SOUTH POLAR REGION
OCEAN
BARENTS SEA
KARA SEA
LAPTEV SEA
EAST SIBERIAN SEA
NORWEGIAN SEA
EUROPE
ASIA
SEA OF OKHOTSK
NORTH
PACIFIC
OCEAN
SAHARA
ARABIAN PENINSULA
AFRICA
ARABIAN SEA
BAY OF BENGAL
PHILIPPINE SEA
AUSTRALIA
INDIAN OCEAN
ENDERBY PLAIN
SOUTH INDIAN BASIN
Queen Maud Land
TARCTICA
WORLD CLIMATE
CLIMATIC ZONES
TROPICAL
DRY
MILD
CONTINENTAL
POLAR
HIGH ELEVATIONS

World Regional Geography

Africa and Asia - GEOG 102

Joseph J. Hobbs | Christopher L. Salter

CENGAGE
Learning™

Australia • Brazil • Japan • Korea • Mexico • Singapore • Spain • United Kingdom • United States

CENGAGE Learning™

World Regional Geography: Africa and Asia - GEOG 102

Joseph J. Hobbs | Christopher L. Salter

Executive Editors:
Michele Baird
Maureen Staudt
Michael Stranz

Project Development Manager:
Linda deStefano

Senior Marketing Coordinators:
Sara Mercurio
Lindsay Shapiro

Production/Manufacturing Manager:
Donna M. Brown

PreMedia Services Supervisor:
Rebecca A. Walker

Rights & Permissions Specialist:
Kalina Hintz

Cover Image:
Getty Images*

* Unless otherwise noted, all cover images used by Custom Solutions, a part of Cengage Learning, have been supplied courtesy of Getty Images with the exception of the Earthview cover image, which has been supplied by the National Aeronautics and Space Administration (NASA).

ISBN-13: 978-0-495-73942-5

ISBN-10: 0-495-73942-1

Cengage Learning
5191 Natorp Boulevard
Mason, Ohio 45040
USA

Printed in the United States of America

CONTENTS

maps

preface

World Regional Geography, Sixth Edition, is a comprehensive introduction to the interactions between people and environments around the world. It uses the tried-and-true convention of organizing the human experience on earth principally around regions rather than themes. This approach reflects my training as a geographer, as well as what I have found to be most useful for students in two decades of teaching geography. I introduce geographic themes to students and then immerse them in the interconnected but different realms on earth where these themes are manifested in distinctive and generally predictable ways.

My personal experience with geography underpins my approach to writing this text. I was trained as a regional geographer (focusing on the Middle East) at the University of Texas at Austin. Dr. Ian Manners was my mentor in all things Middle Eastern, while Dr. Robin Doughty helped develop my thematic and practical abilities in ethnographic research, humanistic geography, and natural history; Robin himself had been tutored by the great Berkeley cultural geographer Carl Sauer. I have been able to put the universal "Doughty skills" to work on research in many regions around the world but have only truly mastered the problems at hand where I can also tap into the regional "Manners skills": near-fluency in the Arabic language; intimate knowledge of history, politics, cultures, and environments of the Middle East; and, above all, long-term on-the-ground fieldwork in that region.

The regional approach is an extremely effective way for students to understand what is happening in the world today. Let us take the example of Iraq. An individual with thematic or systematic expertise in economic or medical geography might be placed on the ground there, or in any number of countries, and quickly come to terms with Iraq's fundamental problems in those areas. He or she might come up with a plan that would logically address or solve the problems, and may even have implemented similar plans successfully elsewhere. But what particular cultural, social, religious, linguistic, political, environmental, or other unique attributes of Iraq (or even within one its particular areas or ethnic groups) would have to be considered for that plan to work effectively in that particular place? Training in the regional context helps ensure the most effective outcomes. If we are to be wise stewards and decision makers in the world, we must have intimate and detailed knowledge about places.

The world is "globalized," but it is not—as the social commentator Thomas Friedman points out—"flat." Its human and physical terrains are extremely rich and varied, and no process will unfold the same way in every location.

This book is the student's "training manual" for the world. It begins with three introductory chapters laying the systematic and thematic foundations of geography, the regional approach, and the overarching human and physical processes at work around the world. It then leads the reader into eight world regions. In each region, the opening chapter lays the systematic and thematic foundations of area and population, physical geography and human adaptations, cultural and historical geographies, economic geography, and geopolitical issues. Subsequent modules give the reader an opportunity to explore the region in greater detail, on a subregional, country-by-country, and topical basis.

This book is comprehensive. It opens the door for the reader to become thoroughly introduced to a region without having to resort to outside readings and other supplements. It allows for flexibility in using the book in a one- or two-semester course (more about this is explained in the section "How to Use This Book").

What Makes This Book Different?

Many professors have lamented that most world regional texts are heavily weighted toward either human or physical geography. This text seeks a balance between the two, offering professors the option of assigning more or less of the cultural or natural material as he or she sees fit.

On the human geography side, readers will notice that I offer exceptional depth in both historical geography and current events. There is nothing contradictory about this. Things that have just happened and things that have yet to occur are seldom unprecedented. I have established the groundwork in both the physical and human realms that will help students picking up today's paper to understand the news in its regional, historical, and cultural contexts. Using the book's foundation, readers should also be able to develop reasonable forecasts of what is to come. Sections dedicated to geopolitical issues are particularly useful for honing those skills.

Underscoring the point that the world is not flat, more attention is paid in this text than in most to the issues of culture, ethnicity, language, religion, and indigenous peoples. Though the world seems to change quickly, leveling the playing field—with more people accessing the Internet, using the English language, and adopting other international cultural attributes—regional and local traditions often endure and even strengthen. As the book points out, the Internet is in many cases helping resuscitate dying languages. And while I do not expect that professors will require students to memorize the names of all the languages spoken in a particular region, I do hope they will ask students to recognize how diverse the world's peoples are by appreciating, in the written text and on the maps, their rich linguistic landscapes.

What's New in This Edition?

For longtime users, the first remarkable feature of the book is that Kit Salter is no longer coauthor. Kit is happily retired and active on other projects. I have missed working with him. For readers, however, the advantage is a more consistent approach, tone, and style throughout the book. It has been a challenge, but a very exciting and informative one, for me to try to keep up with literally everything going on in the world and to distill this into a manageable book for students.

The format is entirely new. The previous exhaustive examination of physical and human processes that operate in each of the various world regions has been cleaved into two chapters, with new attention given to tectonic processes, landforms, and water resources in the physical geography chapter. The book now consists of 11 chapters (3 introductory and 8 regional) rather than 24. There are also 14 more detailed "breakout modules" designed for further exploration or longer courses.

Much of the economic geography in the previous edition had grown long in the tooth. The Sixth Edition still lays the foundation for understanding how regional economies evolved, however the postindustrial characteristics of the more developed countries get the attention they deserve (the final chapter on the United States has been overhauled most extensively). And perhaps most significant of all, the economic "800-pound (and growing) gorilla" of China tramps through every region in this edition.

Andrew Dolan has drafted a number of new maps to reflect these changing economies and other shifts in landscape use and impact. There are also more satellite images to give readers a better feel for natural and cultural landscapes from the extraordinary perspective of space. I have photographed in more than 100 countries and selected images that I think provide a good and accurate illustration of the people and places of the world.

Ancillaries

This text is accompanied by a number of ancillary publications to assist instructors and enhance student learning:

Online Instructor's Manual with Test Bank

Contains chapter outlines, key terms, and themes for general discussion as well as perspectives on selected discussion questions from the text. The Test Bank offers nearly 1,000 multiple-choice, classroom-tested questions and answers keyed to the test.

Places of the World: A Study Guide for Achieving Geographical Literacy

by James W. Lett

Key geographical terms, political and blank maps, review exercises, summary, facts and figures for each region, and much more.

Exam View® Computerized Testing

Create, deliver, and customize tests and study guides (both print and online) in minutes with this easy-to-use assessment and tutorial system.

PowerLecture

This one-stop presentation tool makes it easy to assemble, edit, and present custom lectures using Microsoft™ PowerPoint™. PowerLecture includes animated versions of many figures and maps from the text, allowing you to integrate dynamic visuals into your course in a snap. You'll also find chapter outlines, summaries of key chapter concepts, art and photographs from the text, and JoinIn™ content to use with your Student Response System.

The Comparative World Atlas

This atlas from Hammond features detailed maps for each continent as well as the world so students can see and evaluate regional terrain and political boundaries. Maps, charts, and images help students grasp how the world is physically and culturally interdependent. Also includes a Master Index, Major City Populations Listing, Quick Reference Guide, Global Politics Map, World Statistical Tables, Time Zone Map, and a guide to using the atlas.

Acknowledgments

Many reviewers and advisors helped to bring this sixth edition to completion. Thanks to each and every one of these distinguished geographers and outstanding teachers and students! They include Elizabeth Leppman, Saint Cloud State University; Kefa Otiso, Bowling Green State University; Mark Guizlo, Lakeland Community College; Tarek Joseph, Henry Ford Community College; Joy Adams, Texas State University; Kyong (Keith) Chu, Bergen Community College; Keith Bell, Volunteer State Community College; William Strong, University of North Alabama; Mary Kimsey, James Madison University; Rebecca A. Howze, South Florida Community College; Robert Maguire, Trinity College; Jeffrey Bury, San Francisco State University; J. Duncan Shaeffer, Arizona State University; Kelly Ann Victor, Eastern Michigan University; David Wall, Saint Cloud State University; Norman Carter, California State University—Long Beach; Gail Hobbs, Pierce College; Jim Nemeth, University of Toledo; Fritz Gritzner, South Dakota State University; Wayland Bauer, St. Ambrose University; James Lindberg, University of Iowa; Richard W. Smith, Harford Community College; Ningchuan Xiao, Ohio State University; Audrey Mohan, Texas State University—San Marcos; Thomas Orf, Los Posistas College; Alexander Diener, Pepperdine University; L. Edward Hicks, Faulkner University; Thomas William Hainze, Jr., Trinity Valley Community College; Jia Lu, University of Wisconsin—Stevens Point;and at the University of Missouri, using and advising on improvements from the Fourth Edition, Larry Brown, Jennifer Presberry,

Victor Lozano, Jason Jindrich, Nathaniel Albers, Rob Long, Jeff Pickles, Sunny Stevens, Johnny Finn, Trent Holmes, and Paul Hammond.

Deep thanks and appreciation go to the Brooks/Cole editorial, marketing, and production staff who labored so diligently to help us all meet the tight deadlines and the technical challenges of the Sixth Edition, especially Mindy Newfarmer, Peter Adams, Alex Brady, Rebecca Heider, Joe Rogove, Ashley Pickering, Belinda Krohmer, Michelle Cole, Vernon Boes, Linda Hsu, and Bob Kauser. Kim Adams worked quickly to supply a range of stock photo choices.

I want especially to recognize two persons without whom this book would have suffered a dramatic loss of quality. I first met Jon Peck by email as I worked on the book in a hotel room in the United Arab Emirates. He introduced himself as the book's project manager, and quickly assumed more and more responsibilities for seeing that everything would come together to get the book published. He also brought his extensive knowledge of the world to bear, politely correcting my errors and introducing new pieces of information. I am deeply grateful for his dedication to this effort. I have the same depth of gratitude to Andrew Dolan, formally known as the book's cartographer and research assistant. Andy is a perfectionist in the best sense of the word, never compromising on the accuracy and currency of his maps and data updates. He has also been meticulous in bringing new developments to my attention, and helping to ensuring the trustworthiness of my narrative

Joseph J. Hobbs

about the author

Katie Hobbs

Joseph J. Hobbs received his B.A. at the University of California Santa Cruz in 1978 and his M.A. and Ph.D. at the University of Texas Austin in 1980 and 1986 respectively. He is a professor and chairman of the Department of Geography at the University of Missouriand a geographer of the Middle East with many years of field research on biogeography and Bedouin peoples in the deserts of Egypt. Joe's interest in the region grew from a boyhood lived in Saudi Arabia and India. His research in Egypt has been supported by Fulbright fellowships, the American Council of Learned Societies, the American Research Center in Egypt, and the National Geographic Society Committee for Research and Exploration. In the 1990s he served as the team leader of the Bedouin Support Program, a component of the St. Katherine National Park project in Egypt's Sinai Peninsula. In 2007 he led an effort to establish a national plan for environmental management in the United Arab Emirates. His current research interests are indigenous peoples' participation in protected areas in the Middle East, Southeast Asia and Central America, and the establishment of joint research and education programs between Vietnam and the United States. He is the author of Bedouin Life in the Egyptian Wilderness and Mount Sinai (both University of Texas Press), co-author of The Birds of Egypt (Oxford University Press), and co-editor of Dangerous Harvest: Drug Plants and the Transformation of Indigenous Landscapes (Oxford). He teaches graduate and undergraduate courses in world regional geography, environmental geography, the geography of the Middle East, the geography of caves, and the geography of global current events, the geographies of drugs and terrorism, and a field course on the ancient Maya geography of Belize. He has received the University of Missouri's highest teaching award, the Kemper Fellowship. During summers he has he led "adventure travel" tours to remote areas in Latin America, Africa, the Indian Ocean, Asia, Europe, and the High Arctic, and done his own research abroad. Joe lives in Missouri with his wife, Cindy, daughters Katie and Lily, and an animal menagerie.

how to use this book

Before you begin reading this Sixth Edition of Essentials of World Regional Geography, there are two important features of the book that you need to know about.

Organization. The text is organized with maximum flexibility in mind. It is suited for both one- and two-term courses. For a one-term course, the three thematic introductory chapters and the eight chapters introducing each region are a good match. These may be complemented by a few more in-depth readings about particular issues or places in the modules. For a two-term course, the three introductory thematic chapters, the eight introductory regional chapters, and as many complete or partial modules as wanted may be used. The regions do not need to be covered in any particular order. For example, although they appear last in the book, the North America chapters may be read first if so desired. For current events that deserve discussion—a volcanic eruption in Indonesia or Middle East peace talks, for example—the instructor may easily call attention to those parts of the book that provide useful background.

Cross-Referencing. The book is written with global interconnections in mind. "Globalization" is understandable as a concept, but how exactly does it work? The page and figure numbers in the book's margins serve to tie the diverse strands of global issues together. For example, when you read that Europeans generally dislike genetically modified foods, page numbers in the margin next to that discussion lead you to descriptions of how farmers in Africa and Latin America decide to plant crops accordingly. As you read about China's economic growth and appetite for raw materials, you will likewise be directed to places all around the world where these forces come into play.

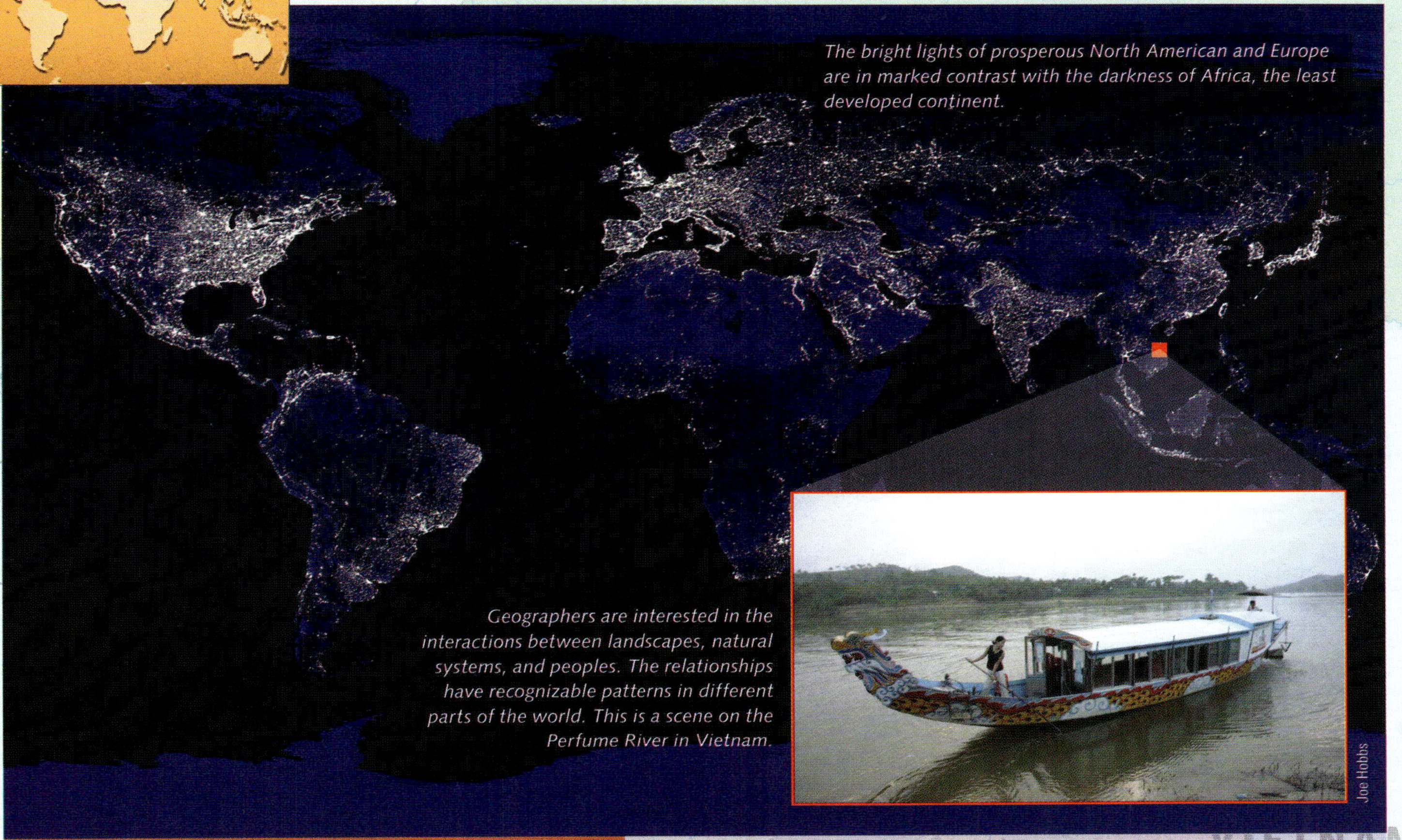

The bright lights of prosperous North American and Europe are in marked contrast with the darkness of Africa, the least developed continent.

Joe Hobbs

Geographers are interested in the interactions between landscapes, natural systems, and peoples. The relationships have recognizable patterns in different parts of the world. This is a scene on the Perfume River in Vietnam.

CHAPTER 1 OBJECTIVES AND TOOLS OF WORLD REGIONAL GEOGRAPHY

1.1 Welcome to World Regional Geography

In its once-a-decade survey conducted in 2002, the National Geographic Society asked a large group of United States citizens aged 18 to 24 to identify on maps some of the world's most important places. Eleven percent of those surveyed could not locate the United States on a blank map of the world. Forty-nine percent could not find New York City, ground zero for the most spectacular of the terrorist attacks of September 11, 2001. Eighty-three percent did not know where Afghanistan is, despite that country's steady presence in the news. Eighty-seven percent did not know where to situate Iraq, which U.S. forces were preparing to invade. The National Geographic survey also tested the geographic awareness of 18- to 24-year-olds from Canada, France, Germany, Italy, Japan, Mexico, Sweden, and Great Britain. The Americans came in next to last, ahead only of the young people of Mexico. (Sweden, incidentally, ranked number one, followed by Germany and Italy).

chapter outline

chapter objectives

This chapter should enable you to

- → Know what geography is
- → Understand the world regional approach to geography
- → Identify the six essential elements of geography
- → Learn some key concepts in geography
- → Appreciate the book's overall objectives
- → Learn the basic language of maps
- → Understand what GIS and remote sensing are
- → Know what geographers do and what kinds of jobs in geography are available

Most of you reading this book are in that group of 18- to 24-year-old Americans. The dismal findings about your peers a few years ago are not meant to shame you or confront you with what you might not know. Instead, they pose a challenge. They make us think about why it is important to know what is going on in the world and where events are taking place. Does it matter if a lot of us don't know where we are on a map? Does it matter if we don't know where other people and countries are? In short, does geography matter?

Years ago, geography earned a reputation for forcing students to memorize long lists of facts. And the truth is, by itself, knowledge of these facts probably means little. In the context of daily life, however, they have great power to transform our lives and contribute to the welfare of our communities and our country. By the end of this chapter, you will know what geography is, recognize the benefits you might gain from learning world regional geography, and understand the organization and objectives of this book.

What Is Geography?

Geography, a term first used by the Greek scholar Eratosthenes in the 3rd century B.C.E.,[1] literally means "description of the earth" but is probably best summed up as "the study of the earth as the home of humankind." Focusing on interactions between people and the environments in which we live, the modern academic discipline of geography has its roots in the Greek and Roman civilizations and emerged from that classical tradition through the Scientific Revolution in Europe (• Figure 1.1). Geography today is carried out today by professionals who have inherited that legacy. They are trained mainly in Western scientific techniques and publish online and in print in a host of languages.

There have always been other geographies, too. Arab geographers in places like Baghdad and Damascus, for example, kept the lights of geographic knowledge burning while Europe experienced its "Dark Ages" before a new dawn around 1000 C.E. All the world's other civilizations carried out explorations and scientific inquiry in fields related to what we call geography today. And for thousands of years, there have also been systems of geographic knowledge used by indigenous peoples who have not had written languages. Such **ethnogeographies,** still found among foraging, agricultural, and other rural peoples today, have gone largely unknown and untapped in the Western geographic tradition. Some geographers are actively collecting this kind of indigenous knowledge in the field, hoping to assemble and publish it before it is lost.

In another important exercise, the National Geographic Society commissioned a team of geographers to identify the core features of the discipline of geography. The team came up with what are known as the **six essential elements of geography.** These elements summarize what geographic research and education involve today and what distinguishes geography from other fields. They are as follows:

1. *The World in Spatial Terms.* Geography studies the relationships between people, places, and environments by mapping information about them into a spatial context.

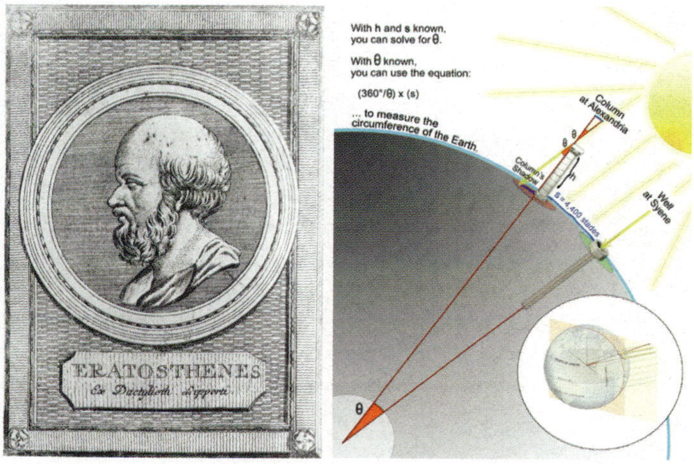

Visual Arts Library (London)/Alamy

• **Figure 1.1** By measuring the lengths of shadows in the Egyptian cities of Alexandria and Aswan (Syene) and applying a simple mathematical formula, the Greek geographer Eratosthenes—the first person known to use the term *geography*—was able in the 3rd century B.C.E. to determine the circumference of the earth as 252,000 stadia—between 24,660 and 28,965 miles (39,690 and 46,620 km; we are not certain of the length of his stadion). The circumference of the earth, measured around the poles, is recognized today as 24,901 miles (40,075 km); Eratosthenes was right "in the ball park"—not bad for 2,000-year-old science!

2. *Places and Regions.* The identities and lives of individuals and peoples are rooted in particular places and in human constructs called regions.
3. *Physical Systems.* Physical processes shape earth's surface and interact with plant and animal life to create, sustain, and modify ecosystems.
4. *Human Systems.* People are central to geography; human activities, settlements, and structures help shape the earth's surface, and humans compete for control of the earth's surface.
5. *Environment and Society.* The physical environment is influenced by the ways in which human societies value and use the earth's physical features and processes.
6. *Uses of Geography.* Knowledge of geography enables people to develop an understanding of the relationships between people, places, and environments over time—that is, of the earth as it was, is, and might be.

While "essential," these six elements are extremely comprehensive. Geography is probably the most all-encompassing of the sciences—a fact of which geographers are duly proud. Broadly, the discipline has two major branches, **physical geography** and **human geography**, each of which has roots and relationships with other disciplines in the social and physical sciences (•Figure 1.2). Geographers very often bridge the social and natural sciences in their research (another fact they are proud of, as it is inappropriate or impossible for most disciplines to do this). Consider, for example, the context of this study, published in 2005 in the *Annals of the Association of American Geographers* (the "flagship" journal for geographers in the United States): "Contemporary Human Impacts on Alpine Ecosystems in the Sagarmatha (Mt. Everest) National Park, Khumbu, Nepal."[2] Alton Byers, the geographer who wrote this article, documented the natural history of one of the world's great national parks, came to understand the human social dynamics bringing about changes in the ecosystems there, and mapped those changes. His work involved every one of geography's six essential elements. It was a classic piece of research in a discipline concerned above all with the theme of **human-environment interaction.**

This interest in the **human agency** (humans' role in changing the face of the earth) has emerged time and again as one of geography's central themes. The great German geographer Alexander von Humboldt (1769–1859) initiated geography's modern era in a series of classic studies on this theme. From field observations in Venezuela, he concluded, "Felling the trees which cover the sides of the mountains provokes in every climate two disasters for future generations: a want of fuel and a scarcity of water."[3] Today, as we observe, for example, the direct connection between deforestation in Nepal's Himalaya Mountains and devastating floods downstream in Bangladesh, we can appreciate the foresight in Humboldt's warnings.

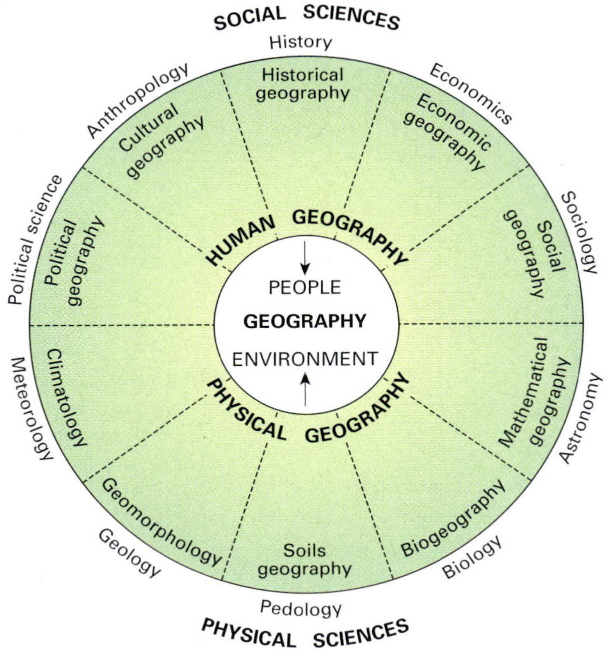

• **Figure 1.2** Selected subfields of geography. These are the main subject areas in human geography and physical geography and their links with the most closely related disciplines in the social and natural sciences.

In Humboldt's wake, other geographers in Europe and the United States wrote about climatic changes resulting from destruction of forest and expansion of farmland. Civilization leads to aridity, they concluded, and they chastised people for violating the harmony and balance that seemed inherent in the natural world. George Perkins Marsh (1801–1882), who served as President Abraham Lincoln's ambassador in Italy, was troubled by the legacy of ancient Greece and Rome on the landscapes of the eastern Mediterranean region. He used the past as a cautionary tale for the future, writing in his book *Man and Nature,* "Man has too long forgotten that the earth was given to him for usufruct alone, not for consumption, still less for profligate waste."[4] *Usufruct* means the nondestructive use of something that belongs to someone else, so Marsh was warning us that we should not be stealing resources and quality of life from one another and from future generations.

A century later, Carl Sauer (1889–1975), another American, wrote:

> We have accustomed ourselves to think of ever expanding productive capacity, of ever fresh spaces of the world to be filled with people, of ever new discoveries of kinds and sources of raw materials, of continuous technical progress operating indefinitely to solve problems of supply. We have lived so long in what we have regarded as an expanding world, that we reject in our contemporary theories of economics and of population the realities that contradict such views. Yet our modern expansion has been effected in large measure at the cost of an actual and permanent impoverishment of the world.[5]

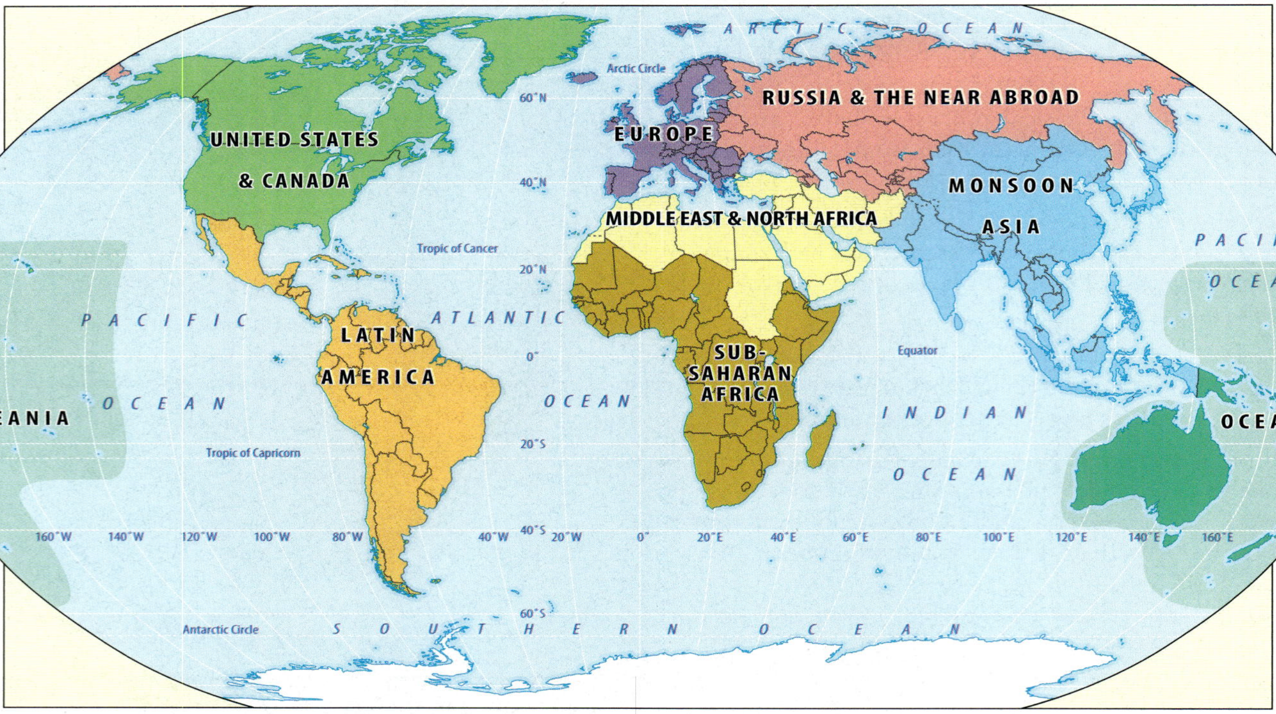

• **Figure 1.3** World regions as identified and used in this book.

These words have a modern ring to them, but Sauer, a geographer at the University of California–Berkeley, wrote them in 1938. Sauer is credited with founding the **landscape perspective** in American geography, based on the method of studying the transformation through time of a **natural landscape** to a **cultural landscape.** People are the agents of that metamorphosis. Sauer defined the geographer as the scientist determined to "grasp in all of its meaning and color the varied terrestrial scene."[6] Geographers continue to be interested in identifying how the forces of nature and culture have been at work in the creation of the **landscape**—the collection of physical and human geographic features on the earth's surface—and in particular the roles that human ideas, activities and cultures play in modifying the landscape. **Culture**—mainly thought of as the realm of anthropology—is an important component in the study of geography and may be defined as a shared, learned, symbolic system of values, beliefs, and attitudes that shapes and influences perception and behavior.[7]

The World Regional Approach to Geography

The **world regional approach** ranges across the human and physical subfields of geography, synthesizing, simplifying, and characterizing the human experiences of earth as home. This text is designed to introduce you in a logical and manageable way to the entire world, first by dividing the earth into eight regions, then by presenting a thematic profile of each region, and finally by examining in greater detail the geographic qualities of countries and other areas within each region.

Because it is impossible to introduce order and logic to something so massive and diverse as our planet without an organized framework built on smaller units, in this book, the world is divided into eight spatial subdivisions, or **regions** (•Figure 1.3 and Table 1.1). Regions are human constructs, not "facts on the ground." People create and draw boundaries around regions in an effort to define distributions of relatively homogeneous characteristics. The Middle East and North Africa, where Europe, Asia, and Africa meet, comprise a region where, very generally, there is a majority Arab population, most of whom are Muslims, and an arid environment. But different analysts draw different boundaries for this region, and those boundaries contain many non-Arab and non-Muslim peoples and non-arid environments. There are often recognizable transition zones between regions; the country of Sudan, for example, is both Middle Eastern and African in its cultures. A region is simply a convenience and a generalization, helping us become acquainted with the world and preparing us for more detailed insight.

Three types of regions are used by geographers and in this book. Each is helpful in its own way in conveying information about different parts of the world:

TABLE 1.1 The Major World Regions: Basic Data

Region	Area (sq mi; thousands)	Area (sq km; thousands)	Estimated Population (millions)	Annual Rate of Increase (%)	Estimated Population Density (sq mi)	Estimated Population Density (sq km)	Human Development Index[a]	Urban Population (%)	Arable Land (%)	GNI PPP Per Capita[b] ($US)
Europe	1,959.2	5,072.2	523.5	0.1	258	100	0.901	74	24	22,611
Russia and the Near Abroad	8,533.2	22,100.8	280.3	−0.1	32	12	0.771	64	8	6,577
Middle East and North Africa	5,916.7	15,324.2	479.0	1.9	160	62	0.690	54	6	5,345
Monsoon Asia	8,013.5	20,755.0	3,504.2	1.2	437	169	0.677	37	20	5,019
Pacific World	3,305.7	8,561.7	32.7	1.0	10	4	0.846	71	5	21,553
Africa South of the Sahara	8,406.5	21,772.5	693.7	2.4	85	33	0.439	31	6	1,767
Latin America	7,946.2	20,580.7	549.4	1.6	69	27	0.773	74	7	7,262
North America	8,403.8	21,765.8	325.6	0.5	38	15	0.939	79	10	37,013
World Regional Totals	52,484.8	135,935.6	6,396.0	1.3	123	47	0.696	48	10	7,590
USA (for comparison)	3,717.8	9,629.1	293.6	0.6	79	30	0.939	79	10	37,800

[a]HDI is a numeric index developed by the United Nations that reflects gross domestic product per capita, literacy rate, level of education on average, and life expectancy. The maximum HDI possible is 1.0, representing the highest level of development.

[b]Per capita gross national income purchasing power parity (per capita GNI PPP) combines gross national income (GNI) and purchasing power parity (PPP) as a method of comparing the real value of output between different countries' economies. This will be fully discussed in Chapter 3. 44

Source: *World Population Data Sheet,* Population Reference Bureau, 2007; *U.N. Human Development Report,* United Nations, 2007; *World Factbook,* CIA, 2007.

- A **formal region** (also called a **uniform** or **homogeneous region**) is one in which all the population shares a defining trait or set of traits. A good example is an administrative political unit such as a county or a state, where the regional boundaries are defined and made explicit on a map.

- A **functional region** (also called a **nodal region**) is a spatial unit characterized by a central focus on some activity (often an economic one). At the center of a functional region—a good example is the distribution area for a metropolitan newspaper—the activity is most intense, while toward the edges of the region, the defining activity diminishes in importance.

- A **vernacular region** (or **perceptual region**) is a region that exists in the mind of a large number of people and may play an important role in cultural identity but does not necessarily have official or clear-cut borders. Good examples are concepts of the Midwest, the Bible Belt, and the Rust Belt in the United States (•Figure 1.4). These regional terms have economic and cultural connotations, but 10 people might have 10 different definitions of the qualities and boundaries of these regions. Vernacular or perceptual regions, constructed by individuals and cultures, serve as shorthand for place and regional identity, helping us organize, simplify, and make sense of the world around us.

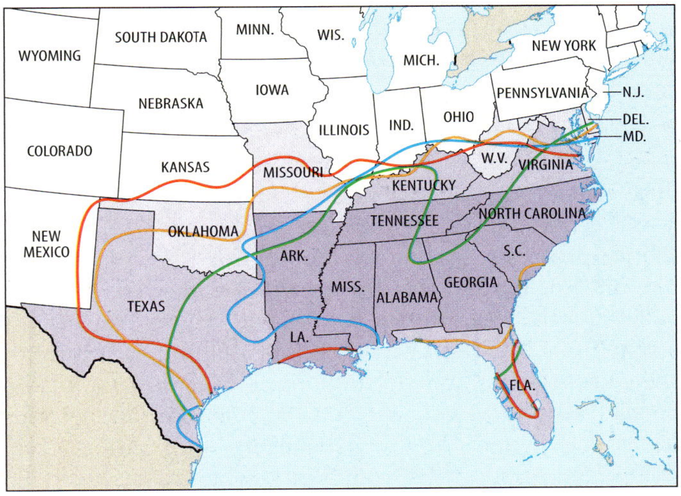

• **Figure 1.4** Definitions of a vernacular region, the American South. Purple shading represents three state-based delineations; colored lines delimit various religious, linguistic, and cultural "Souths." These are just a few of the many different interpretations of the region.

The Objectives of This Book

Mindful of the six essential elements of geography and the comprehensive task of introducing the entire world, this book is written to help you achieve four objectives:

1. *To understand important geographic problems and their potential solutions.* As the home of humankind, the earth requires maintenance; it has some problems that need attention and can be remedied. Many of these problems are environmental, and like geography in general, world regional geography is very concerned with environmental problems and their solutions. Some of these, such as climate change, are global issues that may be addressed with global agreements, like the Kyoto Protocol described in Chapter 2. Others are national and international.

Many regional and international security problems have roots that are environmental. Why can't some countries feed themselves? In some cases, their populations have exceeded their resource bases. What are their prospects then? The least appealing outcome is that famine will lower the population. Another unsavory option is for the country to aggressively seek the resources of others. You will see many examples of countries that have felt pushed into a corner by agricultural failure, debt, or geographic constrictions have lashed out against their neighbors or threatened them (the situation of North Korea comes immediately to mind). Another option is for rich countries to assist the country in need. As the world's sole superpower, what can the United States do to help a country that is in trouble? As a major financial power, the country can help by investing in education, employment, and infrastructure, for example. But should it help? Some authorities (like Garrett Hardin, discussed in Chapter 3) have insisted that Americans should feel no obligation to do so. But if Americans decide to be engaged, there are achievable solutions available. The book will introduce some of these.

2. *To learn to make connections between different kinds of information as a means of understanding the world.* To understand the problems of North Korea, for example, you must consider many variables, including resources, population, economic development, ethnicity, history, and geopolitical interests. Is that too much information for you to take in? Not if you have the right tool with which to filter and synthesize the information. World regional geography is that tool. Geography is all about making connections. Using geography's holistic and integrative approach in a regional framework, you will look at relationships between places and people. You will be using information, techniques, and perspectives from both the natural sciences and the social sciences. You will become familiar with issues in political science, history, economics, anthropology, sociology, geology, atmospheric science, and many other areas. You will pull these issues and perspectives together and find the links between them. In short, you will be doing geography.

This approach will be rewarding for you in both tangible and intangible ways. Your overall university experience should become richer as you become better able to tie together the various courses you are taking. In the future, you will be more likely to interact with the world with new wisdom that can be rewarding both professionally and personally. More complete knowledge of the world—good geography—is good business. American businesses working around the world are doing much better now than in the past in large part because of a greater understanding of local cultures and environments. If you end up working in a business environment, chances are you will be among the best informed of your colleagues when it comes to the cultural and physical settings of a particular place, and that knowledge may offer you new professional opportunities.

3. *To understand current events.* You should become able to pick up a newspaper or turn on the TV news with a greatly enhanced understanding of the issues underlying the world's news. Incidents of violence in the Middle East, earthquakes in the western Pacific, and outbreaks of bird flu are not random, unpredictable events but instead are rooted in consistent, recognizable problems that have geographic dimensions. You should find it very satisfying to feel like something of an expert on a problem that appears in the news.

4. *To develop the ability to interpret places and read landscapes.* Studying geography, you are concerned both with **space**—the precise placement of locations on the face of the earth—and with **place**—the physical and cultural context of a location. Place is much more subjective than location, because it often is defined by the meanings of a particular location; for example, your perceptions of what New York is like may be very different from those of your friend and may be shaped by personal experience in the Big Apple or by photographs or movies you have seen. In this book, there is much discussion of the "sense of place" that individuals and groups have about various locations and regions. Perception of place can have a very strong influence on how we make decisions and interact with others. Perception of place can even have a strong impact on world events. In the Middle East chapters, for example, you will see how Jewish and Muslim perceptions about the sanctity of places within a few meters of each other in Jerusalem play a regular role in both conflict and peacemaking in the Middle East and beyond.

As an example of how you can come to define and identify place, study the photograph in •Figure 1.5, and try to identify the country or region in which this photograph was taken.* What clues in the physical and human geographies of this place help you locate it? As you use this book, you should be increasingly able to look at a photograph of a place or travel to a place with new insight into many of the features that make up its identity: its climate, vegetation, and landforms, for example, and the language, religion, history, and livelihoods of the people who live there.

This book uses photographs that will teach you much about the characteristics of places. There are also aerial and satellite images that will help you identify places on a broader scale and appreciate the world as if you were looking at it from above. (I recommend that you *do* look at the world from above whenever you have the chance. When you fly, ask for a window seat that is not over the airplane's wing.) These images, plus something a book cannot give

• **Figure 1.5** Where is this place?* What clues on the landscape or in the women's appearance might tell you where you are?

you but you should use whenever you can—a globe—are important tools in your gaining the "whole earth" perspective that geography requires.

1.2 The Language of Maps

We turn now to other important geographers' tools that will help you explore and explain relationships on our planet. Although they may at times seem to act like anything from anthropologists to zoologists, geographers reveal their true identities by their preoccupation with space, place, location, landscape, and especially, maps. As geographers study the relationships between people, places, and environments, they usually (but not always) collect and depict information that can be mapped. In other words, they are interested in the **spatial** context of the things; the world in spatial terms is the first essential element of geography. *Spatial* means "pertaining to space" and in the geographic context refers to the distribution of various phenomena on the earth's surface.

The science of making maps is called **cartography.** Cartographers are the geographers and other skilled technicians who create maps through the research and presentation of geographic information. Using maps, the cartographer shows where things and places are located relative to one another. Cartographers today almost always use computers to interpret and display spatial data, but it was not so long ago that they drafted maps laboriously by hand.

Because maps are among the essential tools in the study of world regional geography, it is important to know how to read them. Major elements of their language (beyond the title, date, and north arrow) are *scale, coordinate systems, projections,* and *symbolization.*

Scale

A map is a reducer; it shrinks an area to the manageable size of a chart, piece of paper, or computer monitor. The amount of reduction appears on the map's **scale,** which shows the actual distance on earth as represented by a given linear unit on the map. A common way of denoting scale is to use a representative fraction or ratio, such as 1:10,000 or 1:10,000,000. In other words, one linear unit on the map (for example, 1 inch or 1 centimeter) represents 10,000 or 10,000,000 such real-world units on the ground. A **large-scale map** is one with a relatively large representative fraction (for example 1:10,000 or even 1:100) that portrays a relatively small area in more detail. A **small-scale map** has a relatively small representative fraction (such as 1:1,000,000 or 1:10,000,000) that, in contrast, portrays a relatively large area in more generalized terms. Compare the two maps in • Figure 1.6. Figure 1.6a is a small-scale map showing San Francisco and surrounding parts of the Bay Area. Figure 1.6b is a large-scale map that "zeroes in" on part of San Francisco. Remember this inverse relationship: a small-scale map shows a large area, and a large-scale map shows a small area.

Coordinate Systems

A principal concern in maps is **location.** In addition to core and peripheral locations (see Insights, page 9) there are two general types of locational information discussed in this book: absolute location and relative location.

Relative location defines a place in relationship to other places. It is one of the most basic reference tools of everyday

*This scene is outside a village in Guangxi, China's southernmost province. Agricultural terraces like these are usually associated with high population densities as large numbers of rural people try to get maximum crop returns from steep landscapes, and terrace building requires a lot of labor. These conditions are true of China but could also describe parts of the South American Andes. The clouds suggest a humid climate, found in southern China and also parts of the Andes. Even the women's dark hair could be Asian, Native American, or Hispanic. Their dress is distinctively that of an ethnic minority in China, however.

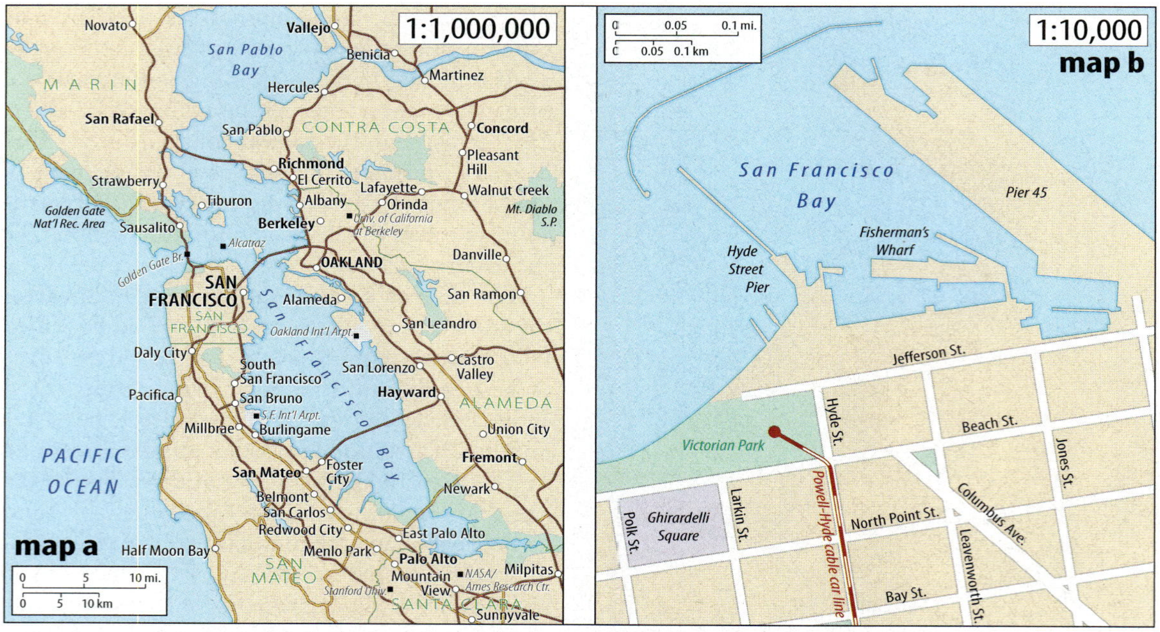

Figure 1.6 Small-scale (a) and large-scale (b) maps of San Francisco and environs.

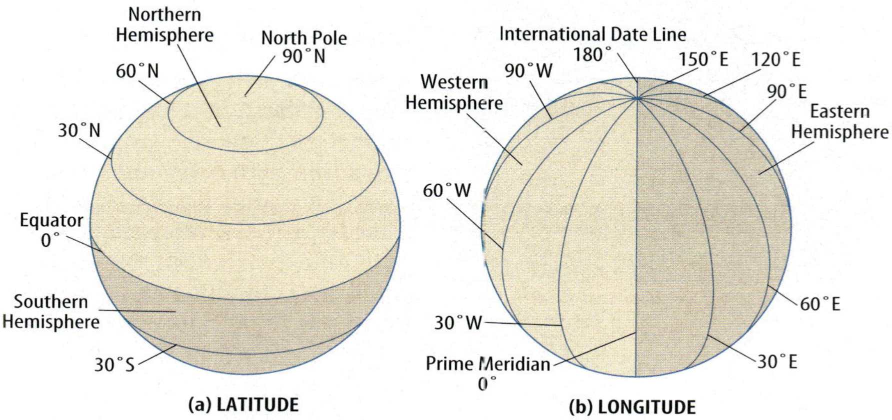

Figure 1.7 (a) Earth's lines of latitude (parallels) in increments of 30 degrees, from the equator (0 degrees) to the North Pole (90 degrees north latitude). (b) Earth's lines of longitude (meridians) in increments of 30 degrees.

life; you might say you live south of the city, just west of the shopping mall, or next door to a good friend. Relative location will become part of your basic geographic literacy as you learn, for example, that Sumatra is northwest of Java and that Bolivia is tantalizingly close to, but cut off from, the Pacific Ocean. Understanding the implications of relative location can be quite useful in following world affairs; be watchful for Bolivia's diplomatic efforts to retrieve its lost access to the sea, for example.

Absolute location, also known as **mathematical location,** uses different ways to label places on the earth so that every place has its own unique location, or "address." **Coordinate systems** are used to determine absolute location. These coordinate systems use grids consisting of horizontal and vertical lines covering the entire globe. The intersections of these lines create the addresses in the global coordinate system, giving each location a specific, unique, and mathematical situation.

The most common coordinate system uses **parallels** of **latitude** and **meridians** of **longitude.** The term *latitude* denotes position with respect to the equator and the poles. Latitude is measured in **degrees** (°), **minutes** (′), and **seconds** (″). Each degree of latitude, which is made up of 60 minutes, is about 69 miles (111 km) apart; these distances vary a little because of earth's slightly ellipsoid shape (see Figure 1.7a). Each minute of latitude, which is made up of 60 sec-onds, is thus roughly a mile apart. The **equator,** which circles the globe east and west midway between the poles, has a

INSIGHTS

Core Location and Peripheral Location

Among the concepts that will prove useful to you in studying world regional geography are **core location** and **peripheral location;** in this book, the region of Europe is subdivided along these lines, for example. Some locales have greater importance in local, regional, or world affairs because they have a central, or core, location relative to others. Other locales are less important because they are situated far from "where the action is." A comparison of two countries, the United Kingdom (U.K.) and New Zealand, provides an example (Figure 1.A). Both are island countries. Their climates are remarkably similar, although they are in opposite hemispheres and are about as far apart as two places on earth can be. Westerly winds blow off the surrounding seas, bringing abundant rain and moderate temperatures throughout the year to both.

But there are important differences. The U.K. is located in the Northern Hemisphere, which has the bulk of the world's land (and may thus be described as the **land hemisphere**) and most of its principal centers of population and industry; New Zealand is on the other side of the equator, in the Southern Hemisphere, where there is less land and less economic activity.

The U.K. is located near the center of the world's landmasses and is separated by a narrow channel from the densely populated industrial areas of western continental Europe; New Zealand is surrounded by vast expanses of ocean in the **water hemisphere.** The U.K. is located in the western seaboard area of Europe, where many major ocean routes of the world converge; New Zealand is far away from the centers of world commerce. For centuries, the United Kingdom has shared in the development of northwestern Europe as an organizing center for the world's economic and political life; New Zealand, meanwhile, has languished in comparative isolation.

The United Kingdom, in other words, has a core or central location in the modern framework of human activity on earth, whereas New Zealand has a peripheral location. Centrality of location is a very important consideration when it comes to assessing the economic and political geographies of places, countries, and regions.

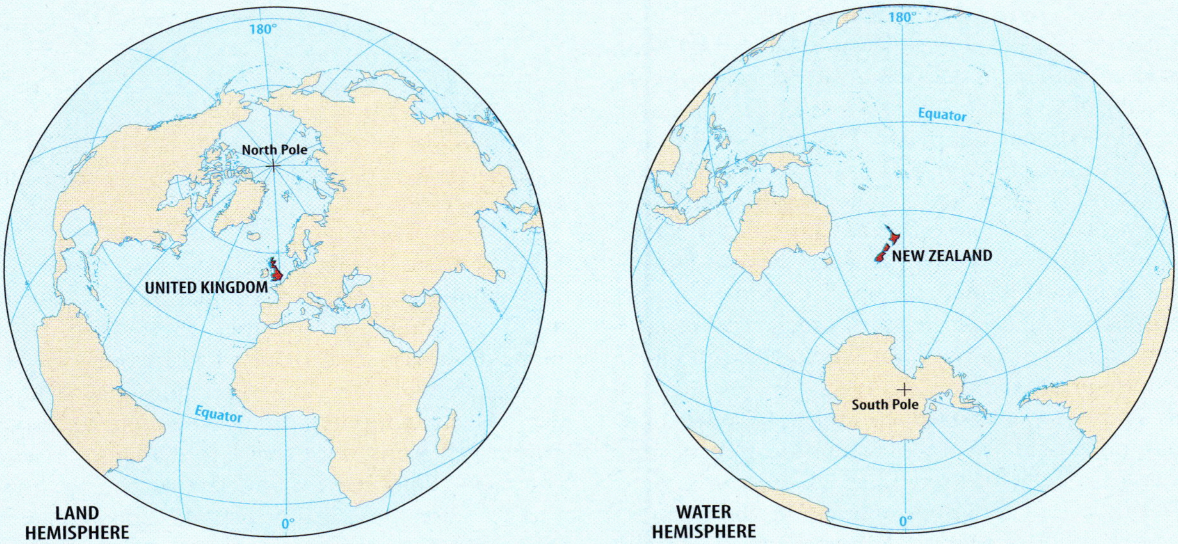

Figure 1.A In the top map, note how the major landmasses are grouped around the margins of the Atlantic and Arctic Oceans. The British Isles and the northwestern coast of Europe lie in the center of the "land hemisphere," which constitutes 80 percent of the world's total land area and has about 90 percent of the world's population. New Zealand lies near the center of the opposite hemisphere, or "water hemisphere," which has only 20 percent of the land and about 10 percent of the population.

latitude of 0°. All other latitudinal lines are parallel to the equator and to each other, which explains why they are called *parallels*.

Every point on a given parallel has the same latitude. Places north of the equator are in **north latitude.** Places south of the equator are in **south latitude.** The highest latitude a place can have is 90°N (the **North Pole**) or 90°S (the **South Pole**). Places near the equator are said to be in **low latitudes;** places near the poles are in **high latitudes.** The **Tropic of Cancer** and the **Tropic of Capricorn,** at 23.44°N and 23.44°S, respectively, and the **Arctic Circle** and **Antarctic Circle,** at 66.56°N and 66.56°S, respectively, form the most commonly recognized boundaries of the low and high latitudes. Places occupying an intermediate position with respect to the poles and the equator are said to be in the **middle latitudes.** Incidentally, there are no universally accepted definitions for the boundaries of the high, middle, and low latitudes. The northern half of the earth between the equator and the North Pole is called the **Northern Hemisphere,** and the southern half between the equator and the South Pole is the **Southern Hemisphere.**

Meridians of longitude are straight lines connecting the poles (see Figure 1.7b). Every meridian is drawn due north and south. All the meridians converge at the poles and are farthest apart at the equator. Longitude, like latitude, is measured in degrees, minutes, and seconds, but because the longitude lines are not equidistant across the globe, their values vary. At the equator, the distance between lines of longitude is about 69.15 statute miles (111.29 km), while at the Arctic Circle, it is only about 27.65 miles (44.50 km). A **statute mile** ("land mile") is the mile you are most likely familiar with, 5,280 feet (1,609 m); a **nautical mile** ("sea mile") is based on one minute of arc of a great circle and equals 6,076 feet (1,852 m). The meridian most used as a base or starting point is the one running through the Royal Astronomical Observatory in Greenwich, England. Known as the **meridian of Greenwich** or the **prime meridian,** it has a longitude of 0°. Places east of the prime meridian are in **east longitude;** places west of it are in **west longitude.**

The meridian of 180°, exactly halfway around the world from the prime meridian, is the basis for the other dividing line between places east and west of Greenwich and is called the **International Date Line.** This line, which has a few zigzags in it for political and practical reasons (see Figure 8.1, page 407), separates two consecutive calendar days. The date west of the line is one day ahead of the date east of the line. All of the earth's surface eastward from the prime meridian to the International Date Line is in the **Eastern Hemisphere,** and all of the earth's surface westward from the prime meridian to the International Date Line is in the **Western Hemisphere.**

Using the simple tools of latitude and longitude notation, you may now establish the absolute location of any given place. Here is a useful exercise to ensure you understand how latitude and longitude work. On the map in •Figure 1.8, the latitude of Madrid, Spain, is approximately 41 degrees north latitude, 4 degrees west longitude (41°N, 4°W). What are the approximate latitude and longitude coordinates of Oslo, Norway? In which hemispheres is Oslo located?

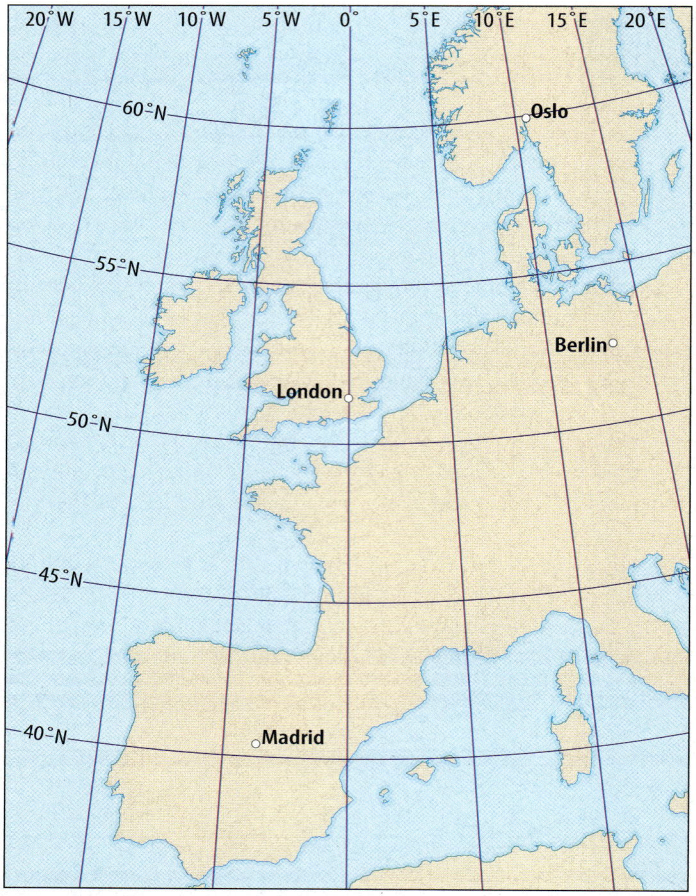

• **Figure 1.8** What are the approximate latitude and longitude coordinates of Oslo, Norway?*

Projections

A **map projection** is a way of depicting the curved surface of the earth—which can be represented accurately only on a globe—on a flat surface such as a piece of paper. There are four metric relationships or properties of objects on a globe: area, shape, distance, and direction. A flat map cannot replicate all of these simultaneously; most projections can preserve only one. As a result, there will inevitably be some distortion on a flat map. On maps depicting very small areas (large-scale maps), the distortion may be minimal enough that it can be disregarded. On small-scale maps, however, the effects of distortion can become very serious. Fortunately, there are thousands of different projections cartographers can choose from to display the geographic area and data they are mapping. There is no one "perfect" projection for any particular map, however.

There are three geometric "developable surfaces" (surfaces that can be flattened without stretching or tearing) that many map projections are based on: the plane, the cylinder, and the cone (•Figure 1.9).

- Projections onto a plane are called **azimuthal projections** and are typically used for maps of the polar regions.
- **Cylindrical projections** are mostly used for areas around the equator or to depict the entire world; regular cylindrical projections have straight meridians, while pseudocylindrical variants have curved ones.
- **Conic projections** are common for middle latitudes; **polyconic projections** can be used for larger areas.

*Using this map, you should determine Oslo's location as about 60 degrees north, 11 degrees east (60°N, 11°E). Oslo is therefore in the Northern Hemisphere, the Eastern Hemisphere, and the land hemisphere.

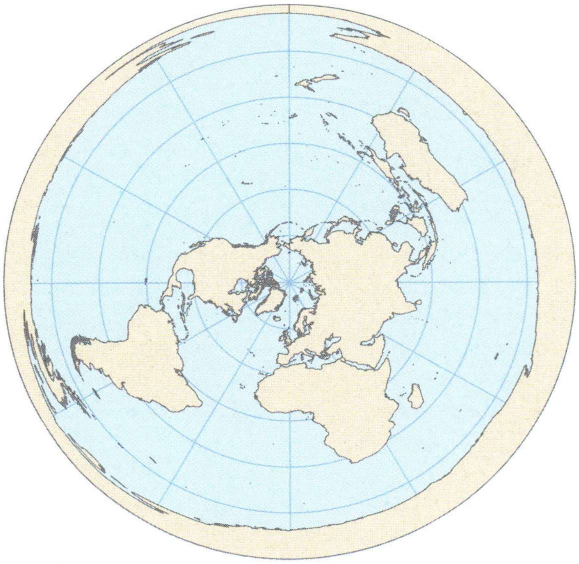

a An azimuthal projection (North Pole Azimuthal Equidistant)

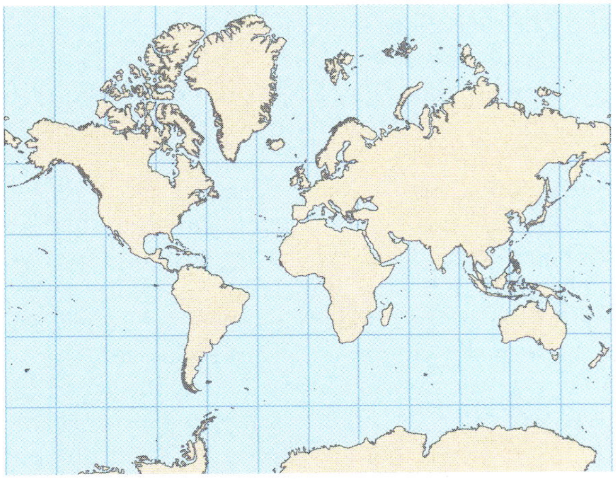

b A cylindrical projection (Mercator)

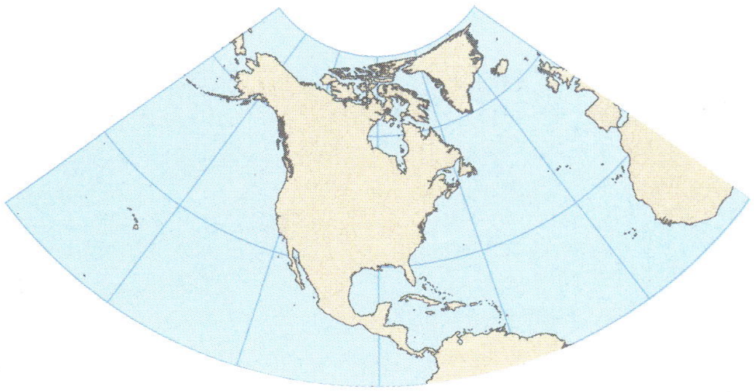

c A conic projection (Albers Equal Area Conic)

• **Figure 1.9** Common map projections.

Projections not based on developable surface geometry are called **compromise projections** (or *mathematical projections*). These projections attempt to create a balance of distortion among the four metric properties to create an aesthetically pleasing map. The **Robinson projection,** used for all the world maps in this book, is an example of a compromise projection.

Map projections are also classified by which metric property they preserve (or distort the least):

- **Equal-area projections** preserve areas consistently across the entire map; each area on the map is proportional to the area it occupies on the earth's surface. Some equal-area projections are segmented into various "lobes" to preserve area size while attempting to minimize distortion of shapes. The result looks something like the peeled skin of an orange laid flat.
- **Equidistant projections** show accurate distances, but only from the center of the projection (in most cases). For example, an equidistant map centered on New York City would accurately show the distance between New York and London and between New York and Los Angeles, but it would not show the correct distance between London and Los Angeles. Flat maps cannot be equidistant and equal-area at the same time.
- **Conformal projections** keep the map's scale the same in every direction from any given point, preserving shapes in very small, localized areas. Sizes are usually distorted, especially toward the edges (no flat map can be both conformal and equal-area). The best-known conformal map is the **Mercator projection,** which was designed for navigation. A straight line drawn between any two points on a Mercator map indicates a true direction (a "rhumb line"). However, to maintain the accuracy of those rhumb lines, Mercator maps increasingly distort the north-south dimension away from the equator and distort the east-west dimension near the poles because of the parallel meridians that conic projections employ (on the three dimensional reality of a globe, the meridians converge at the poles). Therefore, the sizes of areas near the poles on Mercator maps are greatly exaggerated, so much so that Greenland and South America appear to be the same size, even though South America is actually about eight times larger than Greenland! This distortion is clearly visible in the Mercator projection in Figure 1.9b.

Symbolization

Maps allow us to extract certain information from the totality of things, to see patterns of distribution, and to compare these patterns with one another. No map is a complete record of an area. It represents instead a selection of certain details, shown by symbols, that a cartographer uses for a particular purpose. Unprocessed data must be classified into categories that symbols can represent. The details selected for **symbolization** may be categories of physical or cultural forms like rivers or roads, aggregates such as 100,000 people or 1 million barrels of crude oil, or averages such as population density (the number of persons per square mile or kilometer within a defined area).

Cartographers often order aggregates or averages into ranked categories in a graded series (for example, population densities of 0 to 49, 50 to 99, and 100 or more people per square mile) to be portrayed on a map. The categories are usually not self-evident. The cartographer must select them, sometimes through elaborate statistical procedures. To represent different phenomena, the cartographer may choose from a wide range of symbols, including lines, dots, circles, squares, shadings, colors, and typography.

There are two major types of maps that employ symbols: reference maps and thematic maps. **Reference maps** (such as Figure 4.1 on page 69) are concerned mainly with the locations of various features on the earth's surface and their spatial relationships with each other. Highway maps are reference maps. **Thematic maps** also show spatial relationships, but they have a more specific purpose: they often show the distribution of just one phenomenon. Thematic maps are sometimes called *statistical maps* because they typically show distributions of numeric data, such as population density in given area. They can also show qualitative (nonmathematical) information. Several types of thematic maps are used throughout this book:

- On a **choropleth map** (the most common thematic map type in the book), each political unit is filled in with a distinguishing color or pattern representing some derived value, such as per capita income or number of college students per 100,000 people (Figure 3.5 on page 43, showing wealth and poverty by country, is a good example).
- **Isarithmic maps** do not use political boundaries with solid colors or patterns but instead use lines to join points of equal value across the mapped area (for example figure 1.4 on page 5). Topographic maps based on contour lines are also isarithmic maps.
- As another alternative to colored political units, **graduated symbol maps** use a simple symbol, such as a circle, square, or bar graph. The symbol is scaled proportionally to the quantity of the data being mapped. On a graduated symbol map of world population, for example, the circle for China would have an area over four times larger than the circle representing the United States. A graduated symbol map that scales the often highly generalized outlines of political units to the data is called a **cartogram.**
- **Dot maps** use dots to represent a stated amount of some phenomenon within a political unit. Using a scale equating one dot with 1,000 people, 12 dots in one area would represent 12,000 people living in that area.
- **Flow maps** use arrows to detail the movement of people or goods from one area to another (see Figure 3.16 on page 58 for an example).

Isarithmic and graduated symbol maps can display the same data very differently. Figures 3.13 and 3.15, on pages 56 and 58, clearly illustrate that China and India are the most populous nations on earth. Figure 3.13 is a cartogram: it shows that China and India are the world's demographic giants but not where the populations within those two countries are concentrated. Figure 3.15 is an isarithmic map: you cannot tell from this map how many people live in China or India, but the contrast between areas of dense settlement and virtually uninhabited land in both countries is very clear.

Maps need to be used judiciously, and you will gradually acquire the skill of evaluating map quality. A well-known book in the field of geography is titled *How to Lie with Maps,* suggesting, like the motto "let the buyer beware," that the fact that something has been mapped does not guarantee that it has been mapped accurately.

Mental Maps

When someone asks you, "Could you draw me a map of how to get there?" you might quickly draw some lines, write down some street names, point out familiar landmarks, and apologize for how crude your map is. Your map would probably end up looking very different from that of another person asked the same question. Our understanding of location is not completely objective. Each of us has a personal sense of space and place and associations with them.

A **mental map,** rather like a vernacular region, is a collection of personal geographic information that each of us uses to organize spatially the images and facts we have about places, both local and distant. Sometimes we use that information to create actual maps, as when a guest asks you to draw a map showing how to get to your house or when a nomad in the Middle East draws lines in the sand diagramming the territories of his and others' tribes (•Figure 1.10). Such maps are not accurate, precise, or scientific, but they portray useful information and tell us much about the individuals and cultures that create them. They are subjective,

Joe Hobbs

• **Figure 1.10** As we grow, our minds fill with mental maps, which we use consciously or unconsciously to orient ourselves and navigate through daily life. In drawing these maps for others, we often discover the importance of scale, direction, and other map essentials.

just like the various cultural and other regions identified and used in this book.

1.3 New Geographic Technologies and Careers

To close this chapter, we turn to some of the most exciting and innovative developments in geography and consider how you or someone you know might become part of the action in this growing field.

Geographic Information Systems and Remote Sensing

Geographic information systems (GIS) represent the cutting edge of geography today, and are by far the field's leading area of growth and employment. In recent decades GIS has become an essential tool for geographers, and increasingly for people in many other scientific disciplines; it is also widely used by private businesses and governmental and other public agencies. In essence, a GIS is a computerized system designed to help people analyze, manage, and visualize geographic data. By combining many tools including cartography, remote sensing (described below), statistics, and computer science, GIS users can discover relationships between disparate types of spatial information and make informed decisions to solve complex geographic problems. GIS applications are most used for land and resource management, though it has an almost infinite number of other uses (examples include locating the most efficient route for a truck to deliver goods; plotting the locations of various types of crime to detect patterns; determining the best location to open a new store; creating 3-D models of terrain to model the effects of floods; and many more).

GIS data is stored in "layers" (•Figure 1.11), with each layer containing a related set of spatial data. One layer may have a neighborhood's street network, another layer may have the location of every tree in that neighborhood, a third layer might have the routes of electrical wiring or water pipes running below the surface, and yet another layer could contain elevation data. One of the aspects that sets apart each of these layers from a simple map showing the same data are "attributes," or further information, about each item in those layers. For example, a click on a water main on that layer will bring up a box containing information about how long the pipe is and when it was last serviced. The ability to create, select and manipulate all this spatial information has revolutionized both geography itself and all other fields dealing with spatial relationships.

Remote sensing uses various kinds of satellite imagery and photographic coverage to assess land use or other geographic patterns. Many different technologies are associated with remote sensing, and it can be use in an almost countless variety of applications. Aerial photography, one of the earliest geographic uses of remote sensing, allows for the creation of maps of limited areas entirely from photographs. Radar is used to map elevation and detect rainfall, and a newer technology called LIDAR (laser imaging detection and ranging, similar to radar but using light waves instead of radio waves) is often used to create three-dimensional models, for example, of the remains of the World Trade Center and surrounding buildings after 9/11. And since its introduction in 1960, satellite imagery has been used for an extraordinary range of purposes, from detecting ancient cities covered by the sands of Arabia to tracking the shrinking polar ice cap and monitoring weather systems. Remote sensing is an exceptionally good tool for helping geographers in their mission of understanding how people modify the earth. In the remote-sensing

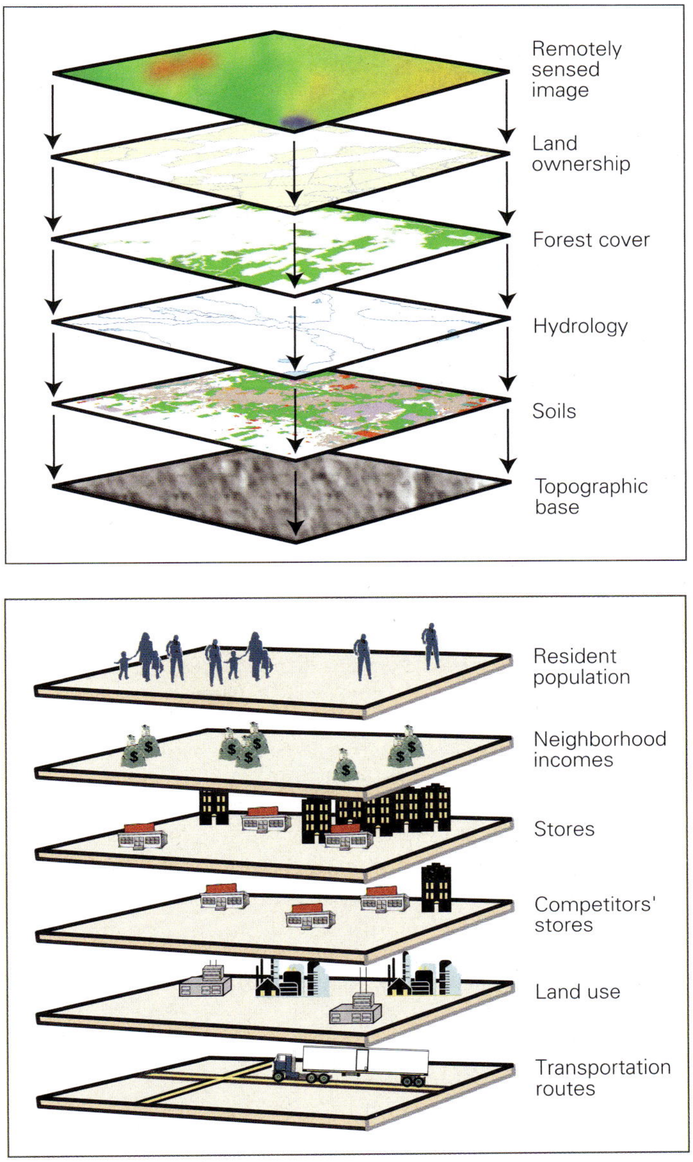

• **Figure 1.11** Geographic information systems (GIS) create and use layers of different types of spatial data. The computer-generated images and models that result have a wide range of academic and practical applications.

Compliments of NASA/USGS

• **Figure 1.12** A remote-sensing image shows one of the few international political boundaries clearly visible from space. High population densities and intensive cultivation have nearly denuded the lush tropical vegetation of southern Mexico, while north-central Guatemala's forests are nearly intact. Guatemalan populations are clustered in the mountain areas farther south. Compare this image with the political map of the
561 region, Figure 10.1.1 on page XXX: the vegetation and political boundaries correlate perfectly. The black area in the wedge on the right side of the photo is area not included in this satellite photo.

Joe Hobbs

• **Figure 1.13** The most in-demand tool of geography today is geographic information systems (GIS). Here the book's cartographer, Andrew Dolan, is using GIS to create a population density map of the United States.

image in •Figure 1.12, for example, human modification of the landscape is clearly apparent.

To see a fascinating and increasingly popular application of remote-sensing and GIS technologies right on your own computer, try Google Earth (http://earth.google.com). Type in your address, any other address, or a place name like "Giza pyramids." You will enjoy a fly-in from outer space to a mosaic of satellite and aerial photographs (from a variety of sources and vintages) of the location you entered. Google Earth acts as a very simple GIS viewer; using the "Layers" sidebar, you can explore various GIS layers of data about the location.

You can see almost anything on the earth with this program. Want to see Iran's nuclear facility of most concern? No problem: type in "Natanz." You can be sure that many governments around the world are very unhappy about this extraordinary software. A Russian intelligence official said of Google Earth, "Terrorists don't need to reconnoiter their target. Now an American company is working for them."[8]

What Do Geographers Do for a Living?

These days, it is appropriate for this question to follow discussion of GIS and remote sensing. More and more, GIS and geography are becoming synonymous, and people with the ability to work with GIS are finding jobs in a number of fields, often at very attractive starting salaries (•Figure 1.13).

Table 1.2 shows a few of the jobs of recent graduates of geography in just one department, the University of Texas at Austin. Note that even with just undergraduate degrees in geography, students are finding work in the GIS and remote-sensing fields. Many graduates with these technical skills find employment in government agencies at the local, state, regional, and federal levels. Firms working with land use decisions at all scales need employees with these skills. Retail firms find the technical and problem-solving know-how of geographers to be useful in management, survey design and implementation, estimation of public response to innovations, and decision making about location expansion or curtailment.

After some years in decline, geography is making a strong comeback as an academic discipline for holders of the doctoral degree. These researchers typically specialize in fields that are either regional (some are Middle East experts, for example), systematic (they study topics of various physical or human geographic elements), or technical (they rely strongly on GIS and remote sensing). Geographers with **regional specialties** in the United States and Europe are most numerous. The relative shortage of U.S. geographers with professional expertise in non-Western and developing countries is a disadvantage, but it does present opportunities for younger geographers willing to undertake the challenges of foreign-language study and fieldwork. As might be expected, there is particularly high demand among intelligence agencies for trained geographers who are acquainted with regions deemed to be strategic in the national interest, such as the Middle East (ironically, though, very few Middle East geographers are currently in training). However, most geographers with foreign-area expertise have university research and teaching positions.

Most academic geographers—many of whom use GIS, remote sensing, and other technical tools of the discipline—practice a **systematic specialty** rather than a regional one. Experts in *physical geography* study spatial patterns and associations of natural features. Prominent subfields include **geomorphology** (the study of landforms and a field in which geography intersects with geology), **climatology** (cli-

TABLE 1.2 Positions Occupied by Recent Graduates of the Department of Geography, University of Texas at Austin[9]

Recipients of a Bachelor of Arts Degree in Geography
Information system analyst, Transportation Planning and Programming Division, Texas Department of Transportation
CFO and geospatial engineer, Spatial Innovations International
Preserve manager, Nature Conservancy
GIS analyst, Environmental Section, Information Solutions Department, PBS&J, Inc.
Natural science educator, Gore Range Natural Science School, Colorado
GIS administrator, Planning and Zoning Department, City of Longview, Texas
Recipients of a Master of Arts Degree in Geography
Senior conservation scientist, Nature Conservancy, Guatemala
Geographer, Texas Water Science Center, United States Geological Survey
Lead environmental scientist, Aquatic Weed Unit, California Department of Boating and Waterways
CEO, Trinidad and Tobago Insurance Consultants, Ltd., Trinidad and Grenada, West Indies
Vice president for market research, Crescent Real Estate Equities, Ltd.
Recipients of a Ph.D. in Geography
Associate vice president, Community Partnerships, San Diego Foundation
Water resources scientist, Gulf States Natural Resource Center, National Wildlife Federation
Managing principal, SWCA Environmental Consultants, Albuquerque, New Mexico
Executive director, Chihuahuan Desert Research Institute
Supervisor, Geographical Information Sciences, Texas Commission on Environmental Quality
Assistant professor, Department of Geography, University of South Carolina

matic processes and patterns), **biogeography** (the geography of plants and animals), and **soils geography.** Closely related to physical geography is the large interdisciplinary field of **environmental studies,** which is concerned with reciprocal relationships between society and the environment. Another allied specialty is **medical geography,** which focuses on spatial associations between the environment and human health and on locational aspects of disease and health care delivery.

Specialists in **economic geography** study spatial aspects of human livelihood. Major occupations and products are their main concerns, and they have an active theoretical component that intersects with the field of economics. **Marketing geography** is a relatively small but active applied offshoot of economic geography that is of particular utility in commercial planning and zoning. The subfields of **agricultural geography** and **manufacturing geography** are concerned, respectively, with the productive management of soil and water resources and with networks and hierarchies of economic production. **Urban geography** considers the locational associations, internal spatial organization, socioeconomic characteristics, and functions of cities. It offers an alternative pathway to the applied field of urban planning.

Geographers specializing in **cultural geography** are concerned today, as they were a century ago, with spatial and other aspects of cultural regions, origins, diffusions and interactions, and the cultural forces behind changing landscapes. Closely allied with cultural geography are the fields of **cultural ecology** and **political ecology,** focusing on the relationships between environments and human adaptations to them and on how social groups organize themselves politically to take advantage of environmental and other opportunities. **Political geography** considers topics such as spatial organization of geopolitical units, international power relationships, nationalism, boundary issues, military conflicts, and regional separatism within states. **Social geography** deals with spatial aspects of human social relationships, especially in urban settings. **Population geography** assesses population composition, distribution, migration, and change. **Historical geography** examines the geography of past periods and the evolution of geographic phenomena such as cities, industries, agricultural systems, and rural settlement patterns.

TABLE 1.3 Specialty Groups of the Association of American Geographers

Africa	Disability	Hazards	Recreation, Tourism, and Sport
Applied Geography	Economic Geography	Historical Geography	Regional Development and Planning
Asian Geography	Energy and Environment	History of Geography	Remote Sensing
Bible Geography	Environmental Perception and Behavioral Geography	Human Dimensions of Global Change	Rural Geography
Biogeography	Ethics, Justice and Human Rights	Indigenous Peoples	Russian, Central Eurasian, and East European Geography
Canadian Studies	Ethnic Geography	Latin America	Sexuality and Space
Cartography	European	Medical Geography	Socialist and Critical Geography
China	Geographic Information Science and Systems	Middle East	Spatial Analysis and Modeling
Climate	Geographic Perspectives on Women	Military Geography	Study of the American South
Coastal and Marine	Geography Education	Mountain Geography	Transportation Geography
Communication Geography	Geography of Religions and Belief Systems	Paleoenvironmental Change	Urban Geography
Cryosphere	Geomorphology	Political Geography	Water Resources
Cultural and Political Ecology		Population	Wine
Cultural Geography		Qualitative Research	
Developing Areas			

Source: Association of American Geographers, http://www.aag.org/sg/sg_display.cfm.

So what exactly do geographers do? As you see in Table 1.3, which lists the 55 specialty groups of the approximately 9,000 members of the **Association of American Geographers (AAG),** almost anything may fall in their realms of interest, activities, and employment.

There are other kinds of employment, too, notably in the travel and tourism industry, for geographers trained at the undergraduate and graduate levels. Even if you are not a geography major or you have not yet seen an advertisement proclaiming "Geographer Wanted," chances are that many professional opportunities will arise if you use geography's tools to grow knowledge of the world in which you live. This book will help you do just that.

SUMMARY

- A recent study showed that U.S. citizens aged 18 to 24 generally have poor knowledge of world geography.
- There are six essential elements of the national geography standards: the world in spatial terms, places and regions, physical systems, human systems, environment and society, and the uses of geography.
- Geography means "description of the earth" and is also defined as "the study of the earth as the home of humankind."
- Four main objectives of the text are for readers (1) to understand important geographic problems and their potential solutions, (2) to become better able to make connections between different kinds of information as a means of understanding the world, (3) to understand current events, and (4) to develop skills in interpreting places and reading landscapes.
- Maps are the geographers' most basic tools. The basic language of maps includes the concepts and terms of scale, coordinate systems, projection, and symbolization. Maps can depict spatial data in a variety of ways.
- Individuals and cultures generate their own unique "mental maps." Regions are in effect mental maps that help us make sense of a complex world. Modern geographic thought derives from a long legacy of interest in how people interact with the environment. The dominant approach has been to understand how people have changed the landscape or face of the earth.
- The discipline of geography may be divided into regional and systematic specialties, with the systematic fields having the most followers. Their concerns overlap many disciplines in the natural and social sciences. Geographers are employed in many private and public capacities. The strongest growth area, with the most jobs, is in geographic information systems (GIS).

KEY TERMS + CONCEPTS

absolute location (p. 8)
agricultural geography (p. 15)
Antarctic Circle (p. 9)
Arctic Circle (p. 9)
Association of American Geographers (AAG) (p. 16)
biogeography (p. 15)
cartogram (p. 12)
cartography (p. 7)
choropleth map (p. 12)
climatology (p. 15)
coordinate systems (p. 8)
core location (p. 9)
cultural ecology (p. 15)
cultural geography (p. 15)
cultural landscape (p. 4)
culture (p. 4)
degrees (p. 8)
dot map (p. 12)
economic geography (p. 15)
environmental studies (p. 15)
equator (p. 8)
ethnogeography (p. 2)
flow map (p. 12)
formal region (p. 5)
functional region (p. 5)
geographic information systems (GIS) (p. 12)
geography (p. 2)
geomorphology (p. 15)
graduated symbol map (p. 12)
hemisphere
- Eastern Hemisphere (p. 10)
- land hemisphere (p. 9)
- Northern Hemisphere (p. 10)
- Southern Hemisphere (p. 10)
- water hemisphere (p. 9)
- Western Hemisphere (p. 10)

historical geography (p. 15)
homogeneous region (p. 5)
human agency (p. 3)
human-environment interaction (p. 3)
human geography (p. 3)
International Date Line (p. 10)
isarithmic map (p. 12)
landscape (p. 4)
landscape perspective (p. 4)
large-scale map (p. 7)
latitude (p. 8)
- high (p. 9)
- low (p. 9)
- middle, (p. 10)
- north (p. 9)
- south (p. 9)

location (p. 7)
longitude (p. 8)
- east (p. 10)
- west (p. 10)

manufacturing geography (p. 15)
map projection (p. 10)
- azimuthal (p. 10)
- compromise (p. 11)
- conformal (p. 11)
- conic (p. 10)
- cylindrical (p. 10)
- equal-area (p. 11)
- equidistant (p. 11)
- Mercator (p. 11)
- polyconic (p. 10)
- Robinson (p. 11)

marketing geography (p. 15)
mathematical location (p. 8)
medical geography (p. 15)
mental map (p. 12)
meridian (p. 8)
meridian of Greenwich (p. 10)
minutes (p. 8)
natural landscape (p. 4)
nautical mile (p. 10)
nodal region (p. 5)
North Pole (p. 9)
parallel (p. 8)
perceptual region (p. 5)
peripheral location (p. 9)
physical geography (p. 3)
place (p. 6)
political ecology (p. 15)
political geography (p. 15)
population geography (p. 15)
prime meridian (p. 10)
reference map (p. 12)
region (p. 4)
regional specialties (p. 14)
relative location (p. 7)
remote sensing (p. 13)
scale (p. 7)
seconds (p. 8)
six essential elements of geography (p. 2)
small-scale map (p. 7)
social geography (p. 15)
soils geography (p. 15)
South Pole (p. 9)
space (p. 6)
spatial (p. 7)
statute mile (p. 10)
symbolization (p. 11)
systematic specialty (p. 14)
thematic map (p. 12)
Tropic of Cancer (p. 9)
Tropic of Capricorn (p. 9)
urban geography (p. 15)
uniform region (p. 5)
vernacular region (p. 5)
world regional approach (p. 4)

REVIEW QUESTIONS

1. What is geography? Is it a natural or a social science? What are some of its characteristic approaches?
2. What are the six essential elements of geography, as defined by the National Geographic Society? What does each element indicate about geography's concern with space, place, or the environment?
3. What does *spatial* mean, and how does geography's interest in space differentiate it from other disciplines?
4. What geographic features make the United Kingdom and New Zealand different?
5. What are the major terms and concepts associated with scale, coordinate systems, projections, and symbolization?
6. Why is a map made with the Mercator projection more suitable for navigation than a map made with the Robinson projection?
7. What is the difference between a dot map and a choropleth map?
8. What is a mental map?
9. What is GIS, and what typically makes it different from old-fashioned manual cartography? What are some applications of remote sensing?
10. What do geographers study, and what do they do for a living?

NOTES

1. The abbreviation B.C.E. stands for "before the Common Era," which is a reference to the dating system invented by European Christians that sets the birth of Jesus Christ as year 1. In Christian cultures, dates before that year are expressed as B.C., meaning "before Christ," and later years are identified as A.D., which stands for *anno Domini* (Latin, "in the year of our Lord"). Religion-neutral dating systems such as the one used in this book employ B.C.E. ("before the Common Era") and C.E. ("Common Era"), respectively, but the years are numbered the same in the two systems.
2. Alton Byers, "Contemporary Human Impacts on Alpine Ecosystems in the Sagarmatha (Mt. Everest) National Park, Khumbu, Nepal." *Annals of the Association of American Geographers, 95,* 2005, pp. 112–140.
3. Quoted in Geoffrey J. Martin and Preston E. James, *All Possible Worlds: A History of Geographical Ideas* (New York: Wiley, 1993), p. 150.
4. Quoted in Ajavit Gupta, *Ecology and Development in the Third World* (New York: Routledge, 1988), p. 2.
5. Quoted in Andrew Goudie, *The Human Impact on the Natural Environment* (Oxford: Blackwell, 1986), p. 6.
6. Quoted in Larry Grossman, "Man-Environment Relations in Anthropology and Geography." *Annals of the Association of American Geographers, 67,* 1977, p. 129.
7. Kathleen A. Dahl, "Culture," http://www2.eou.edu/~kdahl/cultdef.html.
8. Lieutenant General Leonid Sazhin, Federal Security Service, quoted in Katie Hafner and Saritha Rai, "Google Offers a Bird's-Eye View, and Some Governments Tremble." *New York Times,* December 20, 2005, p. 1.
9. Information courtesy of Bill Doolittle, University of Texas at Austin.

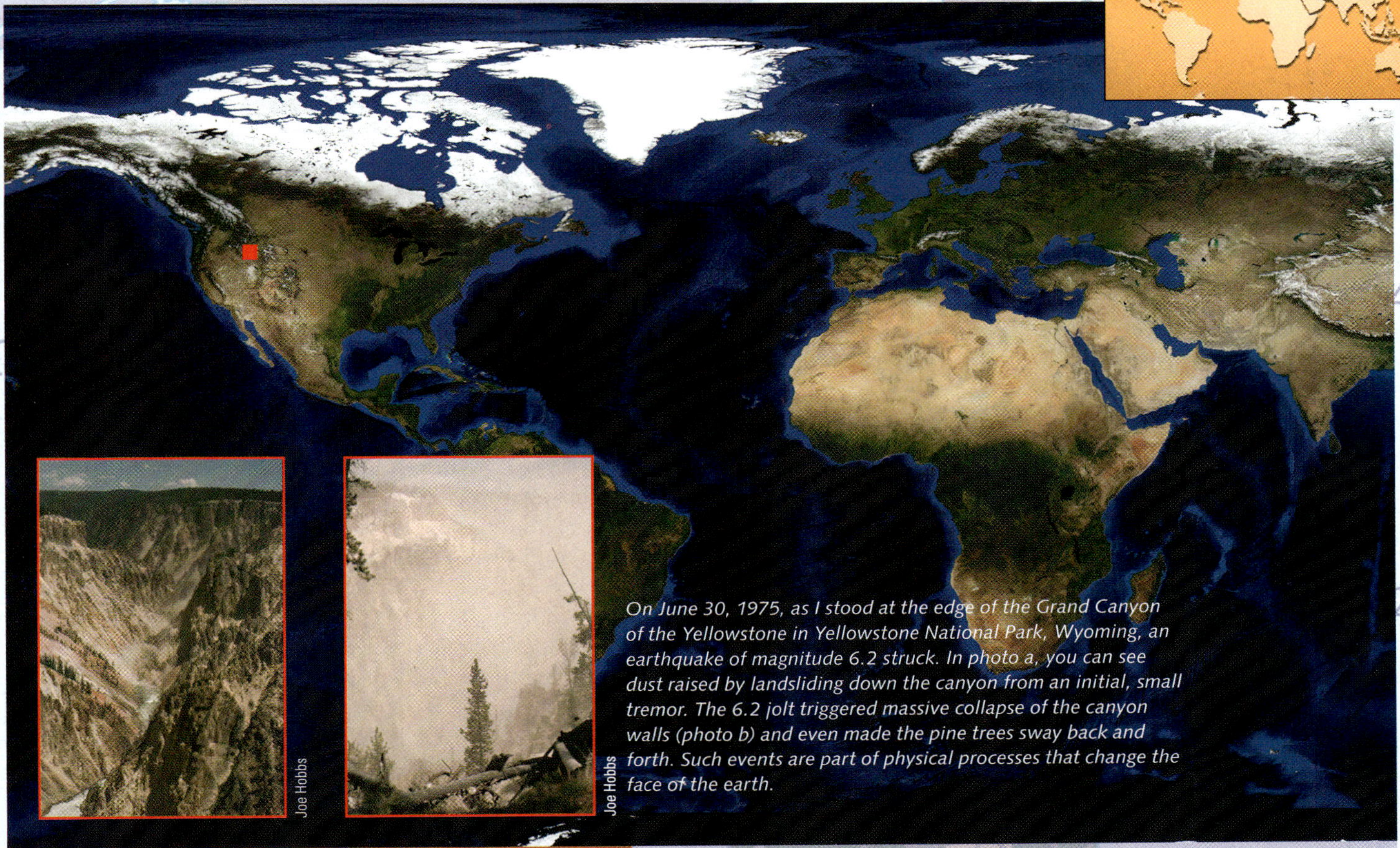

On June 30, 1975, as I stood at the edge of the Grand Canyon of the Yellowstone in Yellowstone National Park, Wyoming, an earthquake of magnitude 6.2 struck. In photo a, you can see dust raised by landsliding down the canyon from an initial, small tremor. The 6.2 jolt triggered massive collapse of the canyon walls (photo b) and even made the pine trees sway back and forth. Such events are part of physical processes that change the face of the earth.

CHAPTER 2 PHYSICAL PROCESSES THAT SHAPE WORLD REGIONS

Many issues in world regional geography relate to the interaction of people and the natural environment. This chapter and the next introduce geography's basic vocabulary about how the natural world works and explain how people have interacted with the environment to change the face of the earth.

This chapter deals with physical geography, introducing the three "spheres" that make up the earth's habitable environment:

- The **lithosphere** is the earth's outer "rind" of rock, varying in thickness from about 50 to 125 miles (60 to 200 km); it is subdivided into the categories of continental and oceanic.
- The **hydrosphere** is made up of the world's oceans and other water features, including lakes and rivers.
- The **atmosphere** is the layer of gases (mainly nitrogen and oxygen) surrounding the earth to roughly 60 miles (100 km) out, where space begins.

chapter outline

chapter objectives

This chapter should enable you to

- → Understand the tectonic forces behind some of the world's major landforms and natural hazards
- → Recognize consistent global patterns in the distribution of temperatures, precipitation, and vegetation types
- → Identify the natural areas most threatened by human activity and explain how natural habitat loss may endanger human welfare
- → Appreciate the important roles of the world's oceans in making the earth habitable
- → Describe the potential impacts of global climate change and international efforts to prevent them

---

Discussion starts with the landforms that set the terrestrial stage for human activities. The earth's lithosphere is a work in progress, and you will see how its changing surfaces provide both opportunities and threats to people. We will consider the climate and vegetation types that play such a large role in human activities and appreciate the rich diversity of wild plant and animal species. We will look briefly at the planet's often overlooked oceans and the resources they hold. Finally, we will examine how the climate may be changing and what this may mean for life on earth.

2.1 Geologic Processes and Landforms

The earth's surface is in motion. Those of you living on the West Coast of the United States have probably had some personal experience of this, the unforgettable sensation of the ground moving beneath you in a mild "temblor" or even a large earthquake. These experiences reflect global processes.

Plate Tectonics

About a century ago (1912), a German geologist named Alfred Wegener came up with an outlandish theory known as **continental drift.** Noting among other things the jigsaw puzzle–like geometry of Africa's west coast and South America's east coast, he proposed that the continents were once joined in a supercontinent (which he named Pangaea) but that they "drifted apart" over time. However, Wegener was not able to explain the forces behind these movements, and he was derided by a scientific community that seemed unwilling to accept anything other than a *terra firma* that was truly firm. Peers pronounced his conclusions "utter damned rot" and "mere geopoetry."[1] Later discoveries in deep-sea science led Wegener's basic proposition to be accepted as fact, and today a good deal is known about how the drift occurs.

The earth's lithosphere is made up of about a dozen giant and several smaller sections called **plates,** and these move in various directions in processes known collectively as **plate tectonics** (•Figure 2.1). On the ocean floors in places such as the Mid-Atlantic Ridge and the East Pacific Rise, new lithosphere is "born" as molten material rises from the earth's mantle and cools into solid rock (•Figure 2.2). Plate tectonics are often explained by the useful analogy of a "conveyor belt" (the convection cell in Figure 2.2) in constant motion. On either side of the long, roughly continuous ridges, the two young plates move away from one another, carrying islands with them; this process is called **seafloor spreading.**

Seafloor spreading has few impacts on people, but when the earth's plates collide, there is cause for great concern: **tectonic forces** are among the planet's greatest **natural hazards.** The **seismic activity** (*seismic* refers to earth vibrations, mainly earthquakes) that causes earthquakes and **tsunamis** (tidal waves) and the **volcanism** (movement of molten earth material) of volcanoes and related features are the most dangerous tectonic forces. 135, 340, 396, 421

Plates collide and converge in different ways and with different consequences. In some parts of the world—offshore from the U.S. Pacific Northwest and the east coast of Japan, for example—one plate "dives" below another in a process known as **subduction** (see Figure 2.2, and an up-close view of the process in Japan in Figure 7.4.A, page 397). The descending lithosphere is melted again as it dives into the earth's mantle along a deep linear feature known as a **trench** (for example, the Mariana Trench off Japan). Subduction is another stage along the "conveyor belt" process that will eventually see this material recycled as newborn lithospheric crust.

This subduction process releases enormous amounts of energy. Periodically, the great stress of one plate pushing beneath another is released in the form of an earthquake. The world's largest recorded earthquakes—registering 9.5 (Chile, 1960), 9.2 (United States, 1964), and 9.1 (Indonesia, 2004), respectively, on the **Richter scale,** which measures the strength of the earthquake at its source—struck along these subduction zones. This sudden displacement of a section of oceanic lithosphere is also what triggers a tsunami and the attendant loss of life and property such a powerful wave can cause (see the feature on the Great Tsunami of 2004 on page 340).

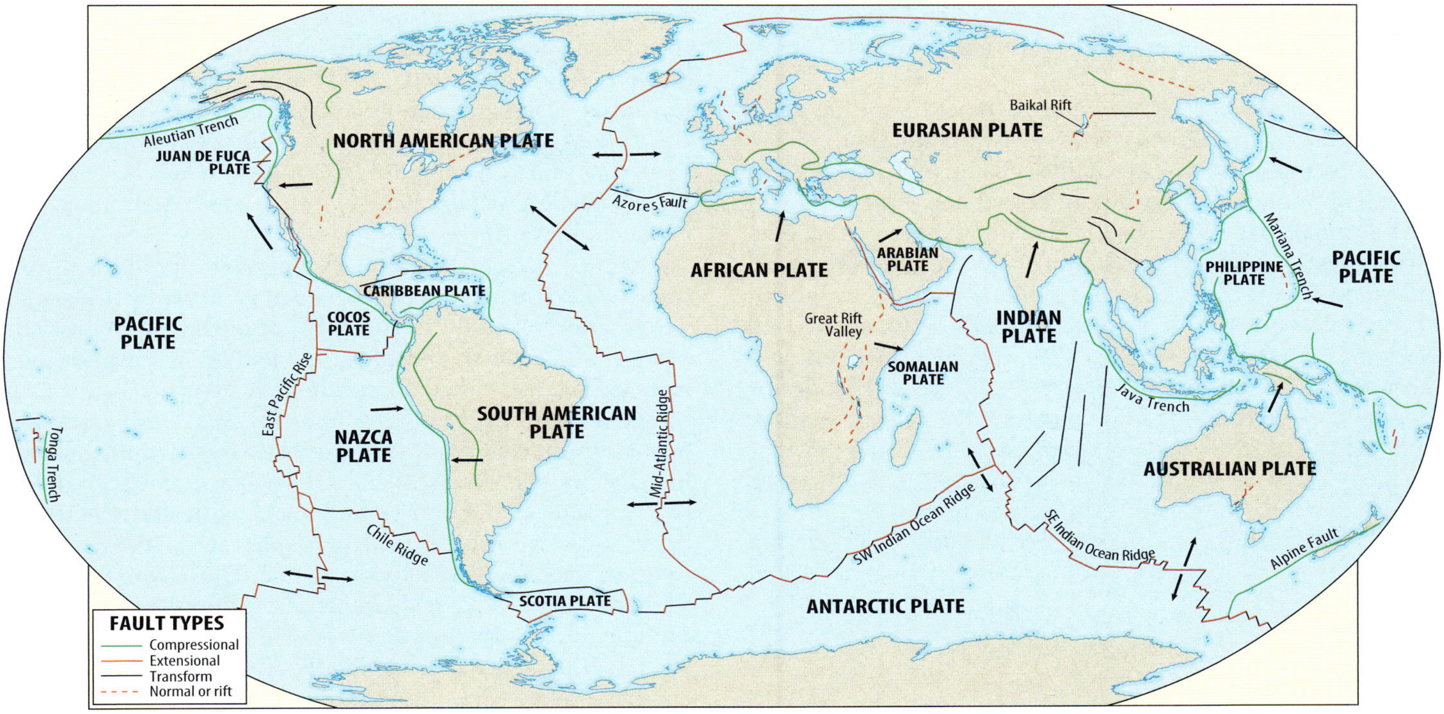

Figure 2.1 Major tectonic plates and their general direction of movement. Earthquakes, volcanoes, and other geologic events are concentrated where plates separate, collide, or slide past one another. Where they separate, **rifting** produces very low land elevations (well below sea level at the Dead Sea of Israel and Jordan, for example) or the emergence of new crust on the ocean floor (in the middle of the Atlantic Ocean, for example).

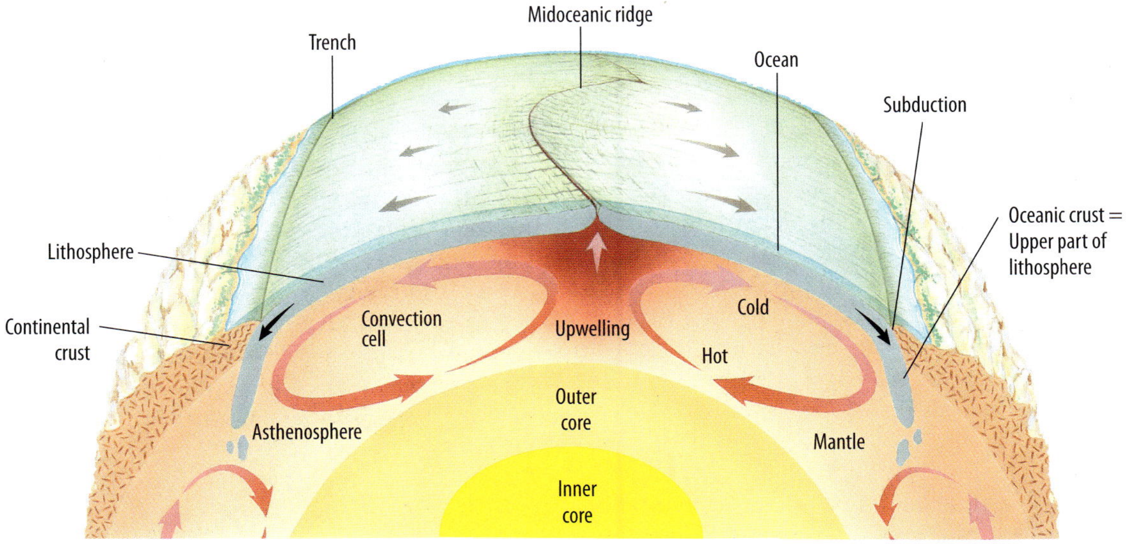

Figure 2.2 Generalized cross section of the earth, showing its main concentric layers and the process by which its lithosphere is recycled.

In other places where they meet, the lithospheric plates grind and slide along one another, as in the state of California and in Turkey.

135, The processes of rock crowding together or pulling apart
212, 336, along these fracture lines is known as **faulting,** with move-
340, ment along various kinds of **faults** causing earthquakes, the
396, 559, emergence of new landforms, and other consequences.

602 Volcanism generally takes place along and near **subduction zones** (see Figure 2.2) and also in the world's several dozen geologic hot spots where molten material has broken through the crust as a *plume* (as in Yellowstone National Park and in Hawaii—see page 420). Despite their posing a host of natural hazards to people living on their slopes or downwind—pyroclastic flows (fast-moving currents of rock fragments, hot gases, and ash), lava flows, ashfalls, and other dangers—volcanoes have beneficial qualities. Volcanic rock generally breaks down to form fertile soils (as in Ethiopia and the Nile Valley), and the range of climate and vegetation
252, types on their slopes poses opportunities for farming and
583, 357 raising livestock.

Major Landform Types

As a student of world regional geography, one of the most basic questions you will ask or answer about places or countries is "What does the landscape look like?" You will probably want to begin by describing the terrain—is it mountainous, flat, or something in between?—and then move on to the climate, vegetation, human activities, and other characteristics that shape the landscape.

In this book, the landforms of each of the world's eight regions are depicted in the "physical geography" map that introduces each region. The planet has an enormous variety of landform types, but to generalize, there are four. Each poses different opportunities for human activities and for natural vegetation and wildlife.

Hill lands and mountains. These are the high and steep features of the earth's surface. There is no formal definition to distinguish hills from mountains, and usage of these terms is usually rooted in local or regional perceptions. In the shadow of the world's highest mountain range, the Himalayas, the people of India call their southernmost high area the "Palni Hills," although the peaks there rise to 6,500 feet (2,000 m). Some people living in the midwestern United States call the rugged region of southern Missouri and northern Arkansas the "Ozark Mountains," although they rise to only 2,500 feet (760 m). Generally, the so-called mountains are higher—typically at least 2,000 to 3,000 feet (600 to 900 m) above sea level—and more rugged than hills and offer fewer possibilities for human settlement and use. Consequently, whereas many areas of hill land support moderate to dense human populations, most mountain areas are sparsely populated. The planet's mountains tend to be along and near zones of convergence of tectonic plates; not just volcanoes but other mountains (including the Himalayas) have risen as a consequence of tectonic forces.

Plains. Most of the world's people live on plains, so they are a very important landscape element to consider in world regional geography. A **plain** is ordinarily a relatively level area of slight elevation, although this usage varies too; the "High Plains" of the United States, for example, rise to over 5,000 feet (about 1,500 m) above sea level. Some plains are almost table-flat, but most are broken by gentle or moderate slopes.

Plateaus. A **plateau** is an elevated plain. Again, while no universal definition exists, a plateau is typically a plain at an elevation of 2,000 feet (610 m) or higher. By some definitions, a plateau should be terminated on at least one side by a steep edge, called an **escarpment,** that marks a sharp boundary with the lower elevation. In areas with abundant precipitation, water may have cut plateaus into hilly terrain, but the "parent" landscape is still a plateau. Depending on their latitude and elevation, and the climatic and vegetation patterns that prevail accordingly, plateaus present a wide range of livelihood options for people. The Tibetan Plateau, for example, is high, arid, and sparsely populated by pastoral nomadic peoples, while the plateaus of East Africa tend to be relatively lush, well watered, and more densely populated by nomads and farmers alike.

We turn now to the climate and vegetation types that give characteristic appearances and patterns to the world's landforms.

2.2 Patterns of Climate and Vegetation

"Everybody talks about the weather, but nobody does anything about it," Mark Twain reportedly said. The day's weather certainly does influence our conversations and decisions. More important for us as students of geography, the cumulative impacts of weather shape landscapes and influence human activities and settlements in patterns recognizable around the world. Once you know these patterns, you will be prepared to step into any world region and appreciate it better.

As you experience a warm, dry, cloudless summer day or a cold, wet, overcast winter day, you are encountering the **weather**—the atmospheric conditions occurring at a given time and place. **Climate** is the average weather of a place over a long time period. Along with surface conditions such as elevation and soil type, climatic patterns have a strong correlation with patterns of natural vegetation and in turn with human opportunities and activities on the landscape. Precipitation and temperature are the key variables in weather and climate.

Precipitation

Water is essential for life on Earth, and it is very useful to understand the processes of weather and climate that dis-

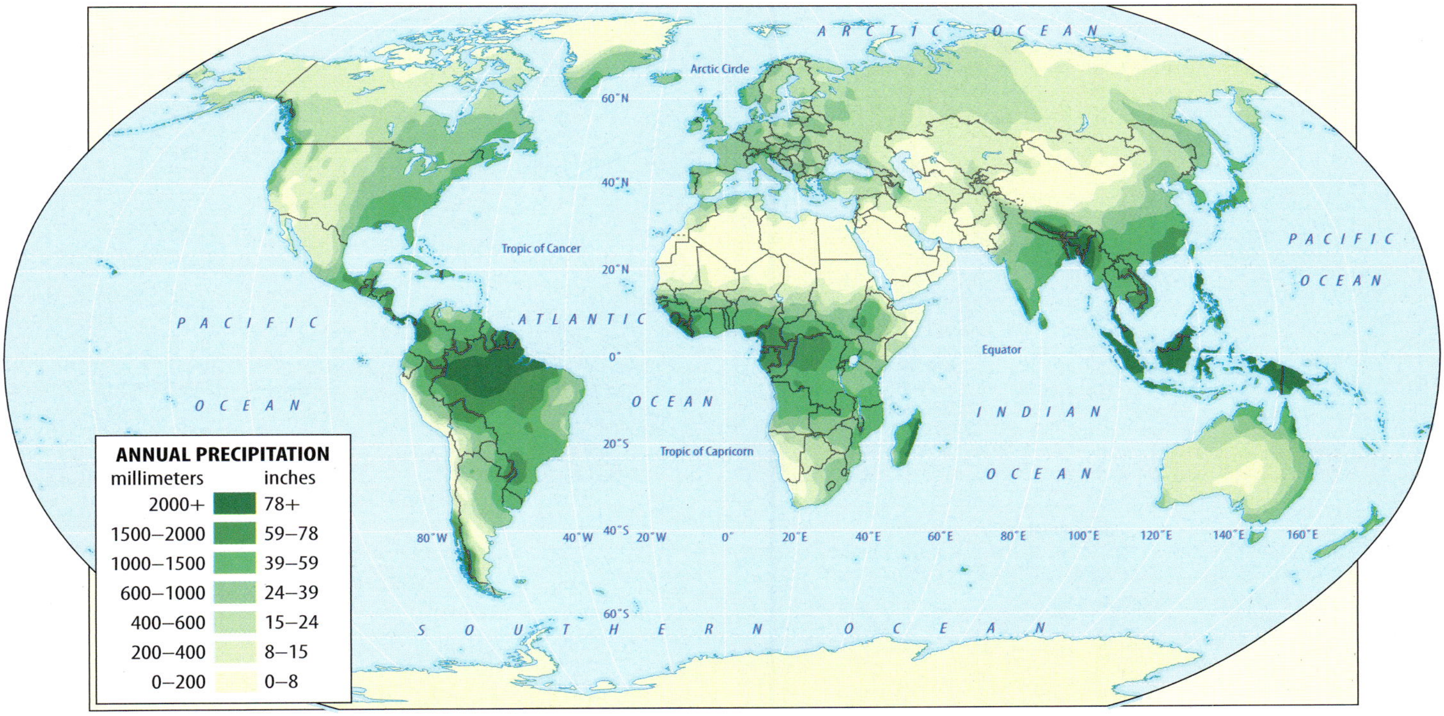

• **Figure 2.3** World precipitation map. Some geographers argue that this is the most important of all maps in understanding life on Earth.

tribute this precious resource (•Figure 2.3). Warm air holds more moisture than cool air, and **precipitation**—rain, snow, sleet, and hail—is best understood as the result of processes that cool the air to release moisture. Precipitation results when water vapor in the atmosphere cools to the point of condensation, changing from a gaseous to a liquid or solid form. The amount of cooling needed for this change depends on the original temperature and the amount of water vapor in the air.

For this cooling and precipitation to occur, generally air must rise in one of several ways. In equatorial latitudes or in the high-sun season (summer, when the sun's rays strike the earth's surface more vertically) elsewhere, air heated by intense surface radiation can rise rapidly, cool, and produce a heavy downpour of rain, an event that occurs often in summer over much of the United States (•Figure 2.4). Precipitation that originates in this way and is released from tall cumuloform clouds is called **convectional precipitation** (this and the other types are depicted in •Figure 2.5).

Orographic (mountain-associated) precipitation results when moving air strikes a topographic barrier and is forced upward. Most of the precipitation falls on the windward side of the barrier, and the leeward (sheltered) side is likely to be very dry.

Cyclonic (frontal) precipitation is generated in traveling low-pressure cells, called **cyclones,** that bring air masses with different characteristics of temperature and moisture into contact (•Figures 2.6 and 2.7). In the atmosphere, air moves from areas of high pressure to areas of low pressure. Air from masses of different temperature and moisture content is drawn into a cyclone, which has low pressure. One air mass is normally cooler, drier, and more stable than the other. These masses do not mix readily but instead come in contact within a boundary zone 3 to 50 miles (c. 5 to 80 km) wide called a **front.** A front is named after which air mass is

Ryan McGinnis/Alamy

• **Figure 2.4** A "supercell" thunderstorm complex can produce heavy rain, hail, and tornadoes.

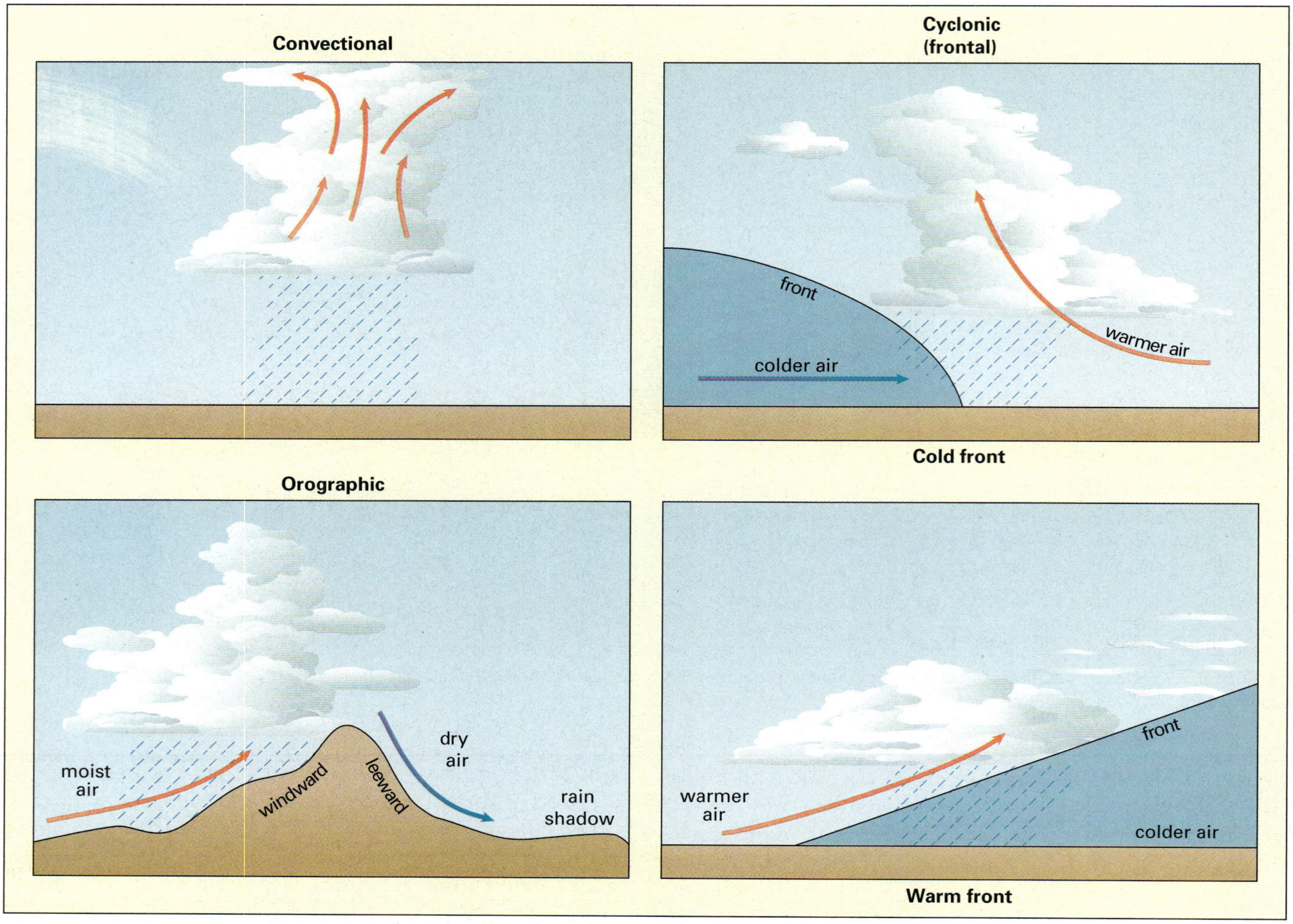

• **Figure 2.5** Diagrams showing the origins of convectional, orographic, and cyclonic precipitation. Note that cyclonic precipitation results from both cold fronts and warm fronts.

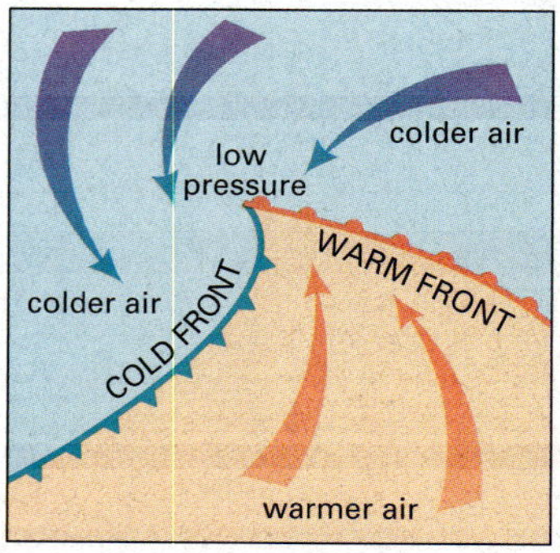

• **Figure 2.6** Diagram of a cyclone in the Northern Hemisphere.

advancing to overtake the other. In a **cold front,** the colder air wedges under the warmer air, forcing it upward and back. In a **warm front,** the warmer air rides up over the colder air, gradually pushing it back. Whether from a warm or a cold front, **frontal precipitation** is likely to result because the warmer air mass rises and condensation takes place.

• **Figure 2.7** Satellite image of Hurricane Fran approaching Florida's east coast in September 1996. Also known as cyclones and typhoons, hurricanes are the largest cyclonic features to form in the earth's atmosphere.

Aridity

The flip side of precipitation is, of course, aridity, and drought or any significant shortages in rainfall are among the natural hazards people around the planet must cope with. There are several causes of low precipitation:

High pressure. Large areas of the earth receive little precipitation because of subsiding air masses having high atmospheric pressure, called **high-pressure cells,** or **anticyclones** (•Figure 2.8). In a high-pressure cell, the air is descending and becoming warmer under the increased pressure (weight) of the air above it. As it warms, its capacity to hold water vapor increases, its relative humidity decreases, and its potential to make precipitation falls. Streams of dry, stable air moving outward from the anticyclones often bring prolonged drought to the areas below their path. Most famous of these air streams are the **trade winds** that originate in semipermanent anticyclones on the margins of the tropics and are attracted equatorward (becoming more moisture-laden as they go) by a semipermanent low-pressure cell, the **equatorial low.** As Figure 2.8 illustrates, high-pressure cells rotate clockwise in the Northern Hemisphere and counterclockwise in the Southern Hemisphere. Conversely, low-pressure systems rotate counterclockwise in the Northern Hemisphere and clockwise in the Southern Hemisphere. This pattern of opposite movements in the hemispheres is caused by the earth's rotation and is known as the **Coriolis effect.**

Atmospheric stability in coastal regions. Cold ocean waters are responsible for coastal deserts in some parts of the world, such as the Atacama Desert in Chile and the
Namib Desert in southwestern Africa. Here, air moving 507
onshore over cooler ocean waters is overlain by warm air being blown seaward from the land. Warmer air is lighter than cooler air, and a layer of warm air sitting on top of
cooler air creates an **inversion.** This **atmospheric stability** 559
prevents the cooler air from rising and thus precludes cloud formation and precipitation.

Rain shadow. As noted earlier, the windward side of mountains often receives abundant orographic precipitation. The leeward side, in contrast, is usually very dry. Nevada, which is on the leeward side of the Sierra Nevada in the
western United States, is in such a **rain shadow.** Rain shad- 247,
ows are the main cause of arid and semiarid lands in some 267, 319,
regions, but other areas with very low precipitation result 522,
from a combination of influences. The Sahara of northern 670
Africa, for example, is mainly a product of high atmospheric pressure, but the rain shadow effect of the Atlas Mountains and the presence of cold Atlantic Ocean waters along its western coast also contribute to the Sahara's dryness.

Many regions that regularly experience long periods of 283,
dry weather are also subject to changes in weather patterns 440, 458
that bring welcome rainfall. These shifts are usually seasonal and are ushered in by changing temperatures.

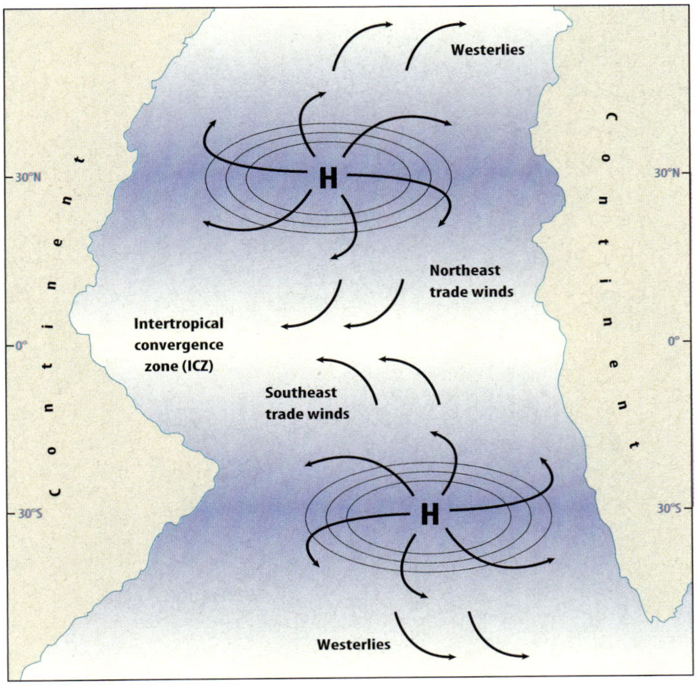

• **Figure 2.8** Idealized wind and pressure systems. Shading indicates wetter areas.

Seasons

In the middle and high latitudes, the most significant factor in determining temperatures is seasonality, which is related to the inclination of the earth's polar axis as the planet orbits the sun over a period of 365 days (•Figure 2.9). On or about June 22, the first day of summer in the Northern Hemisphere, the northern tip of the earth's axis is inclined toward the sun at an angle of 23.44 degrees from a line perpendicular to the plane of the ecliptic (the plane of the planet's orbit around the sun). This is the **summer solstice** in the Northern Hemisphere and the **winter solstice** in the Southern Hemisphere. A larger portion of the Northern Hemisphere than of the Southern Hemisphere remains in daylight, and warmer temperatures prevail. On or about September 23, and again on or about March 20, the earth reaches the **equinox** position. Its axis does not point toward or away from the sun, so days and nights are of equal length at all latitudes on earth. On or about December 22, the first day of winter in the Northern Hemisphere, the southern tip of the earth's axis is inclined toward the sun at an angle of 23.44 degrees from a line perpendicular to the plane of the ecliptic. This is the winter solstice in the Northern Hemisphere and the summer solstice in the Southern Hemisphere. A larger portion of the Southern Hemisphere than of the Northern Hemisphere remains in daylight, and warmer temperatures prevail.

Great regional differences exist in the world's annual and seasonal temperatures. In lowlands near the equator, temperatures remain high throughout the year, while in areas near the poles, temperatures remain low for most of the year. The intermediate (middle) latitudes have well-marked seasonal changes of temperature, with warmer

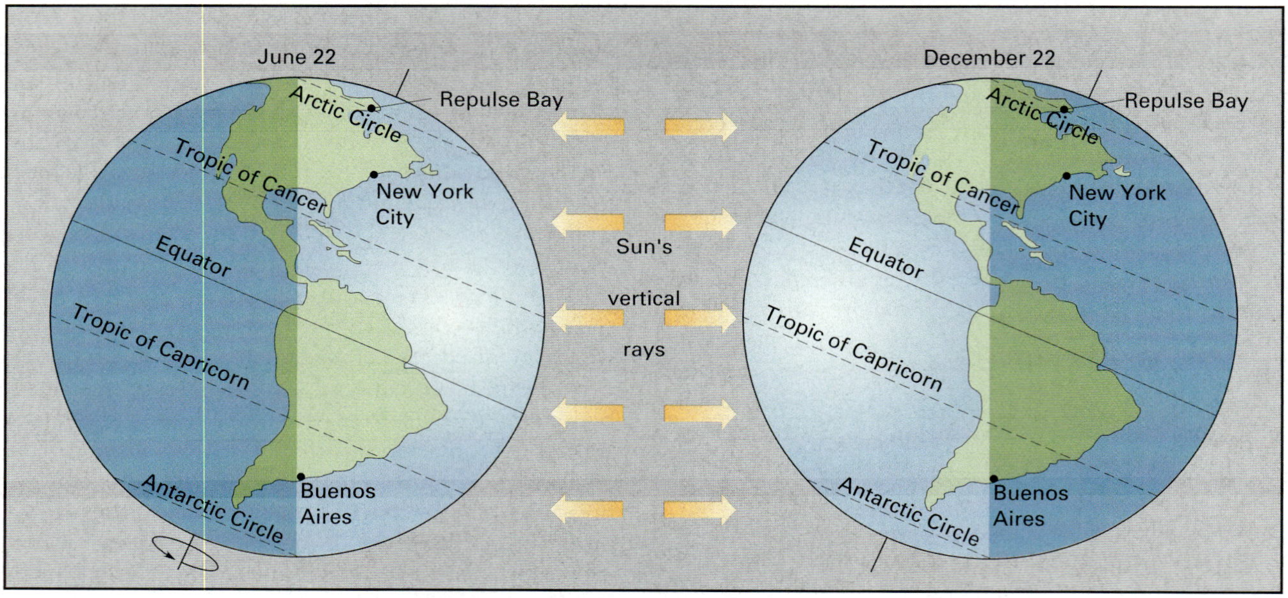

• **Figure 2.9** Geometric relationships between the earth and the sun at the solstices.

temperatures generally in the summer season of high sun and cooler temperatures in the winter season of low sun (when sunlight's angle of impact is more oblique and daylight hours are shorter). Intermittent incursions of polar and tropical air masses increase the variability of temperature in these latitudes, bringing unseasonably cold or warm weather.

Climate and Vegetation Types

The myriad combinations of precipitation, temperature, latitude, and elevation produce a great variety of local climates. Geographers group these local climates into a number of major climate types, each of which occurs in more than one part of the world (•Figure. 2.10) and is associated

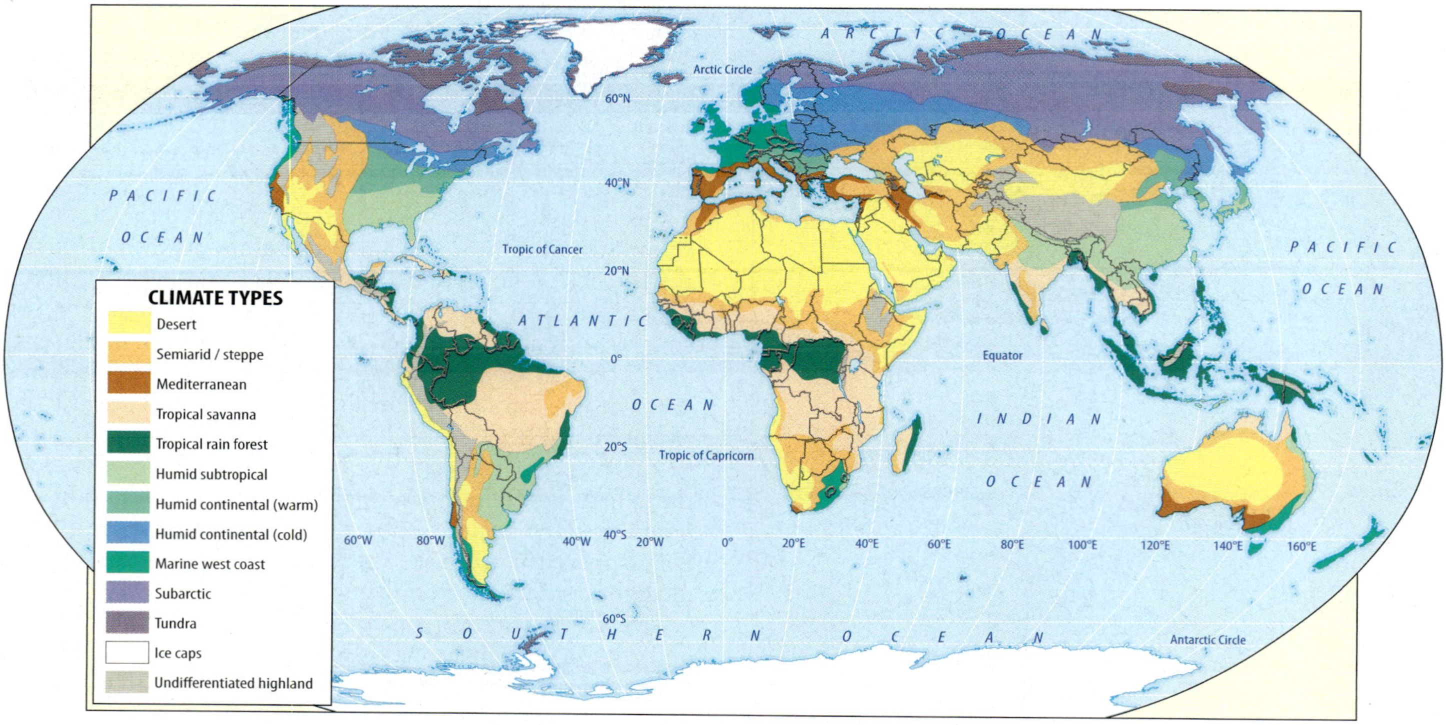

• **Figure 2.10** World climates.

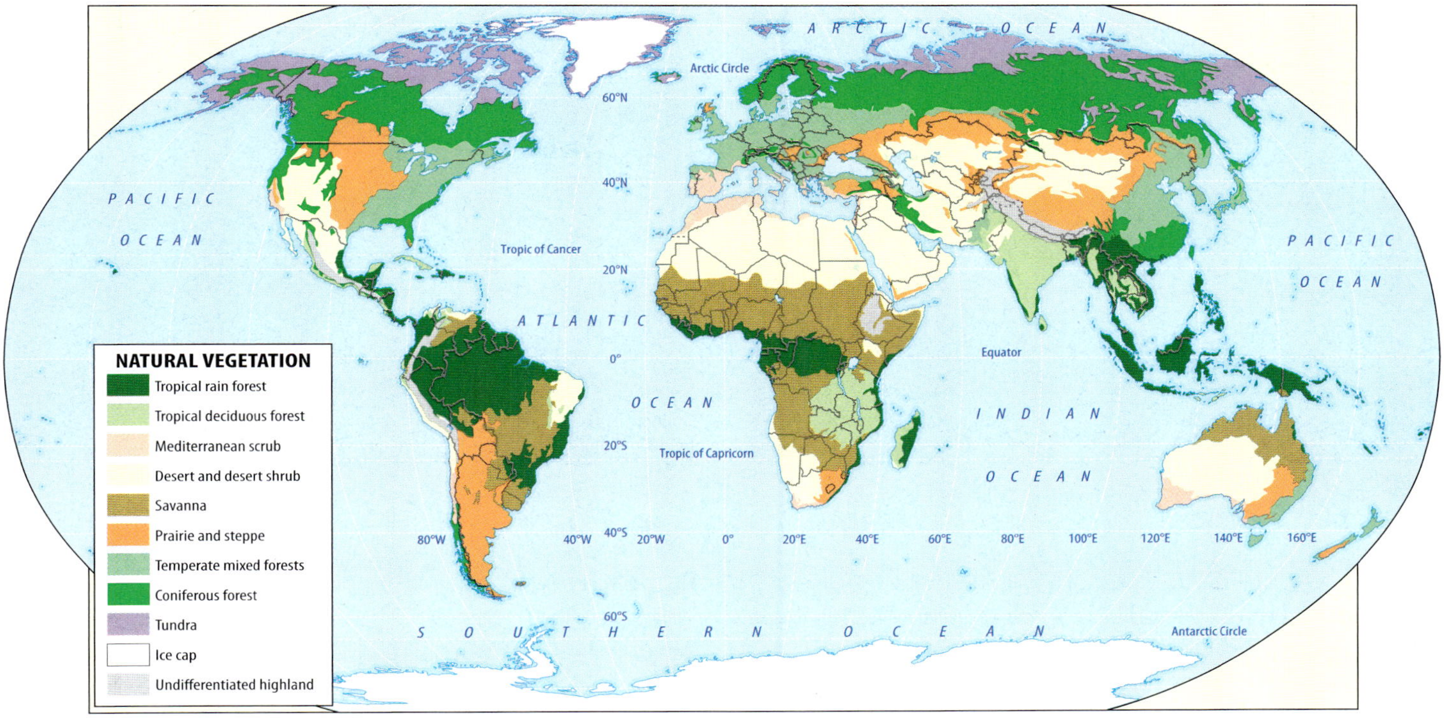

• **Figure 2.11** World biomes (natural vegetation) map.

closely with other types of natural features, especially vegetation. Geographers recognize 10 to 20 major types of terrestrial ecosystems, called **biomes,** which are categorized by dominant types of natural vegetation (•Figures 2.11 and 2.12). Climate plays the main role in determining the distribution of biomes, but differing soils and landforms may promote different types of vegetation where the climatic pattern is essentially the same. Vegetation and climate types are sufficiently related that many climate types take their names from vegetation types—for example, the tropical rain forest climate and the tundra climate. You may easily see the geographic links between climate and vegetation by comparing the maps in Figures 2.10 and 2.11. The spatial distributions of climate and vegetation types do not overlap perfectly, but there is a high degree of correlation.

In the **ice cap, tundra,** and **subarctic** climates, the dominant feature is a long, severely cold winter, making agriculture difficult or impossible. The summer is short and cool. Perpetual ice in the form of glaciers may be found at very high elevations in the lower latitudes (even in equatorial regions) and in the polar realms; the ice cap biome (•Figure 2.12a) is devoid of vegetation, except in those very few spots where enough ice or snow melts in the summer to allow tundra vegetation to grow. Tundra vegetation (•Figure 2.12b)

Joe Hobbs

• **Figure 2.12** (a) Ice cap biome, glacier in British Columbia, Canada

Joe Hobbs

• **Figure 2.12** (b) Tundra, northern Norway

• **Figure 2.12** (c) Coniferous forest, British Columbia, Canada

• **Figure 2.12** (e) Steppe, eastern Turkey

is composed of mosses, lichens, shrubs, dwarfed trees, and some grasses. Needleleaf evergreen **coniferous trees** can stand long periods when the ground is frozen, depriving them of moisture. Thus **coniferous forests,** often called **boreal forests** or **taiga,** their Russian name (•Figure 2.12c), occupy large areas where the climate is subarctic.

In **desert** and **semiarid** or **steppe** climates, the dominant feature is aridity or semiaridity. Deserts and steppes occur in both low and middle latitudes. Agriculture in these areas usually requires irrigation. The earth's largest dry region extends in a broad band across northern Africa and southwestern and central Asia. The deserts of the middle and low latitudes are generally too dry for either trees or grasslands. They have **desert shrub** vegetation (•Figure 2.12d), often only in dry riverbeds, and in many places have no vegetation at all. The bushy desert shrubs are **xerophytic** (from Greek roots meaning "dry plant"), having small leaves, thick bark, large root systems, and other adaptations to absorb and retain moisture. Grasslands dominate in the moister steppe climate, a **transitional zone** between very arid deserts and more humid areas. The biome composed mainly of short grasses is also called the **steppe,** or **temperate grassland** (•Figure 2.12e). The temperate grassland region of the United States and Canada originally supported both tall-grass and short-grass vegetation types, known in those countries as **prairies.**

The **tropical rain forest** and **tropical savanna** climates are rainy, low-latitude climates. The key difference between them is that the tropical savanna type has a pronounced dry season, which is short or absent in the tropical rain forest climate. Heat and moisture are almost always present in the tropical rain forest biome (•Figure 2.12f), where broadleaf evergreen trees (which lose their leaves, but not seasonally or in tandem) dominate the vegetation. In tropical areas with a dry season that still have enough moisture for tree growth, **tropical deciduous forest** (•Figure 2.12g) replaces the rain forest. Here the broadleaf trees are not green throughout the year; they lose their leaves and are dormant during the dry season and then add foliage and resume their growth during the wet season. The tropical

• **Figure 2.12** (d) Desert shrub, southern Sinai Peninsula, Egypt

• **Figure 2.12** (f) Tropical rain forest, Dominica, West Indies

• **Figure 2.12** (g) Tropical deciduous forest, Gir Forest, western India (note Asiatic lions in the center of the photo)

• **Figure 2.12** (i) Savanna, southern Kenya

deciduous forest approaches the luxuriance of the tropical rain forest in wetter areas but thins out to low, sparse scrub and thorn forest (•Figure 2.12h) in drier areas. **Savanna** vegetation (•Figure 2.12i), which has taller grasses than the steppe, occurs in areas of greater overall rainfall and more pronounced wet and dry seasons.

In the **marine west coast** climate, occupying the western sides of continents in the higher middle latitudes (the Pacific Northwest region of the United States, for example), warm ocean currents moderate the winter temperatures, and summers tend to be cool. Coniferous forest dominates some cool, wet areas of marine west coast climate; a good example is the redwood forest of northern California. These humid middle-latitude regions have mild to hot summers and winters ranging from mild to cold, with several other types of climate interspersed.

The **Mediterranean** climate (named after its most prevalent area of distribution, the lands around the Mediterranean Sea) is typically found between a marine west coast climate and a lower-latitude steppe or desert climate. In the summer high-sun period, it lies under high atmospheric pressure and is rainless. In the winter low-sun period, it lies in a westerly wind belt and receives cyclonic or orographic precipitation. **Mediterranean scrub forest** (•Figure 2.12j), known locally by such names as *maquis* and *chaparral*, characterizes Mediterranean climate areas. Because of the hot, dry summers, the natural vegetation consists primarily of xerophytic shrubs.

The **humid subtropical** climate occupies the eastern portion of continents between approximately 20 and 40 degrees of latitude and is characterized by hot summers, mild to cool winters, and ample precipitation for agriculture. The **humid continental** climate lies poleward of the humid subtropical type; it has cold winters, warm to hot summers, and enough rainfall for agriculture, with the greater part of the precipitation falling in the summer. This climate type is often subdivided into warm and cold subtypes, indicating the greater severity of winter in the zones closer to the poles. In middle-latitude areas with these humid subtropical and humid continental climate types, a **temperate mixed forest** with mostly

• **Figure 2.12** (h) Scrub and thorn forest, northern Zimbabwe

• **Figure 2.12** (j) Mediterranean scrub forest, southern California, United States

Joe Hobbs

• **Figure 2.12** (k) Temperate mixed forest, southern Missouri, United States

broadleaf but also coniferous trees (as in the U.S. Midwest and Northeast; •Figure 2.12k) is found. As cold winter temperatures freeze the water within reach of plant roots, broadleaf trees shed their now-colorful leaves and cease to grow, thus reducing water loss. They then produce new foliage and grow vigorously during the hot, wet summer. Coniferous forests can thrive in some hot and moist locations where porous, sandy soil allows water to escape downward, giving conifers (which can withstand drier soil conditions) an advantage over broadleaf trees. Pine forests on the coastal plains of the southern United States are an example.

Undifferentiated highland climates have a range of conditions according to elevation and exposure to wind and sun. The vegetation in these climates (•Figure 2.12l) differs greatly, depending on elevation, degree and direction of slope, and other factors. These climates are "undifferentiated" in the context of world regional geography in that a small mountainous area may contain numerous biomes, and it would be impossible to map them on a small scale.

Joe Hobbs

• **Figure 2.12** (l) Undifferentiated highland vegetation, San Juan Mountains, Colorado, United States

The world's mountain regions have a complex array of natural conditions and opportunities for human use. Increasing elevation lowers temperatures by a predictable **lapse rate,** on average about 3.6°F (2.0°C) for each increase of 1,000 feet (305 m) in elevation. In climbing from sea level to the summit of a high mountain peak near the equator (for example, in western Ecuador), a person would experience many of the major climate and biome types found in a sea-level walk from the equator to the North Pole!

2.3 Biodiversity

Anyone who has admired the nocturnal wonders of the "barren" desert or appreciated the variety of the "monotonous" Arctic can vouch for an astonishing diversity of life even in the biomes that seem least hospitable to it. But geographers and ecologists recognize the exceptional importance of some biomes because of their **biological diversity** (or **biodiversity**)—the number of plant and animal species present and the variety of genetic materials these organisms contain.

The most diverse biome is the tropical rain forest. From a single tree in the Peruvian Amazon region, the entomologist Terry Erwin recovered about 10,000 insect species. From another tree several yards away, he counted another 10,000, many of which differed from those of the first tree. Alwyn Gentry of the Missouri Botanical Garden recorded 300 tree species in a 1-hectare (2.47-acre) plot of the Peruvian rain forest. Less than 50 years ago, scientists calculated that there were 4 to 5 million species of plants and animals on earth. Now, however, their estimates are much higher, in the range of 40 to 80 million. This startling revision is based on research, still in its infancy, on species inhabiting the rain forest.

The Importance of Biodiversity

Such diversity is important in its own right, but it also has vital implications for nature's ongoing evolution and for people's lives on earth. Humankind now relies on a handful of crops as staple foods. In our agricultural systems, the trend in recent decades has been to develop high-yield varieties of grains and to plant them as vast **monocultures** (single-crop plantings). This trend, which is the cornerstone of the so-called **Green Revolution,** is controversial. On the 299, 321, 377
one hand, it puts more food on the global table. But on the other, it may render agriculture more vulnerable to pests and diseases and thus pose long-term risks of famine. In evolutionary terms, the Green Revolution has reduced the natural diversity of crop varieties that allows nature and farmer to turn to alternatives when adversity strikes. At the same time, while we remove tropical rain forests and other natural ecosystems to provide ourselves with timber, agriculture, or living space, we may be eliminating the foods, medicines, and raw materials of tomorrow, even before we

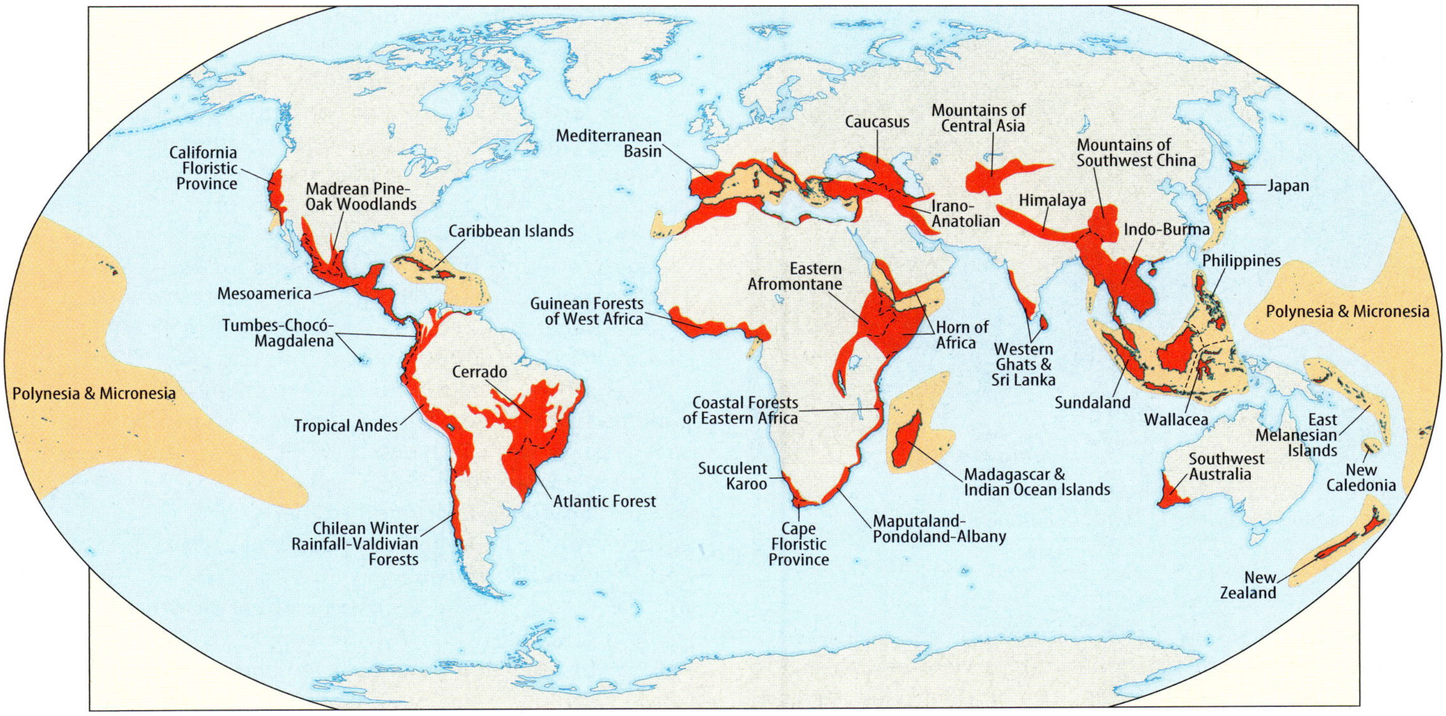

Figure 2.13 World biodiversity hot spots.

have collected them and assigned them scientific names. "We are causing the death of birth," laments the biologist Norman Myers.[2]

Regions where human activities are rapidly depleting the rich variety of plant and animal life are known as **biodiversity hot spots,** which scientists believe deserve immediate attention for study and conservation. Thirty-four priority regions identified by Conservation International are depicted in Figure 2.13. By referring to the map of the earth's biomes in Figure 2.11, you will see that most of these hot spots are in tropical rain forest areas. Many are islands that tend to have high biodiversity because species on them have evolved in isolation to fulfill special roles in these ecosystems and because human pressures on island ecosystems are particularly intense. Efforts are under way in most of these hot spots to establish national parks and other protected areas. Conservationists are also turning their attention increasingly to the state of the world's oceans, which play critical roles in the earth's physical processes.

2.4 The World's Oceans

About 71 percent of the world's surface is comprised of water, but most world regional geography textbooks (including previous editions of this one) devote little, if any, attention to the oceans and their critical resources. Let us not forget them!

Life on earth would be impossible without the roles and resources of the hydrosphere, which includes both oceans and freshwater sources such as lakes and rivers. This book deals often with issues of freshwater, which is appropriate considering the projection by the World Commission on Water that by 2025, half of the world's population will live under conditions of severe water stress.[3] In this section, the focus is on the world's oceans.

Why Should We Care about Oceans?

I live in Missouri, where a day at the beach or a plate of fresh seafood is just a dream. Even those of you living close to the sea may have very little interaction with it. Why bother thinking about the world's oceans, much less worrying about them?

It's a watery world. With so much of the earth's surface covered by seawater, it is vital that we consider how that water shapes life on the planet, including human life. Oceans have the largest role in the **hydrologic cycle,** in which solar radiation evaporates seawater into vapor that is then released onto land in the form of freshwater pre- 23
cipitation. Without the seas, the earth's usable freshwater resources would be extremely limited.

The oceans feed us. Missourians and other landlubbers may not eat fish every day or even every week, but about a billion people, or 15 percent of the world's population, rely mainly on fish for their protein, and even the average American eats 17 pounds (7.7 kg) of fish and shellfish annually (Figure 2.14). Unfortunately, growing human populations and technological improvements in the fishing industry are putting these resources under unprecedented pressure.[4] In

• **Figure 2.14** The Tsukuji fish market in Tokyo. Humanity's appetite for fish and seafood is growing, but what about the supplies?

2006, scientists issued a stern warning on the future of the world's fisheries: unless dramatic steps were taken quickly, there would be a "global collapse" of all species currently fished by 2050.[5]

Defining "collapse" as a population less than 10 percent of its previous levels, these authorities noted that 30 percent of the fish species currently exploited had already collapsed (•Figure 2.15). Their good news was that this gloomy scenario could be averted by action on several fronts, including reducing the number of unwanted fish caught in nets and reducing overfishing where it is recognized to be a problem.
201, 339, 405, 634 More good news is that **aquaculture** (including **fish farming**) has the potential to catch up with and increasingly substitute for wild-caught fish stocks. There has been explosive growth in aquaculture in China within the last decade, and there is unrealized potential in that country and others. Some of this aquaculture is what may be described as "sustainable seafood," including the unlikely harvest of shrimp raised in well water deep in the Arizona desert.[6]

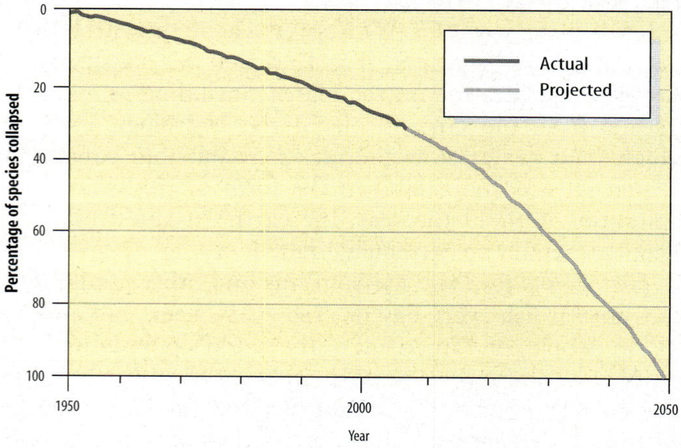

• **Figure 2.15** Percentage of ocean fish species "collapsed" (defined as less than 10 percent remaining).

They provide energy and other raw materials for human use. As terrestrial sources of conventional energy dwindle, our fossil fuel–addicted economies create demand for new supplies. Deep seas are the final frontier for energy exploration. Under the U.S. territorial waters of the Gulf of Mexico, there are as many as 40 billion barrels of petroleum (enough to supply the U.S. oil demand for five years at 2007 levels of consumption). But getting to that oil is going to be a problem: it lies beneath 10,000 feet (3,050 m) of water and then 5 miles (8 km) of rock, salt, and sand. Meanwhile, there is a huge potential to capture unconventional energy supplies from the sea; Scottish engineers have demonstrated that wave power can be used to spin turbines and generate electricity, for example, and in theory, the temperature difference between surface and deeper waters in warm tropical seas can be exploited to produce electricity in a process called *ocean thermal energy conversion* (OTEC). Prospects are increasing for the deep-sea mining of other minerals, including gold, silver, and the copper, cobalt, and nickel held within manganese nodules strewn across much of the world's seafloor.

They play important roles in trade and commerce. The seafaring days of humankind are far from over. A remarkable 90 percent of global trade is seaborne. What about all of that Chinese merchandise finding its way into U.S. retail stores? Some 98 percent of it travels across the Pacific on cargo ships.[7] There is an economic rationale behind this heavy reliance on seaborne trade. Air freight can cost 20 times as much as sea freight. Safeguarding the security of global seaborne commerce is one of the enduring themes of geopolitical strategy making and has all too often been the trigger for regional conflicts and wider wars. 229, 305

2.5 Global Environmental Change

Governments have had great success in recent years in delimiting national parks and other reserves, both on land and at sea, for the protection of biodiversity. Yet even with the most stringent efforts to protect the living things in these wilderness areas, their future is uncertain. Atmospheric changes are occurring that will have profound effects on natural systems and on human uses of the earth. Many of these changes are attributable to human activities, and the final section of this chapter is a prelude to the following chapter on human processes affecting the planet.

Climatic Change

Until about 2000, there was considerable uncertainty in the science of climate change. Most scientists at that time insisted that human activities were responsible for a documented warming of the earth's surface, but a significant minority either insisted that warming was not occurring or, if it was, that a natural climatic cycle was responsible. By 2007, however, most of the uncertainty had evaporated.

In its 2007 report, the atmospheric scientists (representing 154 countries) who make up the United Nations–sponsored **Intergovernmental Panel on Climate Change (IPCC)** concluded that global warming is "unequivocal." The global mean temperate increased by 1.4 degrees Fahrenheit (1.4°F) (about 0.8 degrees Celsius, 0.8°C) since the late 19th century. Human production of **greenhouse gases** such as **carbon dioxide (CO_2)** was, according to the IPCC, "very likely" responsible for this warming (•Figure 2.16). In the scientists' jargon, "very likely" means a more than 90 percent probability; just six years earlier, the IPCC said the human factor was "likely," meaning a 66 to 90 percent probable cause.[8]

The Greenhouse Effect

In 1827, the French mathematician Jean-Baptiste Fourier established the concept of the **greenhouse effect,** noting that the earth's atmosphere acts like the transparent glass cover of a greenhouse (•Figure 2.17)—for modern purposes, think of a car's windshield. Visible sunlight passes through the glass to strike the planet's surface. Ocean and land (the floor of the greenhouse or the car's upholstery) reflect the incoming solar energy as invisible infrared radiation (heat). Acting like the greenhouse glass or car windshield, the earth's atmosphere traps some of that heat.

In the atmosphere, naturally occurring greenhouse gases such as carbon dioxide and water vapor make the earth habitable by trapping heat from sunlight. Concern over global warming focuses on human-derived sources of greenhouse gases, which trap abnormal amounts of heat. Carbon dioxide released into the atmosphere from the burning of coal, oil, and natural gas is the greatest source of concern, but **methane** (from rice paddies and the guts of ruminating animals like cattle), **nitrous oxide** (from the breakdown of nitrogen fertilizers), and **chlorofluorocarbons,** or **CFCs** (used as coolants and refrigerants) are also greenhouse gases resulting from human activities. (CFCs also destroy stratospheric ozone, a gas that has the important effect of preventing much of the sun's harmful ultraviolet radiation from reaching the earth's surface.) In the car-as-earth metaphor, continued production of these greenhouse gases has the effect of rolling up the car windows on a sunny day, with the result of increased temperatures and physical overheating of the occupants.

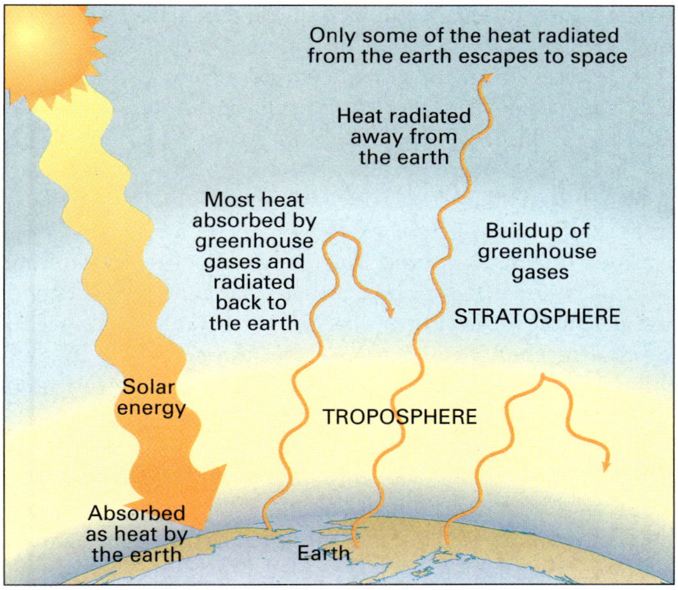

• **Figure 2.17** The greenhouse effect. Some of the solar energy radiated as heat (infrared radiation) from the earth's surface escapes into space, while greenhouse gases trap the rest. Naturally occurring greenhouse gases make the earth habitable, but carbon dioxide and greenhouse gases emitted by human activities accentuate the greenhouse effect, making the planet unnaturally warmer.

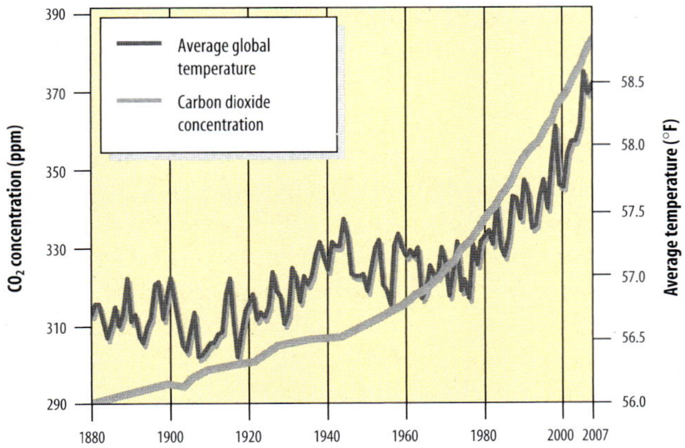

• **Figure 2.16** Industrialization and the burning of tropical forests have produced a steady increase in carbon dioxide emissions. Many scientists believe that these increased emissions explain the corresponding steady increase in the global mean temperature.

Although formal records of meteorological observations began just over a century ago, past climates left evidence in the form of marine fossils, corals, glacial ice, fossilized pollen, and annual growth rings in trees. These indirect sources, along with formal records dating back to 1861, indicate that the 20th century was the warmest of the past six centuries and that the five warmest years in the past 600 years were 1998, 2002, 2003, 2005, and 2007.

The Effects of Global Warming

Scientists have long predicted that global warming due to human activities will have profound and wide-ranging effects. In 1896, the Swedish scientist Svante Arrhenius feared that Europe's growing industrial pollution would eventually double the amount of carbon dioxide in the earth's atmosphere and as a consequence raise the global mean temperature as much as 9°F (5°C). In its 2007 report, the IPCC

reached a remarkably similar conclusion: it calculated the impact of doubling carbon dioxide from the preindustrial 1750 levels as a mean global temperature rise of 3.5°F to 8°F (2°C to 4.4°C). Other climate change models today calculate a doubling of carbon dioxide from year 2000 levels, concluding that the mean global temperatures might warm from 3°F to 11°F (1.7°C to 6.1°C) by 2100. The estimated range and consequences of the increases vary widely because of different assumptions about the little-understood roles of oceans and clouds. Most of these models concur that because of the slowness with which the world's oceans respond to changes in temperature, the effects of this rise
31 will be delayed by decades. Briefly, here are the major predictions:

A warmer climate overall, but . . . Geographers are most concerned with where the anticipated climatic changes will occur and what effects these changes will have. Computer models conclude that there will not be a uniform temperature increase across the entire globe but rather that the increases will vary spatially and seasonally. Differing models sometimes produce contradictory results about the timing and impacts of warming. Models predicting warmer summers generally forecast that crop productivity will decline, while models predicting warmer winters generally forecast an increase in crop yields.

More precipitation overall, but also more pronounced drought. Higher temperatures will mean increased evaporation from the world's oceans. This will result in more precipitation, but it will be distributed irregularly. The 2007 IPCC report forecasts more precipitation in the higher latitudes but less, with intense and longer droughts, in the lower latitudes. What the IPCC calls "heavy precipitation events" will be more common, and the magnitude of hurricanes (whose strength increases with warmer ocean surface waters) will increase.

682 *Pronounced warming in the polar regions.* Geographically, the impacts of global warming appear to be greatest at the higher latitudes. Sailing on a Russian nuclear-powered icebreaker, I saw open water at the North Pole in August 1996 (•Figure 2.18), and subsequent years have seen accelerated shrinkage of polar ice. The 2007 IPCC report concluded that the average coverage of Arctic sea ice has shrunk 2.7 percent per decade since 1978, with summertime ice decreasing 7.4 percent per decade. The trend is self-perpetuating, because the darker ocean waters that replace the white ice cover absorb ever more solar radiation. Some quarters are cheering the trend. A further retreat of polar ice would bode well for maritime shipping through the long-icebound Northwest Passage across Arctic Canada and through the northern sea route across the top of Russia, which in the past has been navigable only in summer and only with the aid of
682 icebreakers.

Shifting biomes, with species extinction. There is general agreement that with global warming, the distribution of climatic conditions typical of biomes will shift poleward worldwide and upward in mountainous regions. Many animal species will be able to migrate to keep pace with changing temperatures, but plants, being stationary, will not. Conservationists who have struggled to maintain islands of habitat as protected areas are particularly concerned about rapidly changing climatic conditions. For example, ecologists have documented that two-thirds of Europe's butterfly species have already shifted their habitat ranges northward by 22 to 150 miles (35 to 240 km), coinciding with the continent's warming trend. The World Wildlife Fund recently issued a report warning that due to global warming, as much as 70 percent of the natural habitat could be lost, and 20 percent of the species could become extinct, in the arctic and in subarctic regions of Canada, Scandinavia, and Russia. The report predicted that more than a third of existing habitats in the American states of Maine, New Hampshire, Oregon, Colorado, Wyoming, Idaho, Utah, Arizona, Kansas, Oklahoma, and Texas would be irrevocably altered by global warming.[9]

Rising sea levels. There is also general agreement that if global temperatures rise, so will sea levels as glacial ice melts and as seawater warms and thus occupies more volume. Sea levels rose 6 to 9 inches (15 to 23 cm) in the 20th century. The 2007 IPCC report forecast a further rise of 7 to 24 inches (18 to 61 cm) rise by 2100, but other scientists call these figures too conservative because they do not account enough for the melting of Greenland and Antarctic land ice. Research published in the journal *Science* in 2007 predicted a rise in sea level of 20 to 55 inches (51 to 140 cm) by 2100, and NASA scientist James Hansen forecast even higher waters.

Decision makers in coastal cities throughout the world, in island countries, and in nations with important lowland

Joe Hobbs

• **Figure 2.18** The passengers and crew of the Russian icebreaker *Yamal* found open water at the North Pole on August 16, 1996. Much of the voyage from northeastern Siberia was ice-free. Similar observations since then have contributed to widespread concern about global warming. The ship's global-positioning instrument indicated that the vessel's bow was at 90 degrees north latitude—the North Pole—when this photo was taken.

areas adjacent to the sea are worried about the implications of rising sea levels. The coastal lowland countries of the Netherlands and Bangladesh have for many years been outspoken advocates for reducing greenhouse gases. The president of the Maldives, an Indian Ocean country comprised entirely of islands barely above sea level, pleaded, "We are an endangered nation!"

Geopolitical instability. There is growing evidence that the poorest and most overcrowded nations will be hit hardest by climate change and that political and economic instability will result. The IPCC forecasts that rising sea levels will flood tens of millions of poor people out of their homes each year and that by 2080, possibly 200 to 600 million people will face starvation because of failing crops. By then, 1 to 3 billion people in developing countries will face shortages of freshwater related to growing aridity. Epidemics of malaria and other killer diseases will be more widespread.

Such environmental crises will likely lead to large movements of refugees and to conflict over resources, upsetting the fragile balance of power between countries. In a report published in 2007, U.S. military thinkers concluded that global warming "presents significant national security challenges to the United States," adding, "Projected climate change will seriously exacerbate already marginal living standards in many Asian, African and Middle Eastern nations, causing widespread political instability and the likelihood of failed states. The chaos that results can be an incubator of civil strife, genocide and the growth of terrorism. The U.S. may be drawn more frequently into these situations." Reacting to the report, Congressman Edward Markey of Massachusetts wrote that "global warming's impacts on natural resources and climate systems may create the fiercest battle our world has ever seen. If we don't cut pollution and head off severe global warming at the pass, we could see extreme geopolitical strain over decreased clean water, environmental refugees and other impacts."[10]

There is a growing international political will to take the strong and costly measures that may be necessary to prevent such dire scenarios. Present efforts to reduce the production of greenhouse gases focus on the world's wealthiest nations, but there is a good reason for the growing pressure on China to join these efforts. China and the United States produced about the same amount of carbon emissions in 2007 and together accounted for roughly 40 percent of the entire global emissions output. As recently as 1994, China was predicted to overtake the United States as the world's largest CO_2 producer by 2019. The fact that China achieved this distinction in 2007 instead is testimony to the dizzying pace of that country's development—a theme that echoes throughout this book. It also underscores the seriousness of the threats posed by global warming. Scientists have begun speaking of a **tipping point** in climate change, a yet unknown point at which feedback effects amplify temperature changes. The tipping point would bring a rapid acceleration in temperature and of effects like more violent storms, dramatic crop losses, and spreading deserts.

What Can Be Done about Global Climate Change?

Some scientists believe that the die has already been cast. No matter how hard we try to reduce greenhouse gases now, the earth's atmosphere will warm dramatically anyway, especially because the oceans are so slow to warm up. Most scientists, however, are more proactive, urging dramatic steps to mitigate global warming. Many policymakers have also been action-oriented, especially in Europe, and even the traditional holdout on climate change action—the United States—has begun to join in the chorus. These are the most commonly prescribed actions:

Negotiate and implement international treaties to reduce greenhouse gas emissions. Most scientists and policymakers agree that the best way to confront global climate change is to implement international treaties to reduce emissions (see Geography of Energy, page 36). There are precedents proving that countries can unite in effective action against global environmental change. Thanks to the **Montreal Protocol** and its amendments signed by 37 countries in the late 1980s, the production of CFCs worldwide has all but ceased; the wealthier countries no longer produce them, and the less developed countries are scheduled to cease production by 2010. The anticipated result is that there will be a marked reduction in the size of stratospheric ozone "holes" that have been observed seasonally over the southern and northern polar regions since about 1985. CFC molecules have very long life spans, but scientists predict that the effects of the Montreal Protocol will be noticeable by 2010, with the ozone layer recovering to pre-1980 levels by 2050. This recovery will also depend on the successful phase-out of another ozone-destroying refrigerant known as HCFC-22 (or R22), which is scheduled to be eliminated worldwide by 2040.

Cut emissions through market-based incentives. Compared with richer nations, the less developed countries other than China and India release relatively little CO_2 into the atmosphere. However, they do contribute to the greenhouse effect, especially because of their heavy reliance on
trees and other vegetation for fuel. When burned, trees re- 42
lease the CO_2 that they stored in the process of photosynthesis. At the same time, they cease to exist as organisms that in the process of photosynthesis remove CO_2 from the atmosphere.

Because the less developed countries produce relatively little CO_2, policymakers are considering innovative ways to keep them from doing so and to encourage them to develop clean industrial technologies. One approach is that of a system of **tradable permits** (also called a **cap-and-trade** or **emission-trading system**). In these schemes, already being used by the United Nations, the European Union, and

GEOGRAPHY OF ENERGY

The Kyoto Protocol

In 1997, at a world conference on climate change, 84 countries (of the 160 countries represented at the conference) signed the **Kyoto Protocol,** a landmark international treaty regarding climate change. The agreement requires 38 more developed countries (known as the Annex I countries) to reduce carbon dioxide emissions to 5 percent or more below their 1990 levels by the year 2012. The United States pledged to cut its emissions to 7 percent below 1990 levels by that date. This would be a huge cut; in 2007, the United States produced over 30 percent more carbon dioxide than it did in 1990. The European Union made a promise of 8 percent below 1990 levels, and Japan promised a 6 percent reduction.

Meeting these pledges would require substantial legislative, economic, and behavioral changes in these more affluent countries. Transportation and other technologies that use fossil fuels would have to become more energy-efficient, making these technologies at least temporarily more expensive. Gasoline prices would rise, so consumers would feel the pinch. Advocates of the protocol argue that the initial sacrifices would soon be rewarded by a more efficient and competitive economy powered by cleaner and cheaper sources of energy derived from the sun. Opponents, however, feel that higher fuel prices will be too costly for the United States and other industrialized economies to bear. Their position made it difficult for some signatories of the Kyoto Protocol to ratify the treaty. For the protocols to take effect, at least 55 countries must have ratified them, and the industrialized Annex I countries that ratify must have collectively produced at least 55 percent of the world's total greenhouse gas emissions in 1990.

By mid-2004, a total of 120 countries—including all the European Union members at that time and Japan—had ratified it, and the critical threshold of 55 percent of the 1990 emissions by Annex I countries was near. Ratification by either the United States or Russia would cross that threshold, putting the treaty into effect.

Much to the dismay of the European Union countries, U.S. President George W. Bush rejected the Kyoto Protocol soon after taking office in 2001. The Bush administration had two objections: the potentially high economic cost of implementing the treaty and the fact that China, along with all the world's less developed countries, was not required by the Kyoto Protocol to take any steps to reduce greenhouse gas emissions. China argued then, as it still does, that it is a developing country whose growth should not be hampered by pressure from countries that became rich by burning fossil fuels.

The European Union and Japan expressed outrage that the United States broke ranks with them on global warming. All eyes turned to Russia, whose ratification of the Kyoto Protocol would fulfill the 55 percent of emissions requirement and therefore finally put the treaty into effect in all ratifying countries. Anticipating financial gains to be earned by selling carbon emissions credits, Russia ratified the treaty late in 2004, and it went into force for its signatory nations in 2005. The European countries seized the initiative in cutting emissions, but tremendous hurdles remained for Kyoto's targets to be met. The Kyoto agreement will expire in 2012, and discussions are already under way about what new—and presumably more dramatic—measures will be needed to succeed it.

---

the unregulated "voluntary market," managers require the richer nations to achieve a net reduction in carbon emissions. It is very expensive for the already highly energy-efficient richer countries to cut their greenhouse gas emissions but relatively inexpensive for rich countries to pay a poorer, energy-inefficient country to cut its emissions. These schemes therefore allow the richer countries either to make the cuts at home or to offset their emissions by buying credits from poorer countries that actually cut carbon emissions. The poorer country selling the credits is obliged to use the income to invest in energy-efficient and nonpolluting technologies.

China is making big profits from emissions trading, using European money to reduce emissions of HFC-23, a potent greenhouse gas produced in many Chinese refrigerator factories. Russia is another big advocate of emission-trading
170 schemes like these. Its industries are producing well below their 1990 carbon dioxide emission levels, so the country has much to gain by selling its unused emission rights. One buyer is Denmark, which is paying to convert coal-fired electricity plants to cleaner-burning gas in Siberia. In such a trading scheme, both countries apparently benefit, as does the global atmosphere. However, this practice is still young, and critics say it ultimately lets the big polluters in the industrialized countries avoid the hard work of cutting emissions further.

Increase carbon sequestration. In the process of photosynthesis, plants absorb carbon dioxide. Collectively, the world's forests, farmlands, and seas (which contain carbon-absorbing phytoplankton) serve as giant **carbon sinks.** Some forest- and farm-rich countries want to receive credit for having these natural buffers against global warming, but the Kyoto Protocol does not allow such so-called **carbon sequestration** to be factored in. The Kyoto treaty negotiators felt that giving credit for carbon sinks would keep big polluters like the United States from attacking the root problem, the greenhouse gases. But not given credit for carbon sinks also means that developing countries lack an important incentive to plant new forests or replace forests that have been cut.

The next chapter discusses many ways in which the world's developed and less developed countries must confront problems like global climate change differently. Like this chapter, it will conclude with some specific ideas about how to deal with some of the most pressing issues of our time.

SUMMARY

→ The earth's three layers of habitable space are the hydrosphere, atmosphere, and lithosphere. The lithosphere is made up of separate plates that are in motion, a process known as plate tectonics. These movements result in mountain building, volcanic activity, earthquakes, and other consequences.

→ The earth's major landform types may be classified as hill lands and mountains, plateaus, and plains.

→ Weather refers to atmospheric conditions prevailing at one time and place. Climate is a typical pattern recognizable in the weather of a region over a long period of time. Climatic patterns have a strong correlation with patterns of vegetation and in turn with human opportunities and activities in the environment.

→ Warm air holds more moisture than cool air, and precipitation is best understood as the result of processes that cool the air to release moisture.

→ Most of the sun's visible short-wave energy that reaches the earth is absorbed, but some of it returns to the atmosphere in the form of infrared long-wave radiation, which generates heat and helps warm the atmosphere. This is the earth's natural greenhouse effect.

→ Geographers group local climates into major climate types, each of which occurs in more than one part of the world and is associated with other natural features, particularly vegetation. Geographers recognize 10 to 20 major types of ecosystems or biomes, which are categorized by the type of natural vegetation. Vegetation and climate types are so sufficiently related that many climate types are named for the vegetation types.

→ Some biomes are particularly important because of their biological diversity—the number of plant and animal species and the variety of genetic materials these organisms contain. Regions where human activities are rapidly depleting a rich variety of plant and animal life are known as biodiversity hot spots, places scientists believe deserve immediate attention for study and conservation.

→ Oceans cover about 71 percent of the earth's surface. They play the key role in the hydrologic cycle, sustain large numbers of people through the protein in fish and seafood, and contain valuable mineral resources.

→ The scientific community is convinced with 90 percent or greater certainty that human activities, particularly the production of carbon dioxide and other greenhouse gases, is responsible for global warming. Computer-based climate change models use the scenario of a doubling of atmospheric carbon dioxide and indicate that the mean global temperature might warm up by an additional 11 degrees Fahrenheit (6.1 degrees Celsius) by the year 2100. The general scientific consensus is that with global warming, the distribution of climatic conditions typical of biomes will shift poleward and upward in elevation, sea levels will rise, and mean global precipitation will increase (but with drought intensified in some areas).

→ The Kyoto Protocol is an international agreement requiring the industrialized countries that ratified it to make substantial cuts in their carbon dioxide emissions to reduce global warming. The United States dropped its support for the treaty. It went into effect for the ratifying countries after Russia ratified it in 2004, and in its present form, it will expire in 2012.

KEY TERMS + CONCEPTS

- aquaculture (p. 32)
- anticyclone (p. 25)
- atmosphere (p. 19)
- atmospheric stability (p. 25)
- biodiversity (p. 30)
- biodiversity hot spots (p. 31)
- biological diversity (p. 30)
- biomes (p. 27)
    - boreal forest (p. 28)
    - coniferous forest (p. 28)
    - desert shrub (p. 28)
    - Mediterranean scrub forest (p. 29)
    - prairie (p. 28)
    - savanna (p. 29)
    - steppe (p. 28)
    - taiga (p. 28)
    - temperate grassland (p. 28)
    - temperate mixed forest (p. 29)
    - tropical deciduous forest (p. 28)
- cap-and-trade system (p. 35)
- carbon sequestration (p. 36)
- carbon sink (p. 36)
- climate (p. 22)
    - desert (p. 28)
    - humid continental (p. 29)
    - humid subtropical (p. 29)
    - ice cap (p. 27)
    - marine west coast (p. 29)
    - Mediterranean (p. 29)
    - semiarid (p. 28)
    - subarctic (p. 27)
    - tropical rain forest (p. 28)
    - tropical savanna (p. 28)
    - tundra (p. 27)
    - undifferentiated highland (p. 30)
- coniferous trees (p. 28)
- continental drift (p. 20)
- Coriolis effect (p. 25)
- cyclone (p. 23)
- emission-trading system (p. 35)
- equatorial low (p. 25)
- equinox (p. 25)
- escarpment (p. 22)
- fault (p. 22)
- faulting (p. 22)
- fish farming (p. 32)
- front (p. 23)
    - cold (p. 24)
    - warm (p. 24)
- Green Revolution (p. 30)
- greenhouse effect (p. 33)

greenhouse gases (p. 33)
- carbon dioxide (CO_2) (p. 33)
- chlorofluorocarbons (CFCs) (p. 33)
- methane (p. 33)
- nitrous oxide (p. 33)

high-pressure cell (p. 25)
hydrologic cycle (p. 31)
hydrosphere (p. 19)
Intergovernmental Panel on Climate Change (IPCC) (p. 33)
inversion (p. 25)
Kyoto Protocol (p. 36)
lapse rate (p. 30)
lithosphere (p. 19)
modified Mercalli scale (p. xxx)
monoculture (p. 30)
Montreal Protocol (p. 35)
natural hazards (p. 20)
plain (p. 22)
plateau (p. 22)
plates (p. 20)
plate tectonics (p. 20)
precipitation (p. 23)
- convectional (p. 23)
- cyclonic (p. 23)
- frontal (p. 23)
- orographic (mountain-associated) (p. 23)

rain shadow (p. 25)
Richter scale (p. 20)
rifting (p. 21)
seafloor spreading (p. 20)
seismic activity (p. 20)
subduction (p. 20)
subduction zone (p. 22)
summer solstice (p. 25)
tectonic forces (p. 20)
tipping point (p. 35)
tradable permits (p. 35)
trade winds (p. 25)
transitional zone (p. 28)
trench (p. 20)
tsunami (p. 20)
volcanism (p. 20)
weather (p. 22)
winter solstice (p. 25)
xerophytic (p. 28)

REVIEW QUESTIONS

1. What are the three "spheres" of habitable life on earth?
2. What is plate tectonics, and what are some of the main consequences of tectonic activity?
3. What is the difference between weather and climate? What are the main forces that produce precipitation and aridity?
4. What are the major climate types and their associated biomes? Where do they tend to occur on earth?
5. Where are the biodiversity hot spots? What kinds of locations and biomes do many of them occur in?
6. What important roles do the world's oceans play, and what are their major resources? In what ways are these resources threatened?
7. What apparent long-term effects on the earth's atmosphere can be attributed to modern technology? What predictions are there for future changes, and what steps are being taken or considered to avert these changes?

NOTES

1. Wolf Roder, "Three Near Misses: Are These Science or Pseudoscience?" *Cincinnati Skeptics Newsletter,* June 1996, http://www.cincinnatiskeptics.org/newsletter/vol5/n5/misses.html. Accessed November 14, 2006.
2. Norman Myers, *Primary Source: Tropical Forests and Our Future* (New York: Norton, 1984), p. x.
3. World Bank, "On World Water Day, World Bank Calls for Investments in Water Infrastructure and Better Governance," press release, March 22, 2007.
4. Review the United Nations' *Annual Report of the State of the World's Fisheries and Aquaculture* at http://www.fao.org/sof/sofia/index_en.htm.
5. Boris Worm et al., "Impacts of Biodiversity Loss on Ocean Ecosystem Services." *Science,* November 3, 2006, pp. 787–790.
6. See the Monterey Bay Aquarium's guide to sustainable seafood, "Seafood Watch," at http://www.mbayaq.org/cr/seafoodwatch.asp.
7. Amy Roach Partridge, "Global Trucking Woes," *Global Logistics,* November 2006, http://www.inboundlogistics.com/articles/global/global1106.shtml. Accessed September 27, 2007.
8. Intergovernment Panel on Climate Change, "Climate Change 2007," http://www.ipcc.ch. Accessed July 4, 2007.
9. Sarah Lyall, "Global Warming Report Predicts Doom for Many Species." *New York Times,* September 1, 2000.
10. Brad Knickerbocker, "Could Warming Cause War?" *Christian Science Monitor,* April 19, 2007, p. 2.

Joe Hobbs

The trail to Mt. Everest, Nepal. The human imprint is almost everywhere on the landscape, even in the most challenging environments.

CHAPTER 3
HUMAN PROCESSES THAT SHAPE WORLD REGIONS

This chapter continues your introduction to geography's basic vocabulary, focusing on how people have interacted with the environment to change the face of the earth, with an emphasis on the human roles. Modern trends in people-land associations are examined as the products of revolutionary changes in the past: the arrival of agriculture and of industrialization. The chapter explains where rich and poor countries are located on the earth's surface and accounts for some of these patterns of prosperity and poverty. It considers where and why populations are increasing and what the implications of that growth are. Finally, it shares ideas about how to solve some of the most important global problems of our time.

chapter outline

chapter objectives

This chapter should enable you to

- Gain a historical perspective on the capacity of human societies to transform environments and landscapes
- Understand why some countries are rich and others poor and recognize the geographic distribution of wealth and poverty
- Explain the simultaneous trends of falling population growth in the richer countries and rapid population growth worldwide
- Explore the principles of sustainable development

Joe Hobbs

• **Figure 3.1** Until the relatively recent past—just a few thousand years ago—people were exclusively hunters and gatherers. This rock art in Egypt's Sinai Peninsula was created over a long time span, as it depicts Neolithic-period hunting of ibex, later uses of camels and horses, and writing from the Nabatean period (1st century D/F/).

3.1 Two Revolutions That Have Changed the Earth

The geographer's approach to understanding a current landscape—in almost all cases, a cultural landscape that has been fashioned by human activity—is sometimes deeply historical, involving study of its development from the prehuman or early human natural landscape. With a perspective of great historical depth, the current spatial patterns of our relationship with the earth may be seen as products of two "revolutions": the Agricultural Revolution that began in the Middle East about 10,000 years ago and the Industrial Revolution that began in 18th-century Europe. Each of these revolutions transformed humanity's relationship with the natural environment. Each increased substantially our capacity to consume resources, modify landscapes, grow in number, and spread in distribution.

Hunting and Gathering

Until about 10,000 years ago, our ancestors lived by **hunting and gathering** (also known as **foraging**). We were apparently quite good at it—it served us well for more than 100,000 years, until we began experimenting with the revolutionary technologies of agriculture.

Foraging was quite different from the farming and industrial ways of living that succeeded it. Joined in small bands of extended family members, hunters and gatherers were nomads with no villages, homes or other fixed dwellings. They moved to take advantage of changing opportunities on the landscape. These foragers scouted large areas to locate foods such as seeds, tubers, foliage, fish, and game animals (•Figure 3.1). Moving their small group from place to place, they had a relatively limited impact on the natural environment, especially compared with the impacts left by agricultural and industrial societies.

Hunters and gatherers may have been the **"original affluent society."**[1] Many scholars have praised these preagricultural people for the apparent harmony they maintained with the natural world in both their economies and their spiritual beliefs. After short periods of work to collect the foods they needed, they enjoyed long stretches of leisure time. Studies of those few hunter-gatherer cultures that lingered into modern times, such as the San (Bushmen) of southern Africa and several Amerindian groups of South America, suggest that although their life expectancy was low, they suffered little from the mental illnesses and broken family structures that characterize industrial societies.

Hunters and gatherers did modify their landscapes. These foragers were not always at peace with one another or with the natural world. With upright posture, stereoscopic vision, opposable thumbs, an especially large brain, and no mating season, *Homo sapiens* became after its emergence in southern Africa about 125,000 years ago an **ecologically dominant species**—one that competes more successfully than other organisms for nutrition and other essentials of life or that exerts a greater influence than other species on the environment. Using fire to flush out or create new pastures for the game animals they hunted, preagricultural people shaped the face of the land on a vast scale relative to their small numbers. Many of the world's prairies, savannas, and steppes where grasses now prevail developed as hunters and gatherers repeatedly set fires.

These people also overhunted and in some cases eliminated animal species. The controversial **Pleistocene overkill hypothesis** states that rather than being at harmony with nature, hunters and gatherers of the Pleistocene Era (2 million to 10,000 years ago) hunted many species to extinction, including the elephantlike mastodon of North America.

The Revolutionary Aspects of Farming

Despite these excesses, the environmental changes that hunters and gatherers could cause were limited. Humans' power to modify landscapes took a giant step with **domestication,** the controlled breeding and cultivation of plants and animals. Domestication brought about the **Agricultural Revolution,** also known as the **Neolithic Revolution** or the **Food-Producing Revolution.**

Why people began to produce rather than continue to hunt and gather plant and animal foods—first in the Middle East and later in Asia, Europe, Africa, and the Americas—is uncertain. Although there are many ideas about this process, surprisingly little hard evidence exists to explain it confidently. Two theories are most often put forward. One is that the climate changed. Increasing drought and reduced plant cover may have forced people and wild plants and animals into smaller areas, where people began to tame wild herbivores and sow wild seeds to produce a more dependable food supply.

A more widely accepted theory is that their own growing populations in areas originally rich in wild foods compelled people to find new food sources, so they began sowing cereal grains and breeding animals. The latter process may have begun about 8000 B.C.E. in the Zagros Mountains of what is now Iran. The culture of domestication spread outward from there but also developed independently in several world regions.

Now that humans were producing as well as consuming foods, their landscape uses, cultures, social organizations, and other characteristics changed dramatically. Among other things, in choosing to breed plants and animals, people settled down. Gradually abandoning the nomadic life and **extensive land use** of hunting and gathering, they came to favor the **intensive land use** of agriculture and animal husbandry. They could sow and harvest crops in specific places year after year. With less need to move around, they began living in fixed dwellings, at least on a seasonal basis. These developed into villages, small settlements with fewer than 5,000 inhabitants.

The settlements raised larger and more reliable stocks of food, making it possible to support their growing populations. Through **dry farming,** which involved planting and harvesting according to the seasonal rainfall cycle, population densities could be 10 to 20 times higher than they were in the hunting and gathering mode. By about 4000 B.C.E., people along the Tigris, Euphrates, and Nile Rivers began **irrigation** of crops—bringing water to the land artificially by using levers, channels, and other technologies—an innovation that allowed them to grow crops year round, independent of seasonal rainfall or river flooding (•Figure 3.2). Irrigation allowed even more people to make a living off the land; irrigated farming yields five to six times more food per unit area than dry farming. In ecological terms, the expanding food surpluses of the Agricultural Revolution raised the earth's **carrying capacity,** the size of a species' population (in this case, humans) that an ecosystem can support.

Joe Hobbs

• **Figure 3.2** The Tigris River in southeastern Turkey. The brown areas are rainless in the long, hot summers and are capable of producing only one crop a year through dry (unirrigated) farming. The green areas along the river are irrigated and can produce two or three crops a year. Irrigation was thus a revolutionary technology that greatly increased the number of people the land could support.

Culture became more complex, and society became more stratified. The steep increase in food production freed more people from the actual work of producing food, and they undertook a wide range of activities unrelated to subsistence needs. Irrigation and the dependable food supplies it provided thus set the stage for the development of **civilization,** the complex culture of urban life characterized by the appearance of writing, economic specialization, social stratification, and high population concentrations. By 3500 B.C.E., for example, 50,000 people lived in the southern Mesopotamian city of Uruk, in what is now Iraq. Other **culture hearths**—regions where civilization followed the domestication of plants and animals—emerged between 8000 and 2500 B.C.E. in China, Southeast Asia, the Indus River Valley, Egypt, West Africa, Mesoamerica, and the Andes.

Human impacts on the natural environment increased. The agriculture-based urban way of life that spread from these culture hearths had larger and more lasting impacts on the natural environment than either hunting and gathering or early agriculture. Acting as agents of humankind, domesticated plants and animals proliferated at the expense of the wild species that people came to regard as pests and competitors. For example, *Bos primigenius,* a wild bovine that was the ancestor of most of the world's domesticated cattle, was hunted to extinction by 1627. Farmers who resented the animals' raid on their crops were probably among the most ardent hunters. Agriculture's permanent and site-specific nature magnified the human imprint on the land, while the pace and distribution of that impact increased with growing numbers of people.

The Industrial Revolution

The human capacity to transform natural landscapes took another giant leap with the **Industrial Revolution,** which began in Europe around 1750 C.E. This new pattern of human-land relations was based on breakthroughs in technology that several factors made possible (•Figure 3.3):

- Western Europe had the economic capital necessary for experimentation, innovation, and risk. Much of this money was derived from the lucrative trade in gold and slaves undertaken initially in the Spanish and Portuguese
84, 85 empires after 1400.
- Significant improvements in agricultural productivity took place in Europe prior to 1500. New tools such as the heavy plow and more intensive and sustainable use of farmland led to increased crop yields. Human populations grew correspondingly.
- Population growth itself was a factor. More people freed from work in the fields could devote their talents and labor to experimentation and innovation. As agricultural innovations spread and industrial productivity improved, more and more of the growing European population was freed from farming, and for the first time in history, a region emerged in which city dwellers outnumbered rural folk. The process of industrialization, having now spread to nearly all parts of the world, continues to promote urbanization today.

Most geographers see population growth today as a drain on resources, but the Industrial Revolution illustrates that given the right conditions, more people do create more resources. Innovations such as the steam engine tapped the vast energy of fossil fuels—initially coal and later oil and natural gas. This energy, the photosynthetic product of ancient ecosystems, allowed the earth's carrying capacity for humankind to be raised again, this time into the billions.

Joe Hobbs

• **Figure 3.3** A tweed mill in Stornoway, on Scotland's Lewis Island. The process of industrialization that began in 18th-century Europe has rapidly transformed the face of the earth.

Industrialization, Colonization, and Environmental Change

As they began to deplete their local supplies of resources needed for industrial production, Europeans started to look for these materials abroad. As early as their **Age of Discovery,** also known as the **Age of Exploration,** which began in the 15th century, Europeans probed ecosystems across the globe to feed a growing appetite for innovation, economic growth, and political power. The process of European **colonization**—the extension of European countries' political and economic control over foreign areas—was thus linked directly to the Industrial Revolution. Mines and plantations from such faraway places as central Africa and India supplied the copper and cotton that fueled economic growth in colonizing countries such as Belgium and England (•Figure 3.4).

No longer dependent on the foods and raw materials they could procure within their own political and ecosystem boundaries, European vanguards of the Industrial Revolution had an impact on the natural environment that was far more extensive and permanent than that of any other people in history. Among the many measures of the unprecedented changes that the Industrial Revolution and its wake have wrought on the earth's landscapes, these are just a few:

- Between 1750 and the present, the total forested area on earth declined by more than 20 percent.
- During the same period, total cropland grew by nearly 500 percent, with more expansion in the period from 1950 to today than in the century from 1750 to 1850.
- Human use of energy increased more than 100-fold from 1750 to now.
- Today, fully 40 percent of the earth's land-based photosynthetic output is dedicated to human uses, especially in agriculture and forestry. Of particular interest to geographers is the unequal distribution of the costs and benefits of such expansion around the globe.

3.2 The Geography of Development

One of the most striking characteristics of human life on earth is the large disparity between wealthy and poor people, both within and between countries. At a high level of generalization, the world's countries can be divided into "haves" and "have-nots" (•Figure 3.5 and Table 3.1). Writers refer to these distinctions variously as "developed" and

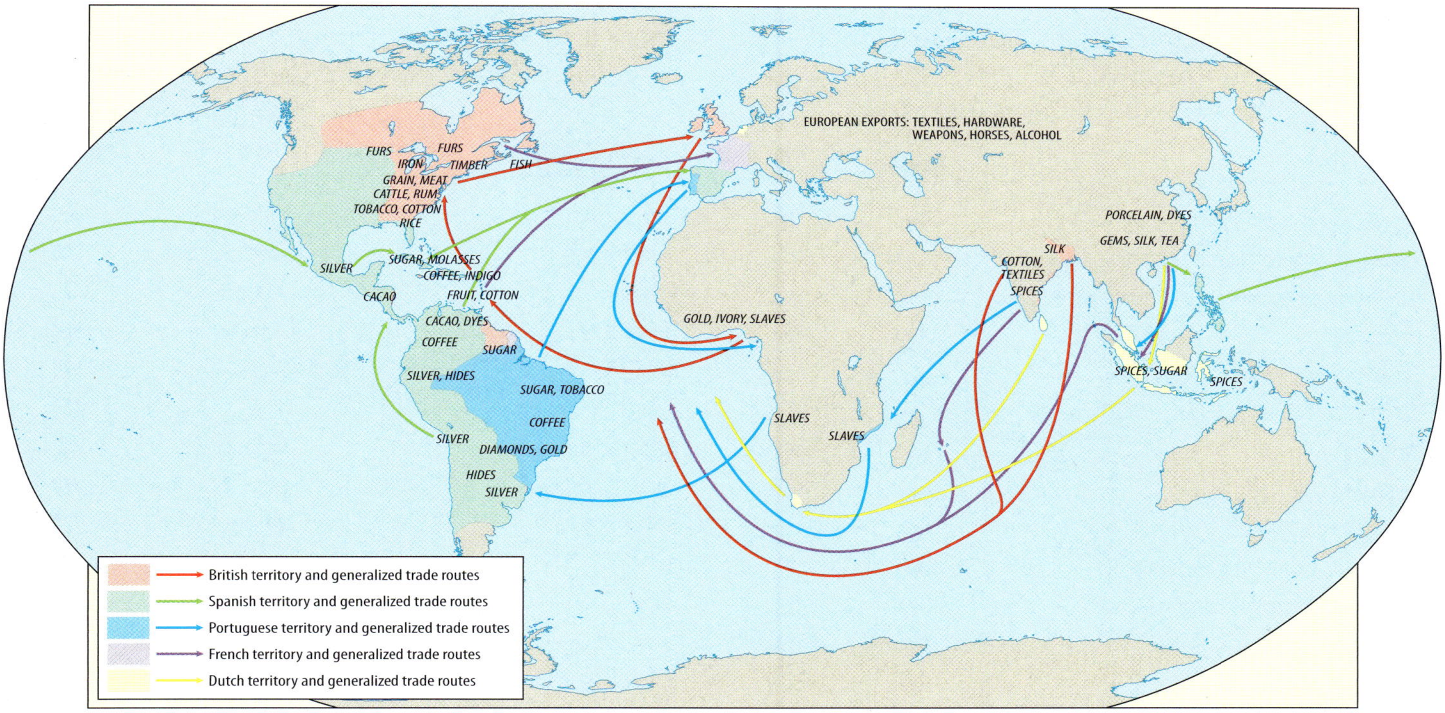

Figure 3.4 In the 18th century, European merchant fleets carried goods, slaves, and information all around the world, profoundly transforming cultures and natural environments.

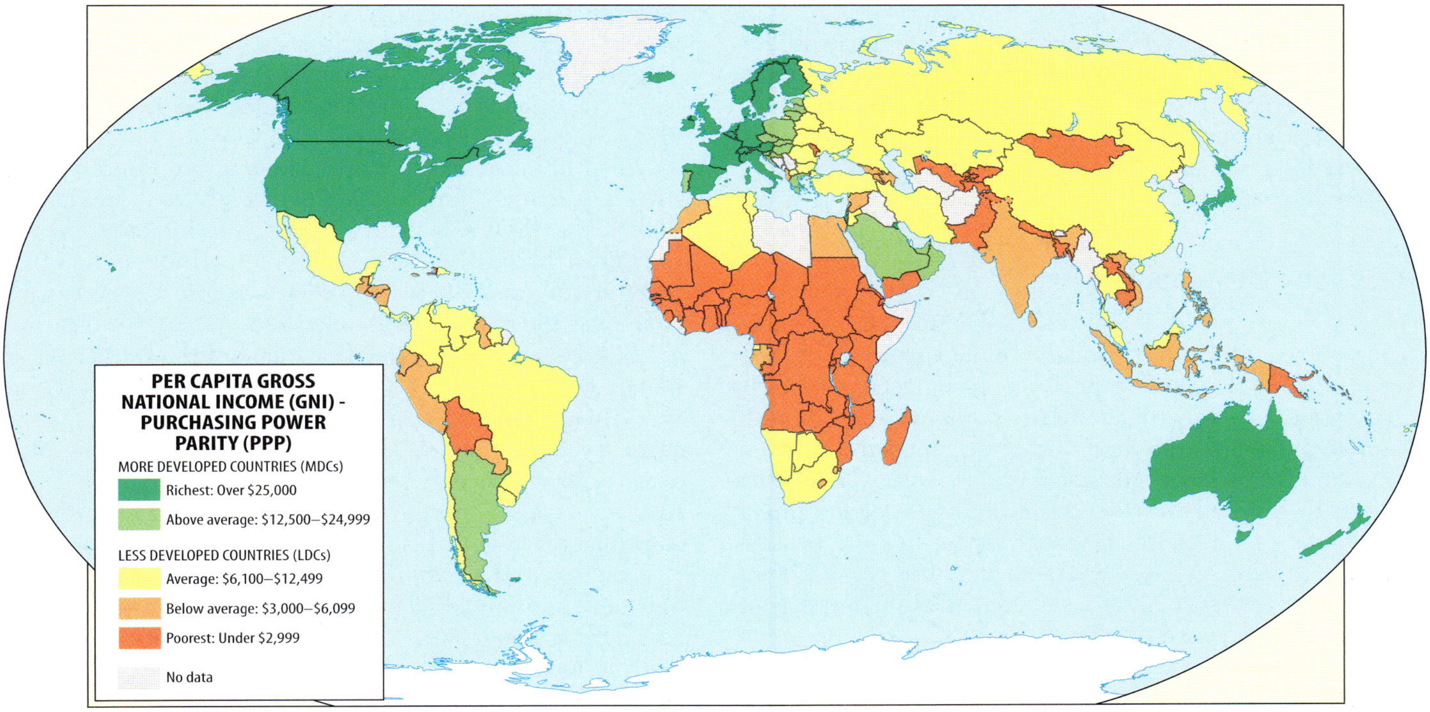

Figure 3.5 Wealth and poverty by country. Note the concentration of wealth in the middle latitudes of the Northern Hemisphere.

"underdeveloped," "developed" and "developing," "more developed" and "less developed," "industrialized" and "nonindustrialized," and "North" and "South," based on the concentration of wealthier countries in the middle latitudes of the Northern Hemisphere and the abundance of poorer nations in the Southern Hemisphere. This text uses the terms **more developed countries (MDCs)** and **less developed countries (LDCs).** It must be emphasized that this framework is an introductory tool and cannot account for the tremendous variations and continuous changes in

TABLE 3.1 Characteristics of More Developed Countries (MDCs) and Less Developed Countries (LDCs)

Characteristic	MDC	LDC
Per capita GDP and income	High	Low
Percentage of population in the middle class	High	Low
Percentage of population involved in manufacturing	High	Low
Energy use	High	Low
Percentage of population living in cities	High	Low
Percentage of population living in rural areas	Low	High
Birth rate	Low	High
Death rate	Low	Low[a]
Population growth rate	Low	High
Percentage of population under age 15	Low	High
Percentage of population that is literate	High	Low
Amount of leisure time available	High	Low
Life expectancy	High	Low

[a]Although death rates are high in the LDCs relative to the MDCs, in most countries they are quite low compared to what they once were in the LDCs.

TABLE 3.2 Top and Bottom Countries in Per Capita GNI PPP (U.S. dollars, 2007)[a]

Top Countries	Per Capita GNI PPP	Bottom Countries	Per Capita GNI PPP
Luxembourg	55,970	Madagascar	960
United States	44,260	Congo, Republic	940
Norway	43,920	Yemen	920
Switzerland	40,630	Sierra Leone	850
Denmark	36,110	Guinea-Bissau	830
Iceland	35,980	Niger	830
Netherlands	35,800	Tanzania	740
United Kingdom	35,690	Congo, Dem. Rep.	720
Ireland	35,540	Malawi	720
Austria	35,300	Burundi	710

[a]This table excludes city-states, territories, colonies, and dependencies. Only countries with available data are listed.
Source: 2007 World Population Sheet, Population Reference Bureau

economic and social welfare that characterize the world today. Some countries, including those known as the "Asian Tigers," are best described as **newly industrializing countries (NICs)** because they do not fit the MDC or LDC idealized types. The relevant regional chapters and modules describe these cases.
228, 298

Measuring Development

There is no single, universally acceptable standard for measuring wealth and poverty on the global scale. However, you are likely to encounter these in your university studies:

Annual per capita gross domestic product. **Gross domestic product (GDP)** is the total output of goods and services that a country produces for home use in a year. Divided by the country's population, the resulting figure of **per capita GDP** is one of the most commonly used measure of economic well-being. A closely related measure is per capita **gross national product (GNP),** which includes foreign output by domestically owned producers.

Annual per capita gross national income purchasing power parity. Although quite a mouthful, **per capita gross national income purchasing power parity (per capita GNI PPP)** is a useful figure in the study of world regional geography. **Gross national income (GNI)** includes gross domestic production plus income from abroad from sources such as rents, profits, and labor. **Purchasing power parity (PPP)** conversion factors consider differences in the relative prices of goods and services, providing a better overall way of comparing the real value of output between different countries' economies. GNI PPP is measured in current "international dollars," which indicate the amount of goods and services one could buy in the United States with a given amount of money. Definitions vary, but in this book, an MDC is a county with an annual per capita GNI PPP of $12,500 or more; all others are considered LDCs.

The gulf between the world's richest and poorest countries is startling (Table 3.2). The average per capita GNI PPP in the MDCs is nearly twelve times greater than in the LDCs. In 2007, with a per capita GNI PPP of $55,970, Luxembourg was the world's richest country, whereas Burundi was the poorest with only $710. By another measure, 1.2 billion people, or about 18 percent of the world's population, were "abjectly poor" (by the World Bank's definition), living on less than $1 per day. These raw numbers suggest that economic productivity and income alone characterize **development,** which, according to a common definition, is a process of improvement in the material conditions of people through diffusion of knowledge and technology.

The Human Development Index. Definitions like these, and statistics like per capita GDP and per capita GNI PPP, reveal little about measures of well-being such as income distribution, gender equality, literacy, and life expectancy. Recognizing the shortcomings of strictly economic definitions, the United Nations Development Programme created the **Human Development Index (HDI),** a scale that considers attributes of quality of life. This book uses HDI in the Basic Data tables of the chapter that introduces each region (for example, Table 4.1 on page 70). On the HDI scale, a measure of 1.0 means "perfect." According to this index, with a rating of 0.965, Norway is "the world's best place to live," although it ranks third in per capita GNI PPP. Following Norway, in descending order, are Iceland, Australia, Ireland, and Sweden. In HDI terms, with a measure of 0. 311, Niger is the world's worst place to live; Sierra Leone, Mali, Burkina Faso, and Guinea-Bissau fare only slightly better. Note that all five of these low-rated countries are in Africa, for reasons discussed in Chapter 9 and Module 9.1.

On the basis of per capita GNI PPP, 1.1 billion, or 16 percent, of the world's people, inhabit the MDCs. Most citizens of these countries, such as the United States, Canada, Japan, Australia, New Zealand, and the nations of Europe, enjoy an affluent lifestyle with freedom from hunger. Employed in industries or services, most of the people live in cities rather than in rural areas. Disposable income, the money that people can spend on goods beyond their subsistence needs, is generally high. There is a large middle class. Population growth is low as a result of low birth rates and low death rates (these demographic terms are explained in Section 3.3 on population). Life expectancy is long, and the literacy rate is high.

Life for the planet's other 84 percent, or about 5.5 billion people, is very different. In the LDCs, including most countries in Latin America, Africa, and Asia, poverty and often hunger prevail. The leading occupation is subsistence agriculture, and the industrial base is small. The middle class tends to be small but growing, with an enormous gulf between the vast majority of poor and a very small wealthy elite, which owns most of the private landholdings. With high birth rates and falling death rates, population growth is high. Life expectancy is short, and the literacy rate is low. However, the fate of LDCs is not predetermined. Wise policies and activities in certain countries or even in certain states or regions of particular countries can make a large difference. For example, while India's overall per capita income is $3,800, its state of Kerala, whose government has long grappled with problems of health and equity, has a per capita income 25 percent higher than the national average, and India's overall life expectancy of 64 years is exceeded in Kerala by 9 years.

With four-fifths of the world's people living in the poorer, less developed countries, it is important to understand the root causes of underdevelopment and to appreciate how wealth and poverty affect the global environment in very different but equally profound ways.

Why Are Some Countries Rich and Others Poor?

Many theories attempt to explain the disparities between MDCs and LDCs. It is important to recognize that there is no single, widely accepted explanation about development in general or in a particular region or country. Probably the best thing a discerning student can do is to weigh the various explanations and see which one, or which combination of them, seem to fit the situation of a given country or region. Here are the main explanations likely to be encountered.

Embraced most strongly in the LDCs, **dependency theory** argues that the worldwide economic pattern established by the Industrial Revolution and the attendant process of colonialism persists today. In his book *Ecological Imperialism,* the historian and geographer Alfred Crosby explains how dependency led to the rich-poor divide, depicting the two very different ways in which European powers used foreign lands during the Industrial Revolution.[2] In the pattern of **settler colonization,** Europeans sought to create new Europes, or **neo-Europes,** in lands much like their own: temperate middle-latitude zones with moderate rainfall and rich soils where they could raise wheat and cattle. Consequently, between 1630 and 1930, more than 50 million Europeans emigrated from their homelands to create European-style settlements in what are now Canada, the United States, Argentina, Uruguay, Brazil, South Africa, Australia, and New Zealand. These lands were destined to become some of the world's wealthier regions and countries.

In contrast to their preference to settle familiar midlatitude environments, Europeans viewed the world's tropical lands mainly as sources of raw materials and markets for their manufactured goods. The environment was too different from home to make settlement attractive. In establishing a pattern of **mercantile colonialism,** Crosby explains, Europeans were less inhabitants than conquering occupiers of the colonies, overseeing indigenous peoples and resettled slaves in the production of primary or unfinished products: sugar in the Caribbean; rubber in Latin America, West Africa, and Southeast Asia; and gold and copper in southern Africa, for example. Colonialism required huge migrations of people to extract the earth's resources, including 30 million slaves and contracted workers from Africa, India, and China to work mines and plantations around 338, 470, 501, 616
the globe.

In the mercantile system, the colony provided raw materials to the ruling country in return for finished goods; this meant that people in India would purchase clothing made in England from the raw cotton they themselves had harvested. The relationship was most advantageous to the colonizer. England, for example, would not allow its colony India to purchase finished goods from any country but England. It prohibited India from producing any raw materials the empire already had in abundance, such as salt (India's Mohandas Gandhi defiantly violated this prohibition in his famous "March to the Sea"). Finished products are **value-added products,** meaning they are worth much more than the raw materials they are made from, so manufacturing in the ruling country concentrated wealth there while limiting industrial and economic development in the colony.

The colony was obliged to contribute to, but was prohibited from competing with, the economy of the ruling country—a relationship that dependency theorists insist continues today. Dependency theory asserts that to participate in the world economy, the former colonies, now independent countries, continue to depend on exports of raw materials to and purchases of finished goods from their former colonizers and other MDCs, and this disadvantageous position keeps them poor. Dependency theorists call this relationship **neocolonialism.** With independence, the former colonies needed revenue. To earn that money, they continued to produce the goods for which markets already existed—generally the same unprocessed primary products they supplied in colonial times. Dependency theorists argue that when the former colonies try to break their dependency by becoming exporters of manufactured goods, the MDCs impose trade barriers and quotas to block that development (see Insights, page 47).

Whether because of neocolonialism or a more complex array of variables, many developing countries continue to rely heavily on income from the export of a handful of raw materials. This makes them vulnerable to the whims of nature and the world economy. The economy of a country heavily dependent on rubber exports, for example, may suffer if an insect pest wipes out the crop or if a foreign laboratory develops a synthetic substitute. When demand for rubber rises, that country may actually harm itself trying to increase its market share by producing more rubber because in doing so, it drives down the price. (Consider oil: members of the OPEC cartel of oil-producing countries drove oil prices to all-time lows in the mid-1980s when they overproduced oil in a bid to earn more revenue.) If the rubber-producing country withholds production to shore up rubber's price, it provides consuming countries with an incentive to look for substitutes and alternative sources. (Again look at oil: after OPEC embargoed shipments of oil to the United States in the 1970s, the United States began developing domestic oil supplies and becoming more energy-efficient, thus temporarily reducing oil prices and OPEC's revenues.) The developing country is in a dependent and disadvantaged position.

Many geographers view dependency theory and neocolonialism as simplistic and politically biased explanations of development. They consider a wider and more complex set of factors, including culture, location, and natural environment, to explain why some countries are wealthy and others are poor. These include the following factors:

Advantageous and disadvantageous location. Does a country's location play a role in how rich or poor it is? In some cases, yes. Location can influence a country's economic fortunes. For example, as discussed in Chapter 1, because it is situated close to a great mainland with which to trade, the island of Great Britain enjoys a core location favorable for economic development. Japan has a similar location relative to the Asian landmass. In contrast, landlocked nations such as Bolivia in South America and numerous nations in Africa have locations unfavorable for trade and economic development, and they have not overcome this disadvantage. But it is important to recognize that geographic location is never the sole decisive factor in development. Like Japan and Britain, Madagascar and Sri Lanka are island nations situated close to large mainlands, but for a variety of (mainly political) reasons discussed in the relevant chapters and modules, neither has experienced prosperity.

Resource wealth or poverty. Having or lacking a diversity or abundance or **natural resources** plays a significant role in development. Superabundance of an especially valuable resource (for example, oil in the Persian/Arabian Gulf countries) or a diversity of natural resources has helped some countries become more developed than others. The former Soviet Union and the United States achieved superpower status in the 20th century in large part by using the enormous natural resources of both countries.

Cultural and historical factors. In some cases, human industriousness has helped compensate for resource limitations and promoted development. For example, Japan has a rather small territory with few natural resources (including almost no petroleum). Yet in the second half of the 20th century, it became an industrial powerhouse largely because the Japanese people united in common purpose to rebuild from wartime devastation, placing priorities on education, technical training, and seaborne trade from their advantageous island location. Conversely, cultural or political 401 problems like corruption and ethnic factionalism can hinder development in a resource-rich nation, as in the mineral-wealthy Democratic Republic of Congo. 480

Environmental Impacts of Underdevelopment

Geographers are very interested in the environmental impacts of relations between MDCs and LDCs, especially on the poorer countries. LDCs generally lack the financial resources needed to build roads, dams, energy grids, and other infrastructure critical to development. They turn to the World Bank, International Monetary Fund (IMF), and other institutions of the MDCs to borrow funds for these projects. Many borrowers are unable to pay even the interest on these loans, which is sometimes huge; debtor nations have been known to spend as much as 40 percent of annual revenues on interest payments to the lenders, or more than they spend on education and health combined.

When lender institutions threaten to cut off assistance, borrowing countries often try to raise cash quickly to avoid this prospect, generally by one or two methods, or both. One method is to dedicate more high-quality land to the production of **cash crops** (also known as **commercial crops**). These are items such as coffee, tea, sugar, coconuts, and bananas exported to the MDCs, where they may be perceived as luxuries or as staple items. Governments or foreign corporations often buy out, force out, or otherwise displace subsistence food farmers in the search for new lands on which to grow these commercial crops. In this process, known as **marginalization,** poor subsistence farmers are pushed onto fragile, inferior, or marginal lands that cannot support crops for long and end up depleted by cultivation. In Brazil's Amazon Basin, for example, peasant migrants arrive from Atlantic coastal regions, where government and wealthy private landowners cultivate the best

INSIGHTS

Globalization: The Process and the Backlash

One of the most remarkable international trends of recent decades has been **globalization**—the spread of free trade, free markets, investments, and ideas across borders and the political and cultural adjustments that accompany this diffusion. There has been much debate and sometimes-violent conflict over the pros and cons of globalization.

Advocates of globalization argue that the newly emerging global economy will bring increased prosperity to the entire world. Innovations in one country will be transferred instantly to another country, productivity will increase, and standards of living will improve. They propose that one obvious solution to the problem of the LDCs' inability to compete in the world economy, and therefore escape their dependency, is to reduce the trade barriers that MDCs have erected against them (some of these barriers are discussed on page 623 in Chapter 11). With free trade, free enterprise will prosper, pumping additional capital into national economies and raising incomes for all. Much of the support for globalization comes from **multinational companies** (also called *transnational companies*, meaning companies with operations outside their home countries) as they increase their investments abroad. Most of the companies are based in the MDCs, but multinational corporations have also grown in LDCs such as China, India, and Mexico.

Opponents of globalization argue that the process will actually increase the gap between rich and poor countries, and even within a country; a selected few developing nations (or people within a country) will prosper from increased foreign investment and resulting industrialization, but the hoped-for "global" wealth will bypass other countries (or people within a country) altogether. And the increasing interdependence of the world economies will make all the players more vulnerable to economic and political instability. The multinational companies will recognize huge profits at the expense of poor wage laborers. In addition, environments will be harmed if environmental regulations are reduced to a lowest common denominator (for example, the high standards of air quality demanded by the U.S. Clean Air Act would be deemed noncompliant with World Trade Organization rules because they make it harder for countries with "dirtier" technologies to compete in the marketplace). Such concerns often lead to massive protests against "corporate-led globalization," especially at meetings of the World Trade Organization, the International Monetary Fund, and the World Bank. Many more confrontations like these can be anticipated. Typical protesters' demands are that working conditions be improved in the foreign "sweatshops," where textiles and other goods are produced at low cost for U.S. corporations, and that corporations like Starbucks should sell only "fair trade" coffee beans bought at a price giving peasant coffee growers a living
354, 540 wage rather than at the "exploitive" price typically paid.

The problems of hot money and the digital divide are also cited as drawbacks to globalization. **Hot money** refers to short-term (and often volatile) flows of investment that can cause serious damage to the "emerging market" economies of less developed countries. Individual and corporate investors in the MDCs can invest such money heavily in stock market securities of the LDCs, reasoning that these developing nations have much greater economic growth potential than the mature economies of the MDCs. This investment can bring rapid wealth to at least some sectors of the economies of LDCs. The problem, however, is that the capital can be withdrawn as quickly as it is pumped in, resulting in huge economic consequences. 344

It is often assumed that the rapid growth and spread of computer and wireless technologies are benefiting all of humankind, but the notion of the **digital divide** challenges this view. The divide is between the handful of countries that are the technology innovators and users and the majority of nations that have little ability to create, purchase, or use new technologies. The United States, most European countries, and Japan are leaders in **information technology (IT)**, and the LDCs are well behind. For example, more than 70 percent of U.S. citizens use the World Wide Web, compared with 20 percent of Russians, 4 percent of Indians, and 10 percent of Middle Easterners (• Figure 3.A). Statistically, an American must save a month's salary to buy a computer, but a Bangladeshi must save eight years' wages—a fact that may change rapidly if the nonprofit organization One Laptop per Child is successful in producing, distributing, and selling its $100 Taiwanese-made "XO" laptop (see http://www.laptop.org).

U.S. companies earn a large share of the revenues from the global Internet business. The fear in the technology-lagging countries is that the growth of e-commerce will concentrate the wealth generated by that commerce in the technology leadership countries. This would enable the leaders to make even further advances in technology and realize even bigger economic gains from a so-called **knowledge economy** based on innovation and services, while the technology laggards fail to catch up and simply become poorer. Will globalization prove to be good or bad? It may depend on where and who you are.

Joe Hobbs

• **Figure 3.A** An Internet café in Mérida, Yucatán, Mexico. Information technology—especially computers, the Internet, and cellular telephones—is spreading rapidly around the world. Some people believe that there is now an unfolding Information Revolution comparable to the Agricultural and Industrial Revolutions in its capacity to change humanity's relationship with the earth.

Joe Hobbs

• **Figure 3.6** The Thu Bon River in Vietnam. In the LDCs, it is very common for people to rely on polluted water sources for drinking, bathing, cooking, and washing their clothes and eating utensils. Health effects can be severe; one of the most tragic is infant diarrhea, typically a waterborne illness that kills an estimated 2 million youngsters worldwide every year.

soils for sugarcane and other cash crops. The newcomers to Amazonia slash and burn the rain forest to grow rice and other crops that exhaust the soil's limited fertility in a few years. Then they move on to cultivate new lands, and in their wake come cattle ranchers, whose land use further
472, 530, 587, 589 degrades the soil.

Another way for debtor countries to raise cash quickly is to sell off their natural assets. National decision makers often face a difficult choice between using the environment to produce immediate or long-term economic rewards. In most cases, they feel compelled to take short-term profits and, by cash cropping and other strategies, initiate a sequence having sometimes-tragic environmental consequences. This is because most LDCs have resource-based economies that rely not on industrial productivity but on stocks of productive soils, forests, and fisheries. The long-term economic health of these countries could be assured by the perpetuation of these natural assets, but to pay off international debts and meet other needs, the LDCs generally draw on their ecological capital faster than nature can replace it. In ecosystem terms, they exceed **sustainable yield** (also known as the **natural replacement rate**), the highest rate at which a **renewable resource** can be used without decreasing its potential for renewal. In tropical biomes, for example, people cut down 10 trees for every tree they plant. In Africa, that ratio is 29 to 1. In 1950, about 30 percent of Ethiopia's land surface was covered with forest. Today, less than 1 percent is forested.

In other words, these countries find themselves in what might be termed **ecological bankruptcy:** they have exhausted their environmental capital. This environmental poverty plays itself out in several damaging ways. First, there are negative effects on the health and well-being of the people living in these countries. People in the LDCs feel the impacts of environmental degradation more directly than people in the MDCs (•Figure 3.6). Many of the world's poor drink directly from untreated water supplies; an estimated 1 billion lack access to clean water. They tend to cook their meals with fuelwood rather than with fossil fuels. They are more dependent on nature's abilities to replenish and cleanse itself, and hence they suffer more when those abilities are diminished (see Insights, below).

In addition, many political and social crises result from this ecological bankruptcy. Revolutions, wars, and refugee migrations in developing nations often have underlying environmental causes. Such problems, as related in the earlier 328, 331, 460
discussion of global warming, are central to the national interests of the United States and other countries far beyond the affected nations. Do you think **deforestation** in Mexico or Haiti is of no concern to you if you are an American? You may want to consider why people from those countries

INSIGHTS

The Fuelwood Crisis

The removal of tree cover in excess of sustainable yield (a characteristic problem in Nepal, for example) illustrates the many detrimental effects that a single human activity can have in the LDCs (•Figure 3.B). These effects cause a complex problem, widespread in the world's LDCs, known as the **fuelwood crisis.** As people remove trees to use as fuel, for construction, or to make room for crops, their existing crop fields lose protection against the erosive force of wind. Less water is available for crops because in the absence of tree roots to funnel water downward into the soil, it runs off quickly. Increased salinity (salt content) generally accompanies increased runoff, causing the quality of irrigation and drinking water downstream to decline. Eroded topsoil resulting from reduced plant cover can choke irrigation channels, reduce water delivery to crops, raise floodplain levels, and increase the chances that floods will destroy fields and settlements. As reservoirs fill with silt, hydroelectric generation, and hence industrial production, is diminished. Upstream, where the problem began, fewer trees are available to use as fuel.

Owing to the depletion of wood, people living in rural areas must now change their behavior (the fuelwood crisis often refers only to this part of the problem). Women, most often the fuelwood collectors, must walk farther to gather fuel. The use of animal dung and crop residues as sources of fuel deprives the soil of the fertilizers these poor people most often use when farming.

INSIGHTS

The Fuelwood Crisis *continued*

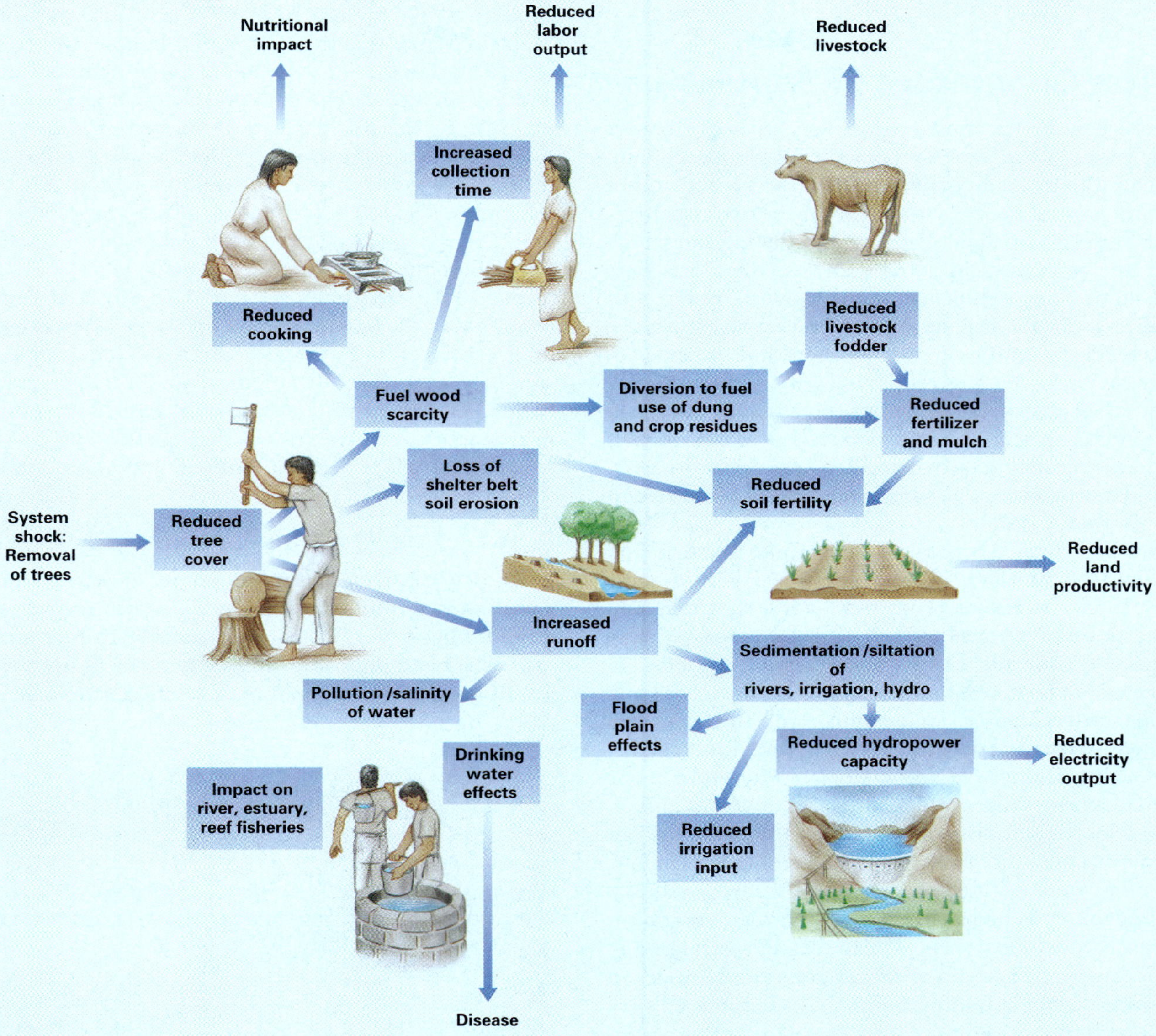

• **Figure 3.B** Impacts of deforestation in the LDCs.

Reduced food output is the result. A family may eventually give up one cooked meal a day or tolerate colder temperatures in their homes because fuel is lacking—steps that have a negative impact on the family's health.

The fuelwood crisis is a problem that governments and societies can confront successfully. A combination of new technologies, government policies, and sound economic growth can halt and even reverse deforestation. The biogas digester, a simple household device usable in warm rural areas, ferments animal manure for the production of methane cooking gas and thus spares fuelwood. Governments can delimit forest reserves but also assist people in earning livelihoods from **nontimber forest products (NTFPs),** such as handicrafts made from bamboo (see http://www.ntfp.org). The process of economic development can itself lead to less pressure on forest resources; the world's wealthier countries have more forest now than they did in 1990, and forest cover actually increased in 22 of world's 50 most forested countries since that time. A key challenge for the wealthier forested countries is to avoid depleting other forests to meet their needs.

try to move to the United States, legally or otherwise. One of the reasons is that the home environments are so degraded that they simply cannot support farming as they once did.

3.3 The Geography of Population

Population may be the most critical issue in geography. It certainly is one of most important issues for human life on earth. The welfare of humanity and of the planet's other species and natural habitats is tied closely to two related issues: the number of people there are and the rates at which we consume resources. In some quarters, there are fears that the human **population explosion** the world has experienced since 1800 will lead to a precipitous crisis: a massive disease outbreak, famine, or some other kind of catastrophe triggered by too many people living too close together without sufficient food and other resources to sustain them. On the other hand, the human population growth rate is slowing, and our numbers are expected to stabilize—ideally at a population that can be sustained with minimal risk of famine or other suffering.

Meanwhile, large numbers of people move across and within political borders, usually by choice but in some cases by force. Migrants bring new cultures, ideas, and opportunities with them, but their reception is not always warm: tension and violence have characterized relations between majority and migrant minority groups in recent years in Europe and Australia, for example, and in the United States there is a heated debate about "immigration reform"
73, 600 directed at illegal migrants from Mexico. **Migration** is one of the major themes discussed throughout this book.

Interest in population is not confined to geography but is shared by many other fields, including biology, sociology, anthropology, and political science. The study of population is known as **demography.** The field of demography is most concerned with patterns of birth, death, marriage, and related issues in themselves, with less attention to issues of migration and population distributions. What most distinguishes population geography is its focus on spatial variations. This section of the book examines how many people have lived on earth and how many we may be in the future, mindful of who these people are and why they have been so few or so many, and paying especially close attention to the fundamentally geographic issue of where they are.

How Many People Have Ever Lived on Earth?

Around 100,000 years ago, our *Homo sapiens* ancestors came out of Africa across the land bridge of Suez and began to populate Eurasia. By around 10,000 years ago, before plants and animals were domesticated, there were probably about 5.3 million humans in the world—roughly the current number of residents of the city of Chicago or the country of Finland. By 1 C.E., humans numbered between 250 and 300 million, about the population of the United States today. The first billion was reached around 1800. Then a staggering population explosion occurred in the wake of the Industrial Revolution. The second billion 42
came in 1930, the fourth in 1975, and the sixth in 1999 (Table 3.3 and •Figure 3.7). At 6.6 billion, *Homo sapiens* is now by far the most populous large mammal on earth and has succeeded where no other animal has in extending its range to the world's farthest corners. Using a 24-hour time period to represent the 125,000 years of our species' history, we reached our first billion within just the last 3 minutes. The 6.6 billion people alive today represent a remarkably large 5.8 percent of all the people who have ever lived on earth!

Figure 3.7 illustrates very dramatically that right after about 1800, the human population surged. What happened after tens of thousands of years that made our numbers suddenly skyrocket? In addition to the factor, addressed earlier, of greater food surpluses that made it possible to feed more people, we answer that question by considering some of the basic vocabulary of population geography.

How Can We Measure Population Changes?

Excluding the issue of migration for now, two principal variables determine population change in a given village, city, or country or the entire planet: birth rate and death rate. The **birth rate** is the annual number of live births per 1,000 people in a population. The **death rate** is the annual

TABLE 3.3 How Many People Have Ever Lived on Earth?

Year	Population
50,000 B.C.E.	few
8,000 B.C.E.	5,000,000
1 C.E.	300,000,000
1200	450,000,000
1650	500,000,000
1750	795,000,000
1850	1,265,000,000
1900	1,656,000,000
1950	2,516,000,000
1995	5,760,000,000
2007	6,625,000,000
Number who have ever been born	106.8 billion
Percent of those ever born who were alive in 2007	6.2%

Source: Modified from Population Reference Bureau.

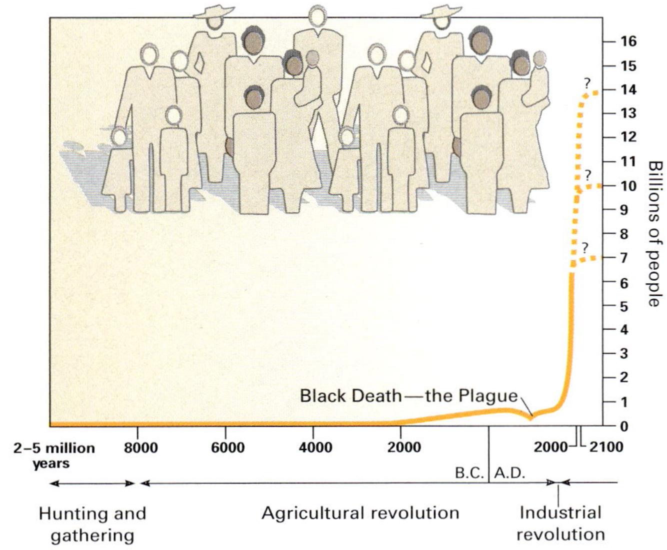

• **Figure 3.7** The human population has exploded since the Industrial Revolution, in a classic J-shaped curve of exponential growth.

number of deaths in that same sample population of 1,000. The **population change rate**—the figure that is often called the "population growth rate" but that may represent either growth or loss—is the birth rate minus the death rate in that population.

We can put these measures to work to appreciate the earth's population today. Supposing a perfect sample of 1,000 people representing the world's population in 2007, the birth rate was 21 per 1,000 and the death rate was 9 per 1,000. This means that by the end of that year, among the 1,000 people, 21 babies had been born and 9 people had died, resulting in a net growth of 12. That figure, 12 per 1,000, or 1.2 percent, represents the 2007 population change rate for the world.

We may now look at some of different combinations of birth rates, death rates, and population change rates and the various forces behind them.

What Determines Family Size?

Many factors affect birth rates, which tend to be much higher in the less developed countries of the world and among the poorer residents of more affluent countries. Some of the motivations and circumstances may seem strange at first glance, but there are good reasons for them:

- Better-educated and wealthier people have fewer children. The parents of most of you reading this book probably considered the economic cost of raising you and sending you to college and made decisions about family size in part because of their education and yours.
- Conversely, less educated and poorer people generally want and have more children. One reason for this is economic: poorer parents are often convinced that "extra" children will help bring more income to the family by working in fields or factories and will help care for them in their old age.
- People in cities tend to have fewer children than those in rural areas.
- Those who marry earlier generally have more children (their reproductive life span is longer).
- Couples with access to and understanding of contraception generally have fewer children.
- Value systems and cultural norms play very important roles. Even where contraception is available and understood, a couple may decide not to interfere with what they perceive as God's will or may seek the social status associated with a larger family.

What Determines Death Rates?

Death rates are correlated mainly with health factors, particularly the level of nutrition and level of medical care available. Improvements in food production and distribution help reduce death rates. Better sanitation, better hygiene, and cleaner drinking water eliminate fatal diseases such as infant diarrhea, a common cause of infant mortality in the less developed countries. Furthermore, the availability of antibiotics, immunizations, insecticides, and other improvements in medical and public health technologies have a marked correlation with declining death rates.

Human death rates overall have been on a steady trend of decline for decades. But death rates sometimes rise, of course, especially with the outbreak of epidemics such as
HIV/AIDS. Large proportions of the world's population 464
have in fact been killed in disease epidemics and natural disasters (Table 3.4). The "Black Death" (caused by bubonic plague) in Europe and Asia killed about 15 percent of the world's population between 1334 and 1349; at least 40 percent of Europe's population died. And we cannot assume that such devastation will never happen again. Estimates for the death toll from the H5NI virus, commonly known as "bird flu," should it become transmissible between humans, range as high as 150 million, or 2.2 percent
of the current world population. 380

Closely related to the measure of death rates is that of *life expectancy*, the number of years a person may expect to live in a given environment (typically defined as a country—see •Figure 3.8—and differentiated between women, who usually live longer, and men). As death rates fall, life expectancy increases, and the reverse is also true. In the United States in 2007, life expectancy for women was 80 years, and for men, 75 years. In Swaziland, the country hit harder than any other by the HIV/AIDS epidemic, a woman could expect to live 34 years and a man 33 years. But as recently as 1986, before the virus was so widespread in Swaziland, the death rate was much lower, and a woman could look forward to 60 years and a man to 53 years.

TABLE 3.4 The World's Most Costly Epidemics and Disasters

Number of Deaths	Place	Date	Cause
unknown	Sumatra (Indonesia)	c. 74,000 years ago	Toba supervolcano eruption [a]
300 million–500 million	worldwide	20th century	smallpox [b]
c. 100 million	Europe	540–590	plague
55 million–75 million	Europe and Asia	1300s	bubonic plague
25 million–100 million	worldwide	1918–1919	influenza
25 million	worldwide	1981–present	AIDS
10 million	China	1892–1896	bubonic plague
3.7 million	China	1931	river flooding, disease, and famine
3 million	India	1900	drought-related famine
3 million	China	1941	drought-related famine
2 million–3 million	worldwide	annually	malaria
more than 2 million	China	1959	river flooding, disease, and famine
2 million	China	1887	river flooding, disease, and famine
1.5 million	India	1965–1967	drought-related famine
1.1 million	Egypt and Syria	1201	earthquakes
1 million	China	1938–1939	river flooding, disease, and famine
1 million	worldwide	1957	influenza
1 million	worldwide	1968	influenza

[a] This single event and its aftermath killed most of our human ancestors. Based on our genetic heritage from that time, estimates vary, but it is believed that 40 to 500 individual female *Homo sapiens* of childbearing age (suggesting a total human population of just a few hundred to several thousand) survived both Toba's initial supervolcanic cataclysm and the subsequent volcanic winter, which caused temperatures to plummet worldwide. (Several competing hominid species were apparently completely wiped out by the Toba eruption and its effects.)

[b] Historically, this disease is thought to have been responsible, at least in Europe, for as many deaths as the plague. It was a great scourge in whole populations during several periods through history, cutting down about one-third in Rome early in the Common Era and decimating the majority of the members of several American Indian tribes in the 16th through 19th centuries. Until the last case was diagnosed in 1977, it was a greater killer by far, in the 20th century, than all wars combined.

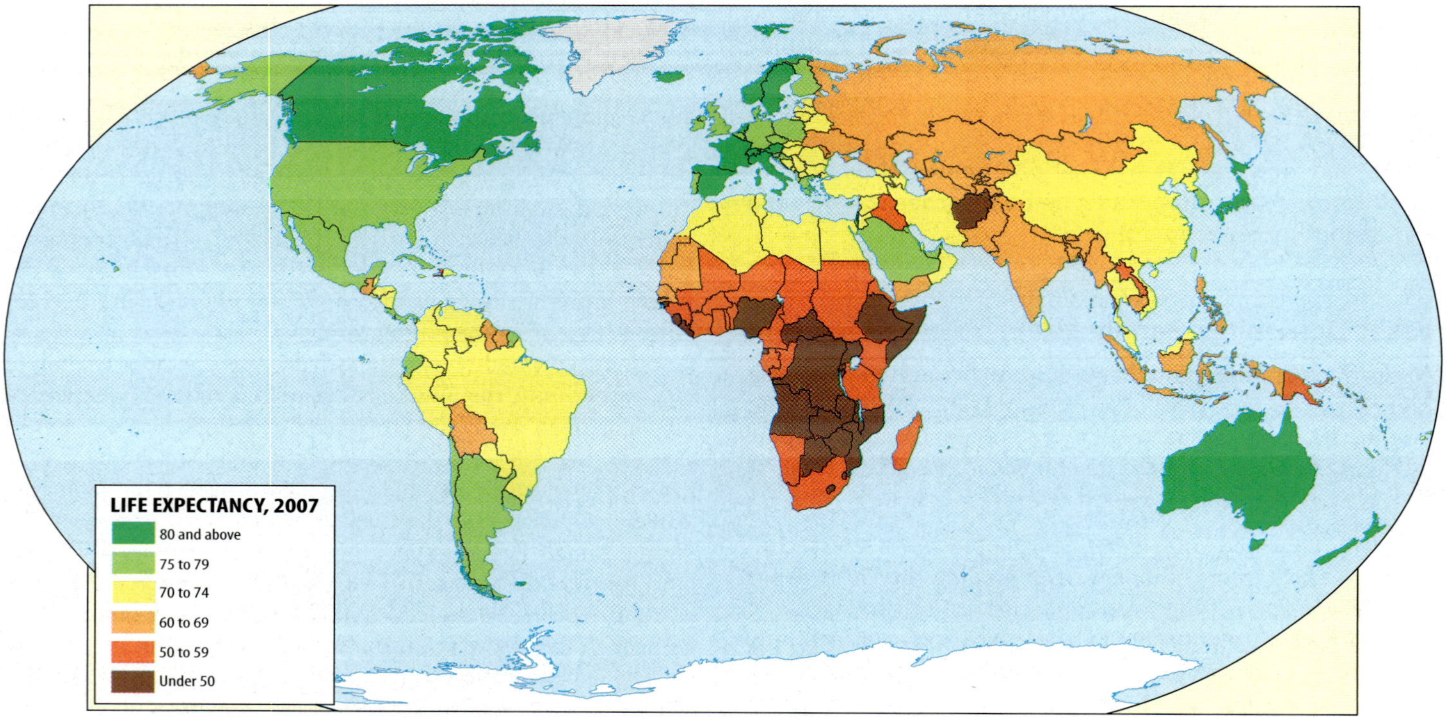

• **Figure 3.8** Life expectancy is closely tied to economic well-being; people live longer where they can afford the medicines and other amenities and technologies that prolong life.

What Determines the Population Change Rate?

Throughout history, natural disasters, diseases, and wars have taken huge bites out of our numbers. Overall, however, with birth rates higher than death rates, the trend line has been one of growth —and since 1800, of spectacular growth. In 1968, the rate of population growth hit an all-time high of 2.0 percent. To appreciate how rapid that growth rate was, you can calculate the doubling time. Applying the often-used rule of 70, in which 70 is divided by the growth rate, the doubling time is the number of years required for the human population to double (assuming that the rate of growth would be unchanged over the entire period). In 1968, the human population was growing at a rate that would, if unchanged, have doubled in 35 years. That rate did slow down, however, illustrating the limited usefulness that doubling time has in projecting future growth. At the 2007 population change rate of 1.2 percent, our numbers would double in 58 years.

Doubling time may be only approximate, but it is a good tool for comparing between countries and also illustrates that human populations have the potential for exponential growth; that is, not an incremental or arithmetic increase from 1 to 2 to 3 to 4, and so on, but geometric growth from 2 to 4 to 8 to 16, for example. (Exponential growth in the context of the famous Malthusian scenario is discussed on page 60.) The annual rate of population change worldwide is depicted by country in •Figure 3.9.

Why Has the Human Population "Exploded?"

If both birth rates and death rates are high (as they were when our ancestors were hunters and gatherers and as recently as the dawn of the Industrial Revolution), population growth is minimal: many people are born to a given population in a given period, but many also die in that same period, so they "cancel each other out." Population growth is also negligible if both birth and death rates are low (as they are today in countries like Japan and Italy). But when the birth rate is high and death rates are low, population surges, as seen in •Figure 3.10. This scenario of high birth rates, plunging death rates, and surging growth is exactly what played out for our species beginning around 1800 in western Europe and after about 1950 in the LDCs.

It is vital to appreciate the fact that the explosive growth in world population since the beginning of the Industrial Revolution is the result not of a rise in birth rates but of a dramatic decline in death rates, particularly in the less developed countries. The death rate has fallen as improvements in agricultural and medical technologies have diffused from the richer to the poor countries. Until recently, however, there were no strong incentives for people in the less developed countries to have fewer children. With birth rates remaining high and death rates falling quickly, the population has grown sharply; the LDCs are generally in stage 2 of an important model that demographers call the **demographic transition** (Figure 3.10).

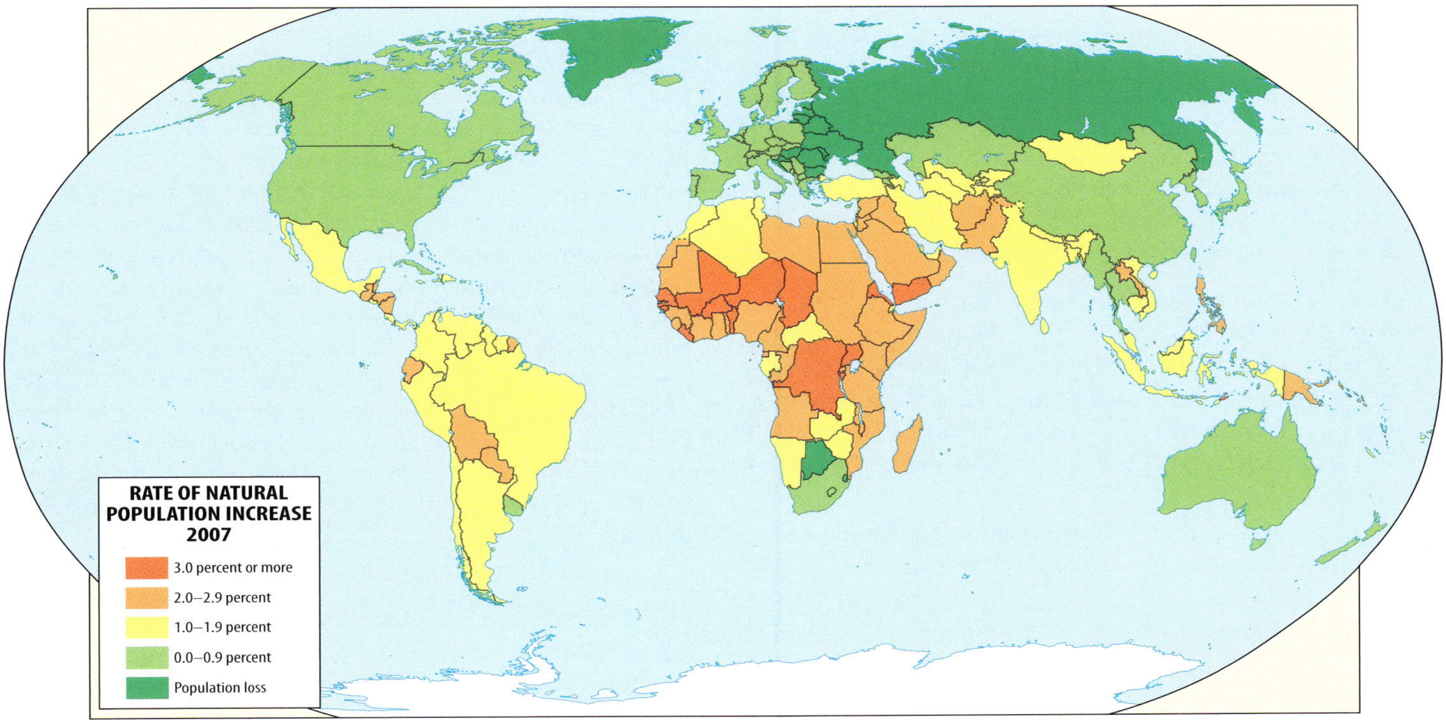

• **Figure 3.9** Population change rates are highest in the countries of Africa and other regions of the developing world and lowest in the more affluent countries.

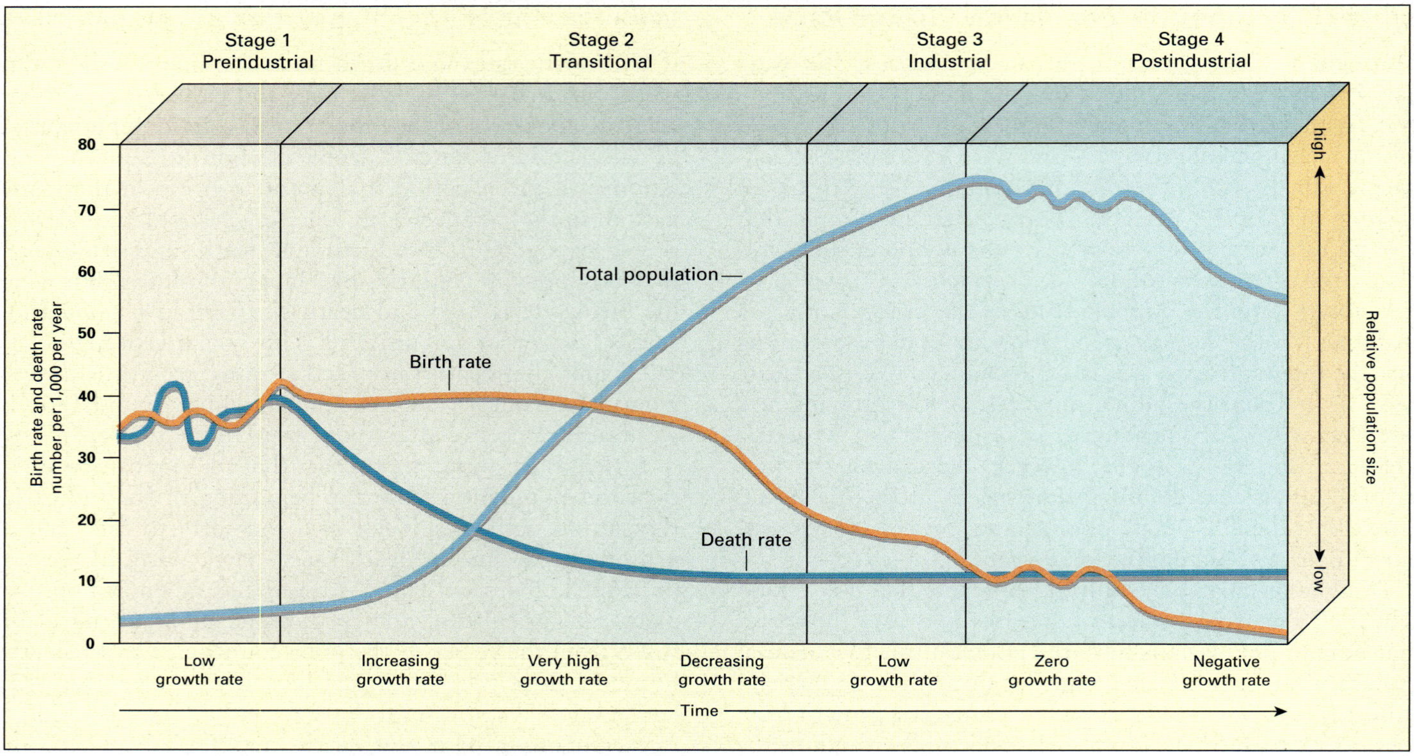

• **Figure 3.10** The demographic transition models population change in the world's wealthier countries. Note how the population surged in the wake of the Industrial Revolution as death rates fell while birth rates remained high but then leveled out and began to decline as economic development advanced.

In its entirety, from stages 1 to 4 as seen in Figure 3.10, the demographic transition model depicts the change from high birth rates and high death rates to low birth rates and low death rates that accompanied economic growth in the more developed countries (for example, western European nations, Japan, and the United States). The first two are the same two stages that humankind as a whole experienced from our earliest days until the present. The latter two have generally been experienced only by people in the wealthier countries. Note how birth rates, death rates, population change rates, and economic development correspond in this model:

- In the **first** or **preindustrial stage** (from the earliest humans to about 1800 C.E.), birth rates and death rates were high, and population growth was negligible.
- In the **second** or **transitional stage,** birth rates remained high, but death rates dropped sharply after about 1800 due to medical and other innovations of the Industrial Revolution.
- In the **third** or **industrial stage,** beginning around 1875, birth rates began to fall as affluence spread.
- Finally, after about 1975, some of the industrialized countries entered the **fourth** or **postindustrial stage,** with both low birth rates and low death rates and therefore, once again, as in stage 1, low population growth.

In this model, the United States, with population growth of about 0.6 percent per year, is in the early years of the postindustrial stage. Other industrialized, affluent countries have made their way well into that stage. Some wealthier countries, including Austria, Portugal, and Greece, have in recent years officially registered **zero population growth (ZPG).** Because of the growing desire of women in Japan and Germany to pursue their own careers and postpone marriage, birth rates in those countries have fallen below death rates, meaning that these countries are actually losing population or even experiencing a **population implosion.** In other words, the fertility rate of Japan and some other postindustrial countries is below the **population replacement level,** the number of new births required to keep the population steady (generally calculated as 2.1 children per woman in the MDCs). Other countries—notably Hungary and Latvia—that qualify as MDCs on the basis of per capita GNI PPP are experiencing population losses unfortunately due as much to rising death rates as to falling birth rates. 156
Trends of this sort can be most easily appreciated by looking at a very useful and informative device, the **age structure diagram.**

The Age Structure Diagram

About 9 of every 10 babies born in the world today are in the poorer countries. Both current and projected rates of population growth are distributed quite unevenly between the poorer and richer nations. This phenomenon is apparent in the age-structure diagrams typical of these countries. An age structure diagram (often called a **population pyramid**) classifies a population by gender and by five-year age increments (•Figure 3.11). One important index these profiles show is the percentage of a population under age 15. A country like Niger in Figure 3.11 is typically poor and faces the prospect of increasing poverty because so many new jobs, food, and other resources will have to be created to meet the demands of those children as they mature and have their own children. The bottom-heavy age structure diagram also suggests a continued surge in population as those children grow to enter their reproductive years. A large, youthful population in which competition for jobs, education, and land is intense is a social environment ripe for discord. A recent study by Population Action International found that 80 percent of the civil conflicts of the 1970s, 1980s, and 1990s took place in countries where at least 60 percent of the population was younger than 30.[3]

The bottom-heavy, pyramid-shaped age structure diagram of Niger contrasts markedly with the more chimney-shaped structures of the United States and Germany in Figure 3.11. These wealthier countries have a much more even distribution of population through age groups, with a modest share under age 15. Such profiles suggest that, not considering migration, their population growth will be low in the near future.

Collectively, about 31 percent of the population of the poorer countries is under age 15, while the corresponding figure for the wealthier countries is 17 percent (•Figure 3.12). As a whole, then, the developing world faces the critical challenge of providing for a burgeoning population in the future, even while struggling to meet the needs of the people alive today.

Where Do We Live?

Where do all these people live? The world population cartogram in •Figure 3.13 shows clearly that China and India are the most populous countries, with 1.3 and 1.1 billion people, respectively; in fact, about 35 percent of all people on earth in 2007 were either Chinese or Indian.

Why are so many people in just two countries? There are several reasons. Both countries are large. China is 20 percent larger than the contiguous 48 U.S. states, and India is about one-third the size of China. Both have been populated since very ancient times, and successful intensive agriculture has been practiced in both for more than 4,000 years. Both have large areas of productive soils and high rainfall, promoting successful farming. Both are developing countries in which birth rates remained high while death rates fell. The resulting population surge prompted China's government to adopt an aggressive population control policy, and India followed suit with less forceful measures—meaning that India will likely overtake China in population by 2030. 279, 280, 286
Overall, Monsoon Asia has about 54 percent of the world's people, and its slice of the population "pie" will continue to grow (•Figure 3.14).

The United States and Indonesia rank third and fourth among the world's most populous countries, but different

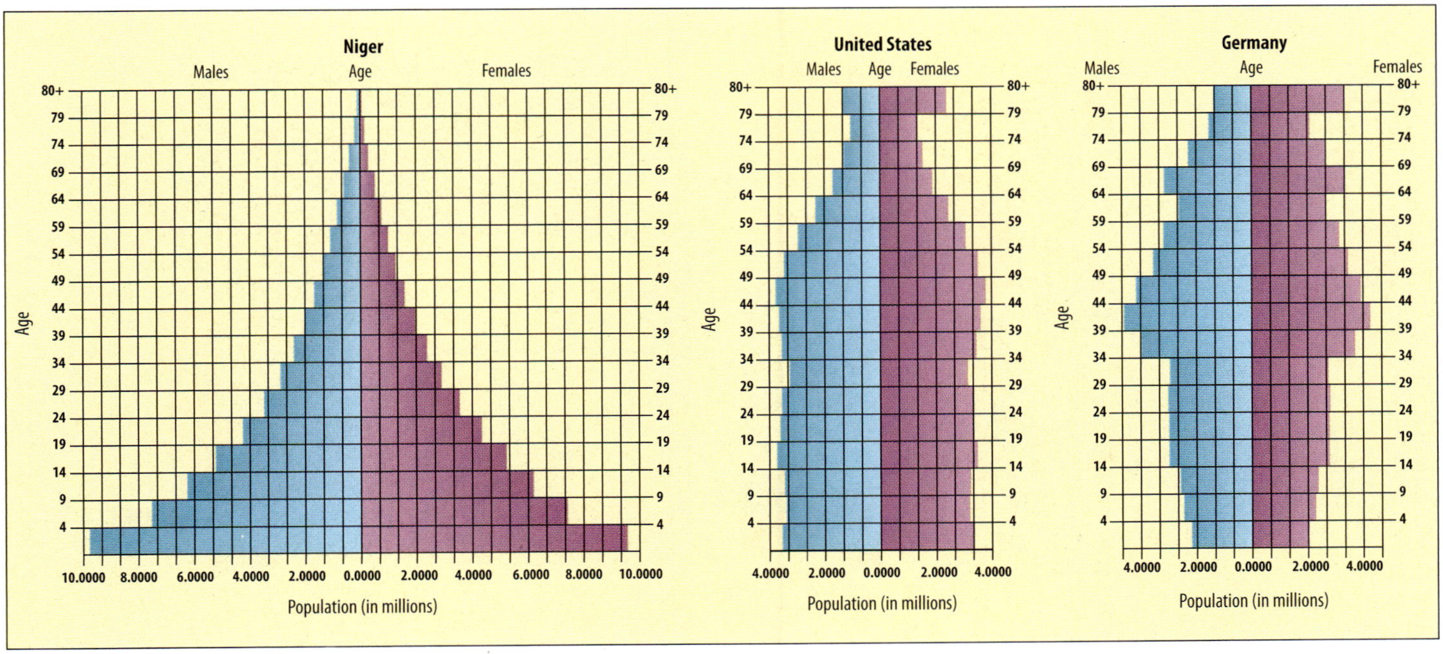

• **Figure 3.11** Population by age and sex. The pyramid-shaped age structure diagram for Niger contrasts remarkably with those of the far more affluent United States and postindustrial Germany, with their chimneylike shapes. A poor country, Niger has a relatively high birth rate, with about 48 percent of the population under age 15. The United States' population is growing slowly, while Germany and some other industrialized nations are losing populations.

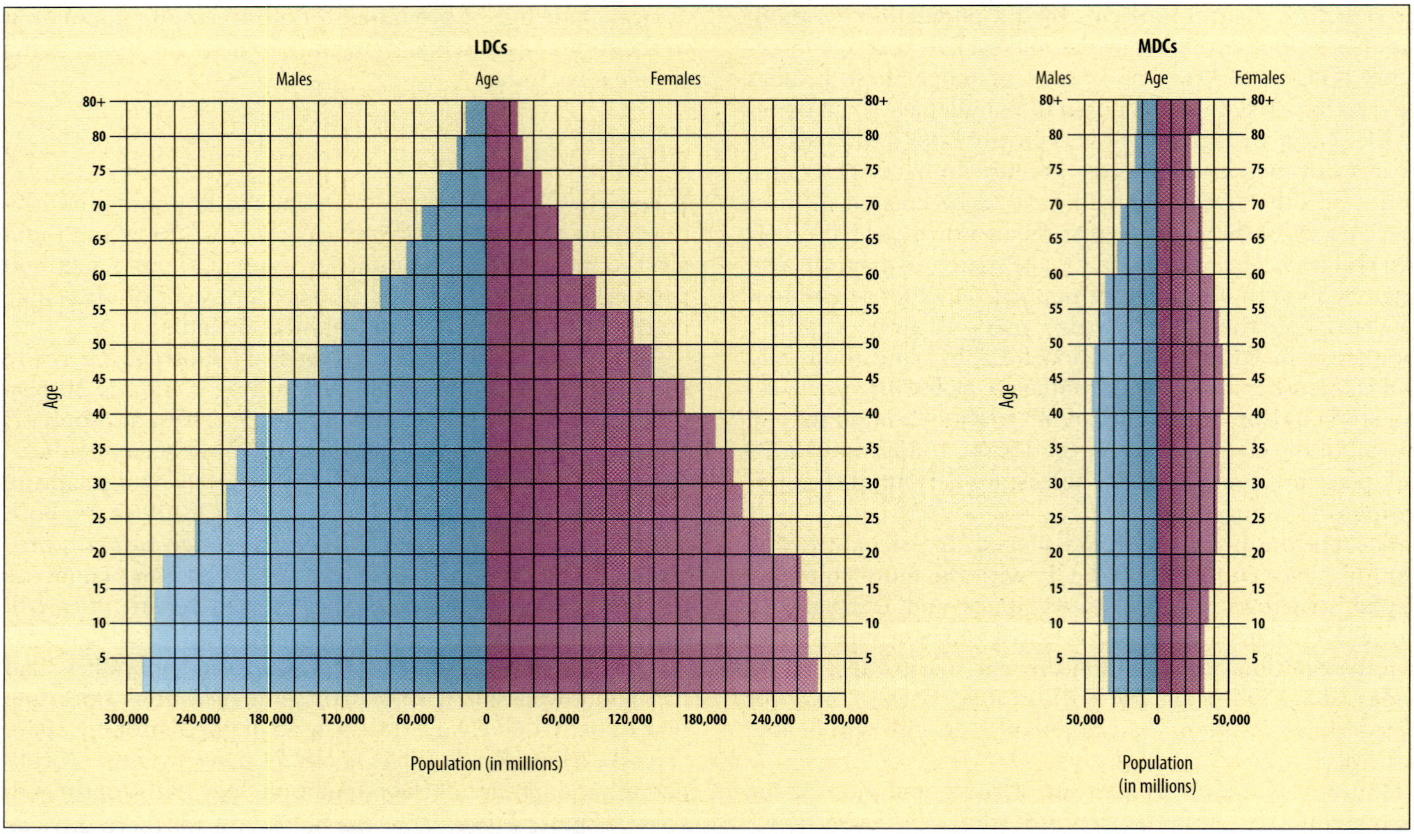

• **Figure 3.12** Population by age and sex. This diagram summarizes some of the most important facts about the human population: the lion's share lives in the poorer countries, and the lion's share of that share is young.

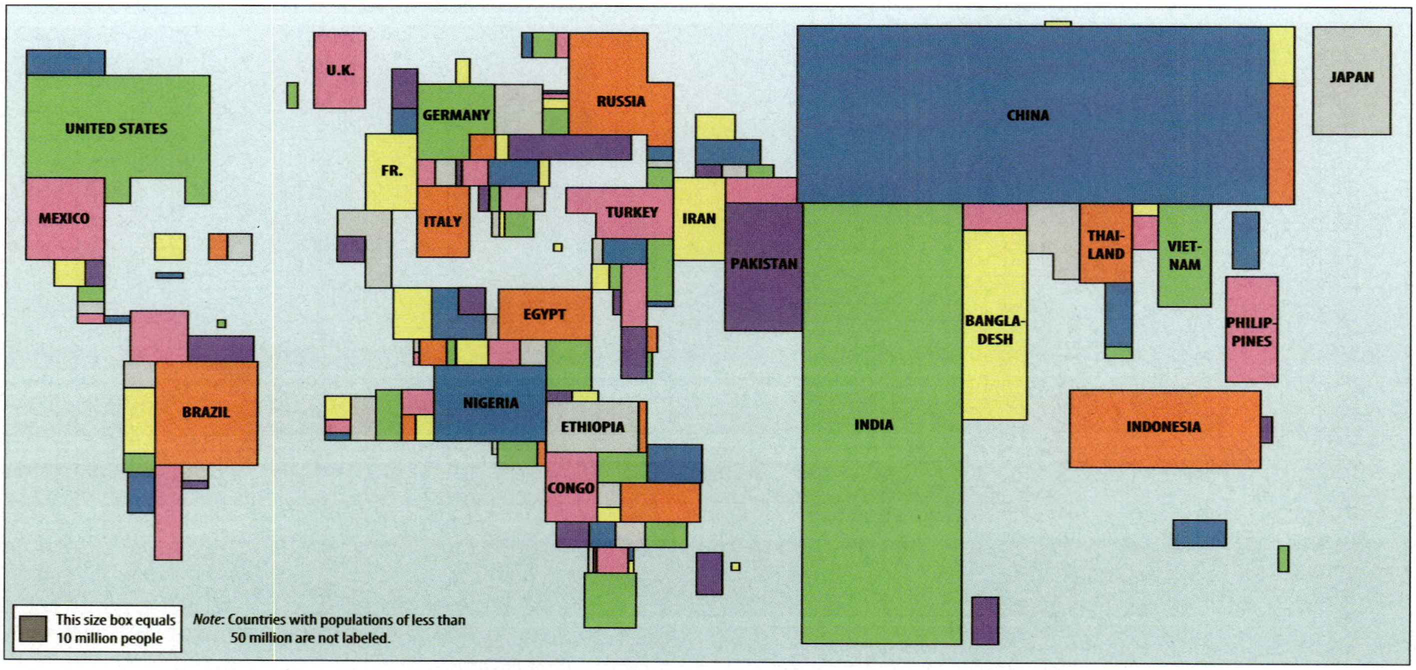

• **Figure 3.13** The demographic heavyweights of China and India stand out in the world population cartogram. The United States and Indonesia, the world's third and fourth most populous countries, are prominent too.

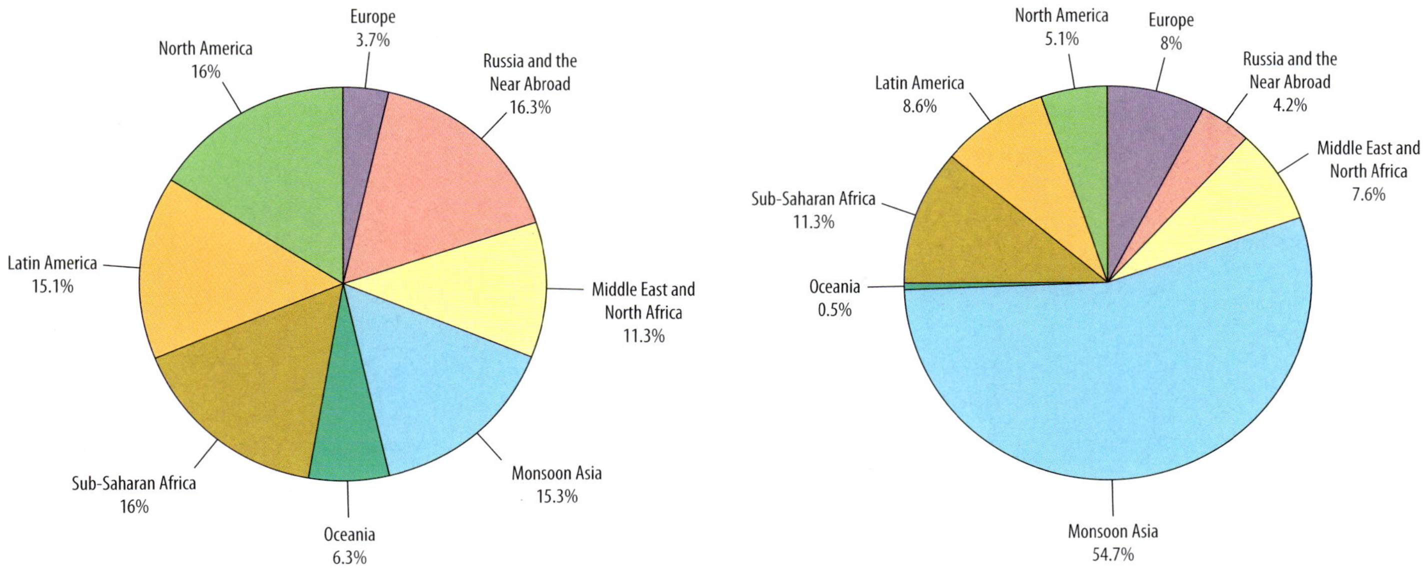

• **Figure 3.14** World regions by land areas (left) and human populations (right).

reasons account for their size. The United States has huge swaths of productive farmland, and so the environment has been able to sustain large numbers of people, but migration in periodic large waves has played a far stronger role than in China, India, or Indonesia in increasing the population. Indonesia, like China and India, is a developing country with an ancient productive agricultural environment and a recent history of high population growth. Its colonial past also played a role in population growth. The Dutch, who colonized the area known as the East Indies in the 19th century, introduced the **culture system**, a scheme to boost the output of valuable food and cash crops by requiring people on the island of Java to contribute their fields and
358 their labors. The agricultural successes of this harsh system contributed to a **positive feedback loop** (in which change in one direction produces more change in that direction) of population growth: more people produced more food, which made it possible to support more people who grew more food. Are there limits to growth of this kind? That question is addressed later in this chapter.

While there are many variables to consider in explaining why large numbers of people are clustered in particular countries, including cultural factors and family planning policies, the natural setting is by far the most important factor. The population densities shown in •Figure 3.15 correlate generally with agricultural and other environmental conditions. The deepest reds showing the highest population densities are in the more humid and fertile regions of both China and India. Conversely, in the very western part of China, including the high Tibetan Plateau, very dry conditions limit agriculture to a few favored areas. The world's highest mountains, the Himalayas, rise just north of the deep red area of high population density in northern India. The moisture-laden winds that bring so much productive rainfall to India cannot cross that mountain barrier, which has been a divide between densely and sparsely populated regions for thousands of years. Looking at the lightly populated areas of the world, you can recognize similar environmental factors at work: in the Sahara of northern Africa, the Arctic of northern Canada, and the Amazon Basin of South America, conditions are too dry, too cold, or too wet and infertile to support large numbers of people.

The Geography of Migration

Migration refers to the movement of people from one location to another in any setting, whether within a community, within a country, or between countries. Migration is one of the most dynamic and most problematic human processes on earth (•Figure 3.16). A migrant is always both an **emigrant** (one who moves from a place) and an **immigrant** (one who moves to a place). Migration is usually associated with either **push factors**, as when hunger or lack of land "pushes" peasants out of rural areas into cities or warfare pushes people from one place to another, or **pull factors**, when, for example, an educated villager takes advantage of a job opportunity in the city or another country. People responding to push factors are often referred to as **nonselective migrants**, and people reacting to pull factors are called **selective migrants.**

Both push and pull forces are behind the **rural-to-urban migration** pattern that is characteristic of most countries. 111,
The growth of cities, known as **urbanization,** is due in part 220, 289,
to the natural growth rate of people already living in urban 297,
areas, but it is particularly strong in the LDCs because of 525
this internal migration of both selective and nonselective migrants.

Within and between countries today, there are considerable movements of **refugees,** the victims of such severe

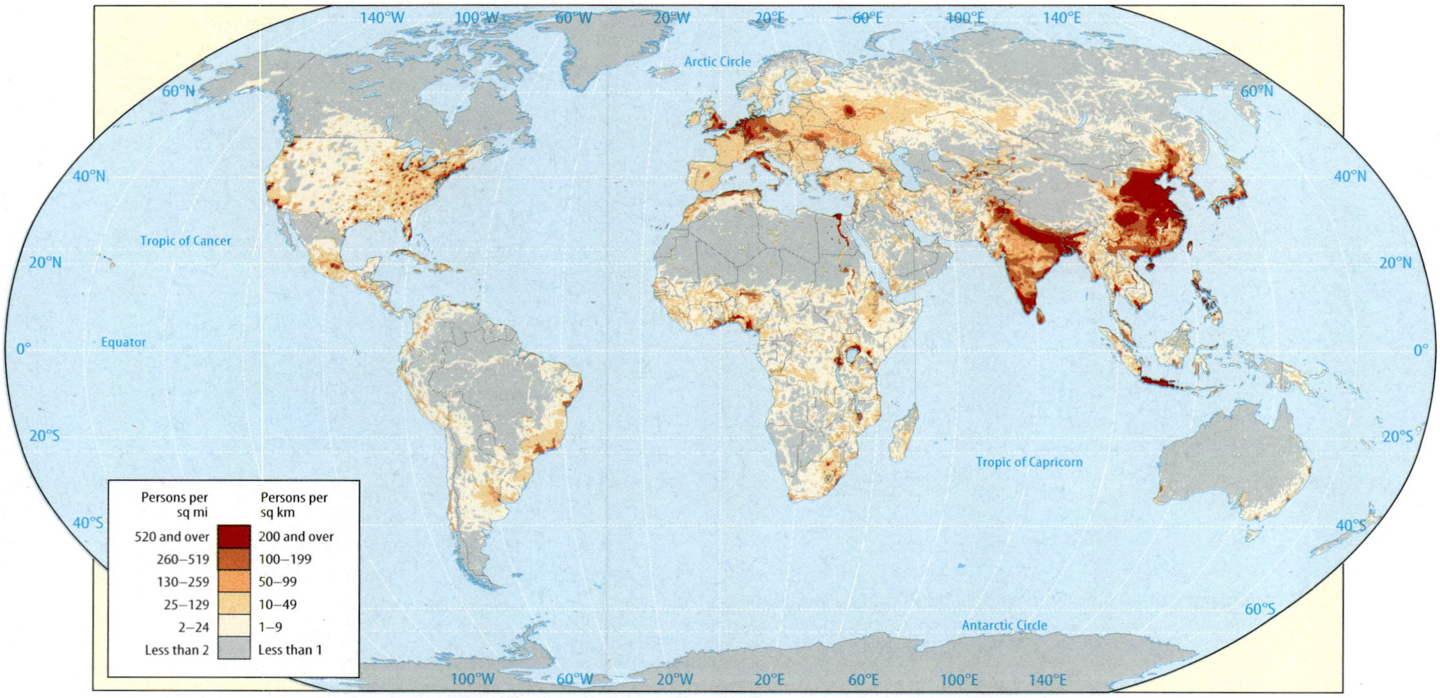

• **Figure 3.15** This isarithmic map of population density shows the approximate distribution of people around the world.

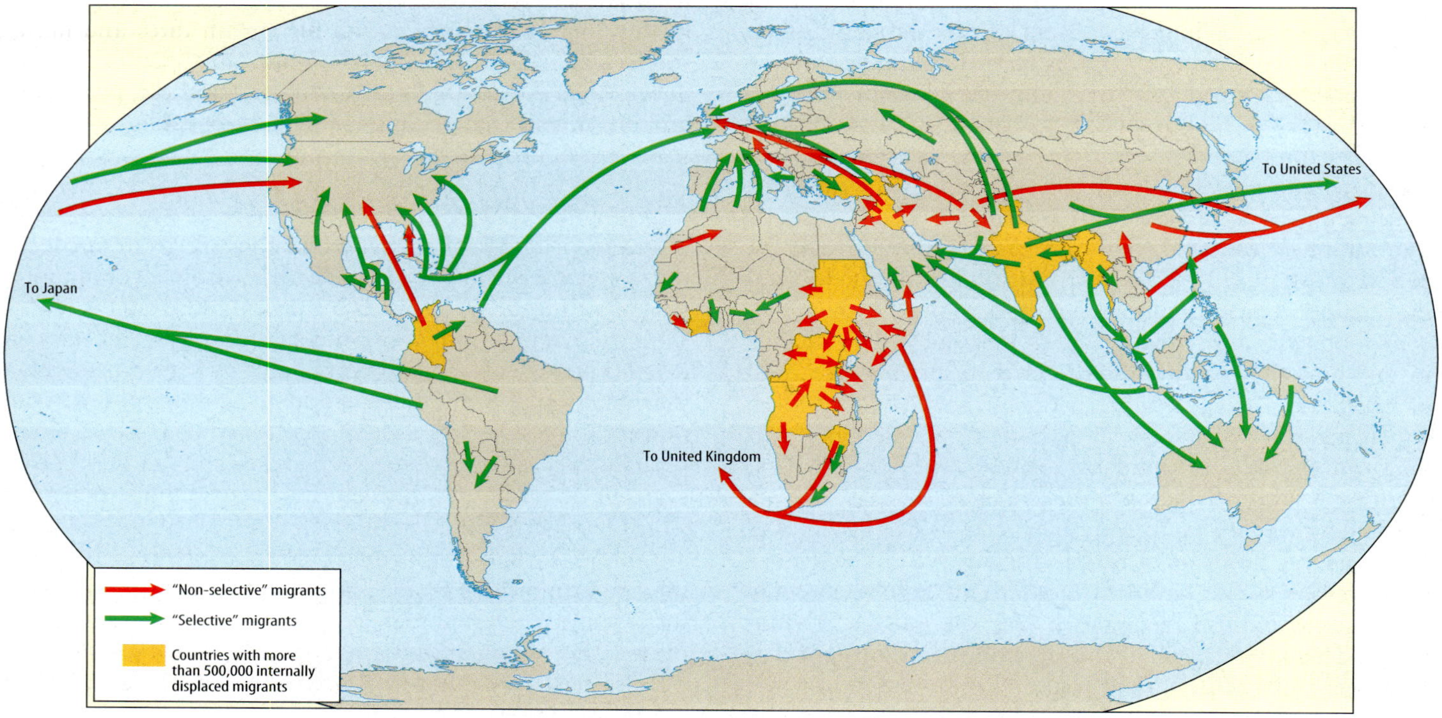

• **Figure 3.16** The global picture of people on the move. The major trends are of migrants in search of work in more affluent countries and of refugees driven by warfare or environmental adversity.

push factors as persecution, political repression, and war. They may be on the move either as illegal immigrants or as people granted **asylum,** or permission to immigrate on the grounds that they would be harmed or persecuted in their country of origin. Immigration and asylum laws and quotas vary widely around the world, depending on host countries' political, economic, and social systems. Among the most disadvantaged of the world's peoples are the **internally displaced persons (IDPs)** who are dislodged and impoverished 255
by strife in their home country but have little prospect of

emigrating. The African nation of Sudan has the world's highest number of internally displaced persons, and in the pages of this book you will read many accounts of others moved by forces beyond their control.

Migration is almost never clearly "good" or "bad" but is almost always a controversial mixed blessing. Migration is bad for the country whose most talented people are leaving but good for the country they go to; the Indian doctor down the street benefits your community in the United States but may be sorely missed in India, for example. The doctor is illustrative of the **brain drain,** the emigration of educated and talented people from a place that needs them. The Mexican immigrant who does low-wage labor in the United States may be perceived as an **illegal alien** threatening to overwhelm social services and take jobs away from local people or as a **guest worker** who performs important services that no one else wants to do. Some host-country peoples are more welcoming of new cultures that enrich their ethnic mosaic (Americans are generally perceived as among the most accommodating cultures in the world), while others fear losing their ethnic majorities and privileges.

How Many People Will Live on Earth?

Population geographers are fairly confident in their calculations of how many people have lived on earth at various times in the past (see Table 3.3). Projecting future numbers is another matter, however. There are many uncertainties. Will birth rates fall faster than anticipated in the developing world? Will death rates surge because of HIV/AIDS or some other epidemic? These are some of the wild cards in the population deck.

In asking how many people will live on earth in the future, we are essentially asking, "Will the poorer countries of the world go through the demographic transition?" Will the current, relatively high birth rates in the less developed countries continue their present slow decline (• Figure 3.17)? In 1970, Kenya had a birth rate of 51 and Bangladesh had a birth rate of 45; by 2007, Kenya's birth rate had dropped to 40 and Bangladesh's to 27. Considering what might have been, given the country's huge population, China's decline may be the most impressive of all, from a rate of 38 in 1965 to 12 in 2007. The Chinese government's generally harsh but successful **one-child policy,** using a combination of in- 286
centives and punishments to encourage couples to bear only one offspring, has been the main reason for the decline. In Bangladesh, growing literacy and the slow but steady economic progress of women have brought down the birth rate. Family planning policies and levels of education and economic well-being have played various roles in the poorer countries, but have collectively combined to bring birth rates down since 1968. If the processes of increasing economic and social development continue in the LDCs as a whole, they should make steady progress through the demographic transition, and the earth's population should cease to grow and should stabilize.

But will the poorer countries complete the transition successfully? An unsavory but possible scenario would see death rates rise dramatically (due to the scourge of HIV/AIDS, for example) in at least some of the countries, in effect pushing them back to stage 1 of the demographic transition, where both birth and death rates are high. Others could remain in stage 2, with high birth rates and falling death rates, long enough to bring unexpectedly high numbers of people into the world. With such different scenarios in mind, the United Nations prefers to use a widely respected model with three projections: high, moderate, and low growth (• Figure 3.18).

In view of declining birth rates worldwide, the United Nations has revised its projection for future population growth downward. The agency predicts 9.2 billion by 2050 (half a billion fewer than it had estimated in earlier

Joe Hobbs

• **Figure 3.17** National family planning programs can have a pronounced impact on birth rates. This sign on the main square in Cairo, Egypt, urges parents to have no more than two children "for the sake of a better life."

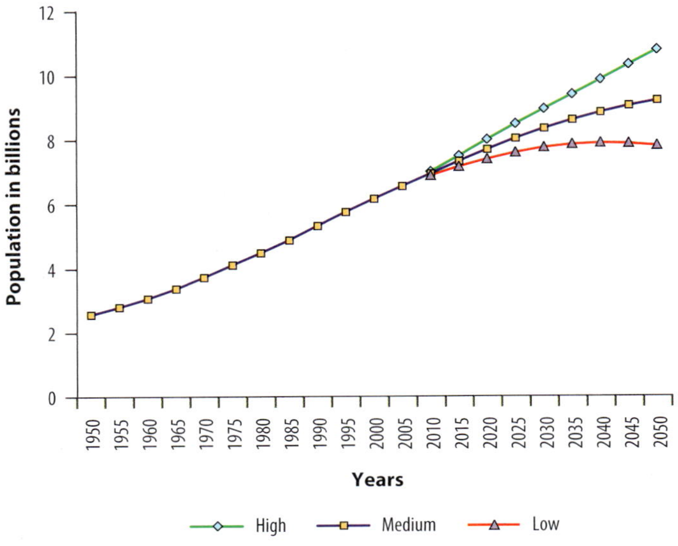

• **Figure 3.18** United Nations projections for population growth.

projections) and stabilization—the maximum number of people that will ever live on earth at one time—at 10.8 billion in 2150 (the earlier projection was 12.6 billion).

This downward revision has prompted a rash of popular and academic articles proclaiming that "the population explosion is over" and even some essays arguing there would soon be too few people on earth, particularly in the richer countries. Other experts have cautioned that it is too soon to declare the population bomb defused. "World population growth turned a little slower," a Population Institute report argues. "The difference, however, is comparable to a tidal wave surging toward one of our coastal cities. Whether the tidal wave is 80 feet or 100 feet high, the impact will be similar."[4] (To put that in more concrete terms, while the world population grew by an all-time high of 88 million people in 1994 and the population growth rate fell after that, it still grew by 70 million in 2007.) Not content with its own projected figure for stabilization, the United Nations has pledged to do its best to help stabilize the global population at no more than 9.8 billion after the year 2050. The organization's plan is to focus on enhancing the education and employment of women in the less developed countries as a means of bringing down birth rates.

The Malthusian Scenario

Even while the birth rate falls, the population increases. Already large and growing numbers pose fundamental questions: Can the earth sustain 9.8 billion or more people? Will we exceed the planet's carrying capacity, and what will happen if we do? Early in the Industrial Revolution, an English clergyman named Thomas Malthus (1766–1834) postulated that human populations, growing geometrically or exponentially, would exceed food supplies, which grow only arithmetically or linearly. He predicted a catastrophic human die-off as a result of this irreconcilable equation (•Figure 3.19). He could not have foreseen that the exploitation of new lands and resources, including tapping the energy of fossil fuels, would permit food production to keep pace with or even outpace population growth for at least the next two centuries.

This **Malthusian scenario** of the lost race between food supplies and mouths to feed remains a source of constant and important debate today. On one side of the debate are optimists, the so-called **technocentrists** or **cornucopians,** who argue that human history provides insight into the future. Thanks to their technological ingenuity, people have always been able to conquer food shortages and other problems and therefore always will (•Figure 3.20). The late Julian Simon, a University of Maryland economist, argued that far from being a drain on resources, additional people create additional resources. Technocentrists therefore insist that people can raise the earth's carrying capacity indefinitely and that the die-off that Malthus predicted will always be averted. Our more numerous descendants will instead enjoy more prosperity than we do.

In contrast, the **neo-Malthusians** (heirs of the reasoning of Thomas Malthus) argue that although we've been successful so far, we cannot increase the earth's carrying capacity indefinitely. There is an upper limit beyond which growth cannot occur, with calculations ranging from 8 to 40 billion people. Neo-Malthusians insist that the poorer countries cannot remain indefinitely in the second stage of the demographic revolution. Either they must intentionally bring birth rates down further and make it successfully through the demographic transition, or they must unwillingly suffer nature's solution, a catastrophic increase in death rates. Thus by either the **birth rate solution** or the **death rate solution** to the problem, the neo-Malthusians argue, the less developed world must confront its population crisis (see pages 62–64).

Whereas technocentrists view the equation between people and resources passively, insisting that no corrective action is needed, neo-Malthusians tend to be activists who describe terrible scenarios of a death rate solution in order to motivate people to adopt the birth rate solution. The biolo-

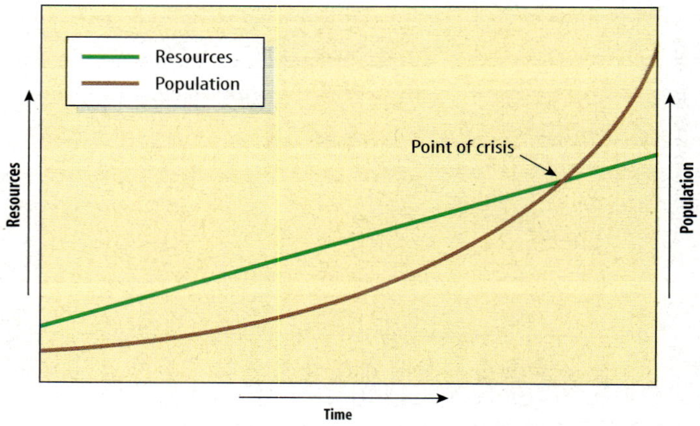

• **Figure 3.19** Malthus envisioned a race between people and resources, in which people lost.

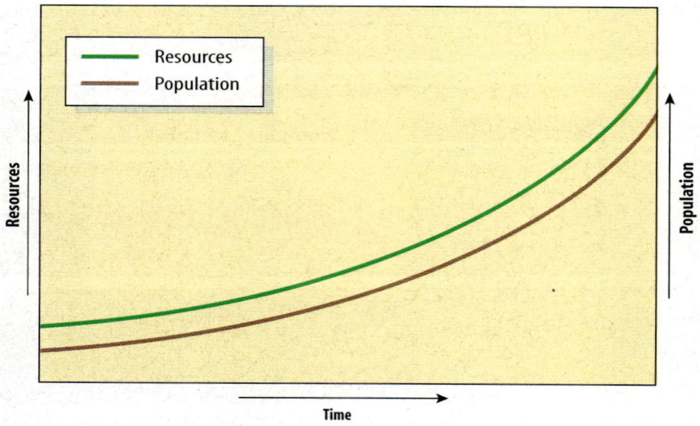

• **Figure 3.20** The technocentrists reason that production of food and other resources will always stay ahead of population growth.

gist Paul Ehrlich thinks that we are increasingly vulnerable to a Malthusian catastrophe, particularly as HIV and other viruses diffuse around the globe with unprecedented speed. "The only big question that remains," Ehrlich writes, "is whether civilization will end with the bang of an all-out nuclear war, or the whimper of famine, pestilence and ecological collapse."[5] Such dire warnings have earned many neo-Malthusians the reputation of being "gloom-and-doom pessimists." The neo-Malthusians insist that there are simply too many people already for the world to support, especially in the LDCs, where there are not enough resources to support them. But what does "too many people" mean? How many are too many, and based on what criteria?

What Is "Overpopulation"?

It is probably most useful to think about two distinct types of overpopulation, one characteristic of the poor countries and the other of the rich (•Figure 3.21). **People overpopulation** is an apparent problem in the poorer countries. The environmental problems characteristic of LDCs are intensified by the relatively high rates of human population growth in those countries. More people cut more trees, hunt more wildlife, and otherwise use more resources. Many persons, each using a small quantity of natural resources daily to sustain life, have a great collective impact on the environment and may add up to too many people for the local environment to support. Common consequences are malnutrition and even the famine emergencies in which richer countries are called on to provide relief. 287, 415, 455, 617

Consumption overpopulation is characteristic of the MDCs. In the wealthier countries, there are fewer persons, but each uses a large quantity of natural resources from ecosystems around the world. Their collective impacts also degrade the environment, and even their smaller numbers may be "too many" at such unsustainably high levels of consumption, particularly with their habit of feeding so high on the food chain (Insights, page 62). With less than 5 percent of the world's population, the United States may be regarded as the world's leading overconsumer, accounting, for example, for about one-quarter of the world's annual energy consumption. According to a useful measure known as the **ecological footprint**—the amount of biologically productive land needed to sustain a person's consumption and absorb his or her wastes—for every acre needed to support an average Ethiopian, 69 acres are needed to support an

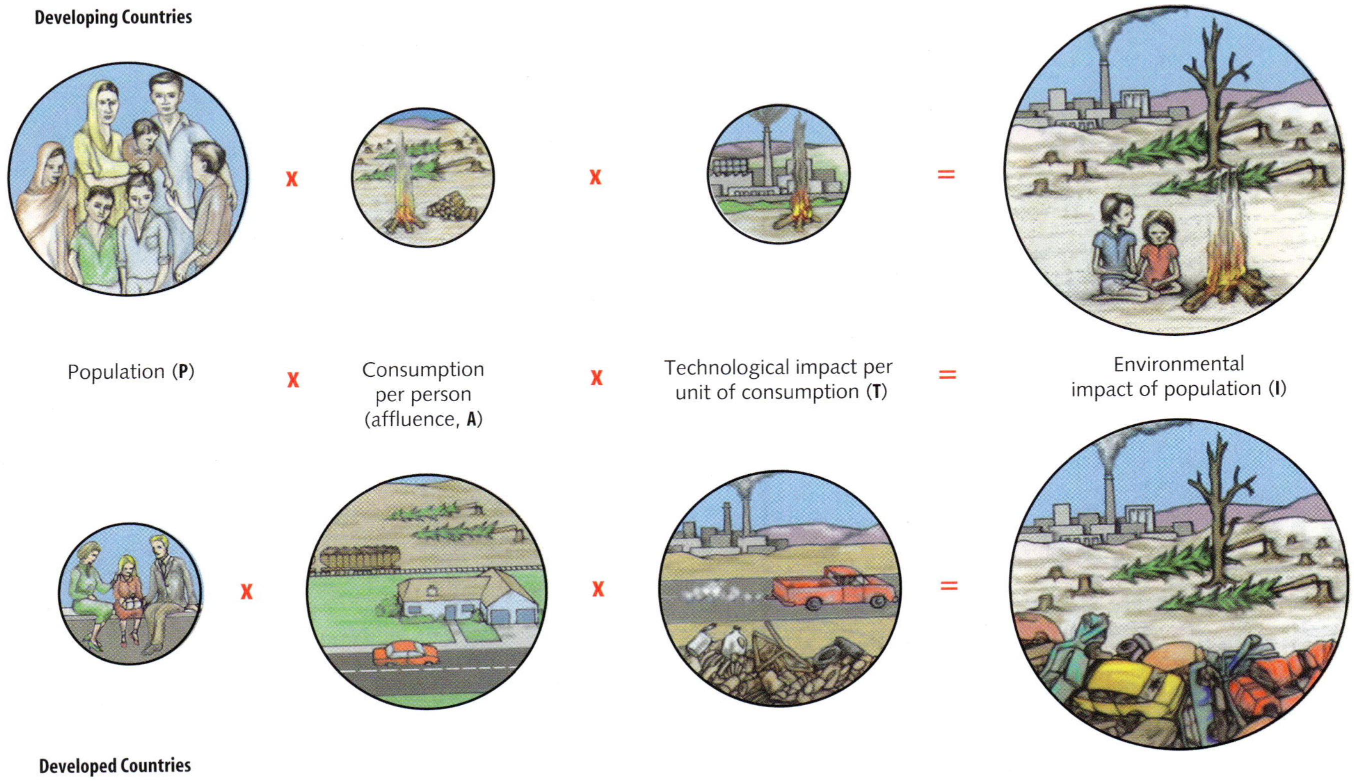

• **Figure 3.21** Two types of overpopulation, calculated according to this formula: *number of people* × *number of units of resources used per person* × *environmental degradation and pollution per unit of resource used* × *environmental impact.* Circle size shows the relative importance of each factor. People overpopulation is caused mostly by growing numbers of people and is typical of LDCs. Consumption overpopulation is caused mostly by growing affluence and is typical of MDCs.

INSIGHTS

The Second Law of Thermodynamics

It is useful to understand how energy flows through ecosystems because this process is central to the question of how many people the earth can support. **Food chains** (for example, in which an antelope eats grass and a leopard eats an antelope) are short, seldom consisting of more than four **trophic (feeding) levels.** The collective weight (known as **biomass**) and absolute number of organisms decline substantially at each successive trophic level (there are fewer leopards than antelopes). The reason is a fundamental rule of nature known as the **second law of thermodynamics,** which states that the amount of high-quality usable energy diminishes dramatically as the energy passes through an ecosystem. In living and dying, organisms use and lose the high-quality, concentrated energy that green plants produce and that is passed up the food chain. As an organism is consumed in any given link in the food chain, about 90 percent of that organism's energy is lost, in the form of heat and feces, to the environment. The amount of animal biomass that can be supported at each successive level thus declines geometrically, and little energy remains to support the top carnivores. So, for example, there are many more deer than there are mountain lions in a natural system.

The second law of thermodynamics has important consequences and implications for the human use of resources. The higher we feed on the food chain (the more meat we eat), the more energy we use. The question of how many people the earth's resources can support thus depends very much on what we eat. The affluent consumer of meat demands a huge expenditure of food energy because that energy flows from grain (producer) to livestock animal (primary consumer, with a 90 percent energy loss) and then from livestock animal to human consumer (secondary consumer, with another 90 percent energy loss). Each year, the average U.S. citizen eats 100 pounds of beef, 50 pounds of pork, and 45 pounds of poultry. By the time it is slaughtered, a cow has eaten about 10 pounds of grain per pound of its body weight; a pig, 5 pounds; and a chicken, 3 pounds. Thus in a year, a single American consumes the energy captured by two-thirds of a ton of grain (1,330 pounds) in meat products alone. By skipping the meat and eating just the corn that fattens these animals, many people could live on the energy required to sustain just one meat-eater. The point is not to feel guilty about eating meat but to realize that there is no answer to the question "How many people can the world support?" The question needs to be rephrased: "How many people can the world support based on people consuming ________?" (with the blank filled in with a specific number of calories per day or with a general lifestyle).

Development means a more affluent lifestyle, including eating more meat (feeding higher on the food chain). Some analysts fear the burden that the changing dietary habits accompanying development will pose to the planet's food energy supply. Lester Brown of the WorldWatch Institute wrote a provocative essay titled "Who Will Feed China?"[6] Keeping in mind the second law of thermodynamics, he argued that if China continues on its present course of growing affluence, with a move away from reliance on rice as a dietary staple to a diet rich in meats and beer, the entire planet will shudder. How could more than a billion people be sustained so high on the food chain?

average American. The disparity suggests that if Ethiopia is "underdeveloped," the United States is "overdeveloped."

If the vast majority of the world's population were to consume resources at the rate that U.S. citizens do, the environmental results might be ruinous (see Insights, above). Even if consumption levels in the LDCs do not rise substantially, the sheer increase in numbers of people in those countries suggests that degradation of the environment will accelerate in the coming decades.

3.4 Addressing Global Problems

The cornucopian view is comforting: we have nothing to worry about. By keeping up the good, ingenious work as we have in the past, our futures will be secure. If, however, one diverges modestly from this premise or even accepts the most dire neo-Malthusian view, it is appropriate to ask what can be done to prevent or solve some of the world's critical problems involving natural resources and human population numbers.

The Death Rate Solution and Lifeboat Ethics

One option is to "let nature take it course" and allow people imperiled by famine or other catastrophe to perish. The neo-Malthusian ecologist Garrett Hardin introduced **lifeboat ethics**—the question of whether or not the wealthy should rescue the "drowning" poor—in a distressing and challenging essay. "People turn to me," wrote Hardin, "and say, 'My children are starving. It's up to you to keep them alive.' And I say, 'The hell it is. I didn't have those children.'"[7] He described the world not as a single "spaceship *Earth*" or "global village" with a single carrying capacity, as many environmentalists do, but as a number of distinct "lifeboats," each occupied by the citizens of single countries and each having its own carrying capacity. Each rich nation is a lifeboat comfortably seating a few people. The world's poor are in lifeboats so overcrowded that many fall overboard. They swim to the rich lifeboats and beg to be brought aboard. What should the passengers of the rich lifeboat do? The choices pose an ethical dilemma (•Figure 3.22).

Hardin set out the following scenario. There are 50 rich passengers in a boat with a capacity of 60. Around them

• **Figure 3.22** An overcrowded wooden boat ferries passengers from Mafia Island to mainland Tanzania. What if the boat capsizes?

are 100 poor swimmers who want to come aboard. The rich boaters have three choices. First, they could take in all the swimmers, capsizing the boat with "complete justice, complete catastrophe." Second, as they enjoy an unused excess capacity of 10, they could admit just 10 from the water. But which 10? And what about the margin of comfort that excess capacity allows them? Finally, the rich could prevent any of the doomed from coming aboard, ensuring their own safety, comfort, and survival.

Translating this metaphor into reality, as an occupant of the rich lifeboat *United States*, for example, what would you do for the drowning refugees from lifeboat *Haiti* or lifeboat *Afghanistan*?

Hardin's choice was the third: drowning. To preserve their own standard of living and ensure the planet's safety, the wealthy countries must cease to extend food and other aid to the poor and must close their doors to immigrants from poor countries. "Every life saved this year in a poor country diminishes the quality of life for subsequent generations," Hardin concluded. "For the foreseeable future, survival demands that we govern our actions by the ethics of a lifeboat."[8] Incidentally, Hardin committed suicide in 2003.

The Birth Rate Solution and Sustainable Development

In recent decades, new concepts and tools for managing the earth and its resources in an effective, long-term way have emerged. Known collectively as **sustainable development** (or **ecodevelopment**), these ideas and techniques consider what both MDCs and LDCs can do to avert the possible Malthusian dilemma and improve life on the planet. By promoting the birth rate solution and other concrete actions, sustainable development offers an activist agenda without the peril of doom depicted by the neo-Malthusians.

The World Conservation Union defines sustainable development as "improving the quality of human life while living within the carrying capacity of supporting ecosystems."[9] Sustainable development refutes what its proponents perceive as the current pattern of unsustainable development, whereby economic growth is based in large part on excessive resource use. Advocates of sustainable development point out that a country that depletes its resource base for short-term profits gained through deforestation increases its gross domestic product (GDP) and appears to be more "developed" than a country that protects its forests for a long-term harvest of sustainable yield. Deforestation appears to be beneficial to a country because it raises GDP through the production of pulp, paper, furniture, and charcoal. However, GDP growth does not measure the negative impacts of deforestation, such as erosion, flooding, siltation, and malnutrition. These consequences are known as **external costs**, or **externalities,** and they are not taken into account in the prices of goods and services. Advocates of sustainable development argue that these externalities should be added or "internalized" before a good or a service is marketed because the true high costs of producing these goods and services would be recognized. The lower true costs of less destructive practices would then be evident, providing stronger incentives for individuals, companies, and nations to invest in sustainable practices and technologies.

Sustainable development is a complex assortment of theories and activities, but its proponents call for eight essential changes in the way people perceive and use their environments:

1. People must change their worldviews and value systems, recognizing the finiteness of resources and reducing their expectations to a level more in keeping with the earth's environmental capabilities. Proponents of sustainable development argue that this change in perspective is needed especially in the MDCs, where instead of trying to "keep up with the Joneses," people should try to enjoy life through more social rather than material pursuits.

2. People should recognize that development and environmental protection are compatible. Rather than viewing environmental conservation as a drain on economies, we should see it as the best guarantor of future economic well-being. This is especially important in the LDCs, with their resource-based economies.

3. People all over the world should consider the needs of future generations more than we do now. Much of the wealth we generate is in effect borrowed or stolen from our descendants. Our economic system values current environmental benefits and costs far more than future benefits and costs, and so we try to improve our standard of living today without regard to tomorrow.

4. Communities and countries should strive for self-reliance, particularly through the use of appropriate technologies. For example, remote villages could rely increasingly on solar power for electricity rather than be linked into national grids of coal-burning plants.

5. LDCs need to limit population growth as a means of avoiding the destructive impacts of people overpopulation. Advances in the status of women, improvements in education and social services, and effective family planning technologies can help limit population growth.

6. Governments need to practice land reform, particularly in the LDCs. Poverty is often not the result of too many people on too little total land area but of a small, wealthy minority holding a disproportionately high share of quality land. To avoid the environmental and economic consequences of marginalization, a more equitable distribution of land is needed.

7. Economic growth in the MDCs should be slowed to reduce the effects of consumption overpopulation. If economic growth, understood as the result of consumption of natural resources, continues at its present rate in excess of sustainable yield, the earth's "environmental capital" will continue to diminish rapidly.

8. Wealth should be redistributed between the MDCs and LDCs. Because poverty is such a fundamental cause of environmental degradation, the spread of a reasonable level of prosperity and security to the LDCs is essential. Proponents argue that this does not mean that rich countries should give cash outright to poor countries. Instead, the lending institutions of MDCs can forgive some existing debts owed by LDCs or use such innovations as **debt-for-nature swaps,** in which a certain portion of debt is forgiven in return for the borrower's pledge to invest that amount in national parks or other conservation programs. Reducing or eliminating **trade barriers** that MDCs impose against products from LDCs would also help redistribute global wealth. 561

Some geographers and other scientists believe that sustainable development (rather than information technology) will bring about the **"Third Revolution,"** a shift in human ways of interacting with the earth so dramatic that it will be compared with the origins of agriculture and industry. The formidable changes called for in sustainable development are attracting increasing attention, perhaps because at present there are no comprehensive alternative strategies for dealing with some of the most critical issues of our time.

SUMMARY

- A useful way to begin to appreciate the current spatial patterns of our relationship with the earth is to view these patterns as products of two "revolutions" in the relatively recent past: the Agricultural or Neolithic (New Stone Age) Revolution and the Industrial Revolution. Each transformed humanity's relationship with the natural environment. After the Industrial Revolution, the human impact on earth grew dramatically.
- The world is markedly divided between the haves and the have-nots, characterized at the largest scale by contrasts between more developed and less developed countries (MDCs and LDCs). Explanations for these disparities include dependency theory, cultural factors, geographic location, and natural resource base.
- LDC economies often rely on the export of a few raw materials or commercial crops and tend to draw down their natural "capital" quickly, with profound impacts on the environments.
- There is disagreement on whether globalization is beneficial or detrimental to poorer countries and to the poorer people who live in those countries.
- Population growth is measured by birth rates, death rates, and migration. Growth rates tend to be higher in the LDCs, due to relatively high but falling birth rates and much lower and falling death rates. The demographic transition, a model of what happened to populations in the MDCs, shows the wealthier countries passing from high birth rates and death rates to low birth rates and death rates.
- The explosive growth in world population since the beginning of the Industrial Revolution is the result not of a rise in birth rates but of a dramatic decline in death rates, particularly in the LDCs. High rates of human population growth intensify environmental problems characteristic of the LDCs. More

people use more resources, a phenomenon that suggests there is a problem of people overpopulation in the poorer countries. Consumption overpopulation is more characteristic of the MDCs, where fewer people use large quantities of natural resources from around the world.

→ Human demands for more food and other photosynthetic products are growing, and it is questionable whether supplies can keep pace. The Malthusian scenario, which has not been realized so far, insists that a catastrophic die-off of people will occur when their numbers exceed food supplies.

→ New concepts and tools for managing the earth and its resources in an effective, long-term way have emerged in the past two decades. Known collectively as sustainable development, or ecodevelopment, these ideas and techniques consider what humanity can do to avert the Malthusian dilemma and improve life on the planet.

KEY TERMS + CONCEPTS

Age of Discovery (p. 42)
Age of Exploration (p. 42)
age structure diagram (p. 54)
Agricultural Revolution (p. 41)
asylum (p. 58)
biomass (p. 62)
birth rate (p. 50)
birth rate solution (p. 60)
brain drain (p. 59)
carrying capacity (p. 41)
cash crops (p. 46)
civilization (p. 41)
commercial crops (p. 46)
consumption overpopulation (p. 61)
cornucopians (p. 60)
culture hearth (p. 41)
culture system (p. 57)
death rate (p. 50)
death rate solution (p. 60)
debt-for-nature swap (p. 64)
deforestation (p. 48)
demographic transition (p. 53)
  first (preindustrial) stage (p. 54)
  second (transitional) stage (p. 54)
  third (industrial) stage (p. 54)
  fourth (postindustrial) stage (p. 54)
demography (p. 50)
dependency theory (p. 45)
development (p. 44)
digital divide (p. 47)
domestication (p. 41)
dry farming (p. 41)
ecodevelopment (p. 63)
ecological bankruptcy (p. 48)
ecological footprint (p. 61)
ecologically dominant species (p. 40)
emigrant (p. 57)
extensive land use (p. 41)
external costs (p. 63)
externalities (p. 63)
feeding levels (p. 62)
food chain (p. 62)
Food-Producing Revolution (p. 41)
foraging (p. 40)
fuelwood crisis (p. 48)
globalization (p. 47)
gross domestic product (GDP) (p. 44)
gross national income (GNI) (p. 44)
gross national product (GNP) (p. 44)
guest worker (p. 59)
hot money (p. 47)
Human Development Index (HDI) (p. 44)
hunting and gathering (p. 40)
illegal alien (p. 59)
immigrant (p. 57)
Industrial Revolution (p. 42)
information technology (IT) (p. 47)
intensive land use (p. 41)
internally displaced persons (IDPs) (p. 58)
irrigation (p. 41)
knowledge economy (p. 47)
less developed countries (LDCs) (p. 43)
lifeboat ethics (p. 62)
Malthusian scenario (p. 60)
marginalization (p. 46)
mercantile colonialism (p. 45)
migration (p. 50)
more developed countries (MDCs) (p. 43)
multinational companies (p. 47)
natural replacement rate (p. 48)
natural resource (p. 46)
neocolonialism (p. 46)
neo-Europes (p. 45)
Neolithic Revolution (p. 41)
neo-Malthusians (p. 60)
newly industrializing countries (NICs) (p. 44)
nontimber forest products (NTFPs) (p. 49)
nonselective migrants (p. 57)
"original affluent society" (p. 40)
one-child policy (p. 59)
people overpopulation (p. 61)
per capita GDP (p. 44)
per capita gross national income purchasing power parity (per capita GNI PPP) (p. 44)
Pleistocene overkill hypothesis (p. 40)
population change rate (p. 51)
population explosion (p. 50)
population implosion (p. 54)
population pyramid (p. 55)
population replacement level (p. 54)
positive feedback loop (p. 57)
pull factors (p. 57)
purchasing power parity (PPP) (p. 44)
push factors (p. 57)
refugees (p. 57)
renewable resource (p. 48)
rural-to-urban migration (p. 57)
second law of thermodynamics (p. 62)
selective migrants (p. 57)
settler colonization (p. 45)
sustainable development (p. 63)
sustainable yield (p. 48)
technocentrists (p. 60)
"Third Revolution" (p. 64)
trade barriers (p. 64)
trophic levels (p. 62)
urbanization (p. 57)
value-added products (p. 45)
zero population growth (ZPG) (p. 54)

REVIEW QUESTIONS

1. What were the Agricultural and Industrial Revolutions? In what ways did they initiate important changes in human-earth relationships?
2. What are the typical differences between MDCs and LDCs?
3. According to dependency theory, what are the causes of disparities between MDCs and LDCs? What other factors may explain global wealth and poverty?
4. What is the fuelwood crisis? What are some of the effects of deforestation on human lives and economies in the LDCs?
5. What are the two types of overpopulation, and how do they differ?
6. What has been the main cause of the world's explosive population growth since the beginning of the Industrial Revolution?
7. What variables distinguish the four stages of the demographic transition, and what explains them?
8. Which world regions have the most and the fewest people, and what factors might account for these differences?
9. How do technocentrists and neo-Malthusians view the balance between people and resources?
10. What do Hardin's lifeboats represent—what is his metaphor?
11. What are the goals and methods of sustainable development?

NOTES

1. Marshall B. Sahlins, *Stone Age Economics* (Chicago: Aldine-Atherton, 1972).
2. Alfred W. Crosby, *Ecological Imperialism: The Biological Expansion of Europe, 900–1900* (New York: Cambridge University Press, 1986).
3. Celia W. Dugger, "Very Young Populations Contribute to Strife, Study Concludes." *New York Times,* April 4, 2007, p. A6.
4. Cited in Steven A. Holmes, "Global Crisis in Population Still Serious, Group Warns." *New York Times,* December 31, 1997, p. A7.
5. Paul R. Ehrlich, "Populations of People and Other Living Things." In *Earth '88: Changing Geographic Perspectives,* ed. Harm De Blij (Washington, D.C.: National Geographic Society, 1988), p. 309.
6. Lester R. Brown, "Who Will Feed China?" *WorldWatch Magazine,* September-October 1994, p. 10.
7. Garrett Hardin, "The Tragedy of the Commons." *Science, 162,* 1968, pp. 1243–1248.
8. Garrett Hardin, "Lifeboat Ethics: The Case against Helping the Poor." *Psychology Today,* September 1974, p. 6; and "Living on a Lifeboat," *BioScience 24,* 1974, p. 568.
9. World Wildlife Fund, *Sustainable Use of Natural Resources: Concepts, Issues and Criteria* (Gland, Switzerland: World Wildlife Fund, 1995), p. 5.

Egypt's Saint Katherine Monastery, built in the 6th century, lies at the foot of Jebel Musa—a mountain believed by many Christians and Muslims to be Mount Sinai of the Bible and the Qur'an.

CHAPTER 6
A GEOGRAPHIC PROFILE OF THE MIDDLE EAST AND NORTH AFRICA

The *Middle East* is a fitting designation for the places and the cultures of this vital world region because they are literally in the middle. This is a physical crossroads, where the continents of Africa, Asia, and Europe meet and the waters of the Mediterranean Sea and the Indian Ocean mingle. Its peoples—Arab, Jew, Persian, Turk, Kurd, Berber, and others—express in their cultures and ethnicities the coming together of these diverse influences. Occupying as they do this strategic location, the nations of the Middle East and North Africa have through five millennia been unwilling hosts to occupiers and empires originating far beyond their borders. They have also bestowed on humankind a rich legacy that includes the ancient civilizations of Egypt and Mesopotamia and the world's three great monotheistic faiths of Judaism, Christianity, and Islam.

chapter outline

chapter objectives

This chapter should enable you to

- → Understand and explain the mostly beneficial relationships between villagers, pastoral nomads, and city dwellers in an environmentally challenging region
- → Know the basic beliefs and sacred places of Jews, Christians, and Muslims
- → Recognize the importance of petroleum to this region and the world economy
- → Identify the geographic chokepoints and oil pipelines that are among the world's most strategically important places and routes
- → Appreciate the problems of control over freshwater in this arid region
- → Know what al-Qa'ida and other Islamist terrorist groups are and what they want

People outside the region tend to forget about such contributions because they associate the Middle East and North Africa with war and terrorism. These negative connotations are often accompanied by muddled understanding. This is perhaps the most inaccurately perceived region in the world. Does sand cover most of the area? Are there camels everywhere? Does everyone speak Arabic? Are Turks and Persians Arabs? Are all Arabs Muslims? The answer to each of these questions is no, yet popular Western media suggest otherwise. This chapter and its accompanying module attempt to make the misunderstood, complex, sometimes bloody, but often hopeful events and circumstances in the Middle East and North Africa more intelligible by illuminating the geographic context within which they occur.

6.1 Area and Population

What is the Middle East, and where is it? The term itself is Eurocentric, invented by the British, who placed themselves in the figurative center of the world. They began to use the term prior to the outbreak of World War I, when the *Near East* referred to the territories of the Ottoman Empire in the eastern Mediterranean region, the *East* to India, and the *Far East* to China, Japan, and the western Pacific Rim. With *Middle East*, they designated as a separate region the countries around the Persian Gulf (known to Arabs as the Arabian Gulf and in this text as the Persian-Arabian Gulf or simply the Gulf). Gradually, the perceived boundaries of the region grew.

Sources today vary widely in their interpretation of which countries are in the Middle East. For some, the Middle East includes only the countries clustered around the Arabian Peninsula. For others, it spans a vast 6,000 miles (9,700 km) west to east from Morocco in northwest Africa to Afghanistan in central Asia, and a north-south distance of about 3,000 miles (4,800 km) from Turkey, on Europe's southeastern corner, to Sudan, which adjoins East Africa (•Figure 6.1). The larger area is the region covered in this chapter and its module, where it is referred to as "the Middle East and North Africa." The North African peoples of Morocco, Algeria, and Tunisia generally do not consider themselves Middle Easterners; they are, rather, from what they call the *Maghreb*, meaning "western land." Many geographers would place Sudan in sub-Saharan Africa and Afghanistan in central Asia or South Asia. Both are border or transitional countries in regional terms, and in this text, they are placed in the Middle East and North Africa with consideration given to their characteristics reflective of the other regions.

Thus defined, the Middle East and North Africa include 21 countries, the Palestinian territories of the West Bank and Gaza Strip, and the disputed Western Sahara (Table 6.1), occupying 5.9 million square miles (15.3 million sq km) and inhabited by about just over 500 million people in 2007. This area is about 1.8 times the size of the lower 48 United States and is generally situated at latitudes equivalent to those between Boston, Massachusetts, and Bogotá, Colombia (•Figure 6.2).

These half-billion people are not distributed evenly across the region but are concentrated in major clusters (•Figure 6.3). Three countries contain the lion's share of the region's population: Turkey, Iran, and Egypt, each with more than 70 million people. One look at a map of precipitation or vegetation explains why people are clustered this way (see Figure 2.3, page 23, and Figure 6.5). Where water is abundant in this generally arid region, so are people. Egypt has the Nile River, and parts of Iran and Turkey have bountiful rain and snow. Conversely, where rain seldom falls, as in the Sahara of North Africa and in the Arabian Peninsula, people are few.

The Middle East and North Africa as a whole have a high rate of population growth. The rapid growth is a general indication that this is a develsoping rather than industrialized region, and also reflects the majority Muslim culture
that favors larger families. The average annual rate of pop- 50
ulation change for the 21 countries, the Palestinian territories, and Western Sahara was 1.9 percent in 2007. The lowest rate of population growth (1.1 percent) is in Tunisia.

• **Figure 6.1** The Middle East and North Africa.

The highest is 3.2 percent in Yemen, followed closely by the Palestinian territories (consisting of the West Bank and Gaza Strip) at 2.9 percent. These are some of the highest population growth rates in the world. In the Palestinian territories, such rapid growth may be ascribed in large part to the Palestinians' poverty and perhaps to the wishes of many Palestinians to have more children to counterbalance the demographic weight of their perceived Israeli foe.

Between these extremes are countries with modest rates of population growth of 1.1 to 1.5 percent per year; these include Israel, Morocco, Iran, Turkey, the United Arab Emirates, Lebanon, and Qatar. Generally, their governments have regarded most of these countries as too populous for their resources and economic base and have encouraged family planning. They have been successful in lowering birth rates. On the other hand, oil-rich Saudi Arabia, with its 2.7 percent annual growth rate, is a good example of how rapid population growth is not always a sign of poverty. In this case, an oil-rich nation has encouraged its citizens to give birth to more citizens so that in the future they will not need to import foreign laborers and technicians and will be more self-sufficient in their development. Some oil-rich countries of the Gulf region have more foreigners than citizens living in them; about 75 percent of the working-
261 age population of the United Arab Emirates, for example, is nonnative, mainly from India and other South Asian countries. Many of these workers complain of dreams broken by poor living conditions and low wages.

Many developing countries have economies largely dependent on subsistence agriculture and have low percentages of urban inhabitants. Perhaps surprisingly, however, the Middle East and North Africa have more urbanites than country folk. The average urban population among the 23 countries and territories is 56 percent. The most prosperous countries are also the most urban. Essentially a city-state, Kuwait is 98 percent urban. The other oil-wealthy Gulf countries also have urban populations over 70 percent. Consistent with its profile as a Western-style industrialized country without oil resources, Israel is 92 percent urban. At the other end of the spectrum, desperately poor Afghanistan and Yemen have urban populations of less than 30 percent.

6.2 Physical Geography and Human Adaptations

The margins of the Middle East and North Africa are mainly oceans, seas, high mountains, and deserts (•Figure 6.4). To the west lies the Atlantic Ocean; to the south, the Sahara and the highlands of East Africa; to the north, the Mediterranean, Black, and Caspian Seas, together with mountains and deserts lining the southern land frontiers of Russia and the Near Abroad; and to the east, the Hindu Kush mountains on the Afghanistan-Pakistan frontier and the Baluchistan Desert straddling Iran and Pakistan. The land is composed mainly of arid plains and plateaus, together with large areas of rugged mountains and isolated "seas" of sand. Despite the environmental challenges, this region has given rise to some of the world's oldest and most influential ways of living.

TABLE 6.1 Middle East and North Africa: Basic Data

Political Unit	Area (thousands; sq mi)	Area (thousands; sq km)	Estimated Population (millions)	Estimated Population Density (sq mi)	Estimated Population Density (sq km)	Annual rate of Natural Increase (%)	Human Development Index	Urban Population (%)	Arable Land (%)	Per Capita GNI PPP ($US)
Middle East	**2,630.7**	**6,813.5**	**308.5**	**117**	**45**	**1.9**	**0.738**	**60**	**8**	**9,540**
Afghanistan	251.8	652.2	31.9	127	49	2.6	N/A	20	12	N/A
Bahrain	0.3	0.8	0.8	2,667	1,030	1.8	0.859	100	3	18,770
Iran	630.6	1,633.3	71.2	113	44	1.2	0.746	67	8	8,480
Iraq	169.2	438.2	29.0	171	66	2.5	N/A	67	13	N/A
Israel	8.1	21.0	7.3	901	348	1.5	0.927	92	16	25,470
Jordan	34.4	89.1	5.7	166	64	2.4	0.760	82	2	6,200
Kuwait	6.9	17.9	2.8	406	157	1.9	0.871	98	1	29,200
Lebanon	4.0	10.4	3.9	975	376	1.5	0.774	87	16	5,460
Oman	82.0	212.4	2.7	33	13	2.2	0.810	71	0	14,570
Palestinian Territories	2.4	6.2	4.0	1,667	644	2.9	0.736	72	17	N/A
Qatar	4.2	10.9	0.9	214	83	1.5	0.844	100	1	N/A
Saudi Arabia	830.0	2,149.7	27.6	33	13	2.7	0.777	81	1	16,620
Syria	71.5	185.2	19.9	278	107	2.5	0.716	50	25	3,920
Turkey	299.2	774.9	74.0	247	95	1.2	0.757	66	31	9,060
United Arab Emirates	32.3	83.7	4.4	136	53	1.5	0.839	74	0	23,990
Yemen	203.8	527.8	22.4	110	42	3.2	0.492	26	2	920
North Africa	**3,286.0**	**8,510.7**	**194.7**	**59**	**23**	**1.9**	**0.664**	**50**	**5**	**5,500**
Algeria	919.6	2,381.8	34.1	37	14	1.7	0.728	58	3	6,900
Egypt	386.7	1,001.6	73.4	190	73	2.1	0.702	43	3	6,900
Libya	679.4	1,759.6	6.2	9	4	2.0	0.798	85	1	N/A
Morocco	172.4	446.5	31.7	184	71	1.5	0.640	55	19	5,000
Sudan	967.5	2,505.8	38.6	40	15	2.2	0.516	41	7	2,160
Tunisia	63.2	163.7	10.2	161	62	1.1	0.760	65	18	8,490
Western Sahara	97.2	251.7	0.5	5	2	2.0	N/A	92	0	N/A
Summary Total	**5,916.7**	**15,324.2**	**503.2**	**85**	**33**	**1.9**	**0.709**	**56**	**6**	**7,980**

Sources: World Population Data Sheet, Population Reference Bureau, 2007; Human Development Report, United Nations, 2006; World Factbook, CIA, 2007.

A Region of Stark Geographic Contrasts

Aridity dominates the Middle East and North Africa (• Figure 6.5). At least three-fourths of the region has average yearly precipitation of less than 10 inches (25 cm), an amount too small for most types of dry farming (unirrigated agriculture). Sometimes, however, localized cloudbursts release moisture that allows plants, animals, and small populations of people—the Bedouin, Tuareg, and other pastoral nomads—to live in the desert. Even the vast Sahara, the world's largest desert, supports a surprising diversity and abundance of life. Plants, animals, and even people have developed strategies of **drought avoidance** and **drought endurance.** Migrating to avoid drought is a coping strategy that pastoral nomads and many animals use. Other organisms, such as trees, must endure drought by such adaptations as having very deep roots and small leaves. Populations of people, plants, and animals are all but nonexistent in the region's vast **sand seas,** including the Great Sand Sea of western Egypt (•Figure 6.6) and the Empty Quarter of the Arabian Peninsula.

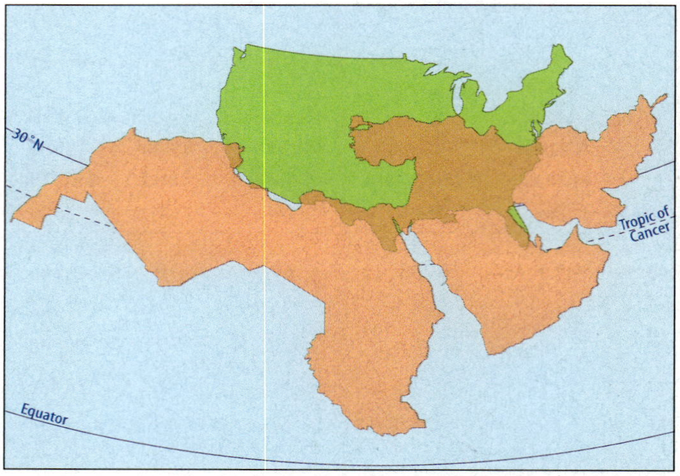

• **Figure 6.2** The Middle East and North Africa compared in latitude and area with the conterminous United States.

The region's climates have the comparatively large daily and seasonal ranges of temperature characteristic of dry lands. Desert nights can be surprisingly cool. Most days and nights are cloudless, so the heat absorbed on the desert surface during the day is lost by radiational cooling to the heights of the atmosphere at night. Summers in the lowlands are very hot almost everywhere. The hottest shade temperature ever recorded on earth, 136 degrees F (58 degrees C), occurred in Libya in September 1922. Many places regularly experience daily maximum temperatures over 100°F (38°C) for weeks at a time.

Human settlements located near the sand seas often experience the unpleasant combination of high temperatures and hot, sand-laden winds, creating the sandstorms known locally by such names as *simuum* ("poison") and *sirocco*. Only in mountainous sections and in some places near the sea do higher elevations or sea breezes temper the intense midsummer heat. The population of Alexandria, on the Mediterranean Sea coast, explodes in summer as Egyptians flee from Cairo and other hot inland locations. In Saudi Arabia, the government relocates from Riyadh to the highland summer capital of Taif to escape the lowland furnace.

Lower winter temperatures bring relief from the summer heat, and the more favored places receive enough precipitation for dry farming of winter wheat, barley, and other cool season crops (•Figure 6.7). In general, winters are cool to mild. But very cold winters and snowfalls occur in the high interior basins and plateaus of Iran, Afghanistan, and Turkey. These locales generally have a steppe climate. Only in the southernmost reaches of the region, notably Sudan, do temperatures remain consistently high throughout the year. A savanna climate and biome prevail there.

Most areas bordering the Mediterranean Sea have 15 to 40 inches (38 to 102 cm) of precipitation a year, falling almost exclusively in winter, while the summer is dry and warm—a typical Mediterranean climate pattern. Throughout history, people without access to perennial streams have stored this moisture to make it available later for growing crops that require the higher temperatures of the summer months. The Nabateans, for example, who were contemporaries of the Romans in what is now Jordan, had a sophisticated network of limestone cisterns and irrigation channels (•Figure 6.8). Rainfall sufficient for dry farming during the summer is concentrated in areas along the southern and northern margins of the region. The Black Sea side of Turkey's Pontic (also known as Kuzey) Mountains is lush and moist in the summer, and tea grows well there (•Figure 6.9). In the southwestern Arabian Peninsula, a monsoonal climate brings summer rainfall and autumn harvests to Yemen and Oman, probably accounting for the Roman name for the area: *Arabia Felix* ("Happy Arabia"). Mountainous areas in the region, like the river valleys and the margins of the Mediterranean, play a vital role in supporting human populations and national economies. Due to orographic or elevation-induced precipitation, the mountains tend to receive much more rainfall than surrounding lowland areas.

There are three principal mountainous regions of the Middle East and North Africa (see Figure 6.4). In northwestern Africa between the Mediterranean Sea and the Sahara, the Atlas Mountains of Morocco, Algeria, and Tunisia reach over 13,000 feet (3,965 m) in elevation. Mountains also rise on both sides of the Red Sea, with peaks up to 12,336 feet (3,760 m) in Yemen. These are the result of tectonic processes that are pulling the African and Arabian plates apart, creating the northern part of the
Great Rift Valley. The hinge of this crustal movement is 20, 21
the Bekaa Valley of Lebanon, where the widening fault line follows the Jordan River Valley southward to the Dead Sea (• Figure 6.10). This valley is the deepest depression on the earth's land surface, lying about 600 feet (183 m) below sea level at Lake Kinneret (also known as the Sea of Galilee or Lake Tiberius) and nearly 1,300 feet (400 m) below sea level at the shore of the Dead Sea, the lowest point of land on the planet. The rift then continues southward to the very
deep Gulf of Aqaba and Red Sea before turning inland into 459, 466, 482
Africa at Djibouti and Ethiopia.

In tectonic processes, one consequence of rifting in one place is the subsequent collision of the earth's crustal plates elsewhere. There are several such collision zones in the Middle East and North Africa, particularly in Turkey and Iran, where mountain building has resulted. These are seismically active zones—meaning earthquakes occur there—and rarely does a year go by without a devastating quake rocking Turkey, Iran, or Afghanistan.

A large area of mountains, including the region's highest peaks, stretches across Turkey, Iran, and Afghanistan. These mountains are products of the collision of continental plates. On the eastern border with Pakistan, the Hindu Kush Range has peaks over 25,000 feet (7,600 m) high. The loftiest mountain ranges in Turkey are the Pontic, Taurus

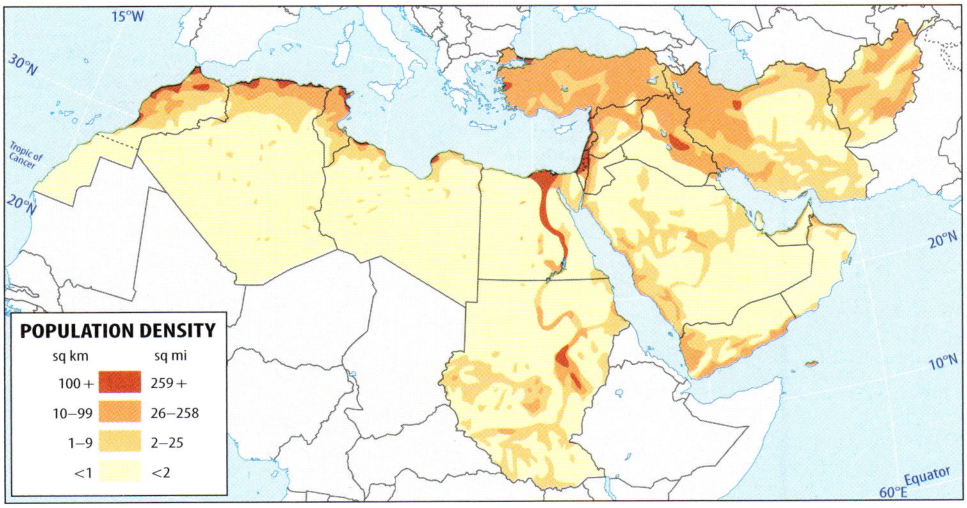

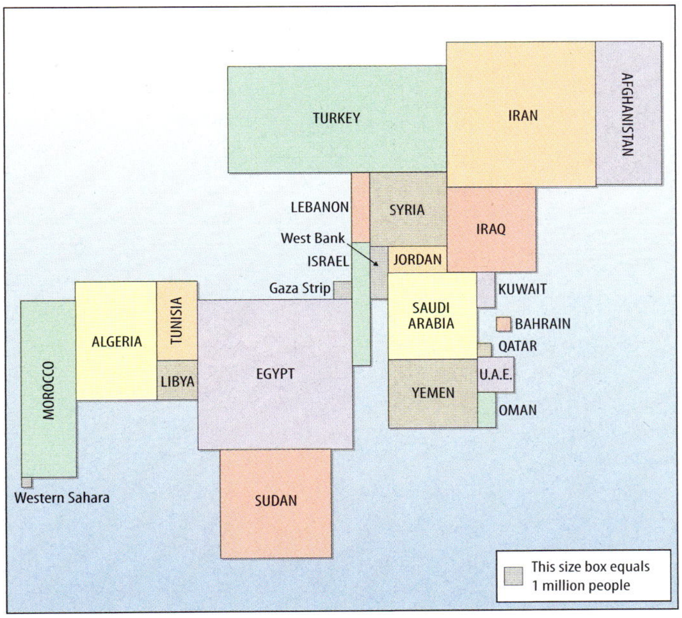

• **Figure 6.3** Population distribution (top) and population cartogram (bottom) of the Middle East and North Africa.

(•Figure 6.11), and Anti-Taurus, and in Iran, the Elburz and Zagros Mountains. These chains radiate outward from the rugged Armenian Knot in the tangled border country where Turkey, Iran, and the countries of the Caucasus meet. Mount Ararat is an extinct, glacier-covered volcano of 16,804 feet (5,122 m) towering over the border region between Turkey and Armenia. Many biblical scholars and explorers think the ark of Noah lies high on the mountain (and some go in search of it), for in the book of Genesis, this boat was said to have come to rest "in the mountains of Ararat."

Extensive forests existed in early historical times in the Middle East and North Africa, particularly in these mountainous areas, but overcutting and overgrazing have almost eliminated them. Since the dawn of civilization in this area, around 3000 B.C.E., people have cut timber for construction and fuel faster than nature could replace it. Egyptian King Tutankhamen's funerary shrines and Solomon's Temple in Jerusalem were built of cedar from Lebanon (•Figure 6.12). So prized has this wood been through the millennia that only a few isolated groves of cedar remain in Lebanon. Described in ancient times as "an oasis of green with running creeks" and "a vast forest whose branches hide the sky," Lebanon is now largely barren. Lumber is still harvested commercially in a few mountain areas such as the Atlas region of Morocco and Algeria, the Taurus Mountains of Turkey, and the Elburz Mountains of Iran, but supply falls far short of demand.

Villager, Pastoral Nomad, Urbanite

In the 1960s, the American geographer Paul English developed a useful model for understanding relationships between the three ancient ways of life that still prevail in the Middle East and North Africa today: villager, pastoral nomad, and urbanite. Each of these modes of living is rooted in a particular physical environment. Villagers are the subsistence farmers of rural areas where dry farming or irrigation is possible; pastoral nomads are the desert peoples who migrate through arid lands with their livestock, following patterns of rainfall and vegetation; and urbanites are the inhabitants of the large towns and cities, generally located near bountiful water sources but sometimes placed for particular trade, religious, or other reasons. Describing these ways of life as components of the **Middle Eastern ecological trilogy,** English explained how each of them has a characteristic, usually mutually beneficial, pattern of interaction with the other two (•Figure 6.13).

The peasant farmers of Middle Eastern and North African villages (the **villagers**) represent the cornerstone of the trilogy. They grow the staple food crops such as wheat

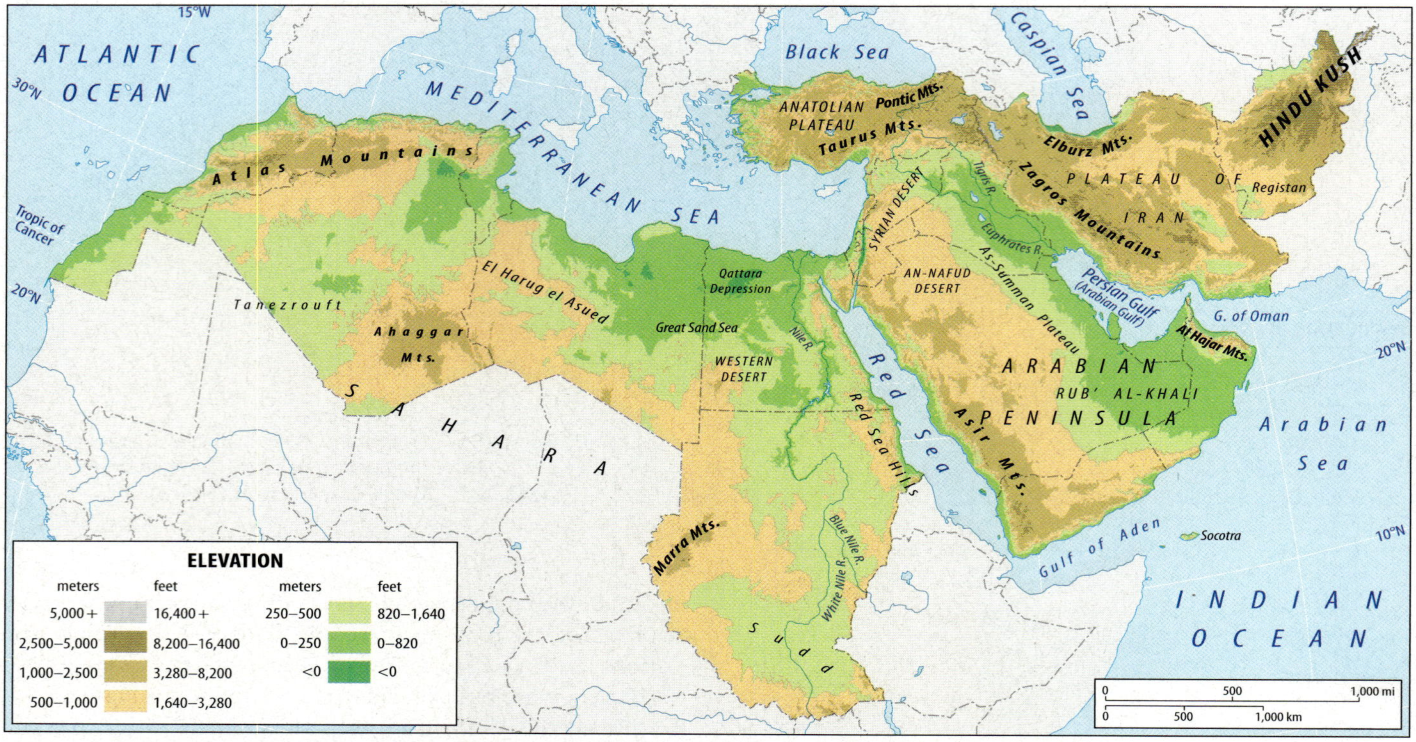

• **Figure 6.4** Physical geography of the Middle East and North Africa.

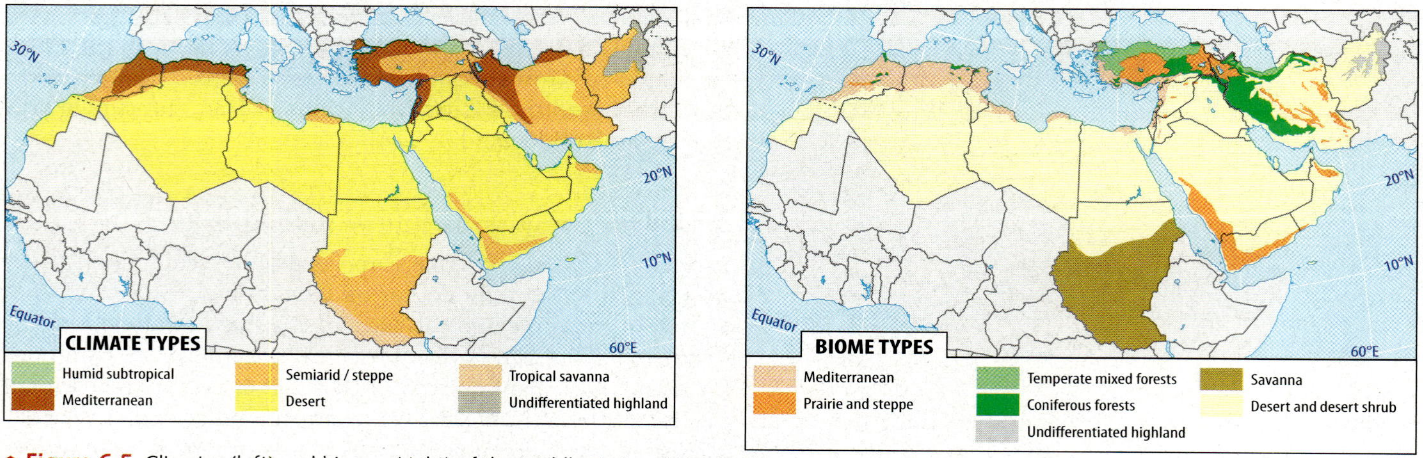

• **Figure 6.5** Climates (left) and biomes (right) of the Middle East and North Africa.

and barley that feed both the city dweller and the pastoral nomad of the desert. Neither urbanite nor nomad could live without them. The village also provides the city, often unwillingly, with tax revenue, soldiers, and workers. And before the mid-20th century, villages provided **pastoral nomads** with plunder as the desert dwellers raided their settlements and caravan supply lines. Generally, however, the exchange is beneficial.

The nomads provide villagers with livestock products, including live animals, meat, milk, cheese, hides, and wool, and with desert herbs and medicines. Educated and progressive **urbanites** provide technological innovations, manufactured goods, religious and secular education and training, and cultural amenities (today, including films and music).

There is little direct interaction between urbanites and pastoral nomads, although some manufactured goods such as clothing travel from city to desert and some desert folk medicines pass from desert to city. Historically, the exchange has been violent, as urban-based governments have sought to control the movements and military capabilities of the elusive and sometimes hostile nomads. Pastoral nomads once plundered rich caravans plying the major overland

Joe Hobbs

• **Figure 6.6** Seas of sand cover large areas in Saudi Arabia, Iran, and parts of the Sahara. This is the edge of the Great Sand Sea near Siwa Oasis in Egypt's Western Desert.

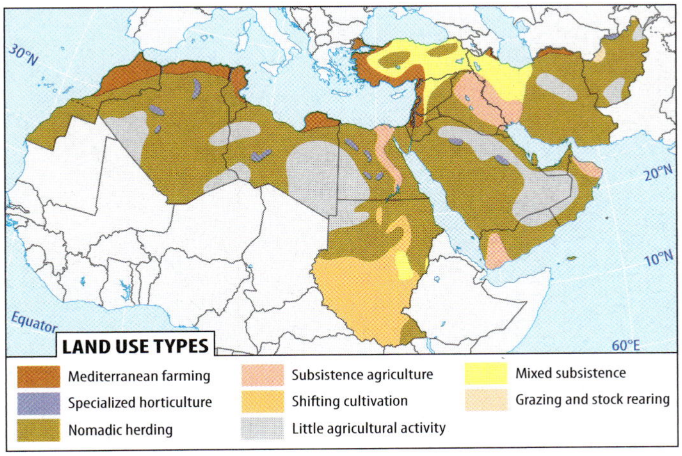

• **Figure 6.7** Land use in the Middle East and North Africa.

Joe Hobbs

• **Figure 6.8** The Treasury, a temple carved in red sandstone, probably in the 1st century D/F/, by the Nabateans at their capital of Petra in southern Jordan. The English poet Dean Burgen called Petra "a rose red city, half as old as time." The Nabateans built sophisticated networks for water storage and distribution. For scale, note the man at the lower right.

trade routes of the Middle East and North Africa. Governments did not tolerate such activities and often cracked down hard on the nomads they were able to catch.

Later, in the 1970s, Paul English wrote an article marking the "passing of the ecological trilogy." He noted that cities were encroaching on villages, villagers were migrating into cities and giving some neighborhoods a rural aspect, and pastoral nomads were settling down—thus the trilogy no longer existed. In reality, although the makeup and interactions of its parts have changed somewhat, the trilogy model is still valid and useful as an introduction to the major ways of life in the Middle East and North Africa. It is especially significant that a given man or woman in the region strongly identifies as either a villager, a pastoral nomad, or an urbanite. This perception of self has an important bearing on how these people of very different backgrounds interact, even when they live in close proximity. Urban officials may work in rural village areas, but they remain at heart and in their perspectives city people and usually live apart from farmers. Extended families of pastoral nomads may settle down and become farmers, but they continue to identify themselves by affiliation with the nomadic tribe. Many continue to harvest desert resources on a seasonal basis and retain marriage and other ties with desert-dwelling relatives.

The Village Way of Life

Agricultural villagers historically represented by far the majority populations in the Middle East and North Africa; only in recent decades have urbanites begun to outnumber them. In this generally dry environment, the villages are located near reliable water sources with cultivable lands nearby. They are usually made up of closely related family

Joe Hobbs

• Figure 6.9 Rainfall is heavy on the Black Sea side of Turkey's Pontic (Kuzey) Mountains. Note that roofs are pitched to shed precipitation; in the region's drier areas, most roofs are flat. The crop growing here is corn.

Joe Hobbs

• Figure 6.10 The surface of the Dead Sea (upper left) is the lowest point on earth, a consequence of the rifting of continental crust. The escarpment on the right is in the Israeli-occupied West Bank, not far from Jerusalem; to the left, out of view, is a counterpart escarpment in Jordanian territory. This photo reveals the strategic advantage Israel gained by capturing the West Bank in 1967: any force invading Israel from the east
242 would have to travel through a steep, easily-defended pass.

Joe Hobbs

• Figure 6.11 The Taurus Mountains of southeastern Turkey.

Joe Hobbs

• Figure 6.12 Ancient Egyptians believed that the solar boat of King Cheops of Egypt (c. 2500 C/D/F/), builder of the Great Pyramid, would carry the pharaoh's spirit through the firmament. It had to be made of the best wood—cedar from Lebanon. The solar boat was interred next to the pyramid and now stands in a specially constructed museum.

groups, with many fields owned by an absentee landlord. Villagers typically live in closely spaced homes made of mud brick or concrete blocks. Production and consumption focus on a staple grain such as wheat, barley, or rice. As land for growing fodder is often in short supply, villagers keep only a small number of sheep and goats and rely in part on nomads for pastoral produce. Residents of a given village usually share common ties of kinship, religion, ritual, and custom, and the changing demands of agricultural seasons regulate their patterns of activity.

These patterns of village life have been increasingly exposed to outside influences since the mid-18th century. European colonialism brought significant economic changes, including the introduction of cash crops and modern facilities to ship them. Improved and expanded irrigation, financed initially with capital from the West, brought more

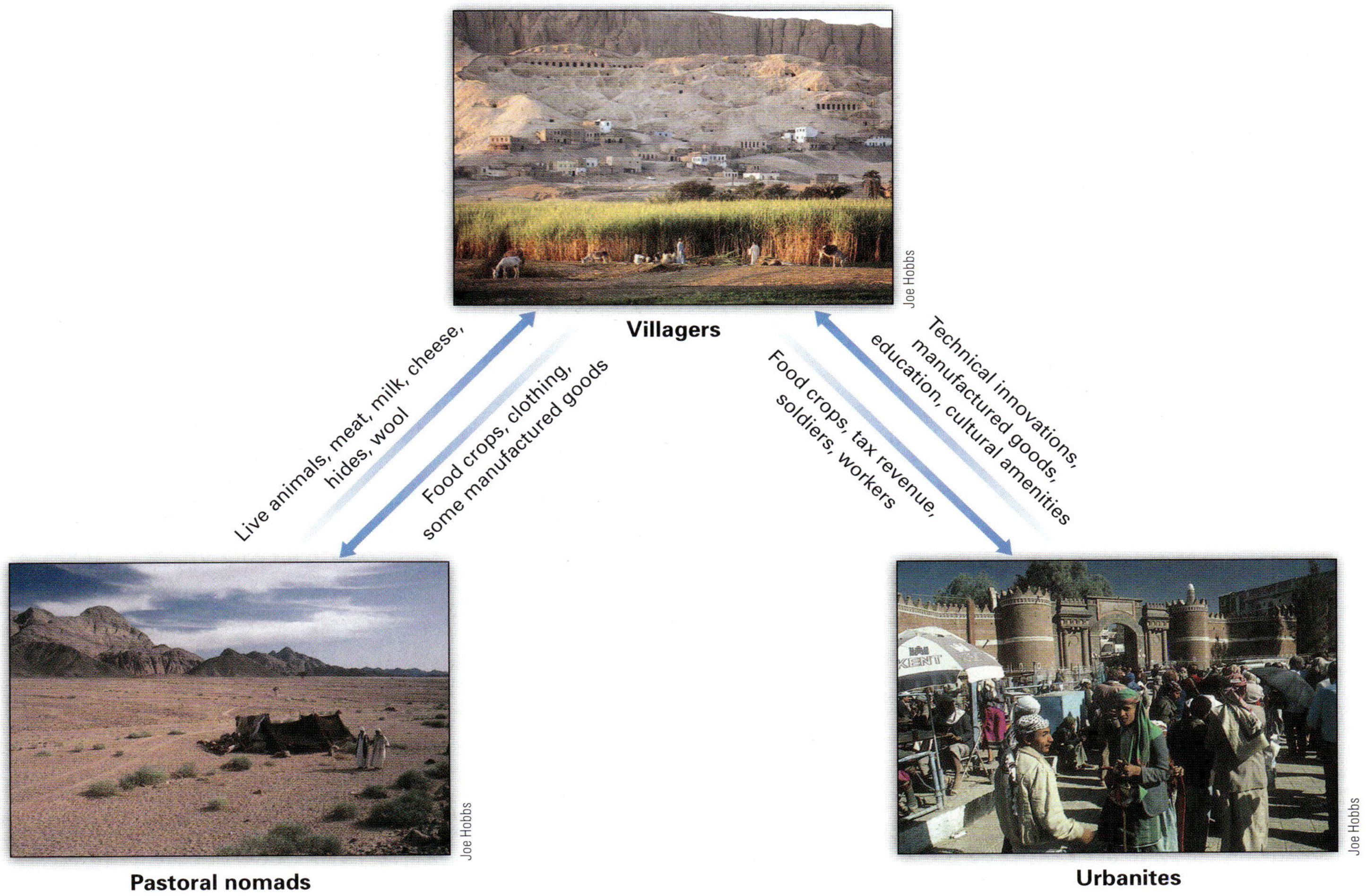

• **Figure 6.13** People, environments, and interactions of the ecological trilogy (here, villagers harvesting sugarcane in upper Egypt, Bedouin at camp in Egypt's Eastern Desert, and shoppers at the main gate to the historic city of San'a, Yemen). This relationship is generally symbiotic, although historically both urbanites and pastoral nomads preyed on the villagers, who are the trilogy's cornerstone.

land under cultivation. Recent agents of change have been the countries' own government-supported doctors, teachers, and land reform officers. Modern technologies such as sewing machines, motor vehicles, gasoline-powered water pumps, radio, television, and most recently, the Internet and cell phones have modified old patterns of living. The educated and ambitious, as well as the unskilled and desperately poor—motivated, respectively, by pull and push factors (see Chapter 3, page 57)—have been drawn to urban areas. Improved roads and communications have in turn carried urban influences to villages, prompting villagers to become more integrated into the national society.

The Pastoral Nomadic Way of Life

Pastoral nomadism emerged as an offshoot of the village agricultural way of life not long after plants and animals were first domesticated in the Middle East (about 7000 B.C.E.). Rainfall and the wild fodder it brings forth, although scattered, are sufficient resources to support small groups of people who migrate with their sheep, goats, and camels (and in some locales, cattle) to take advantage of this changing resource base (see Perspectives from the Field, page 218). In mountainous areas, they follow a pattern of **vertical migration** (sometimes referred to as *transhumance*), moving with their flocks from lowland winter to highland summer pastures. In the flatter expanses that comprise most of the region, the nomads practice a pattern of **horizontal migration** over much larger areas where rainfall is typically far less reliable than in the mountains. In addition to selling or trading livestock to obtain food, tea, sugar, clothing, and other essentials from settled communities, pastoral nomads hunt, gather, work for wages, and where possible, grow crops. Many now work in the tourist industry, as Westerners hungry for insight into traditional cultures seek them out. Their multifaceted livelihood has been described as a strategy of **risk minimization** based on the exploitation of multiple resources so that some will support them if others fail.

PERSPECTIVES FROM THE FIELD

Way-Finding in the Desert

The research for my doctoral dissertation in geography at the University of Texas–Austin was an 18-month journey with the Khushmaan Ma'aza, a clan of Bedouin nomads living in the northern part of Egypt's Eastern Desert. I studied their perceptions, knowledge, and uses of natural resources and tried to understand how their worldviews and kinship patterns apparently helped them devise ways of protecting these scarce resources. I have worked with these people from the early 1980s to the present and have always been astonished by how different they are from me and how much better they are in their ability to "read" the ground and recall landscape details. In the following passage from my book *Bedouin Life in the Egyptian Wilderness,* I marvel at their way-finding abilities.

The process by which the Khushmaan nomads have developed roots in their landscape, fashioning subjective "place" from anonymous "space," and the means by which they orient themselves and use places on a daily basis deserve special attention: these are essential parts of the Bedouins' identity and profoundly influence how they use resources and affect the desert ecosystem.

The Bedouins have little need or knowledge of maps. On the other hand, when shown a map or aerial image of countryside they know, the nomads accurately orient and interpret it, naming the mountains and drainages depicted. Saalih Ali (•Figure 6.A*) especially enjoyed my star chart. He would tell me what time the Pleiades would rise, for example, and ask me what time the chart indicated. The two were almost always in agreement.*

Many desert travelers have been astonished by the nomads' navigational and tracking ability, calling it a "sixth sense." The Khushmaan are exceptional way-finders and topographical interpreters able, for instance, to tell from tracks whether a camel was carrying baggage or a man; whether gazelle tracks were made by a male or female; which way a car was traveling and what make it was; which man left a set of footprints, even if he wore sandals; and how old the tracks are. Bedouins are proud of their geographical skills, which, they believe, distinguish them from settled people. Saalih told me, "Your people don't need to know the country but we do, to know exactly where things are in order to live." Pointing to his head, he said, "My map is here."

The difference in the way-finding abilities of nomads and settled persons may be due to the greater survival value of these skills for nomads and to the more complex meanings they attribute to locations. Bedouin places are rich in ideological and practical significance. The nomads interpret and interact with these meanings on a regular basis and create new places in their lifetimes. Theirs is an experience of belonging and becoming with the landscape, whereas settled people are more likely to inherit and accommodate themselves to a given set of places.

Joe Hobbs

• **Figure 6.A** Saalih Ali, a Bedouin of the Khushmaan Ma'aza, with the author in 2005, 25 years after they began fieldwork together.

*The nomads' homeland is so vast, and the margin of survivability in it so narrow, that topographic knowledge must be encyclopedic. Places either are the resources that allow human life in the desert or are the signposts that lead to resources. A Khushmaan man pinpointed the role of places in desert survival: "Places have names so that people do not get lost. They can learn where water and other things are by using place names." Tragedy may result if the nomad has insufficient or incorrect information about places: many deaths by thirst are attributed to faulty directions for finding water.**

*From Joseph Hobbs, *Bedouin Life in the Egyptian Wilderness* (Austin: University of Texas Press, 1989), pp. 81–82. Copyright © 1989 University of Texas Press. Reprinted with permission.

Although renowned in Middle Eastern legends, pastoral nomads have been described as "more glamorous than numerous." It is still impossible to obtain adequate census figures on the number living in the deserts of the Middle East and North Africa, though estimates range from 5 to 13 million. Recent decades have witnessed the rapid and progressive settling down, or **sedentarization,** of the nomads—a process attributed to a variety of causes. In some cases, prolonged drought virtually eliminated the resource base on which the nomads depended. Traditionally, they were able to migrate far enough to find new pastures, but modern national boundaries now inhibit such movements. Some have returned with the rains to their desert homelands, but others have chosen to remain as farmers or wage laborers in villages and towns. On the Arabian Peninsula, the prosperity and technological changes prompted by oil revenues made rapid inroads into the material culture—and then the livelihood preferences—of the desert people; many preferred the comforts of settled life. Some governments, notably those of Israel and prerevolutionary Iran, were frustrated by their inability to count, tax, conscript, and control a sizable migrant population and therefore compelled the nomads to settle.

Pastoral nomads of the Middle East and North Africa identify themselves primarily by their tribe, not by their nationality. The major ethnic groups from which these tribes draw are the Arabic-speaking Bedouin of the Arabian Peninsula and adjacent lands, the Berber and Tuareg of North Africa, the Kababish and Bisharin of Sudan, the Yoruk and Kurds of Turkey, the Qashai and Bakhtiari of Iran, and the Pashtun of Afghanistan. Members of a tribe claim common descent from a single male ancestor who lived countless generations ago; their kinship organization is thus a patrilineal descent system. It is also a segmentary kinship system, so called because there are smaller subsections of the tribe, known as clans and lineages, that are functionally important in daily life. Members of the most closely related families comprising the lineage, for example, share livestock, wells, trees, and other resources. Both the larger clans, made up of numerous lineages, and the tribes possess territories. Members of a clan or tribe typically allow members of another clan or tribe to use the resources within its territory on the basis of **usufruct**, or nondestructive mutual use.

Although some detractors have depicted pastoral nomads as the "fathers" rather than "sons" of the desert, blaming them for wanton destruction of game animals and vegetation, there are numerous examples of pastoral nomadic groups who have developed indigenous and very effective systems of resource conservation. Most of these practices depend on the kinship groups of family, lineage, clan, and tribe to assume responsibility for protecting plants and animals.

The Urban Way of Life

The city was the final component to emerge in the ecological trilogy, beginning in about 4000 B.C.E. in Mesopotamia (modern Iraq) and 3000 B.C.E. in Egypt. Though they resembled villages in many ways, the early cities were distinguished by their larger populations (more than 5,000 people), the use of written languages, and the presence of monumental temples and other ceremonial centers. The early Mesopotamian city and, after the 7th century C.E., the classic Islamic city, called the **medina**, had several structural elements in common (•Figure 6.14). The medina had a high surrounding wall built for defensive purposes. The congregational mosque and often an attached administrative and educational complex dominated the city center. Although Islam is often characterized as a faith of the desert, religious life has always been focused in, and diffused from, the cities. The importance of the city's congregational mosque in religious and everyday life is often emphasized by its large size and outstanding artistic execution.

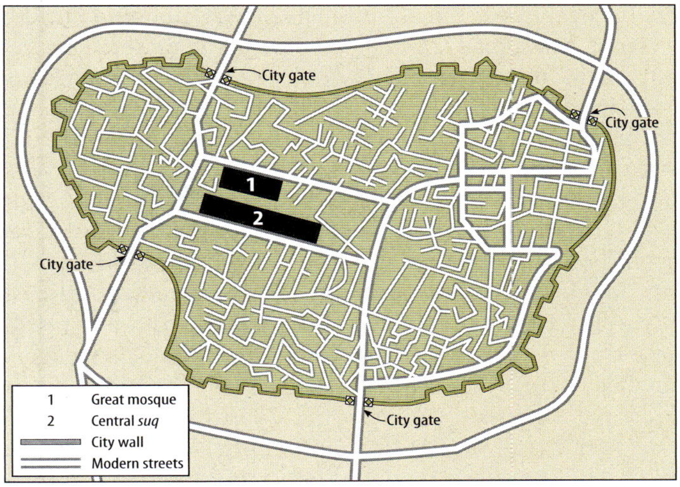

• **Figure 6.14** An idealized model of the classic medina, or Muslim Middle Eastern city.

A large commercial zone, known as a ***bazaar*** in Persian and a ***suq*** in Arabic and recognizable as the ancestor of the modern shopping mall, typically adjoined the ceremonial and administrative heart of the city (•Figure 6.15). Merchants and craftspeople selling various commodities occupied separate spaces within this complex, and visitors to an old medina today can still find sections devoted exclusively to the sale of spices, carpets, gold, silver, traditional medicines, or other particular goods. Smaller clusters of shops and workshops were located at the city gates.

Residential areas were differentiated as quarters not by income group but by ethnicity; the medina of Jerusalem, for example, still has distinct Jewish, Arab, Armenian, and non-Armenian Christian quarters (see Figure 6.B, page 223). Homes tended to face inward toward a quiet central courtyard, buffering the occupants from the noise and bustle of the street. The narrow, winding streets of the medina were intended for foot traffic and small animal-drawn carts, not for large motor vehicles, a fact that accounts for the traffic jams in some old Middle Eastern and North African cities today and for the wholesale destruction of the medina in others.

The medinas that survive today are gently decaying vestiges of a forgotten urban pattern. Periods of European colonialism and subsequent nationalism changed the face and orientation of the city. During the colonial age, resident Europeans preferred to live in more spacious settings at the outer edges of the city, and later, the national elite followed this pattern. In recent times, independent governments have adopted Western building styles, with broad traffic arteries cutting through the old quarters and large central squares near government buildings. This opening up of the cityscape has spread commercial activity along the wide avenues, diluting the prime importance of the central bazaar as the focus of trade.

Rural-to-urban migration and the city's own internal growth contribute to a rapid rate of urbanization that puts enormous pressure on services in the region's poorer countries. Governments often build high-rise public housing to accommodate the growing population, contributing to a cycle in which the urban poor move into the new dwellings, only to leave their old quarters as a vacuum to draw in still more rural migrants. In Cairo, millions of former villagers now live in the "City of the Dead," an extraordinary urban landscape composed of multistory dwellings erected above graves—a last resort for the poor who have no other place
253 to go. The overwhelmingly largest city, or primate city so characteristic of Middle Eastern and North African capitals, thus grows at the expense of the smaller original city.

Much of the rural-to-urban migration and subsequent urban gridlock and squalor could probably be avoided if governments invested more in the development of villages and smaller cities. The oil-rich countries with relatively small populations generally enjoy an urban standard of living equaling that of affluent Western countries. Modern industrial cities such as Saudi Arabia's Jubail and others founded on oil wealth were built virtually overnight, providing fascinating contrast to the region's colorful, complex ancient cities. With the recent resurgence of oil revenue into Saudi Arabia and the other Gulf countries, some of the cities boast the most elegant, sophisticated, and expensive amenities and architecture to be found anywhere in the 258–261
world (•Figure 6.16)

Joe Hobbs

• **Figure 6.15** A typical Middle Eastern *suq* or *bazaar* (market). This one is in Cairo, Egypt.

Joe Hobbs

• **Figure 6.16** July in Dubai, one of the city states of the United Arab Emirates. Wealth built initially on oil revenues has made it possible for some Gulf countries to build almost anything, including a ski resort located in a shopping mall.

6.3 Cultural and Historical Geographies

Cultures of the Middle East and North Africa have made many fundamental contributions to humanity. Many of the plants and animals on which the world's agriculture is based were first domesticated in the Middle East between 5,000 and 10,000 years ago in the course of the Agricultural Rev-
41 olution. The list includes wheat, barley, sheep, goats, cattle, and pigs, whose wild ancestors were processed, manipulated, and bred until their physical makeup and behavior changed to suit human needs. The interaction between people and the wild plants and animals they eventually domesticated took place mainly in the well-watered **Fertile Crescent**, the arc of land stretching from Israel to western Iran.

By about 6,000 years ago, people sought higher yields by irrigating crops in the rich but often dry soils of the Tigris, Euphrates, and Nile River Valleys. Their efforts produced
41 the enormous crop surpluses that allowed civilization—a cultural complex based on an urban way of life—to emerge in Mesopotamia (literally, "the land between the rivers," the Tigris and Euphrates) and in Egypt. Accomplishments in science, technology, art, architecture, language, mathematics, and other areas diffused outward from these centers of civilization. Egypt and Mesopotamia are thus among the world's great culture hearths.

The Middle East and North Africa are sometimes mistakenly referred to as the "Arab world"; in fact, the region has huge populations of non-Arabs. It is true that a majority of the region's inhabitants are Arabs. An **Arab** is best defined as a person of Semitic Arab ethnicity whose ancestral language is **Arabic**, a **Semitic language** spoken by about 245 million, or 48 percent, of the region's people. Originally, the Arabs were inhabitants of the Arabian Peninsula, but conquests after their majority conversion to Islam took them, their language, and their Islamic culture as far west as Morocco and Spain.

84 243 The region is also the homeland of the **Jews**, whose definition today is complex. Originally, Jews were both a distinct ethnic and linguistic group of the Middle East who practiced the religion of Judaism. It is possible (although difficult) for non-Jews to convert to Judaism (the convert is known as a proselyte, or "immigrant"), so a strictly ethnic definition does not apply today. Many Jews do not practice their religion but still consider themselves ethnically or culturally Jewish. Whatever debate exits about what defines a Jew, Jewish identity is strong and resilient.

It is important to recognize that although political circumstances made them enemies in the 20th century, Jews and Arabs lived in peace for centuries, and they share many cultural traits. Both recognize Abraham as their patriarch. Arabic is a Semitic language in the same **Afro-Asiatic language family** as **Hebrew**, which is spoken by most of the 5.5 million Jewish inhabitants of Israel (see the language map, •Figure 6.17).

There are other very large populations of non-Semitic ethnic groups and languages in the region. The greatest are the 60 million **Turks** of Turkey, who speak **Turkish**, a member of the **Altaic language family**, and the 33 million **Persians** of Iran, who speak **Farsi** (also called Persian), in the **Indo-European language family.** Both Persian and Arabic are written in Arabic script and so appear related, but they are not. Turkish was also written in Arabic script until early in the 20th century, but since then, it has been written in a Latin script. The ethnic **Pashtun** majority of Afghanistan speaks **Pashto**, a language closely related to Persian, and the country's official language is **Dari** (Afghan Persian). About 30 million **Kurds**—a people living in Turkey, Iraq, Iran, and Syria—speak **Kurdish**, which is also an Indo-European language. Many people in North Africa speak **Berber** and **Tuareg** (in the Afro-Asiatic language family). Sudan is ethnically and linguistically a transition zone between the Middle East and North Africa region and sub-Saharan Africa. In Sudan, particularly in the south, there are many speakers of **Chari-Nile languages**, in the **Nilo-Saharan language family.**

The Middle East also gave the world the closely related monotheistic faiths of Judaism, Christianity, and Islam. It is impossible to consider the human and political geographies of this region without attention to these religions, the ways of life associated with them, and their holy places, so here is an introduction (•Figure 6.18).

The Promised Land of the Jews

Judaism, the first significant monotheistic faith as far as we know, is today practiced by about 14 million people worldwide, mostly in Israel, Europe, and North America. September 2007 marked the start of the year 5768 in the Jewish calendar, but unlike its kindred faiths—Christianity and Islam—Judaism does not have an acknowledged starting point in time. Also unlike Christianity, Judaism does not have a fixed creed or doctrine. Jews are encouraged to behave in this life according to God's laws, which, according to the Torah (the Jewish holy scripture, which is known as the Old Testament in the Christian Bible), God gave to Moses on Mount Sinai as a covenant with God's "chosen people." The coming of a savior known as the Messiah ("Anointed One" in Hebrew) is prophesied in the Torah; Christians believe that Jesus was that savior, as described in the New Testament; Jews do not recognize Jesus as the fulfillment of that prophecy and so do not accept the New Testament.

Another distinction that sets Christianity and Islam apart from Judaism is that Judaism is not a proselytizing religion; it does not seek converts. Jewish identity is based strongly on a common historical experience shared over thousands of years. That historical experience has included deep-seated geographic associations with particular sacred places in the Middle East—particularly with places in Jerusalem, capital of ancient Judah (Judea), the province from

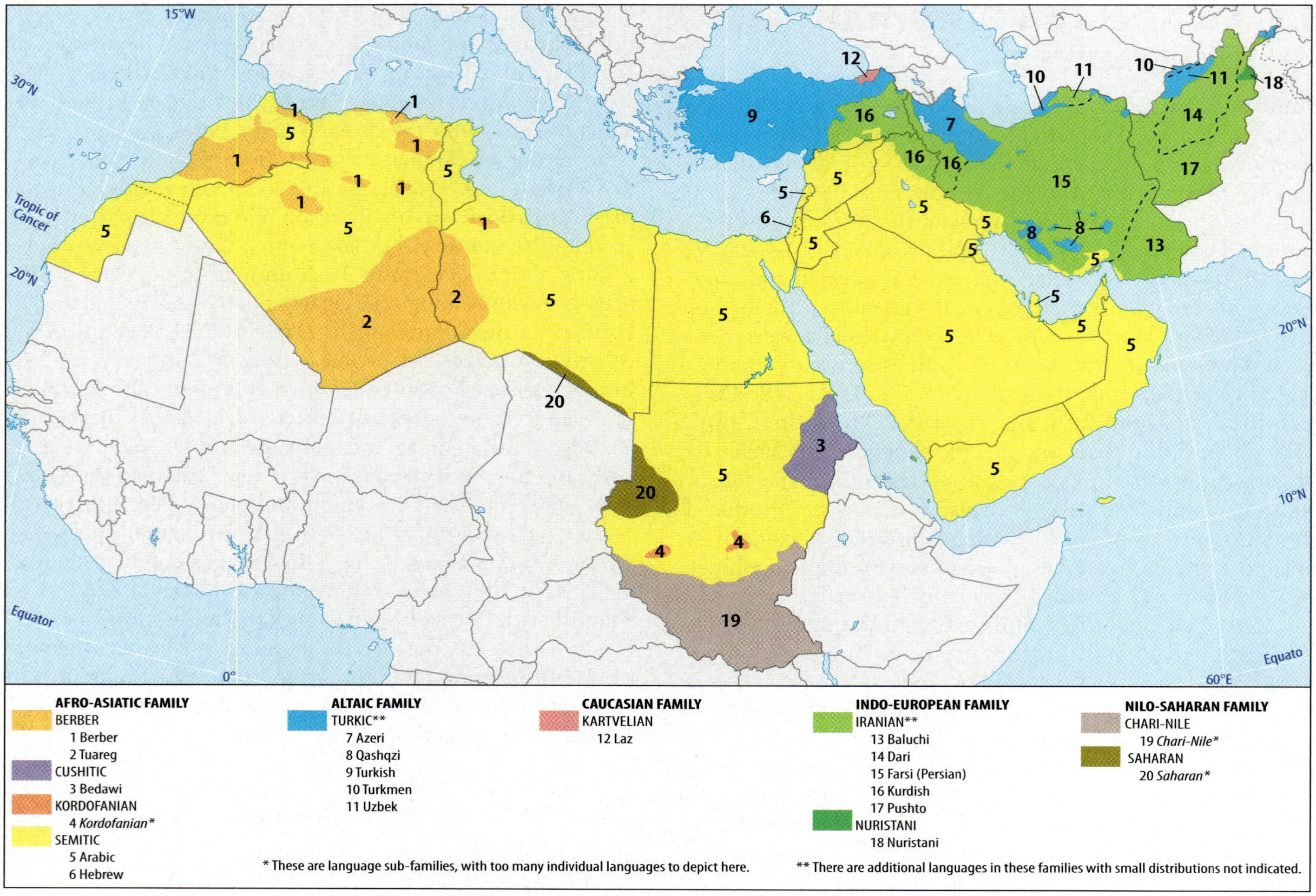

• **Figure 6.17** Languages of the Middle East and North Africa.

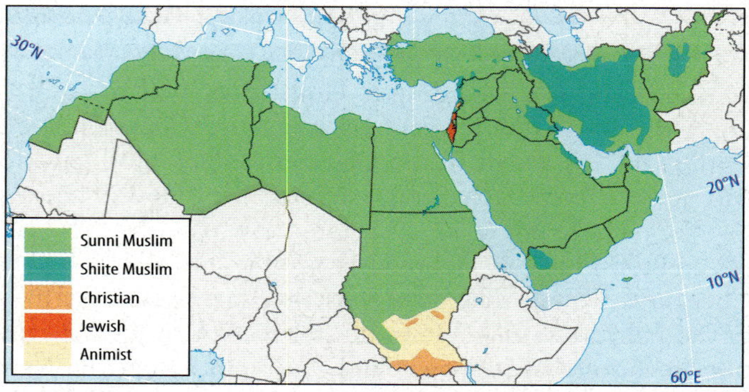

• **Figure 6.18** Religions of the Middle East and North Africa

which Jews take their name. Tragically, the Jewish history also has included unparalleled persecution.

Depending on one's perspective, the Jewish connection with the geographic region known as Palestine, essentially the area now composed of Israel, the West Bank, and the Gaza Strip, is most significant on a time scale of either about 4,000 years or about 100 years. According to the Bible, around 2000 B.C.E., God commanded Abraham and his kinspeople, known as **Hebrews** (later as Jews), to leave their home in what is now southern Iraq and settle in Canaan. God told Abraham that this land of Canaan—geographic Palestine—would belong to the Hebrews after a long period of persecution. The Bible says that the Hebrews did settle in Canaan until famine struck that land. At the command of Abraham's grandson Jacob, the Hebrews—known then as **Israelites**—relocated to Egypt, where grain was plentiful. That began the long sojourn of the Israelites in Egypt, which, according to the Bible, ended in about 1200 B.C.E. when Moses led them out in the journey known as the **Exodus.**

According to Jewish history, the prophecy of Abraham was first fulfilled when the Israelites settled once again in Canaan, their **Promised Land.** The Jewish King Saul unified the 12 tribes that descended from Jacob into the first united Kingdom of Israel in about 1020 B.C.E. In about 950 B.C.E. in Jerusalem—the capital of a kingdom enlarged by Saul's successor, David—King Solomon built Judaism's **First**

GEOGRAPHY OF THE SACRED

Jerusalem

The old city of Jerusalem is filled with sacred places and is one of the world's premier pilgrimage destinations. Over thousands of years, Jerusalem has been coveted and conquered by people of many different cultures and faiths. In the process, a place held sacred by one group has often come to be held sacred by a second and even a third.

The most important thing to notice on this map is that the location of the First and Second Jewish Temples is identical to that of the Muslims' Dome of the Rock. Judaism, Islam, and Christianity are very close in their origins and many of their basic precepts, but their faiths are not conjoined or syncretized in practice, as, for example, the Maya and Catholic faiths have been in Mesoamerica (see page XXX) or Buddhism and other traditions have been in East Asia (see page 294). In each faith, there has been a long tradition of quarreling over ownership of sacred space. Nowhere in the world does contested sacred space have more potential to ignite violence and outright war than in Jerusalem (which means, in Hebrew, "City of Peace"). Module 6.1 offers more insight into role of Jerusalem's holy places in the region's turbulent history. •Figure 6.B depicts the major places sacred to Jews, Christians, and Muslims and also the city's ethnic quarters; the walled old city
537 is a classic medina.

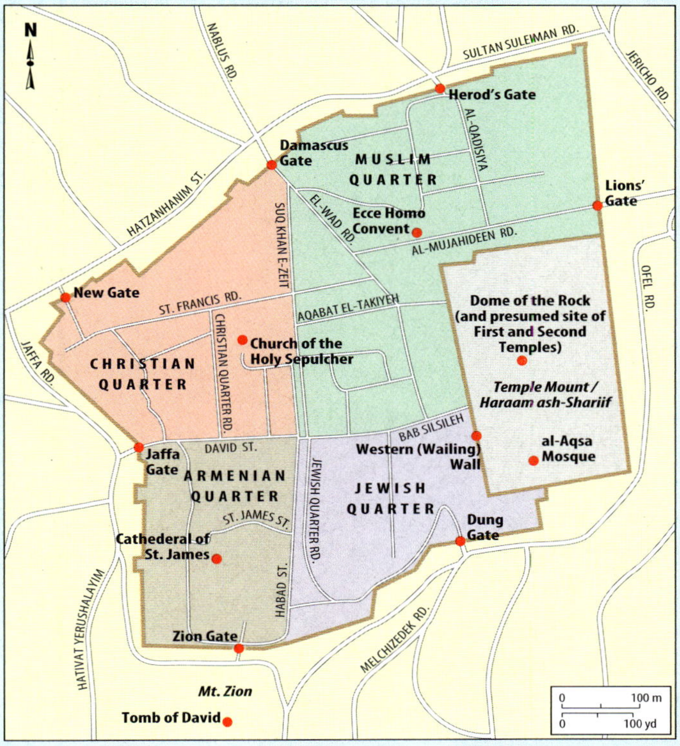

• **Figure 6.B** Sacred sites and ethnic quarters of the old city of Jerusalem.

219 **Temple.** He located it atop a great rock known to the Jews as *Even HaShetiyah*, the **Foundation Stone,** plucked from beneath the throne of God to become the center of the world and the core from which the entire world was created (see Geography of the Sacred, above). The Ark of the Covenant, containing the commandments that God gave to Moses atop Mount Sinai, was placed in the Temple's Holy of Holies.

The united Kingdom of Israel lasted only about 200 years before splitting into the states of Israel and Judah. Empires based in Mesopotamia destroyed these states: the Assyrians attacked Israel in 721 B.C.E., and the Babylonians sacked Judah in 586 B.C.E. The Babylonians destroyed the First Temple (at which point the Ark of the Covenant disappeared) and exiled the Jewish people to Mesopotamia, where they remained until conquering Persians allowed them to return to their homeland. In about 520 B.C.E., the Jews who returned to Judah rebuilt the temple (the **Second Temple**) on its original site. A succession of foreign empires came to rule the Jews and Arabs of Palestine: Persian, Macedonian, Ptolemaic, Seleucid, and around the time of Jesus, Roman. Herod, the Jewish king who ruled under Roman authority and was a contemporary of Jesus, greatly enlarged the temple complex.

The Jews of Palestine revolted against Roman rule three times between 64 and 135. The first revolt broke out as the profoundly monotheistic Jews refused to acknowledge the Roman emperor as a god. The Romans quashed these rebellions in a series of famous sieges, including those of Masada and Jerusalem. The Romans destroyed the Second Temple, and a third has never been built. All that remains of the Second Temple complex is a portion of the surrounding wall built by Herod. Today, this **Western Wall,** known to non-Jews as the Wailing Wall, is the most sacred site in the world accessible to Jews (•Figure 6.19). Some religious traditions prohibit Jews from ascending the **Temple Mount** above, the area where the temple actually stood, because it is too sacred. After the temple's destruction, that site was occupied by a Roman temple and then in 691 by the Muslim shrine called the **Dome of the Rock,** which still stands today (also in Figure 6.19). The mostly Muslim Arabs know the Temple Mount as *al-Haraam ash-Shariif,* the **Noble Sanctuary.** Supercharged with meaning, this place

Joe Hobbs

• **Figure 6.19** Jerusalem's Western Wall, Temple Mount, and Noble Sanctuary. In this view are some of Judaism's and Islam's holiest places. At left, below the golden dome, is the Western Wall, which is all that remains of the structure that surrounded the Jews' Second Temple. The dome is the Muslims' Dome of the Rock. At far right, with the black dome, is the al-Aqsa Mosque, another very holy place in Islam.

has in modern times often been the spark of conflagration between Jews and Palestinian Arabs.

The victorious Romans scattered the defeated Jews to the far corners of the Roman world. Thus began the Jewish exile, or **Diaspora.** In their exile, the Jews never forgot their attachment to the Promised Land. One Jewish prayer, prayed during the weeklong Passover holiday, ends with the words "Next year in Jerusalem!" In Europe, where their numbers grew to be greatest, Jews were subjected to systematic discrimination and persecution and were forbidden to own land or engage in a number of professions. Known as **anti-Semitism,** hatred of Jews developed deep roots in Europe. This sentiment in part grew out of the long-simmering perception that Jews were responsible for the murder of Jesus Christ and in part out of the fact that Christian Europeans were generally prohibited from practicing moneylending. Jews were permitted to loan money to the Christians, who accepted the loans but then resented having to pay the Jews interest.

In the 1930s, anti-Semitism became state policy in Germany under the Nazis, led by Adolf Hitler. Many German Jews fled to the United States, and others emigrated to Palestine in support of the **Zionist movement,** which had since the late 1890s aimed at establishing a Jewish homeland in Palestine with **Zion** (Jerusalem) as its capital. Most Jews were not as fortunate as the Zionist emigrants. Within the boundaries of the Nazi empire that conquered most of continental Europe during World War II, Hitler's regime implemented its "final solution" to the "Jewish problem" by shipping to prison camps and eventually murdering all the Jews they could round up. The Nazi Germans and their allies killed an estimated 6 million Jews, along with millions of other minorities they deemed "inferior," including Roma (Gypsies) and homosexuals. It was this Holocaust that 84, 163
prompted the victorious allies of World War II, from their powerful position in the newly formed United Nations, to create a permanent homeland for the Jewish people in Palestine. A seemingly endless cycle of violence, discussed beginning on page 241, ensued.

Christianity: Death and Resurrection in Jerusalem

Nearly 1,000 years after Solomon established the Jews' First Temple, a new but closely related monotheistic faith emerged in Palestine. This was **Christianity,** named for Jesus Christ (*Christ* is Greek for "Anointed One," the equivalent of the Hebrew word for Messiah). Jesus, a Jew, was born near Jerusalem in Bethlehem, probably around 4 B.C.E. Tradition relates that when he was about 30 years old, Jesus began spreading the word that he was the Messiah, the deliverer of humankind long prophesied in Jewish doctrine. A small group of disciples accepted that he was the Messiah and followed him for several years as he preached his message. He taught that love, sacrifice, and faith were the keys to salvation. He had come to redeem humanity's sins through his own death. To his followers, he was the Son of God, a living manifestation of God Himself, and the only path to eternal life was by accepting his divinity.

Jesus Christ's teachings denied the validity of many Jewish doctrines, and around 29 C.E., a growing chorus of Jewish protesters called for his death. Palestine was then under

Roman rule, and Roman administrators in Jerusalem placated the mob by ordering that he be put on trial. He was found guilty of being a claimant to Jewish kingship, and Roman soldiers put him to death by the particularly degrading and painful method of crucifixion. The cornerstone of Christian faith is that Jesus Christ was resurrected from the dead two days later and ascended into heaven. Christians believe that he continues to intercede with his Father on their behalf and that he will come again on Judgment Day, at the end of time.

After a period of relative tolerance, the Romans began actively persecuting Christians. Nevertheless, Christian ranks and influence grew. The turning point in Christianity's career came after 324 C.E., when the Roman Emperor Constantine embraced Christianity and established it as the official religion of the empire. The Christian Byzantine civilization that developed in the **"New Rome"** that Constantine established—Constantinople, now Istanbul, Turkey—created fine monuments at places associated with the life and death of Jesus Christ. These include the place 133, 162 long acknowledged as the center of the Christian world, Jerusalem's **Church of the Holy Sepulcher** (•Figure 6.20). This extraordinary, sprawling building, now administered by numerous separate Christian sects, contains the locations where tradition says Jesus Christ was crucified and buried. Incidentally, several of these Christian denominations have a long history of squabbling over turf within the church, and the key to this place is in the hands of a trusted, neutral party: a Muslim family.

Christianity has seldom been the majority religion in the land where it was born; only until Islam arrived in Palestine in 638 C.E. was the region primarily Christian. From then until the 20th century, most of Palestine's inhabitants were Muslim, and since the mid-20th century, Muslims and Jews have been the major groups. Between the 11th and 14th centuries, European Christians dispatched military expeditions to recapture Jerusalem and the rest of the Holy Land from the Muslims. These bloody campaigns, known as the **Crusades,** resulted in a series of short-lived Christian administrations in the region.

There are significant minority populations of Christians throughout the Middle East, including members of distinct sects such as the **Copts** of Egypt and the **Maronites** of Lebanon. Their share of the population has generally been declining, both because of emigration and lower birth rates than those of the majority Muslims. Nevertheless, the Middle East remains the cradle of their faith for Christians the world over, and Jerusalem and nearby Bethlehem are the world's premier Christian pilgrimage sites.

Joe Hobbs

• **Figure 6.20** Jerusalem's Church of the Holy Sepulcher, containing the locations where many Christians believe Jesus Christ was crucified and buried.

The Message of Islam

Islam is by far the dominant religion in the Middle East and North Africa; only Israel within its pre-1967 borders has a non-Muslim majority. Because of Islam's powerful influence not merely as a set of religious practices but as a way of life, an understanding of the religious tenets, culture, and diffusion of Islam is vital for appreciating the region's cultural geography.

Islam is a monotheistic faith built on the foundations of the region's earliest monotheistic faith, Judaism, and its offspring, Christianity. Indeed, **Muslims** (people who practice Islam) call Jews and Christians **People of the Book,** and their faith obliges them to be tolerant of these special peoples. Muslims believe that their prophet, Muhammad, was the very last in a series of prophets who brought the Word of God to humankind. Thus they perceive the Bible as incomplete but not entirely wrong—Jews and Christians merely missed receiving the entire message. Muslims do not accept the Christian concept of the divine Trinity (God manifested in the form of the Father, his son Jesus, and the Holy Spirit) and regard Jesus as a prophet rather than as a manifestation of God.

Muhammad was born in 570 C.E. to a poor family in the western Arabian (now Saudi Arabian) city of Mecca. Located on an important north-south caravan route linking the frankincense-producing area of southern Arabia (now Yemen and Oman) with markets in Palestine (now Israel) and Syria, Mecca was a prosperous city at the time. It was also a pilgrimage destination because more than 300 deities were venerated in a shrine there called the **Ka'aba** (the Cube; •Figure 6.21). Muhammad married into a wealthy family and worked in the caravan trade. Muslim tradition holds that when he was about 40 years old, Muhammad was meditating in a cave outside Mecca when the Angel Gabriel appeared to him and ordered him to repeat the words of God that the angel would recite to him. Over the next 22 years, the prophet related these words of God (whom Muslims call **Allah**) to scribes who wrote them down as the **Qur'an** (or **Koran**), the holy book of Islam.

During this time, Muhammad began preaching the new message, "There is no god but Allah," which the polytheistic people of Mecca viewed as heresy. As much of their income depended on pilgrimage traffic to the Ka'aba, they also viewed Muhammad and his small band of followers as an economic threat. They forced the Muslims to flee from Mecca and take refuge in Yathrib (modern Medina), where a largely Jewish population had invited them to settle. There were subsequent skirmishes between the Meccans and Muslims, but in 630, the Muslims prevailed and peacefully occupied Mecca. The Muslims destroyed the idols enshrined in the Ka'aba, which became a pilgrimage center for their one God.

The Ka'aba is Islam's holiest place, and Mecca and Medina are its holiest cities. Jerusalem is also sacred to Muslims. Muslim tradition relates that on his **Night Journey,** the Prophet Muhammad ascended briefly into heaven from the great rock now beneath the Dome of the Rock (the same rock Jews regard as the Foundation Stone). Nearby on the Temple Mount (*al-Haraam ash-Shariif*) is **al-Aqsa Mosque,** a sacred congregational site. The proximity, even the duplication, of holy places between Islam and Judaism came to be 223, 224 the most difficult issue in peace negotiations between Palestinians and Israelis, as explained on page 245 in Module 6.1.

After Muhammad's death in 632, Arabian armies carried the new faith far and quickly. The two decaying empires that then prevailed in the Middle East and North Africa—the Byzantine or Eastern Roman Empire, based in Constantinople (Istanbul), and the Sassanian Empire, based in Persia (Iran) and adjacent Mesopotamia (Iraq)—put up only limited military resistance to the Muslim armies before capitulating. Local inhabitants generally welcomed the new faith, in part because administrators of the previous empires had not treated them well, whereas the Muslims promised tolerance. Soon the Syrian city of Damascus became the center of a Muslim empire. Baghdad assumed this role in 750 C.E.

Arab science and civilization flourished in the Baghdad immortalized in the legends of *The Thousand and One Nights*. Important accomplishments and discoveries were made in mathematics, astronomy, and geography. Scholars translated the Greek and Roman classics, and if not for their efforts, many of these works would never have survived to become part of the modern European legacy. It was an age of exploration, when Arab merchants and voyagers visited China and the remote lands of southern Africa. Many important discoveries by the Arab geographers were recorded in Arabic, a language unfamiliar to contemporary

Nabeel Turner/Stone/Getty Images

• **Figure 6.21** The black-shrouded cubical shrine known as the Ka'aba (just right of center) in Mecca's Great Mosque is the object toward which all Muslims face when they pray and is the centerpiece of the pilgrimage to Mecca required of all able Muslims. Note the carpet of humanity spread across this image.

INSIGHTS

Sunni and Shiite Muslims

A schism occurred very early in the development of Islam, and it persists today. The split developed because the Prophet Muhammad had named no successor to take his place as the leader (caliph) of all Muslims. Some of his followers argued that the person with the strongest leadership skills and greatest piety was best qualified to assume this role. These followers became known as **Sunni,** or orthodox, Muslims. Others argued that only direct descendants of Muhammad, specifically through descent from his cousin and son-in-law Ali, could qualify as successors. They became known as **Shia,** or **Shiite,** Muslims.

The military forces of the two camps engaged in battle south of Baghdad at Karbala in 680 D/F/, and in the encounter, Sunni troops caught and brutally murdered Hussein, a son of Ali (Karbala is thus sacred to Shiites, as is nearby al-Najaf, where Ali is buried). The rift thereafter was deep and permanent. The martyrdom of Hussein became an important symbol for Shiites, many of whom still today regard themselves as oppressed peoples struggling against cruel tyrants, including some Sunni Muslims.

Today, only two Muslim countries, Iran and Iraq, have Shiite 264, 268 majority populations. Significant minority populations of Shiites are in Syria, Lebanon, Yemen, and the Arab states of the Persian-Arabian Gulf (see Figure 6.18 on page 222).

Europeans, and had to be rediscovered centuries later by the Portuguese and Spaniards. Arab merchants carried their faith on the spice routes to the East. One result, surprising to many today, is that the world's most populous Muslim country is not in the Middle East and North Africa, and its people are not Arabs; it is Indonesia, 5,000 thousand miles (8,000 km) east of Arabia.

Whether in Arabia or Indonesia, whether they be Sunni Muslims or Shiite Muslims (see Insights, page 227), all believers are united in support of the five fundamental precepts, or **Pillars of Islam.** The first of these is the **profession of faith:** "There is no god but Allah, and Muhammad is His Messenger." This expression is often on the lips of the devout Muslim, both in prayer and as a prelude to everyday activities.

The second pillar is **prayer,** required five times daily at prescribed intervals. Two of these prayers mark dawn and sunset. Business comes to a halt as the faithful prostrate themselves before God. Muslims may pray anywhere, but wherever they are, they must turn toward Mecca. There also is a congregational prayer at noon on Friday, the Muslim Sabbath.

The third pillar is **almsgiving.** In earlier times, Muslims were required to give a fixed proportion of their income as charity, similar to the concept of the tithe in the Christian church. Today, the donations are voluntary. Even Muslims of very modest means give what they can to those in need.

The fourth pillar is **fasting** during Ramadan, the ninth month of the Muslim lunar calendar. Muslims are required to abstain from food, liquids, smoking, and sexual activity from dawn to sunset throughout Ramadan. The lunar month of Ramadan falls earlier each year in the solar calendar and thus periodically occurs in summer. In the torrid Middle East and North Africa, that timing imposes special hardships on the faithful, who, even if they are performing manual labor, must resist the urge to drink water during the long, hot days.

The final pillar is the ***hajj,*** or **pilgrimage to Mecca,** Islam's holiest city. Every Muslim who is physically and financially able is required to make the journey once in his or her lifetime. A lesser pilgrimage may be performed at any time, but the prescribed season is in the 12th month of the Muslim calendar. Those days witness one of earth's greatest annual human migrations as about 3 million Muslims from all over the world converge on Mecca. Hosting these throngs is an obligation the government of Saudi Arabia fulfills proudly and at considerable expense. However, there has been some trepidation in recent years because of the security threat foreign visitors may pose to the host country and because accidents such as stampedes and tent city fires have cost numerous lives. Many pilgrims also visit the nearby city of Medina, where Muhammad is buried. Most Muslims regard the *hajj* as one of the most significant events of their lifetimes. All are required to wear simple seamless garments, and for a few days, the barriers separating groups by income, ethnicity, and nationality are broken. Pilgrims return home with the new stature and title of *hajji* but also with humility and renewed devotion.

All Muslims share the Five Pillars and other tenets, but they vary widely in other cultural practices related to their faith, depending on the country they live in; whether they are from the desert, village, or city; and their education and income. The governments and associated clerical authorities in Saudi Arabia and Iran insist on strict application of **Islamic law** (known as ***sharia***) to civil life; in effect, there is no separation between church and state. The Qur'an does not state that women are required to wear veils, but it does urge them to be modest, and it portrays their roles as different from those of men. Clerics in Saudi Arabia insist that women wear floor-length, long-sleeved black robes and black veils in public, that they travel accompanied by a male member of their families, and that they not drive cars. In Egypt, by contrast, Muslim women are free to appear in public unveiled if they choose. However, in most Muslim countries, conservative ideas about the role of women are still very strong: they should be modest, retiring, good mothers, and keepers of the home. The Qur'an portrays women as equal to men in the sight of God, and in principle, Islamic teachings guarantee the right of women to hold and inherit property.

Most Muslim women argue that what others often see as "backward" cultural practices are in fact progressive. For example, they say, their modest dress compels men to evaluate them on the basis of their character and performance, not their attractiveness. They argue that segregation of the sexes in the classroom makes it easier for both women and men to develop their confidence and skills. Sexual assault is rare. A married woman retains her maiden name. These apparent advantages can be weighed against the drawbacks that in many cases women are generally subordinate to men in public affairs and have fewer opportunities for education and for work outside the home.

6.4 Economic Geography

Overall, the Middle East and North Africa is a poor region; per capita GNI PPP for the 23 countries and territories averages only $7,980 (see Table 6.1). This may seem surprising in view of the "rich Arab" stereotype. Only the oil-endowed states around the Persian-Arabian Gulf deserve the reputation for wealth, and by most measures other than per capita wealth, only non-Arab Israel is truly a more developed country (MDC). Israel's prosperity comes from its innovation in computer and other high-technology industries, the processing and sale of diamonds, large amounts of foreign (mostly U.S.) aid, and investment and assistance by Jews and Jewish organizations around the world.

Vital to the industrialized countries as a source of fuels, lubricants, and chemical raw materials, petroleum is one of the world's most important natural resources. A crucial feature of world geography is the concentration of approximately two-thirds of the world's proven petroleum reserves in a few countries that ring the Persian Gulf (Arabian Gulf).

By coincidence, the countries rich in oil tend to have relatively small populations, whereas the most populous nations have few oil reserves; Iran is an exception. All but about 1 percent of the Persian Gulf (Arabian Gulf) oil region's proven reserves of crude oil are located in Saudi Arabia, Iraq, Kuwait, Iran, and the United Arab Emirates (UAE), with smaller reserves in Oman and Qatar. Saudi Arabia, by far the world leader in reserves, has about 20 percent of the proven crude oil reserves on the globe. Canada is in second place with almost 14 percent of the world's proven oil; the next four are all along the Persian Gulf (Arabian Gulf): Iran (10 percent), Iraq (9 percent), Kuwait (8 percent), and the UAE (8 percent). Venezuela has 6 percent of world reserves, Russia another 5 percent, and the world's largest oil consumer, the United States, only 2 percent.

In the Gulf region, the great thickness of the region's oil-bearing strata and high reservoir pressures have made it possible to secure an immense amount of oil from a small number of wells. The productivity of each well makes each barrel inexpensive to extract and makes it simple to increase or reduce production quickly in response to world market conditions. Gulf oil is thus "cheap" to produce unless expenditures to maintain huge military forces to defend it are factored in. In 2007, the United States maintained about 190,000 military personnel in the region (including Iraq) at a cost of $100 billion per year. Some analysts—including a former U.S. Navy secretary—like to calculate the "real price" of oil with military expenditures factored in. If one accepts the premise that maintaining a costly military presence in the region is part of the United States' long-term strategy to maintain the flow of Gulf oil, that oil is no longer cheap by any measure. With these external costs factored in, Gulf oil would be priced at over $250 per barrel in September 2007, not the roughly $80 per barrel that was 63
the world market price at that time.

Production, export, and profits of Middle Eastern and North African oil were once firmly in the hands of foreign companies. That situation changed after 1960 when most of the Gulf countries and other exporting nations formed the **Organization of Petroleum Exporting Countries (OPEC)** with the aim of taking joint action to demand higher profits from oil. It changed again after 1972 when the oil-producing countries began to nationalize the foreign oil companies. OPEC was relatively obscure until the Arab-Israeli war of 1973, after which the organization began a series of dramatic price increases. In 1980, the organization's price reached $37 (U.S.) per barrel, compared with $2 a barrel in early 1973. Adjusted for inflation, that $37 would have been the equivalent of $100 per barrel in 2007, so the price hikes were truly titanic.

These events had enormous repercussions for the world economy. Immense wealth was transferred from the more developed countries to the OPEC countries to pay for indispensable oil supplies. The skyrocketing cost of gasoline and other oil products helped cause serious inflation in the United States and many other countries and contributed to the 1973 **energy crisis** in the United States. Desperately poor, less developed countries (LDCs) found that high oil prices not only hindered the development of their industries and transportation but also reduced food production because of high prices for fertilizer made from oil and natural gas. In the Gulf countries, the oil bonanza produced a wave of spending on military hardware, showy buildings, luxuries for the elite, and ambitious development projects of many kinds. Per capita benefits to the general populace were greatest in the Arabian Peninsula, where small populations and immense inflows of oil money made possible the abolition of taxes, the establishment of comprehensive social programs, and heavily subsidized amenities such as low-cost housing and utilities, including freshwater distilled from the salty Gulf in desalination plants.

Then, as the 1980s began, the era of continually expanding OPEC oil production, sales, and profits seemed to come to an end. After 1973, the high price of oil stimulated oil development in countries outside OPEC. Oil conservation measures such as a shift to more fuel-efficient vehicles and furnaces were instituted. Substitution of cheaper fuels for oil increased. Coal replaced oil in many electricity-generating stations. Oil refineries were converted to make gasoline from cheaper "heavy" oils rather than the more expensive "light" oils previously used. Meanwhile, the world entered a period of economic recession due in part to high oil prices. Decreased business activity reduced the demand for oil. Profits of the world oil industry (and taxes paid to governments) were severely cut, large numbers of refineries had to close, and much of the world tanker fleet was idled.

Oil prices rose temporarily in 1990–1991 when the flow of Iraqi and Kuwaiti oil was cut off following Iraq's military
262 takeover of Kuwait, but the prices soon fell again. In the 1990s, Saudi Arabia and other oil-rich Gulf states adopted economic austerity measures for the first time. However, the immense oil and gas reserves still in the ground guaranteed that the Gulf region would continue to have a major long-term impact on the world and would remain relatively prosperous as long as these finite resources are in demand in the MDCs. The lasting economic clout of the region was apparent early in the 21st century as the price of oil once again climbed to record levels, due especially to OPEC decisions to reduce production, and to explosive economic growth
297 and energy demands in China and India.

There are of course other resources and industries in the region's economic geography—remittances (earned income) sent home by guest workers in the oil-rich countries, revenues from ship traffic through the Suez Canal, and exports of cotton, rice, and other commercial crops, for example—but oil dominates the region's economy and is central to the global economy. The economies of the respective countries are described in more detail in Module 6.1. Here the focus remains on Middle Eastern oil and its crucial role in geopolitical affairs.

6.5 Geopolitical Issues

This has long been a vital region in world affairs and a target of outside interests. Its strategic crossroads location often has made it a cauldron of conflict. From very early times, overland caravan routes, including the famous Silk Road,
371 crossed the Middle East and North Africa with highly prized commodities traded between Europe and Asia. The security of these routes was vital, and countries at either end could not tolerate any threat to them. In more recent times, geopolitical concerns have focused on narrow waterways, access to oil, access to freshwater, and terrorism.

Chokepoints

One of the striking characteristics of the geography of the Middle East and North Africa is how many seas border and penetrate the region. In many cases, these seas are connected to one another though narrow straits and other passageways. In geopolitical terms, such constrictions are known as **chokepoints**—strategic narrow passageways on land or sea that may be easily closed off by force or even the threat of force (•Figure 6.22). Chokepoints must be unimpeded if world commerce is to carry on normally. Keeping them open is therefore usually one of the top priorities of regional and external governments. Similarly, closing them is a priority to a combatant nation or a terrorist entity seeking to gain a strategic advantage. Many notable events in military history and the formation of foreign policy in the Middle East focus on these strategic places.

One of the world's most important chokepoints is the Suez Canal, which opened in 1869. Slicing 107 miles (172 km) through the narrow Isthmus of Suez, the British- and French-owned Suez Canal linked the Mediterranean Sea with the Indian Ocean, saving cargo, military, and passenger ships a journey of many thousands of miles around the southern tip of Africa (•Figure 6.23). Keeping the Suez Canal in friendly hands was one of Britain's major military concerns during World War II. That objective led to a hard-fought and eventually successful British and Allied campaign against Nazi Germany in North Africa. Egyptian President Gamal Abdel Nasser's nationalization of the canal in 1956 led immediately to a British, French, and Israeli invasion of Egypt and a conflict known as the **Suez Crisis**

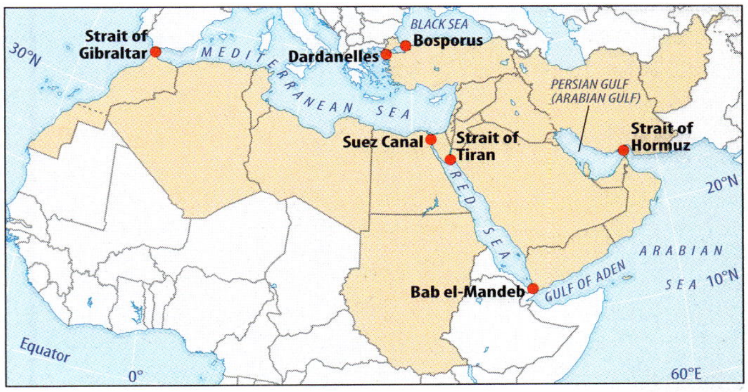

• **Figure 6.22** Chokepoints in the Middle East and North Africa.

Joe Hobbs

• **Figure 6.23** The strategically vital Suez Canal zone saw bitter fighting in the Middle East wars of 1956, 1967, and 1973.

GEOGRAPHY OF ENERGY

Middle Eastern Oil Pipelines

The oil-exporting countries of the Middle East, often with the financial support of the United States and other leading oil-consuming countries, have invested enormous resources to ensure the safe passage of oil to world markets. The Middle East is not landlocked like the oil-rich "stans" of central Asia are, but some of the same kinds of geopolitical issues are involved in oil export
176–177 planning. These concerns often point toward pipelines as the best means of routing oil shipments (• Figure 6.C). Pipelines are attractive because they shorten the time and expense involved in seaborne transport, and they bypass chokepoints. However, they are vulnerable to disruption. Weapons as small as grenades can disrupt supplies. A major challenge is how to route a pipeline so that it will not cross through a potential enemy's territory, and another is how to maintain friendships with the countries the oil crosses.

One of the first major pipelines in the region was the 1,100-mile (1,760-km) Trans-Arabian Pipeline (Tapline), leading from the oil-producing eastern province of Saudi Arabia to a terminal on the Mediterranean coast of Lebanon. It has the advantage of bypassing three chokepoints (Strait of Hormuz, Bab el-Mandeb, and Suez Canal). However, at the time it was completed (1950), it could not have been anticipated that it would cross two of the worst conflict zones in the Middle East: the Golan Heights of Syria (which fell to Israel in the 1967 war) and southern Lebanon (a major theater of the Lebanese civil war of the 1970s and the Israeli–Lebanese Shiite struggles of the 1980s and 1990s). Tapline was knocked out early by its unfortunate geography.

Another Middle East conflict, the 1967 Six-Day War between Israel and its neighbors, led to the closure of the Suez Canal (for about a decade) and thus to the construction of two new pipelines to bypass Suez: one across southern Israel from the Gulf of Aqaba to the Mediterranean Sea and another across Egypt from the Gulf of Suez to the Mediterranean (thus its acronym, SUMED).

No Middle Eastern country has been more dependent on pipelines than Iraq, and none has so systematically experienced the
liabilities of pipelines. The Iran-Iraq war of 1980–1988 had ma- 264
jor impacts on pipeline geography. Saudi Arabia supported Iraq in

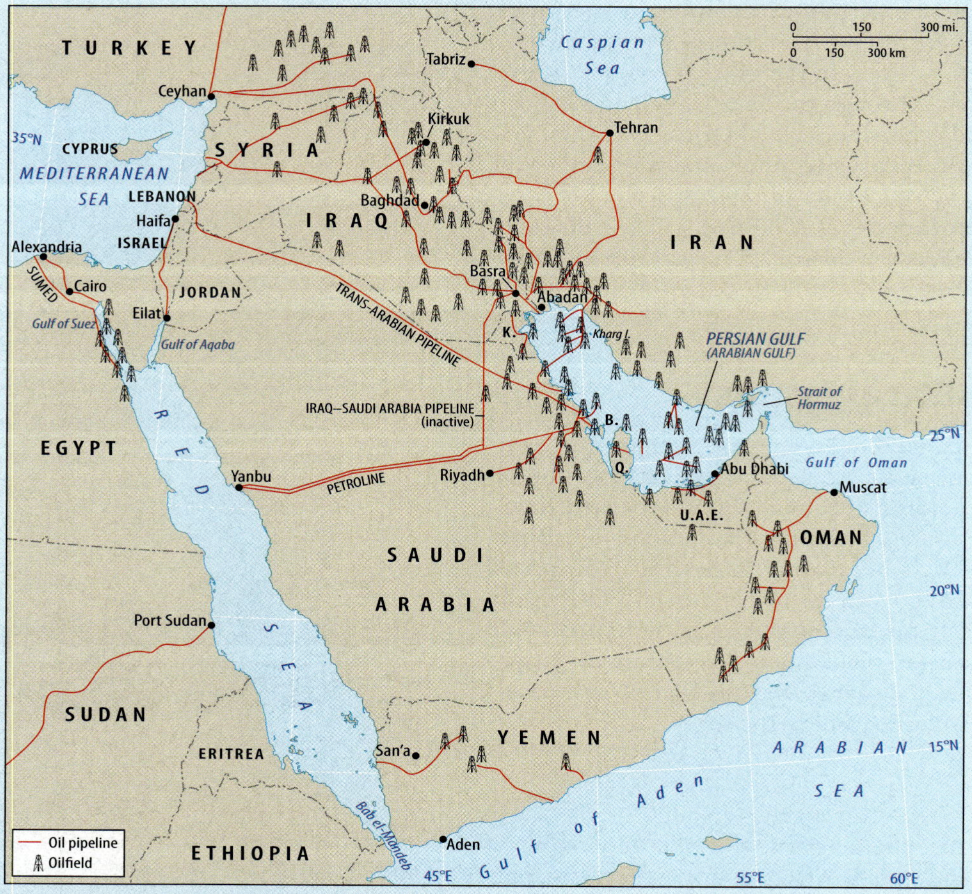

• **Figure 6.C** Principal oilfields and pipelines in the heart of the Middle East. Vulnerable chokepoints and volatile political relations have led to the construction and often indirect routing of many pipelines.

GEOGRAPHY OF ENERGY

Middle Eastern Oil Pipelines *continued*

the war and so tried to ensure it could get oil to market without being threatened by Iran; this meant bypassing the Strait of Hormuz. That led to the construction of Petroline (opened in 1981), the east-west pipeline across the Arabian Peninsula. Iraq built a pipeline to ship some of its oil south to Petroline, thereby bypassing the Strait of Hormuz as well. Beginning in 1961, Iraq was also able to ship oil through pipelines almost due west though Syria to the Mediterranean Sea. But when it attacked Iran in 1980, Iraq lost that route because Syria was Iran's ally.

Iraq quickly responded to that setback by building a new pipeline leading almost due north to bypass Syrian territory and then making a sharp 90-degree turn westward through southern Turkey to the Mediterranean Sea. But in 1990, Iraq attacked Kuwait, threatened Saudi Arabia, and prompted a U.S.-led counterattack. Saudi Arabia responded by closing the southern link with Petroline. Turkey, a NATO ally of the United States, reacted by closing the northern link. The United Nations responded by issuing strict controls on Iraqi oil exports. In sum, by its military actions, Iraq did almost everything imaginable to deprive itself of oil export capabilities.

and the **Arab-Israeli War of 1956.** Egypt effectively won that war when international pressure caused the invading forces to withdraw, leaving the canal in Egypt's hands.

Nearby, another chokepoint played a critical role in the
242 most important Arab-Israeli war, the Six-Day War of 1967. One of the events that precipitated the war was President Nasser's closure of the Strait of Tiran (at the southern end of the Gulf of Aqaba) to Israeli shipping. Israel had won the right of navigation through the strait after its war of independence and would not accept its closure; therefore, it attacked Egypt.

On either side of the Arabian Peninsula are two more critical chokepoints. One is extraordinarily vital to the world's economy: the Strait of Hormuz, connecting the Persian-Arabian Gulf with the Gulf of Oman and the Arabian Sea. Much of the world's oil supply passes through here in the holds of giant supertanker ships. Closure of the Strait of Hormuz would have devastating effects on the world's industrial and financial systems. Iran's plans to station Chinese-made missiles on the Strait of Hormuz and thus threaten international oil shipments led to a new level
264 of U.S. involvement late in the Iran-Iraq War of 1980–1988. That war also prompted a flurry of new oil pipeline construction designed to bypass the Strait of Hormuz chokepoint (see Geography of Energy, page 230). One of these pipelines, the Petroline route running east-west across the Arabian Peninsula, was built to bypass another important chokepoint, the Bab el-Mandeb, which connects the Red Sea with the Gulf of Aden and the Indian Ocean.

Turkey controls two more chokepoints that together are known as the Turkish Straits. The northernmost strait is the Bosporus, which cleaves the city of Istanbul into western (European) and eastern (Asian) sides, and the southern strait is called the Dardanelles. Their security has long been critical to the successful passage of goods between Europe and Asia and even more so to the successful passage of vessels between Russia (and the Soviet Union) and the rest of the world. Throughout the 20th century, one of the Soviet Union's constant strategic priorities was the right of navigation through the Turkish Straits.

Finally, it should be noted that the Strait of Gibraltar, connecting the Mediterranean Sea with the Atlantic Ocean, is also a chokepoint. Here, too, maintaining and monitoring the flow of maritime traffic have long been important concerns. Britain's insistence on maintaining control over its enclave of Gibraltar, decades after the decolonization of 254–255
most of the world, is an excellent indicator of how critical this chokepoint is.

Access to Oil

The Suez Canal and other chokepoints, the cotton of the Egypt's Nile Delta, and the strategic location of the region were important during colonial times and have remained so ever since. But oil has been and will remain (as long as fossil fuels drive the world's economies) what keeps the rest of the world interested in the Middle East and North Africa. The region's oil flows to many countries, but most of it is marketed in western Europe and Japan. The United States also imports large amounts of Gulf oil, but has a smaller relative dependence on this source than Japan and Europe do; 22 percent of its imported oil came from the Gulf in 2007. However, the Gulf region is very important to the United States because of the heavy dependence of close American allies on Gulf oil and because of the importance of the oil as a future reserve. American companies are also heavily involved in oil operations and oil-financed development in the Gulf countries. Gulf "petrodollars" are spent, banked, and invested in the United States, contributing significantly to the U.S. economy. Maintaining a secure supply of Gulf oil has therefore been one of the long-standing pillars of U.S. policy in the Middle East.

The United States has long had a precarious relationship with the key players in the Middle Eastern arena. The United States has pledged unwavering support for Israel while at the same time courting Israel's traditional enemies

such as oil-rich Saudi Arabia. That Arab kingdom and its neighbors around the Persian-Arabian Gulf possess more than 60 percent of the world's proven oil reserves. Thus they are vital to the long-term economic security of the Western industrial powers and Japan. The United States and its Western allies made it clear they would not tolerate any disruption of access to this supply when Iraqi troops, directed by Iraqi President Saddam Hussein, occupied Kuwait on August 2, 1990. U.S. President George Bush drew a "line in the sand," proclaiming, "We cannot permit a resource so vital to be dominated by one so ruthless—and we won't." In what came to be known as the **Gulf War,** the United States and a coalition of Western and Arab allies mounted a massive array of military might that ousted the Iraqi invaders within months and secured the vital oil supplies for Western markets. When the United States invaded and oc-
265 cupied Iraq in 2003, ostensibly to extinguish Iraq's ability to develop and deploy **weapons of mass destruction (WMD),** including chemical, biological and nuclear weapons, many critics insisted that this was just another example of Ameri-
265 ca's determination to control Middle Eastern oil.

The Gulf War was not the first time the United States expressed its willingness to use force to maintain access to Middle Eastern oil. In the wake of the revolution in Iran in
268 1979, the Soviet Union invaded neighboring Afghanistan. U.S. military analysts feared that the Soviets might use Afghanistan as a launch pad to invade oil-rich Iran. The United States deemed this prospect unacceptable, and President Jimmy Carter issued the policy statement that came to be known as the **Carter Doctrine:** the United States would use any means necessary to defend its vital interests in the region. "Vital interests" meant oil, and "any means necessary" meant that the United States was willing to go to war with the Soviet Union, presumably nuclear war, to defend those interests.

Middle Eastern wars had already become proxy wars for the superpowers, with oil always looming as the prize. In the 1967 and 1973 Arab-Israeli wars, for example, Soviet-backed Syrian forces fought U.S.-backed Israeli troops. American support of Israel in this war prompted Arab members of OPEC to impose an **oil embargo,** refusing to sell their oil to the United States, thus precipitating the nation's first energy crisis. During the 1973 war, the United States put its forces on an advanced state of readiness to take on the Soviets in a nuclear exchange if necessary. All of these events illustrate that the Middle East and North
138 Africa form, as eastern Europe did, a shatter belt—a large, strategically located region composed of conflicting states caught between the conflicting interests of great powers.

Access to Freshwater

Some of the most serious geopolitical issues in the Middle East and North Africa relate to **hydropolitics,** or political leverage and control over water. In this arid region, where most water is available either from rivers or from underground aquifers that cross national boundaries, control over water is an especially difficult and potentially explosive issue. An estimated 90 percent of the usable freshwater in the Middle East crosses one or more international borders.

Water is one of the most problematic issues in the Palestinian-Israeli conflict. Freshwater aquifers underneath the West Bank supply about 40 percent of Israel's water. Palestinians point to Israel's control over West Bank water as one of the most troubling elements of its occupation. The average Jewish settler on the West Bank uses 74 gallons (278 l) per day, whereas the average West Bank Palestinian uses 19 (72 l). (The World Health Organization calculates that 13 gallons or 50 liters per person per day is needed for minimal health and sanitation standards.) Israeli policies prohibit Palestinians from increasing their water usage. Many Israeli policymakers insist that water resources in the West Bank must remain under strict Israeli control and on these grounds oppose the creation of a Palestinian state in the West Bank. Critically, it is estimated that the West Bank aquifers will not contain enough water to support the region's population at current levels of consumption for more than a few more years (even taking into account anticipated replenishment from rainfall).

More promisingly, Jordan and Israel are working on agreements to share waters from the Jordan River (which forms a portion of their common border) and its tributary, the Yarmuk River. They are discussing a joint venture to build the **Red-Dead Peace Conduit,** which would send seawater from the Gulf of Aqaba to the Dead Sea via a network of canals and pipelines (see Figure 6.1.4, page 244). This would replenish the Dead Sea, which has retreated by about 3 feet (c. 1 m) per year for the last 25 years. The gravity flow of seawater to the Dead Sea would run generators to produce electricity, some of which would be used to desalinate the water. The two nations, and potentially the Palestinian Authority, could share this water and power.

Water is a critical issue blocking a peace treaty between Israel and Syria. If Syria were to recover all of the Golan Heights (which it lost to Israel in the 1967 war), it would have shorefront on Lake Kinneret and therefore, presumably, rights to use its water. That prospect is unacceptable to Israel. This body of water, also known as Lake Tiberias and the Sea of Galilee, is Israel's principal supply of freshwater, feeding the National Water Carrier system that transports water south to the Negev Desert. In occupying the Golan Heights, Israel also controls some of the northern bank of the Yarmuk River on the border with Jordan. For many years, Israel has stated it would never allow Syria and Jordan to construct their proposed Unity Dam, intended to store waters of the Yarmuk River to be shared between those countries. Israel thus implied it would bomb the dam rather than allow it to deprive Israel of Jordan River water.

A useful way to think about the geography of hydropolitics is in terms of **upstream** and **downstream countries.** Simply because water flows downhill, an upstream country is usually able to maximize its water use at the expense of a downstream country (a situation described as a *zero-sum game,* where any gain by one party represents an equivalent loss to the other). However, the situation between Israel and the countries upstream on the Yarmuk shows that this is not always true. Although downstream, Israel is far more powerful militarily and can use the threat of force to wrest more water out of the system.

Historically, the same has been true of Egypt. It is the ultimate downstream country, at the mouth of a great river that runs through five countries and sustains about 160 million people (a population that is expected to double in about 20 years; •Figure 6.24). However, it has long been the strongest country in the Nile Basin and has threatened to use its greater force if it does not get the water it wants. In 1926, when the British ruled Egypt and many other colonies in Africa, 10 countries located on the Nile or its tributaries upstream of Egypt were compelled to sign the **Nile Water Agreement.** This guaranteed Egyptian access to 56 billion cubic meters of the Nile's water, or fully two-thirds of its 84 billion cubic meters—even though barely a drop of the Nile's waters actually originates in Egypt. The treaty forbids any projects that might threaten the volume of water reaching Egypt, prohibits use of Lake Victoria's water without Egypt's permission, and gives Egypt the right to inspect the entire length of the Nile to ensure compliance. In recent years, however, one country after another has defied the treaty, calling it an outmoded legacy of colonialism. Kenya and Tanzania have plans to build pipelines to carry Lake Victoria waters to thirsty towns and villages inland. Uganda is building its controversial Bujagali Dam on the Nile, mainly for hydroelectricity production. With Chinese assistance, Ethiopia is building the huge Tekaze Dam, for hydropower and irrigation, on a tributary of the Blue Nile. Sudan is building the Merowe (Hamdab) and Kajbar Dams on its northern stretch of the Nile. Predictably, Egypt has had a bellicose response to these developments. For example, Egypt called Kenya's stated intention to withdraw from the Nile Water Agreement an "act of war."

Meanwhile, Egypt's demands on Nile waters are increasing. Egypt recently built the multibillion-dollar Toshka Canal, which transports water from Lake Nasser over a distance of 100 miles (160 km) to the Kharga Oasis of the Western Desert (see Figure 6.24). Proponents of the canal insist that it will result in the cultivation of nearly 2,350 square miles (c. 6,100 sq km) of "new" land and provide a living for hundreds of thousands of people. Critics argue that it is a waste of money and that salinization and evaporation will take a huge toll on the cultivated land and the country's water supply.

Turkey—the source of four-fifths of Syria's water and two-thirds of Iraq's—exercises its upstream advantage on the Tigris and Euphrates Rivers (•Figure 6.25). Turkey's position has long been that water in Turkey belongs to Turkey, just as Saudi Arabian oil belongs to Saudi Arabia. Not surprisingly, downstream Syria and Iraq reject this position and are distraught by the diminished flow and quality of water resulting from Turkey's comprehensive Southeast Anatolia Project. When completed, the project is expected to reduce Syria's share of the Euphrates waters by 40 percent and Iraq's by 60 percent. Also increasing the likelihood of serious future tension is a history of strained relations among Turkey, Syria, and Iraq, accompanied by the fact that no commonly accepted body-of-water law governs the allocation of water in such international situations.

270

316, 353

Turkish leaders have said they will never use water as a political weapon, but Turkey has already wielded water to its advantage. In 1987, for example, Turkey increased the Euphrates flow into Syria in exchange for a Syrian pledge to stop support of Kurdish rebels inside Turkey. Turkey now says it wants to use water to promote peace in the Middle East by shipping it in converted supertankers for sale to such thirsty (and therefore potentially combative) countries as Israel, Jordan, Saudi Arabia, Libya, and Algeria.

• **Figure 6.24** Recent and proposed water developments in the Nile Basin. Note that this map is oriented such that north is at the left rather than at the top.

Joe Hobbs

• **Figure 6.25** The Tigris and Euphrates Rivers rise in Turkey, giving this non-Arab country control over a resource vital to the lives of millions of Arabs in downstream Syria and Iraq. This waterfall is on a tributary of the Tigris in far eastern Turkey.

Terrorism

Viewed from the perspective of the U.S. administrations of President George W. Bush, almost all of the geopolitical issues related to this region—oil, economic development, trade, aid, the Arab-Israeli conflict, and more—were subsumed beneath the broader rubric of the president's declared **war on terror.** Even the 2003 invasion of Iraq was explained in part as essential to that new kind of war (see that discussion in Module 6.1, pages 265–266). How the United States pursues that war in the coming years will have enormous effects on societies and economies in the Middle East and North
303 Africa, South Asia, and perhaps in the United States as well.

The terrorists pursued by the United States are almost without exception Islamist militants—best known as **Islamists**—and so it is useful to understand the nature of Islamic "fundamentalism" and radicalism. Not all Islamists are militant or terrorist, but they all reject what they view as the materialism and moral corruption of Western countries and the political and military support these countries lend to Israel. Both Sunni and Shiite Muslims have advanced a wide range of Islamic movements, notably in Iran, Lebanon, Egypt, Afghanistan, Sudan, and Algeria.

Although nominally religious, the more radical of these movements have political and cultural aims, particularly the destabilization or removal of U.S. and Israeli interests in the region and abroad. In 1993, followers of the radical Egyptian cleric Sheikh Umar Abdel-Rahman bombed New York City's World Trade Center as a protest against American support of Israel and Egypt's pro-Western government. In an attempt to destabilize and replace Egypt's government, which it viewed as an illegitimate regime too supportive of the United States, another Egyptian Islamist group attacked and killed foreign tourists in Egypt in the 1990s. In the 1980s, members of the pro-Iranian **Hizbul-**
250 **lah,** or Party of God, in Lebanon kidnapped foreign civilians to use as bargaining chips for the release of comrades jailed in other Middle Eastern countries.

Within Israel and the autonomous Palestinian territories of the West Bank and Gaza Strip, Palestinian members of **Hamas** (an Arabic acronym for the Islamic Resistance Movement) and another organization called the al-Aqsa Martyrs Brigade carried out terrorist attacks on Israeli civilians and soldiers in an effort (apparently successful) to derail implementation of the peace agreements reached in
the 1990s between the Israeli government and the Palestine 246
Liberation Organization (PLO).

In Algeria, years of bloodshed followed the government's annulment of 1991 election results that would have given the **Islamic Salvation Front (FIS)** majority control in the parliament. Muslim sympathizers carried the battle to France, bombing civilian targets in protest against the French government's support for the Algerian regime.

In Western capitals, concern about "state-sponsored terrorism" has long focused on Iran, one of the three coun-
tries (along with Iraq and North Korea) that President Bush 246,
in 2002 dubbed the **"axis of evil."** Iran has extended direct 304
or clandestine assistance to a variety of Islamist terrorist groups, including Hizbullah and Hamas. There is great concern about Iran's nuclear weapons potential because such weapons might find their way to terrorist groups or be delivered by Iran itself on its own missiles against Israel or another target. Iran denies that it is developing the weapons, but U.S. and Israeli intelligence agencies believe Iran is on course to create them some time before 2016. Israeli officials have said publicly that Israel (which has its own nuclear weapons) regards Iran as an "existential threat" and will prevent that development, presumably by an air strike like the one Israel carried out against Iraq's nuclear facility in 1981. The United States, Israel, and the United Nations' **International Atomic Energy Agency (IAEA)** are particularly watchful of Iran's acquisition of centrifuges, which are used to enrich uranium. The enrichment process is necessary for the production of fuel for civilian use in nuclear power production, which Iran is permitted to do, but can also be expanded for the production of nuclear weapons.

Prior to a late 2007 release of a U.S. intelligence finding that Iran suspended its nuclear weapons development program in 2003, the United States was weighing two choices. It could engage in diplomatic dialogue with Iran, as it had done with North Korea, offering a package of incentives and security guarantees in exchange for Iran's pledge to
halt weapons development. This is the option that Rus- 303
sia, China, and European allies of the United States favored. The United States and Iran do share some common interests, notably stability in Iraq; Iran does not want to deal with strong new Sunni or Kurdish states there or with the tide of refugees that more Iraqi conflict would send its way. Alternatively, the United States could launch military strikes on Iran's suspected nuclear weapons development sites, beginning at Natanz. Most analysts believed that the military option would have far-ranging and dangerous consequences. Iran's nuclear program would be set back a few years, but the country would portray itself as a victim of U.S. aggression, galvanizing support across the Muslim

world (even among Iran's traditional enemies, the Sunni Arabs), perhaps inciting its Hizbullah allies in Lebanon and Hamas allies in the Palestinian territories to attack Israel, encouraging al-Qa'ida-style terrorism against the West, and discouraging Iranians (particularly the younger ones) who favor peaceful relations with the United States. A military strike would probably also lead to a sharp reduction in oil exports from the Gulf, perhaps triggering a global economic slowdown. Russia and China would meanwhile have an opportunity to enhance their geopolitical interests by offering at least diplomatic support to Iran throughout such a crisis. Both are already gaining more leverage in oil-rich Iran.

In 1998, the world began to hear about Osama bin Laden, a former Saudi businessman living in exile in Afghanistan, whose **al-Qa'ida** organization bombed U.S. embassies in Kenya and Tanzania as part of a worldwide armed struggle, or ***jihad,*** against American imperialism and immorality (see Geography of Terrorism, pages 236–237). Bin Laden's organization was also responsible for the 2000 bombing of the American naval destroyer U.S.S. *Cole* in Yemen's harbor of Aden. However shocking those assaults were, they pale in comparison to al-Qa'ida's attacks against targets in the United States on September 11, 2001. In the most ferocious terrorist actions ever undertaken to that date, members of al-Qa'ida "cells" (small groups of agents) in the United States hijacked four civilian jetliners and succeeded in piloting two of them into the twin towers of New York City's World Trade Center and one into the Pentagon, home of the U.S. Department of Defense, outside Washington, D.C. More than 3,000 people, mainly civilians, perished. Bin Laden and his followers cheered the carnage as justifiable combat against an infidel nation whose military troops occupied the holy land of Arabia, where Mecca and Medina are located.

As the U.S. counterattacked, ousting al-Qa'ida from its bases in Afghanistan and cracking down hard on its leadership around the world, al-Qa'ida evolved into a far more geographically diffuse organization. It carried out or supported attacks in Indonesia, Morocco, Algeria, Tunisia, Turkey, Spain, Uzbekistan, Egypt, Saudi Arabia, Jordan, and the United Kingdom (•Figure 6.26). A succession of al-Qa'ida videotapes and audiotapes promised an unrelenting and costly continuation of *jihad* against the United States and its allies. It will be many years before Americans in particular might be able to emerge from the shadow of this threat. Al-Qa'ida clearly will not hesitate to use the most devastating weapons, even against large numbers of civilians; this com-

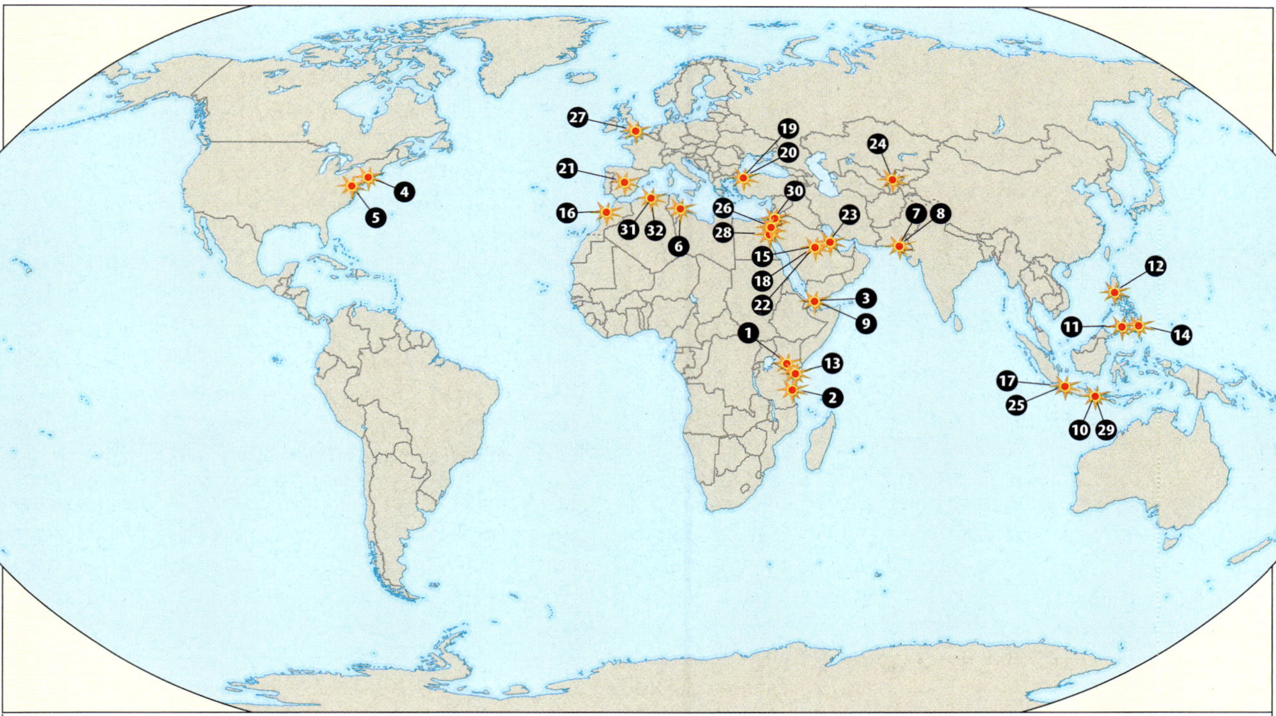

1 NAIROBI, KENYA 8/07/98
2 DAR ES SALAAM, TANZANIA 8/07/98
3 OFFSHORE OF ADEN, YEMEN 10/12/00
4 NEW YORK, UNITED STATES 9/11/01
5 WASHINGTON, UNITED STATES 9/11/01
6 DJERBA ISLAND, TUNISIA 4/11/02
7 KARACHI, PAKISTAN 5/08/02
8 KARACHI, PAKISTAN 6/14/02
9 OFFSHORE OF ADEN, YEMEN 10/06/02
10 KUTA, INDONESIA 10/12/02
11 ZAMBOANGA, PHILIPPINES 10/17/02
12 MANILA, PHILIPPINES 10/18/02
13 MOMBASA, KENYA 11/28/02
14 DAVAO, PHILIPPINES 3/04/03
15 RIYADH, SAUDI ARABIA 5/12/03
16 CASABLANCA, MOROCCO 5/16/03
17 JAKARTA, INDONESIA 8/05/03
18 RIYADH, SAUDI ARABIA 11/08/03
19 ISTANBUL, TURKEY 11/15/03
20 ISTANBUL, TURKEY 11/20/03
21 MADRID, SPAIN 3/11/04
22 RIYADH, SAUDI ARABIA 4/21/04
23 KHOBAR, SAUDI ARABIA 5/29/04
24 TASHKENT, UZBEKISTAN 7/30/04
25 JAKARTA, INDONESIA 9/09/04
26 TABA, EGYPT 10/08/04
27 LONDON, UNITED KINGDOM 7/7/05
28 SHARM EL-SHEIKH, EGYPT 7/23/05
29 KUTA, INDONESIA 10/01/05
30 AMMAN, JORDAN 11/09/05
31 ALGIERS, ALGERIA 4/10/07
32 ALGIERS, ALGERIA 12/11/07

• **Figure 6.26** Locations of terrorist attacks attributed to al-Qa'ida and affiliated organizations, 1998–2007.

GEOGRAPHY OF TERRORISM

What Does al-Qa'ida Want?

This is an important question. Perhaps if an answer could be found, there would be a way either to defeat this organization or to address the root causes of its existence in such a way that it would no longer have a reason to exist. Here we will look at the question mainly from the inside, examining what al-Qa'ida says it wants. The answer may be fundamentally a geographic one.

To begin, it is useful to know what al-Qa'ida is. Al-Qa'ida ("the Base" in Arabic) is a transnational organization that seeks to unite Islamist militant groups worldwide in a common effort to achieve its goals. An Islamist or *jihadist* organization is one that employs Islamic faith, culture, and history to legitimize its philosophy and actions.[*] The principal sources Islamists cite are the Qur'an, the Hadith (sayings of the Prophet Muhammad), and the writings of earlier Islamic militants such as Ibn Taymiyya and Sayyid Qutb. According to Rohan Gunaratna, the leading academic authority on the organization, al-Qa'ida is "above all else a secret, almost virtual, organization, one that denies its existence in order to remain in the shadows."[†] This desire for secrecy explains why al-Qa'ida seldom takes direct responsibility for terrorist acts. Instead, these are attributed to the other names and identities employed by al-Qa'ida, particularly the "World Islamic Front for the Jihad against the Jews and the Crusaders," a coalition of seven Islamist militant groups (three Egyptian, two Pakistani, one Bangladeshi, and one Afghan). Osama bin Laden, formerly a Saudi national, emerged as the figurehead of this organization.

Al-Qa'ida certainly is a terrorist organization if one employs the U.S. Department of State's definition of **terrorism:** "premeditated, politically motivated violence perpetrated against noncombatant targets by subnational groups or clandestine agents, usually intended to influence an audience."[‡] Ayman al-Zawahiri, al-Qa'ida's second in command, called for an escalation of attacks with "the need to inflict the maximum casualties against the opponent, for this is the language understood by the West, no matter how much time and effort such operations take."[§] Osama bin Laden, as evident in his own words, has called for the killing of American and other Western civilians. Gunaratna regards al-Qa'ida as an extremely unusual terrorist group in the category he calls "apocalyptic," one that believes "it has been divinely ordained to commit violent acts, and likely to engage in mass casualty, catastrophic terrorism."[**] He warns that al-Qa'ida "will have no compunction about employing chemical, biological, radiological and nuclear weapons against population centers."[††]

But why? Why do they want to kill Western, and particularly American, civilians? Al-Qa'ida leaders have been very explicit in their rationale. Osama bin Laden explained that the main reason for the terrorist attacks is to convince the United States that it should withdraw its military forces and other interests from the Islamic Holy Land. In the following statement, the "Land of the Two Holy Places" means Saudi Arabia. The two holy places are 246 the Saudi Arabian cities of Mecca (site of the Ka'aba, and Islam's most sacred city) and Medina (Islam's second holiest place, where the Prophet Muhammad is buried). The Dome of the Rock is the shrine in Jerusalem, described earlier, containing the sacred rock from which the Prophet Muhammad was said to have ascended into heaven on the Night Journey.

> The Arabian Peninsula has never—since God made it flat, created its desert, and encircled it with seas—been stormed by any forces like the crusader armies spreading in it like locusts, eating its riches and wiping out its plantations. . . . The latest and greatest of these aggressions incurred by the Muslims since the death of the Prophet . . . is the occupation of the Land of the Two Holy Places—the foundation of the house of Islam, the place of the revelation, the source of the message, and the place of the noble Ka'aba, the *qibla* [direction faced during prayer] of all Muslims—by the armies of the American crusaders and their allies. We bemoan this and can only say "No power and power acquiring except through Allah." . . . To push the enemy—the greatest *kufr* [infidel]—out of the country is a prime duty. No other duty after Belief is more important than this duty. Utmost effort should be made to prepare and instigate the *umma* [Islamic community] against the enemy, the American-Israeli alliance—occupying the country of the two Holy Places . . . to the al-Aqsa mosque in Jerusalem. . . . The crusaders and the Jews have joined together to invade the heart of Dar al-Islam—the Abode of Islam: our most sacred places in Saudi Arabia, Mecca and Medina, including the prophet's mosque and Dome of the Rock in Jerusalem, *al-Quds*.[‡‡]

In what was arguably his most important policy statement, announcing the formation of the Islamic World Front for the Jihad against the Jews and the Crusaders, bin Laden used the occupation of Islamic sacred space as the principal justification for war against the United States:

bination is in fact its tactical priority. There is also growing consensus that al-Qa'ida is no longer a solitary severe threat to U.S. interests; instead, the *jihadist* cause has spread to numerous smaller or less renowned groups that are also are keen to take on the West.

In 9/11 and other atrocities, a tiny minority of Muslims carried out terrorist actions that the great majority of Muslims condemned. Islamic scholars and clerics pointed out in each case that the murder of civilians is prohibited in Islamic law and that the attacks had no legitimate religious grounds. Mainstream Islamic movements are not military or terrorist organizations but have distinguished themselves through public service to the needy and through encouragement of strong moral and family values. For most Muslims, the growing Islamist trend means a reembrace of traditional values like piety, generosity, care for others, and Islamic legal systems, which have proved effective for centuries. For them, these values pose a reasonable alternative

GEOGRAPHY OF TERRORISM

What Does al-Qa'ida Want? *continued*

> In compliance with God's order, we issue the following *fatwa* [religious injunction] to all Muslims: The ruling to kill the Americans and their allies—civilians and military—is an individual duty for every Muslim who can do it in any country in which it is possible to do it, in order to liberate the al-Aqsa mosque and the holy mosque [in Mecca] from their grip, and in order for their armies to move out of all the lands of Islam, defeated and unable to threaten any Muslim.[§§]

In an earlier *fatwa* (1996), bin Laden laid out these goals for al-Qa'ida: to drive U.S. forces out of the Arabian Peninsula, overthrow the Saudi government, and liberate the holy places of Mecca and Medina. Overthrowing the Saudi government and the other autocratic dynasties of the Persian Gulf (Arabian Gulf) region, along with other secular and pro-Western regimes of the Middle East—notably Mubarak's Egypt—is a theme that is very often articulated in al-Qa'ida ideology.

What is al-Qa'ida's ultimate goal? Is the organization satisfied now that U.S. troops actually have been withdrawn from Saudi Arabia? What would be achieved if all Western interests were driven from the Islamic Holy Land and if all of the governments sympathetic to the West were overthrown? According to Gunaratna, the ultimate aim is to reestablish the **caliphate**—the empire of Islam's early golden age—and thereby empower a formidable array of truly Islamic states to wage war on the United States and its allies.[***]

With these goals in mind, what should the United States do? Al-Qa'ida's strategists clearly hope that terrorism will inflict unacceptable losses of American lives, forcing the United States to withdraw its troops from Iraq and neighboring countries. That conviction may stem from the withdrawal of U.S. troops from Somalia after the loss of 18 American soldiers in Mogadishu in 1993 (the "Blackhawk Down" episode; see page 506) and the withdrawal of U.S. forces from Lebanon following the suicide bombing of the Marine barracks in Beirut in 1983 (see page 250). Al-Qa'ida thinks the United States has no stomach for sustained sacrifice.

Al-Qa'ida poses a dilemma for the United States. If the United States were to withdraw its troops from the region, it might only reaffirm al-Qa'ida's belief that its adversary is weak and vulnerable, thus encouraging more attacks. And if al-Qa'ida's ultimate goal is to take on the United States once it has established an Islamic empire, there is no reason to believe that a unilateral withdrawal would bring a cessation of hostilities.

On the other hand, if the United States continues to conduct a war on al-Qa'ida and a broader war on terror in a host of Muslim countries, al-Qa'ida can use the American presence, and especially unintended civilian losses, to incite widespread hatred and violence directed against the United States. Al-Qa'ida used the U.S. occupation of Iraq to argue its case that "the crusaders" *are* waging a war against Muslims and the Islamic world. President Bush argued that Iraq is "the central front on the war on terror,"[†††] but many analysts argue that the U.S. invasion of Iraq served as a catalyst to recruit uncounted numbers to the *jihadist* cause, turning Iraq into a "superbowl of terror" where new militant groups could cut their teeth on American soldiers and civilian contractors and their Iraqi allies.

Using geographic information systems and a variety of other tools, many geographers are interpreting the geographic dimensions of terrorism. One of the most profound questions for geographers is where future attacks will take place. Gunaratna estimates that al-Qa'ida maintains a reserve of at least 100 targets worldwide.[‡‡‡] Where are most of these? Suleiman Abu Ghaith, one of bin Laden's top aides, offered a clue, along with a geographic answer to the question of what al-Qa'ida wants: "Let the United States know that with God's permission, the battle will continue to be waged on its territory until it leaves our lands."[§§§]

*John L. Esposito, *Unholy War: Terror in the Name of Islam* (Oxford: Oxford University Press, 2002), p. 28.

†Rohan Gunaratna, *Inside al Qaeda: Global Network of Terror* (New York: Cambridge University Press, 2002), p. 3.

‡Quoted in Rex A. Hudson, *Who Becomes a Terrorist and Why: The 1999 Government Report on Profiling Terrorists* (Guilford, Conn.: Lyons Press, 1999), p. 18.

§Ayman al-Zawahiri, "Why Attack America? (January 2002)." In *Anti-American Terrorism and the Middle East,* Barry Rubin and Judith Colp Rubin, eds. (Oxford University Press, 2002), p. 133.

**Gunaratna, *Inside al Qaeda,* p. 93.

††Ibid., p. 11.

‡‡Osama bin Ladin in three selections from *Anti-American Terrorism and the Middle East:* "Declaration of War (August 1996)," pp. 137, 139; "Statement: Jihad against Jews and Crusaders (February 23, 1998)," p. 149; and "Al-Qa'ida Recruitment Video (2000)," p. 174.

§§Bin Laden, "Statement," note vii, p. 150.

***Gunaratna, *Inside al Qaeda,* note ii, p. 55; p. 89.

†††George W. Bush, "President Discusses War on Terror and Operation Iraqi Freedom," March 20, 2006, http://www.whitehouse.gov/news/releases/2006/03/20060320-7.html.

‡‡‡Gunaratna, *Inside al Qaeda,* p. 188.

§§§Suleiman Abu Ghaith, "Al-Qa'ida Statement (October 10, 2001)." In *Anti-American Terrorism and the Middle East,* p. 251.

to Western cultural influences and often repressive political and administrative systems. For many people outside the region, however, "Muslim" and "terrorist" have become synonymous—an erroneous association that can be overcome in part by careful study of the complex Middle East and North Africa.

This concludes our overview of land and life in the Middle East and North Africa. Module 6.1 offers more insight into the peoples and nations of this vital region. It begins with a continuation of the geopolitical theme, examining one of the world's most problematic, persistent, and influential conflicts, the one between Israel and its Arab neighbors.

SUMMARY

→ Misleading stereotypes about the environment and people of the Middle East and North Africa are common, as people outside the region often associate the region solely with military conflict and terrorism.

→ The region has bestowed on humanity a rich legacy of ancient civilizations and the three great monotheistic faiths of Judaism, Christianity, and Islam.

→ Middle Easterners include Jews, Arabs, Turks, Persians, Pashtuns, Berbers, people of sub-Saharan African origin, and other ethnic groups who practice a wide variety of ancient and modern livelihoods.

→ Arabs are the largest ethnic group in the Middle East and North Africa, and there are also large populations of ethnic Turks, Persians (Iranians), and Kurds. Islam is by far the largest religion. Jews live almost exclusively in Israel, and there are minority Christian populations in several countries.

→ Population growth rates in the region are moderate to high. Oil wealth is concentrated in a handful of countries, and as a whole, this is a developing region.

→ The Middle East has served as a pivotal global crossroads, linking Asia, Europe, Africa, and the Mediterranean Sea with the Indian Ocean. These countries have historically been unwilling hosts to occupiers and empires originating far beyond their borders.

→ The margins of this region are occupied by oceans, high mountains, and deserts. The land is composed mainly of arid and semiarid plains and plateaus, together with considerable areas of rugged mountains and isolated "seas" of sand.

→ Aridity dominates the environment, with at least three-fourths of the region receiving less than 10 inches (25 cm) of yearly precipitation. Great river systems and freshwater aquifers have sustained large human populations.

→ Many of the plants and animals on which the world's agriculture depends were first domesticated in the Middle East.

→ The Middle Eastern "ecological trilogy" consists of peasant villagers, pastoral nomads, and city dwellers. The relationships among them have been mainly symbiotic and peaceful, but city dwellers have often dominated the relationship, and both pastoral nomads and urbanites have sometimes preyed on the villagers, who are the trilogy's cornerstone.

→ About two-thirds of the world's oil is here, making this one of the world's most vital economic and strategic regions.

→ Since World War II, several international crises and wars have been precipitated by events in the Middle East. Strong outside powers depend heavily on this region for their current and future industrial needs. Unimpeded access to Persian-Arabian Gulf oil is one of the pillars of U.S. foreign policy.

→ The region of the Middle East and North Africa is characterized by a high number of chokepoints, strategic waterways that may be shut off by force, triggering conflict and economic disruption.

→ Oil pipelines in the Middle East are routed both to shorten sea tanker voyages and to reduce the threat to sea tanker traffic through chokepoints, but are themselves vulnerable to disruption.

→ Access to freshwater is a major problem in relations between Turkey and its downstream neighbors, Egypt and its upstream neighbors, and Israel and its Palestinian, Jordanian, and Syrian neighbors.

→ Al-Qa'ida and affiliated Islamist terrorist groups aim to drive the United States and its allied governments from the region and to replace them with an Islamic caliphate. Al-Qa'ida is an apocalyptic group that seeks to inflict mass casualties on its enemies, particularly on Americans in their home country.

KEY TERMS + CONCEPTS

hydropolitics (p. 232)
Indo-European language family (p. 221)
- Dari (Afghan Persian) (p. 221)
- Farsi (Persian) (p. 221)
- Kurdish (p. 221)
- Pashto (p. 221)

International Atomic Energy Agency (IAEA) (p. 234)
Islam (p. 235)
Islamic law (*sharia*) (p. 227)
Islamic Salvation Front (FIS) (p. 234)
Islamists (p. 234)
Israelites (p. 222)
Jews (p. 221)
jihad (p. 235)
Ka'aba (p. 226)
Kurds (p. 221)
Maronites (p. 225)
medina (p. 219)
Middle Eastern ecological trilogy (p. 213)
- pastoral nomads (p. 214)
- urbanites (p. 214)
- villagers (p. 213)

Muslims (p. 225)
"New Rome" (p. 225)
Night Journey (p. 226)
Nile Water Agreement (p. 233)
Nilo-Saharan language family (p. 221)
- Chari-Nile languages (p. 221)

Noble Sanctuary (*al-Haraam ash-Shariif*) (p. 223)
oil embargo (p. 232)
Organization of Petroleum Exporting Countries (OPEC) (p. 228)
Pashtun (p. 221)
People of the Book (p. 225)
Persian (p. 221)
Pillars of Islam (p. 228)
- almsgiving (p. 227)
- fasting (p. 227)
- pilgrimage to Mecca (*hajj*) (p. 227)
- prayer (p. 227)
- profession of faith (p. 228)

Promised Land (p. 222)
Qur'an (Koran) (p. 226)
Red-Dead Peace Conduit (p. 232)
risk minimization (p. 217)
sand sea (p. 211)
Second Temple (p. 223)
sedentarization (p. 218)
Shia (p. 227)
Shiite (p. 227)
Suez Crisis (p. 229)
Sunni (p. 227)
suq (p. 219)
Temple Mount (p. 223)
terrorism (p. 236)
Turks (p. 221)
upstream countries (p. 232)
usufruct (p. 219)
vertical migration (p. 217)
war on terror (p. 234)
weapons of mass destruction (WMD) (p. 232)
Western Wall (p. 223)
Zion (p. 224)
Zionist movement (p. 224)

REVIEW QUESTIONS

1. What countries constitute the Middle East and North Africa? Which are the three most populous? Which encourage and discourage population growth?
2. What are the major climatic patterns of the Middle East? Where are the principal mountains, deserts, rivers, and areas of high rainfall?
3. Why can this region be described as a culture hearth? What major ideas, commodities, and cultures originated there?
4. What are the major ethnic groups, and in which countries are they found? What is an Arab? A Jew? A Turk? A Kurd? A Persian? A Muslim?
5. What are the principal beliefs and historical geographic milestones of Jews, Christians, and Muslims?
6. What is the difference between Shiite and Sunni Islam?
7. Where is oil concentrated in this region?
8. What is the Carter Doctrine?
9. What options does the United States weigh in dealing with Iran's suspected nuclear weapons development program?
10. What are upstream and downstream countries, and which are usually the more powerful? What are the exceptions to this rule?
11. What are Hizbullah, Hamas, and al-Qa'ida?

Joe Hobbs

A sheikh (headman) of the Bisharin, camel-raising pastoral nomads of Egypt's southern Eastern Desert. Bisharin men carry their swords for ornamentation, and the photographer asked to see it.

MODULE 6.1
THE MIDDLE EAST AND NORTH AFRICA: MODERN STRUGGLES IN AN ANCIENT LAND

Conflict between Israel and its neighbors has been one of the most persistent and most dangerous problems in global affairs since the end of World War II. This module begins with an introduction to the key players and issues in the conflict. This depiction reveals why the Palestinian-Israeli dispute is so difficult to solve and how beneficial its resolution would be. The module continues with a survey of the major subregions, nations, and geographic problems of the diverse, fascinating Middle East.

module outline

module objectives

This module should enable you to

- → Understand the major issues of the Arab-Israeli conflict and the obstacles to their resolution
- → Appreciate how Lebanon's political system failed to represent the country's ethnic diversity, setting the stage for civil war
- → Balance the pros and cons of Egypt's Aswan High Dam, Libya's Great Manmade River, and other large water engineering projects
- → View the Sudanese civil war as an effort by one ethnic group to dominate others
- → Recognize how the U.S. war on terrorism has transformed political systems and international relations
- → Trace the trajectory of Persian Gulf (Arabian Gulf) countries from underpopulated, wealthy welfare states to populous, vulnerable states overly dependent on a single commodity
- → See the U.S. wars on Iraq as the result of tragic miscalculations by Saddam Hussein, combined with strategic American interests in the region
- → Recognize Turkey as an "in-between" country, a less developed Asian nation aspiring to be a European power
- → Consider how both superpower preoccupation and profound neglect sowed the seeds of conflict in Afghanistan

6.1.1 The Arab-Israeli Conflict and Its Setting

The Arab-Israeli conflict persists as one of the world's most intractable disputes. It has not been resolved in part because the central issues are closely tied to such life-giving resources as land and water and to deeply held religious beliefs.

The Arab-Israeli conflict is above all a conflict over who owns the land—sometimes very small pieces of land—and is therefore of extreme interest in the study of geography. It is also a conflict that has repercussions far beyond the boundaries of the small countries and territories involved. As long as it simmers or boils, there are other countries and entities that will use the unresolved Arab-Israeli conflict to advance their interests at the expense of others; both al-Qa'ida and Iran, for example, derive much beneficial propaganda value from it. A United Nations–sponsored group called the Alliance of Civilizations concluded in 2006 that the Palestinian-Israeli conflict is the largest force behind global tensions.

A geographic understanding of the Middle East today requires familiarity with the events leading up to the creation of the state of Israel and with the wars that have followed, particularly as they have rearranged the boundaries of nations and territories.

The modern state of Israel was carved from lands whose fate had been undetermined since the end of World War I. The Ottoman Empire, based in what is now Turkey, had ruled Palestine (roughly the area now made up of Israel and the Palestinian territories) and surrounding lands in the eastern Mediterranean since the 16th century. After the British and French defeated the Ottoman Turks in World War I and destroyed their empire, they divided the region between themselves (•Figure 6.1.1). The British received the "mandate" (authority to establish a government) for Palestine, Transjordan (modern Jordan), and Mesopotamia (Iraq), while the French received the mandate for Syria (now Syria and Lebanon).

During World War I, British administrators of Palestine had made conflicting promises to Jews and Arabs. They implied that they would create an independent Arab state in Palestine and yet at the same time vowed to promote Jewish immigration to Palestine with an eye to the eventual establishment of a Jewish state there. The **Palestinians**—Arabs who historically formed the vast majority of the region's inhabitants—did not welcome the ensuing Jewish immigration and rioted against both the migrants and the British administration. Militant Jews attacked British interests in Palestine, hoping to precipitate a British withdrawal.

Placing themselves in a no-win position with these conflicting promises and under increasing pressure from both Jews and Arabs, in 1947 the British decided to withdraw from Palestine and leave the young United Nations with the task of determining the region's future. The United Nations responded in 1947 with the **two-state solution** to the problem of Palestine. It established an Arab state (which would have been called Palestine) and a Jewish state (Israel). The plan was deeply flawed. The states' territories were long, narrow, and fragmented, giving each side a sense of vulnerability and insecurity (•Figure 6.1.2). When Israel declared itself into existence in May 1948, the armies of the neighboring Arab countries of Transjordan, Egypt, Iraq, Syria, and Lebanon mobilized. In what Israelis call the War of Independence and Palestinians call the Catastrophe (*al-Nakba*), the smaller but better-organized and more highly motivated Israeli army defeated the Arab armies, and Israel acquired what have come to be known as its **pre-1967 borders** (the 1949 Armistice Agreement borders shown in Figure 6.1.2).

Prior to and during the fighting of the **1948–1949 Arab-Israeli war,** approximately 800,000 Palestinian Arabs chose or were forced to flee from the new officially Jewish state

• **Figure 6.1.1** The Middle East and North Africa in 1920. The victorious allies of World War I carved up the Middle East among themselves. Growing difficulties of administration would drive them from the region within a few decades.

of Israel to neighboring Arab countries (•Figure 6.1.3). The United Nations established refugee camps for these displaced persons in Jordan, Egypt, Lebanon, and Syria. Little was done to resettle them in permanent homes, and both the local Arab governments and the refugees themselves continued to insist on the right of the refugees to return to Israel and the restoration of their property there. This Palestinian **right of return** is one of the central problems of the ongoing peace process.

Other countries assumed control of those parts of the proposed Palestine that Israel did not absorb. Egypt occupied the Gaza Strip, a piece of land on the Mediterranean shore adjacent to Egypt's Sinai Peninsula that was inhabited mostly by Palestinian Arabs. Transjordan, renaming itself simply Jordan, occupied the predominantly Arab hilly region of central Palestine on the west side of the Jordan River, known as the West Bank, and the entire old city of Jerusalem, including the Western Wall and Temple Mount,
223 Judaism's holiest sites. The Palestinian Arab state envisioned in the UN partition plan was thus stillborn.

The **Six-Day War** of 1967 fundamentally rearranged the region's political landscape in Israel's favor, setting the stage for subsequent struggles and the peace process. This conflict was precipitated in part when Egypt, by positioning arms at the Strait of Tiran chokepoint, closed the Gulf of Aqaba to Israeli shipping. Egypt's President Nasser and his Arab allies took several other belligerent but nonviolent steps toward a war they were ill prepared to fight. Israel elected to make a preemptive strike on its Arab neighbors, virtually destroying the Egyptian and Syrian air forces on the ground. Israel gave Jordan's King Hussein an opportunity to stay out of the conflict. However, Jordan went to war and quickly lost the entire West Bank and the historic and sacred Old City of Jerusalem. The entire nation of Israel was transfixed by the news that Jewish soldiers were praying at the Western Wall. The Israeli army (Israeli Defense Forces, or IDF) also seized the Gaza Strip, Egypt's Sinai Peninsula (shutting down the Suez Canal), and the strategic Golan Heights section of Syria overlooking Israel's Galilee region. These three pieces of land would henceforth be known to the world as the **Occupied Territories.** Israel had tripled its territory in six days of fighting (•Figure 6.1.4).

The Palestinian refugee situation became more complex as a result of the war. Israel took over the areas where most of the camps were located. Many persons in the camps again took flight, and Palestinians fleeing villages and towns in the newly Occupied Territories joined them as refugees (see Figure 6.1.3). Some later returned to their homes, but an estimated 116,000 either did not attempt to return or were denied permission by Israel authorities to do so.

The **1973 Arab-Israeli War** was a multifront Arab attempt to reverse the humiliating losses of 1967. On October 6, 1973, the Jewish holy day of Yom Kippur, Egypt and Syria launched a surprise attack on Israel with the hope of rearranging the stalemated political map of the Middle East. Egypt's army initially showed surprising strength, penetrat-

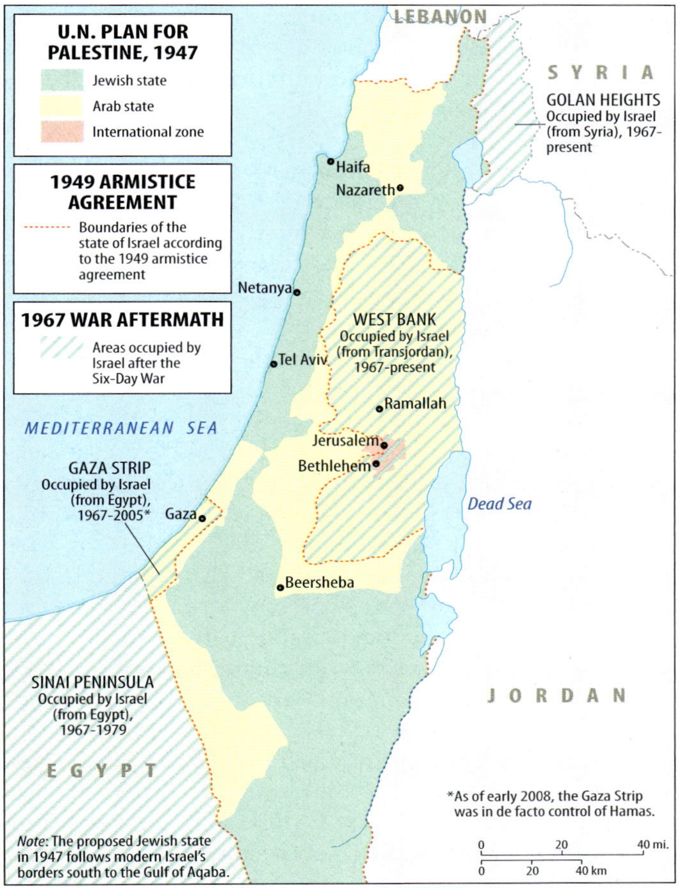

• **Figure 6.1.2** The 1947 UN partition plan for Palestine and Israel's original ("pre-1967") borders. The 1948–1949 war, which began as soon as Britain withdrew from Palestine and Israel proclaimed its existence, aborted the UN plan and created a tense new political dynamic in the region.

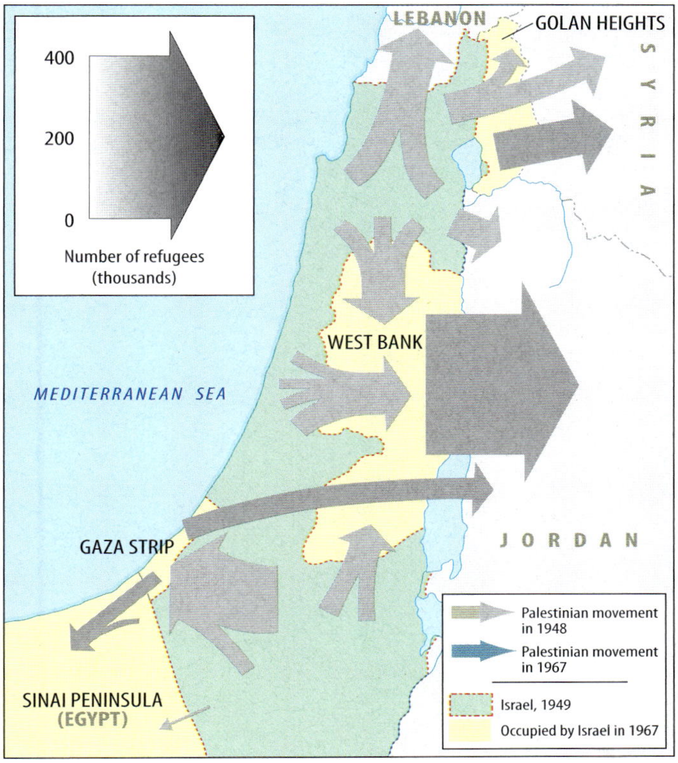

• **Figure 6.1.3** Palestinian refugee movements in 1948 and 1967. Many who fled in the first conflict relocated in the second.

ing deep into the Sinai Peninsula and overturning Israel's image of invulnerability. Israeli troops soon reversed the tide and surrounded Egypt's army, but in the ensuing disengagement talks, Egypt won back the eastern side of the Suez Canal and by 1975 was able to reopen it to commercial traffic. In 1979, Israel agreed to return Sinai to Egypt under the U.S.-sponsored **Camp David Accords,** and Egypt recognized Israel's rights as a sovereign state. The Gaza Strip, West Bank, and Golan Heights continued to be held by Israel. Different Israeli administrations regarded them as lands that rightfully belong to Jews or as cards to be played in the regional game of peacemaking.

Arabs and Jews: The Demographic Dimension

In addition to issues of land, water, politics, and ideology, the Palestinian-Israeli conflict is about sheer numbers of people. Each side has wanted to maximize its numbers to the disadvantage of the other. To realize the Zionist dream of establishing a Jewish state, Jews began immigrating to Palestine around the start of the 20th century. Jews made up 11 percent of Palestine's population in 1922, 16 percent in 1931, and 31 percent in 1946, on the eve of Israel's creation. In keeping with national legislation known as the **Law of Return,** the state of Israel has always granted citizenship to any Jew who wishes to live there. Following the 1948–1949 war, waves of new immigrants from Europe (**Ashkenazi Jews**) joined the Middle Eastern **Sephardic Jews** 130 who had inhabited Palestine and other parts of the Middle East (and until 1492, Spain) since early times.

In the 1990s, more waves of Jewish immigration followed the dissolution of the Soviet Union and a change of government in Ethiopia, where an ancient Jewish group called **Falashas** had lived in isolation for thousands of years. In 2007, Jews made up 76 percent of the population of 7.3 million people living within Israel's pre-1967 borders. About 1.6 million Palestinian and non-Palestinian Arabs, or 23 percent of the country's population, are Israeli citizens and residents; they are known as "Israeli Arabs." Palestinian Arabs without Israeli citizenship, along with a variety of other ethnic groups, comprise the balance of the population within Israel's pre-1967 borders.

Another 4 million people live in the West Bank, Gaza Strip, and Golan Heights—areas occupied by Israel in the wars of 1967 and 1973. About 187,000 of those are Jewish settlers, and most of the rest are Palestinian Arabs (mainly in the West Bank and Gaza Strip), **Druze** (members of a small offshoot of Shiite Islam, living mainly in the Golan

• **Figure 6.1.4** Israel and Arab territories as of 2007. Israel occupied the Sinai Peninsula in the 1967 war but returned it to Egypt following the Camp David Accords of 1979. Israel withdrew settlers and soldiers from the Gaza Strip in 2005 but kept them in the West Bank and the Golan Heights.

Heights), and other Arabs. Another 177,000 Jewish settlers live in East Jerusalem, which is also part of the territories Israel conquered in 1967. There are approximately 9 million more Palestinians living outside Israel and the Occupied Territories. Jordan has the largest number (3 million), followed by Syria and Lebanon (with 400,000 each). The remaining 1.5 million are in other Arab countries and in numerous nations around the world, comprising a sizable diaspora of their own that Palestinians call the ***ghurba.*** In this respect, the Palestinians are in much the same situation as the Jews prior to Israel's creation.

After 1967, and especially after the election of its conservative **Likud Party** to power in 1977, Israel moved to strengthen its grip on the Occupied Territories through security measures and the government-sponsored establishment of new **Jewish settlements.** Most of these settlements were, and still are, spread through the West Bank (see Figure 6.1.B, page 246). Much smaller numbers of Jews settled in the troubled Golan Heights, where about 15,000 remained in 2007. These are generally not frontier farming settlements but communities inhabited by relatively prosperous middle-class Jews who commute to jobs in Israel. They were not built as temporary encampments but as permanent fixtures meant to create what have been described as **"facts on the ground"**—an Israeli presence so entrenched that its withdrawal would be almost inconceivable (•Figure 6.1.5). Some of the motivation for this settlement has been ideological, as the West Bank is composed of the biblical lands of **Judaea and Samaria,** which many devout Jews regard as part of Israel's historical homeland.

The proliferation of Jewish settlements worsened relations between Israel and its Arab neighbors and even drove a wedge into the Jewish population of Israel itself. Many Israelis strongly oppose further settlement in the Occupied Territories and annexation of these territories to the Jewish state because if the territories were annexed, Israel

Jodi Cobb/National Geographic Image Collection

• **Figure 6.1.5** The Jewish settlement of Maaleh Adumim has a commanding position overlooking the strategic West Bank road linking Jerusalem with the Dead Sea rift valley.

PROBLEM LANDSCAPE

Land for Peace

They loom large in the headlines and in world affairs, but a closer look reveals that Israel and the Occupied Territories are almost unfathomably small. Within its legally recognized, pre-1967 borders, Israel's area is only 8,019 square miles (20,770 sq km), about the size of New Jersey or Slovenia (• Figure 6.1.A). Israelis often cite the small size of their country, nestled within the vastness of the Arab Middle East and North Africa, to highlight their vulnerability on the world stage. When and if Palestinians acquire their country, Palestine will be even smaller. The question "Whose lands are these?" has always been at the heart of the conflict between Israel and its neighbors.

President Anwar Sadat of Egypt set the precedent in 1979: Arabs could make peace with Israel on the formula of "land for peace," with Israel swapping Arab lands it occupied in 1967 in exchange for peace with its neighbors. The United States, Russia, and Norway initiated negotiations in the 1990s that made such a prospect appear possible. The process began in earnest in 1993, when Israel recognized the legitimacy of and began to negotiate with the **Palestine Liberation Organization (PLO),** long recognized by the Palestinians as their legitimate governing body. In return, the PLO, under the leadership of Yasser Arafat, recognized Israel's right to exist and renounced its long-standing use of military and terrorist force against Israel. U.S. President Bill Clinton orchestrated a historic handshake between Arafat and the Israeli prime minister, Yitzhak Rabin, and the leaders signed an agreement known as the **Gaza-Jericho Accord** (in its implementation, it came to be known as the **Oslo I Accord,** after the Norwegian capital where it was negotiated). It was designed to pave a pathway to peace by Israel's granting of limited autonomy, or self-rule, in the Gaza Strip and in the West Bank town of Jericho. Autonomy meant that the Palestinians were responsible for their own affairs in matters of education, culture, health, taxation, and tourism. Israel also pulled its troops out of these areas. Palestinians were allowed to form their own government, known as the **Palestinian Authority (PA),** and they elected Yasser Arafat as its first president. The accord established a five-year timetable for the resolution of much more difficult matters, the so-called **final status issues.** These included the political status of Jewish and Muslim holy places in Jerusalem and of the city itself, the possible return of Palestinian refugees, and Palestinian statehood.

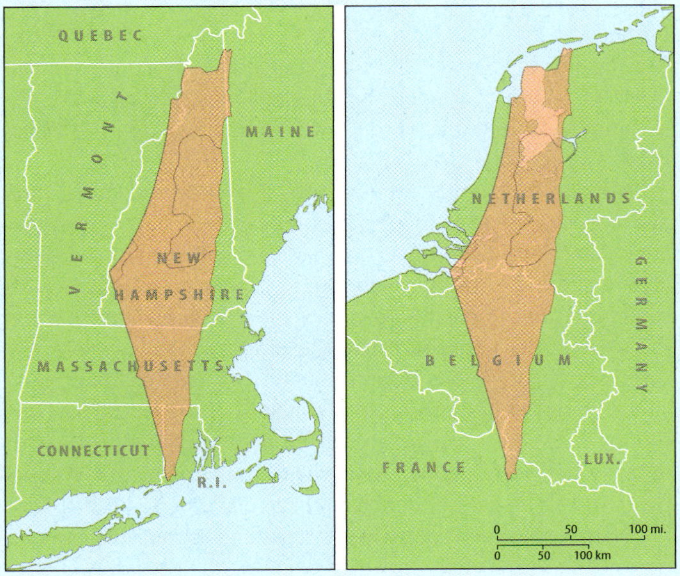

• **Figure 6.1.A** Israel and the Occupied Territories compared in size with New England and the Benelux countries.

In a deal known as the Wye Agreement or **Oslo II Accord,** three successive Israeli prime ministers (Yitzhak Rabin, Benjamin Netanyahu, and Ehud Barak) promised to transfer more West Bank lands from Israeli to Palestinian control, and Arafat promised increased Palestinian efforts to crack down on Palestinian terrorists and so guarantee the security of Israelis in the Palestinian territories and in Israel. If finally implemented (it had not been as of 2007), this arrangement would have brought the total of West Bank lands under complete Palestinian control to 17 percent, leaving 57 percent completely in Israeli hands and 26 percent under joint control (• Figure 6.1.B).

During his final year in office in 2000, U.S. President Clinton sought to solidify his legacy as peacemaker by brokering a historic final settlement between Israelis and Palestinians. PLO Chairman Arafat and Israeli Prime Minister Barak huddled with Clinton and his advisers in the presidential retreat at Camp David, a site chosen because of its historical significance in Middle East peacemaking. Over weeks of tough negotiations, the two sides came close to a deal. It would have included the creation of an independent Palestinian country in the Gaza Strip and approximately 95 percent of the West Bank, with the other 5 percent representing clusters of Jewish settlements that would remain under Israeli control while Palestine received 5 percent of Israel's land in exchange. However, the talks broke down over two thorny issues. One was the status of Palestinian refugees abroad; Israel was unwilling to permit the right of return of the large numbers demanded by the Palestinians. But the deal breaker was Jerusalem, which both Israelis and Palestinians proclaim as their capital. Israel agreed to have the Old City of Jerusalem divided, with Israel assuming sovereignty over the Jewish and Armenian quarters and Palestinians assuming sovereignty over the Arab and Christian quarters (see Figure 6.B, page 223). These issues were not as problematic as those involving the places held holy by each side. Israel agreed to allow the Palestinians "soft sovereignty" over the Temple Mount (*al-Haraam ash-Shariif*), meaning that Palestinians could administer the area and fly a flag over it, but Israel would technically "own" it. Israel saw this point as essential because the First and Second Temples had stood on this spot. The Palestinians, however, insisted on full sovereignty over the Temple Mount and walked out on the negotiations.

PROBLEM LANDSCAPE

Land for Peace *continued*

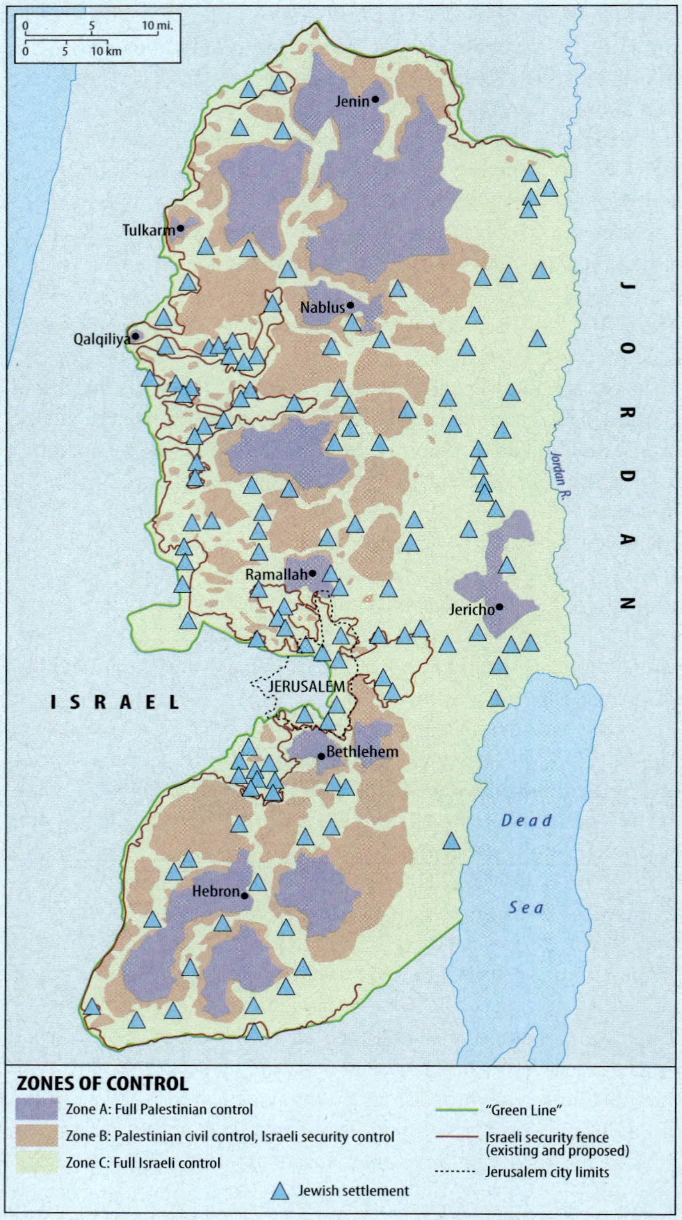

• **Figure 6.1.B** The West Bank. Israeli and Palestinian areas of control were delimited in the peace process of the 1990s, but because of recurrent violence, almost all areas are effectively controlled by Israel. Also depicted are Jewish settlements, the completed and planned portions of the Israeli-built security fence, and the Green Line delimiting the internationally recognized border between sovereign Israel and the occupied West Bank.

Tragically, within weeks, this historic opportunity for peace evaporated and was replaced by a state of war. On September 28, 2000, the right-wing Israeli opposition leader (and subsequently prime minister) Ariel Sharon walked up to the Temple Mount under heavily armed escort. He meant to make the point that this sacred precinct belonged to Israel. Many observers claim he also meant to disrupt the peace process. Whether or not it was his intention, a disruption in the peace process resulted. Beginning that day and for the next few years, Palestinians took to the streets in the **al-Aqsa *Intifada*** to challenge Israeli forces with rocks, bottles, and guns. Hundreds of Palestinians and Israelis died.

Following the collapse of the Camp David talks and an escalation of suicide bombings of Israeli civilians, Israel chose to effectively suspend the peace process. Prime Minister Sharon branded Arafat an untrustworthy partner for peace negotiations, isolating him physically in the West Bank town of Ramallah and systematically dismantling the Palestinian Authority he headed. Israel also persuaded the United States to stop negotiating with Arafat, who died a forlorn figure in 2004. Israel and the United States insisted on the creation of the new post of Palestinian prime minister, hoping it would be filled by someone they could negotiate with. The United States led an effort joined by the other members of the so-called Quartet—the European Union, the United Nations, and Russia—to establish what they called a **"road map for peace"** between the Israelis and the Palestinian Authority. It was based largely on the 2000 Camp David formula and, if successful, would presumably lead to an independent Palestinian state consisting of the West Bank and the Gaza Strip.

Prime Minister Sharon, while resuming dialogue with the Palestinian Authority, decided unilaterally to withdraw Israeli settlers and troops from the Gaza Strip in August 2005 and to build an immense **security fence** separating the West Bank from Israel. Israel's stated goal in constructing the new barrier was to prevent Palestinian suicide bombers from reaching Israel. Noting that the security fence did not follow the internationally recognized **Green Line** of Israel's pre-1967 border but instead penetrated significant portions of the West Bank to pick up Jewish settlements there, Palestinians condemned it as a "land grab." The International Court of Justice concluded that the wall violated customary international law and several conventions on human rights to which Israel is a signatory.

Work on the barrier continued after Sharon lapsed into a coma and was succeeded as prime minister by a Likud Party colleague, Ehud Olmert, in 2006. Olmert vowed to continue Sharon's consolidation of Israeli territory, holding on to three major blocs of Jewish settlements in the West Bank and to "united Jerusalem." On the Palestinian side, a political earthquake occurred that year with the surprising majority vote for parliamentary candidates from the Islamist group Hamas, giving Hamas control of the Palestinian Authority. Palestinians had grown disillusioned with the showy wealth 234
of many elected Palestinian Authority officials and by the PA's failure to deliver an independent country or any other tangible benefits. Hamas, meanwhile, won hearts and minds through its network of social services, often providing what the Palestinian Authority failed to, including education, medicine, and food rations.

Hamas's stunning victory in the parliamentary elections added another layer of uncertainty and discord to the region's political

PROBLEM LANDSCAPE

Land for Peace *continued*

landscape. In the subsequent power-sharing agreement, Palestinian president Mahmoud Abbas of the PLO retained his post, but a Hamas leader, Ismail Haniyeh, took office as the Palestinian prime minister presiding over the Hamas-dominated parliament. Because the Hamas charter does not recognize Israel and in fact calls for its destruction, Israel and the Quartet suspended financial aid and discontinued dialogue with the Palestinian Authority. These sanctions would be lifted on three conditions: the Hamas-led PA must recognize Israel's right to exist, renounce violence, and abide by previous agreements reached between Israel and the Palestinians.

As the Palestinian economy deteriorated even further, clashes erupted between Hamas and PLO factions led by Abbas's Fatah Party. Hamas emerged victorious in fighting to control the Gaza Strip in 2007. Expelled from the territory, Abbas withdrew his recognition of Haniyeh as prime minister, and the Palestinian Authority government was dissolved. Abbas's Fatah Party consolidated its hold in the West Bank while the international community including numerous Arab countries, convened in Annapolis, Maryland, in an effort to resurrect the peace process.

would become a country whose population was only about 60 percent Jewish. With an Arab minority that has a far higher birth rate than that of the Jews, Jews could eventually become the minority population in their own country. In 2006, the Israeli advocacy group Peace Now leaked official Israeli government documents showing that 39 percent of the land on which Jewish settlements in the West Bank were built is privately owned by Palestinians. There is also international pressure against the settlements because they violate Geneva Conventions on activities allowed in territories captured in war. The International Court of Justice in the Hague has declared them illegal. **United Nations Resolutions 242 and 338** (passed after the wars of 1967 and 1973, respectively) called on Israel to withdraw completely from the Occupied Territories.

The settlements and other aspects of Israeli occupation, including the periodic closing off of the territories that prevents many Palestinians from reaching their jobs inside Israel proper, have fostered widespread and long-lasting Palestinian resistance against Israel. In 1987, Palestinians instigated a popular uprising in the Occupied Territories known as the ***Intifada*** (Arabic for "shaking"). Initially, they relied on rocks and bottles to engage Israeli troops, who answered with rubber and metal bullets. International media coverage of a Palestinian "David" fighting an Israeli "Goliath" did much to damage Israel's image abroad, and many Israelis came to question the justice and importance to security of Israel's continued occupation. Popular opinion and, by 1992, a new Israeli government led by the liberal **Labor Party** began to consider the previously unthinkable: giving the Palestinians at least some control over land and internal affairs in the Occupied Territories (see Problem Landscape, pages 245–247).

A Geographic Sketch of Israel

Israel is much more than a country in a perpetual state of conflict. It is a physically and culturally diverse and vibrant nation with a promising future, especially if relations with its Arab neighbors improve.

One of Israel's greatest achievements since independence has been the expansion and intensification of agriculture within its small territory; it is almost self-sufficient in agriculture, excluding grains. Production concentrates on citrus fruits (including the famous Jaffa oranges), which provide export revenue, and on dairy, beef, and poultry products, as well as flowers, vegetables, and animal feeds. Israel's intensive, mechanized agriculture resembles the agriculture of densely populated areas in western Europe. Collectivized settlements called ***kibbutzim*** (sing., *kibbutz*) are a distinctively Israeli feature on the agricultural landscape. Many of these lie near the frontiers and have defensive as well as agricultural and industrial functions. They have also been important in the development of Israeli identity, instilling a sense of work for the common good and reminding Israelis that their ancestors were often deprived of the right to
own and till the land. Far more numerous and important 222
in Israel's agricultural economy are other types of villages, including the small farmers' cooperatives called ***moshavim*** (sing., *moshav*) and villages of private farmers.

Agricultural development is concentrated in the northern half of the country, with its heavier rainfall (concentrated during the winter in this Mediterranean climate; see Figure 6.5, page 214) and more ample supplies of surface and underground water for irrigation. Annual precipitation averages 20 inches (50 cm) or more in most sections north of the city of Tel Aviv and in the interior hills where Jerusalem lies. Due to the rain shadow effect, very little rain falls in the deep rift valley of the Jordan River. From Tel Aviv southward to Beersheba, annual rainfall decreases to about 8 inches (20 cm), and in the central and southern Negev Desert, it drops to 4 inches (10 cm) or less. To expand agriculture along the coastal plain and in the northern Negev, Israel transfers large quantities of water from the north by the aqueduct and pipeline network of the National Water Carrier (see Figure 6.14, page 244). The largest source for this network is Lake Kinneret (Lake Tiberias, or Sea of Galilee), which is fed by the upper Jordan River. Since occupying the West Bank in 1967, Israel has intensively developed irrigated agriculture along the lower Jordan River and in scattered areas of the drier West Bank hills. Israel would lose most of this production to any future Palestinian state.

Jerusalem (population 700,000, excluding East Jerusalem, which international law regards as occupied territory) is the country's capital, but few countries recognize it as such because of the political sensibilities involved. Tel Aviv–Jaffa (population 3.2 million), on the Mediterranean coast, is Israel's greatest industrial center. The leading seaport is Haifa in northern Israel. The ancient town of Beersheba is the administrative and industrial center of the Negev Desert region. To the south lies Israel's small Red Sea port and resort town of Eilat. Thanks to the successful peace accords with Egypt and Jordan, open doors now link Eilat with the neighboring resort towns of Taba in the Egyptian Sinai and Aqaba in Jordan. Whenever conflict subsides, tourism booms, and developers are building and promoting a three-nation "Red Sea Riviera" to satisfy mainly European sun worshipers. Repeated episodes of conflict since 2000 have taken a significant toll on Israel's economy, which relies heavily on the exports of high-technology equipment, cut diamonds, and agricultural products and on assistance from the United States.

Jordan: Between Iraq and Israel

The economic and political situation of Jordan (•Figure 6.1.6) has always been precarious. Throughout his long reign, ending with his death in 1999, the country's King Hussein withstood various crises. Many of these related to the large numbers of Palestinian refugees in Jordan. Palestinians still make up about 52 percent of Jordan's population of 5.7 million, but most are refugees without citizenship. After the 1967 war, Jordan was the chief base for military operations of the PLO against Israel, but in 1970 and 1971, the king expelled the armed Palestinian forces, which relocated to Lebanon. Jordan's monarchy has always walked a tightrope between its pro-Western stance and the need to satisfy its anti-Israeli and anti-American inhabitants and neighbors. Its support of the U.S. war in Iraq put Jordan on al-Qa'ida's target list; in 2005, the organization carried out merciless bombings (including one on a wedding party) at three Western-owned hotels in Amman.

• **Figure 6.1.6** Principal features of Jordan, Syria, and Lebanon.

Jordan is a land of desert and semidesert. The main agricultural areas and settlement centers, including the largest city and capital, Amman (population 1.3 million), are in the northwest. Prior to the 1967 war with Israel, northwestern Jordan included two upland areas—the West Bank on the west (captured during the war) and the East Bank, or Transjordanian Plateau, on the east—separated by the deep rift valley occupied by the Jordan River, Lake Kinneret, and the Dead Sea (see Figure 6.1.4). Most of the valley bottom lies in deep rain shadow and receives less than 4 inches (10 cm) of precipitation a year. The bordering uplands, rising to more than 2,000 feet (610 m) above sea level and lying in the path of moisture-bearing winds from the Mediterranean, receive precipitation averaging 20 inches (50 cm) or more annually. But overall, rainfall is so scarce that only small areas can be cultivated without irrigation, and the water available for irrigation is limited. Only about 2 percent of Jordan is cultivated, but larger areas are devoted to the grazing of sheep and goats. Irrigated vegetables, olives, citrus fruits, and bananas provide substantial exports, but some foods must be imported.

Jordan's natural resources are limited, and the country has little manufacturing, though it does export textiles. Rock phosphate and potash (both used for fertilizers) extracted from the Dead Sea are shipped out of Jordan's small port of Aqaba on the Gulf of Aqaba (known to Israelis as the Gulf of Eilat). Tourism is a growing industry, with the Nabatean rose-red city of Petra the chief attraction (see Figure 6.8, page 215). Al-Qa'ida probably sought to destabilize Jordan's regime by attacking hotels frequented by foreign tourists.

Lebanon and Syria

Israel's immediate neighbors on the north are Lebanon and Syria (see Figure 6.1.6). Their physical and climatic features are similar to those of Israel and Jordan. As in Israel, narrow coastal plains backed by highlands front the sea. The highlands of Lebanon and Syria are loftier than those of Israel and Jordan. Mount Hermon and the Lebanon and Anti-Lebanon Ranges rise above 10,000 feet (c. 3,000 m). Cyclonic precipitation brought by air masses off the Mediterranean is supplemented by orographic precipitation as rain and snow fall along the mountain chains paralleling the coast.

Lebanon and Syria have agricultural patterns suited to the Mediterranean climatic pattern of winter rain and summer drought. The coastal plains and seaward-facing mountain slopes are settled by villagers growing Mediterranean crops such as winter grains, vegetables, and drought-resistant trees and vines. Lebanon's Bekaa Valley, drained in the south by the Litani River, lies between the Lebanon and Anti-Lebanon Ranges and is a northward continuation of the Jordan rift valley and the African Great Rift Valley.
459, 466, 482 Baalbek, its chief city and the site of an important ancient Roman temple, became a Syrian military stronghold in 1976 and a center of pro-Iranian Hizbullah activities against Israel after 1982. The political and military volatility of the Bekaa Valley was conducive to the emergence of a flourishing drug industry, producing both hashish (from marijuana plants) and opium (from poppies), from the mid-1970s until the 1990s. Marijuana cultivation made a comeback in the early 2000s as alternative crops proved to bring farmers far
548 less income.

Syria is more industrialized than Lebanon, but neither country compares with Israel industrially. Textile milling, agricultural processing, and other light industries predominate. Both countries are deficient in minerals, although Syria has an oilfield in the east. Its reserves are expected to be exhausted by 2010.

There are three large, historically important cities in Syria and Lebanon. Damascus (population 2.6 million), Syria's capital, is in a large irrigated district at the foot of the Anti-Lebanon Range. Its Umayyad Mosque, originally a Byzantine cathedral, is one of the great monuments of the Middle East. The main city of northern Syria, Aleppo, is an ancient caravan center located in the Syrian Saddle between the Euphrates Valley and the Mediterranean. The old Phoenician city of Beirut (population 2 million), on the Mediterranean, is Lebanon's capital, largest city, and main seaport. Equipped with modern harbor facilities and an important international airport, Beirut before 1975 was one of the busiest centers of transportation, commerce, finance, and tourism in the Middle East.

For its prosperity and its natural beauty, Lebanon was once known as the **"Switzerland of the Middle East."** Then a precarious political balance broke down, and this Connecticut-size country was plunged into civil war in 1975 and 1976. The origins of that war and the country's suffering ever since are rooted in Lebanon's mixture of strongly localized religious communities, an ethnogeographic recipe for tragedy.

What the French carved out of greater Syria and put together as Lebanon in 1920 was a collection of small geographic areas, each dominated by an ethnic or religious group. Seventeen separate groups, known as *confessions*, are recognized officially (•Figure 6.1.7). Major confessions include the Maronite Christians, who predominate in the Lebanon Range north of Beirut; Sunni Muslims close to the northern border with Syria; Shiite Muslims in the Bekaa Valley and along Lebanon's southern border with Israel; and Druze Muslims in the Shouf Mountains, part of the Lebanon Range south of Beirut. Eastern Orthodox, Greek Catholic, and nearly a dozen other Christian sects are also present in Lebanon. Beirut itself is divided between the predominantly Muslim west and Christian east.

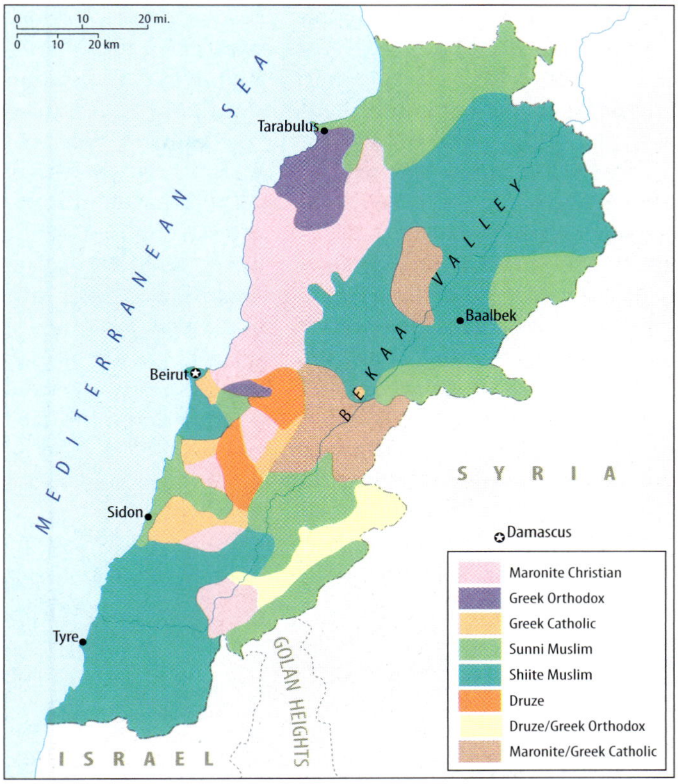

• **Figure 6.1.7** Generalized map of the distribution of religious groups in Lebanon.

A 1932 census revealed a majority of Christians in Lebanon, and the country's 1946 constitution distributed power according to those census figures. Maronite Christians held a majority of seats in parliament, followed by Sunni Muslims and Shiite Muslims. Top leadership positions were allocated by a formula still followed today, with the president a Maronite Christian, the prime minister a Sunni Muslim, and the speaker of parliament a Shiite Muslim.

The central government did not try to unify the disparate geographic and ethnic enclaves of this mountainous land. Muslim populations grew and eventually outnumbered the Christians, who were unwilling to relinquish their political
and economic privileges. Palestinians became part of the 248
volatile mix. Pushed from Jordan in 1971, PLO guerrillas established a virtual "state within a state" in the southern slums of Beirut and in Shiite-controlled areas of southern Lebanon, from which they launched attacks on Israel.

The **Lebanese Civil War** started in 1975 as the have-not Sunnis, soon joined by Shiites and Palestinians, took on the more prosperous Maronite Christians. Syria and Israel supported opposing sides. By the end of the war the following year, as many as 100,000 people, most of them civilians, had been killed (•Figure 6.1.8). The war accomplished little. Lebanon effectively split into a number of geographically distinct microstates or fiefdoms—a politically unstable situation that set the stage for the next war in 1982.

The **1982 Lebanese war** was one facet of a larger war in Lebanon involving both civil strife and international intervention. Lebanon had become the main stronghold of the Palestine Liberation Organization, which used the country as a base for guerrilla operations against Israel. In June 1982, Israel invaded Lebanon with the avowed intention of smashing the PLO and establishing security for the northern Israeli frontier. Syria resisted the Israeli advance and gave support to the PLO. Israeli troops routed and inflicted heavy casualties on both the Syrians and the PLO. The Israeli push continued northward to the suburbs and vicinity of Beirut, culminating in a ferocious aerial and artillery bombardment of the city—the **Battle of Beirut**—that ended only when U.S. President Ronald Reagan exerted pressure on Israel to declare a cease-fire. Under supervision by U.S. and other international peacekeeping troops, the PLO withdrew from Beirut to Tunisia. After the multinational troops withdrew, Muslim assailants assassinated Lebanon's Christian president, Bashir Gemayal. Angry Christian forces then massacred hundreds of Palestinians and other civilians in the Sabra and Shatila refugee camps south of Beirut. U.S. and other multinational forces were again deployed to Beirut, and American ships used firepower to support the new Christian president in solidifying his authority.

• **Figure 6.1.8** Much of the fighting in the Lebanese civil war of 1975–1976 took place on streets like this, along the line separating Muslim west from Christian east Beirut. This portion of what had been a thriving, prosperous city became a desolate moonscape. A gunfight was under way here as the photographer took this picture in 1976; a taxi driver dumped him here and sped off.

Lebanese Shiite Muslim guerrillas, seeking to avenge U.S. attacks on Lebanon and its support of Israel, soon bombed American and French targets in Beirut, killing over 200 U.S. Marines in a barracks near Beirut's airport in 1983. The multinational forces withdrew, but Syrian troops remained in Lebanon and allowed Iranian forces and Iran's Shiite Muslim allies in Lebanon to establish a stronghold in the Bekaa Valley town of Baalbek. Subjected to persistent guerrilla attacks, Israeli troops withdrew from all of Lebanon except a self-declared "security zone" in the south, inhabited by both hostile Shiite Muslims and Israel's de facto Christian Lebanese allies. Israel exerted control over this zone and engaged in regular skirmishes with Hizbullah, a 234
Shiite Muslim organization sympathetic to Iran and hostile to the peace process between Israel and its Arab neighbors. With growing numbers of casualties among Israeli forces in the security zone, the occupation became increasing unpopular in Israel, and troops were finally withdrawn in 2000.

Buoyed by a peace pact signed between the warring factions in 1989, Lebanon embarked on a steady path of reconciliation and reconstruction. Beirut rose from the ashes in an ambitious, expensive reconstruction program that was the brainchild of its wealthy Sunni prime minister, Rafiq al-Hariri. Political power was redistributed to reflect the new demographic composition in which Muslims outnumber Christians, although a new, official census was such a sensitive issue that none was conducted. Once dominated by Christians, the Lebanese army enrolled soldiers from all of Lebanon's ethnic confessions. Social and education programs got under way to mix children from different ethnic groups, with the hope they would learn tolerance. Remarkably, however, Lebanon's textbooks still avoid any mention of the ethnic divisions and civil strife of the country, as if not discussing these rifts might help them go away.

Its latest war over, Lebanon was still not in control of its own destiny, however. The Syrians had never accepted the 1920 French decision to cleave the separate state of Lebanon from greater Syria, and since the civil war of the 1970s had stationed tens of thousands of soldiers in Lebanon. In 2005, a powerful suicide bomb killed former Lebanese Prime Minister al-Hariri, who had been outspokenly critical of Syria's presence in and influence over Lebanon. Many Lebanese concluded that Syria was behind the assassination, and hundreds of thousands of them took to the streets in the so-called **Cedar Revolution.** Their loud calls for Syria's withdrawal were answered within weeks. But the call for Syria's withdrawal was not unanimous. Supporters of Lebanon's Shiite Hizbullah—traditionally aligned with both Syria and Iran—massed in the streets to protest Syria's pullout. With Syria gone, Iran's direct influence on Hizbullah grew.

The stage was set for another conflagration in Lebanon. In July 2006, Hizbullah forces kidnapped two Israeli soldiers. Israel's usual response to this situation would have been to arrange a prisoner swap, but this time, Israeli Prime Minister Ehud Olmert decided to crush Hizbullah in Lebanon. The Israeli Defense Force unleashed a ferocious 34-day assault on Hizbullah and other Shiite targets, attacking more than 400 cities, towns, and villages across Lebanon. Israeli warplanes concentrated on targets in the southern part of the country but also hit the mainly Shiite southern suburbs of Beirut hard. Extensive damage was done to the country's bridges and other infrastructure that had been rebuilt following years of warfare. Israel's superior military forces proved incapable of halting a steady barrage of Hizbullah rocket fire into northern Israel, where Israeli civilian casualties mounted. A rising drumbeat of Israeli public opinion against the fighting led to a cease-fire between Israel and Hizbullah.

Israel had failed to defeat or even substantially weaken Hizbullah, which declared itself the victor of the **2006 Israeli-Hizbullah war.** Hizbullah sought to build on its momentum, moving quickly through Lebanon's devastated Shiite communities to help rebuild services. Rallying behind Hizbullah, Lebanon's long-marginalized Shiites sensed an opportunity to increase their political presence, insisting that they now outnumbered both Sunnis and Christians and beginning to challenge the prominence of the country's Sunni Muslim and Christian establishments. A divide reminiscent of Lebanon's civil war days emerged, with the Shiite Hizbullah, backed by its proxies Iran and Syria (and joined by one Christian faction in Lebanon) squaring off against Sunni, Druze, and other Christian groups. Supporters of each side gathered in huge demonstrations in Beirut. There were fears that an all-out civil war might reignite.

There were some silver linings in all this regional tension, including the prospect of a long-delayed peace agreement between Syria and Israel. The countries revealed that they had been conducting "back door" negotiations since 2004. Their formula for peace would likely include a Syrian cessation of support for Hizbullah in Lebanon in exchange for an Israeli withdrawal from the Golan Heights, occupied since 1967. A buffer zone designated as a peace park to be shared by Israelis and Syrians would run down the eastern shore of Lake Kinneret.

Syria, like Lebanon, is a country beset by religious factionalism. In Syria, Muslims outnumber Christians nine to one, and the major rifts are between Sunni and Shiite Muslims. In 1970, a coup overthrew the ruling Sunni Muslim establishment and brought to power Hafez al-Assad, a member of the **Alawite** sect, an offshoot of Shiite Islam. Under his authoritarian rule as president, dominance of the government quickly passed to the Shiite Alawites, who make up only 12 to 15 percent of the country's population. The police and army crushed opposition in other communities, notably Sunni Islamists, who reject Alawites as heretics.

For three decades, until his death in 2000, President Assad was a skilled politician who played Middle Eastern affairs cautiously. His regime survived the economic and political setbacks accompanying the downfall of its benefactor, the Soviet Union. During Iran-Iraq War of the 1980s, Alawite-controlled Syria supported its kindred 264 Shiite Iran, and Syria promoted the rise of the Shiite Hizbullah organization in Lebanon. These alignments earned Syria pariah status in neighboring Arab countries and especially in the United States, which views Syria as a proxy for Iran's interests in the region. Unlike the PLO leaders, President Assad did not pursue formal reconciliation with Israel. His son and successor, Bashar al-Assad, has continued the country's generally unaccommodating approach to Israel and the West. There is always a danger that Israeli or U.S. confrontations with Shiite interests in Middle East will spark a wider conflict, as the governments in Syria and Iran and Shiite factions in Lebanon could all be drawn in.

6.1.2 Egypt: The Gift of the Nile

Southwest of Israel lies the Arab Republic of Egypt, the most populous Arab country (73 million) (•Figure 6.1.9). This ancient land, strategically situated at Africa's northeastern

• **Figure 6.1.9** Principal features of Egypt.

corner between the Mediterranean and Red Seas, is utterly dependent on a single river. The Nile has created the floodplain and delta on which more than 95 percent of all Egyptians live (see Figure 6.24, page 233). The Greek historian and geographer Herodotus aptly called Egypt "**the gift of the Nile**." The river's bountiful waters and fertile silt helped Egypt become a culture hearth that produced remarkable achievements in technology and cultural expression over a period of almost 3,000 years until about 500 B.C.E., when a succession of foreign empires came to rule the land.

The Nile Valley may appropriately be described as a "river oasis," for stark, almost waterless desert borders this lush ribbon. Only 3 percent of Egypt is cultivated, and nearly all this land lies along and is watered by the great river. The conversion of the original papyrus marshes and other wetlands along the Nile to the thickly settled, irrigated landscape of today is a process that has been unfolding for more than 50 centuries. It is difficult to imagine that in the time of the pharaohs, crocodiles and hippopotamuses swam in the Nile and game animals typical of East Africa roamed the nearby plateaus.

Ancient Egyptian civilization was based on a system of **basin irrigation**, in which the people captured and stored in built-up embankments the floodwaters that spread over the Nile floodplain each September. Egyptian farmers allowed the water to stand for several weeks in the basins, where it supersaturated the soil and left a beneficial deposit of fertile volcanic silt washed down by the Blue Nile from Ethiopia. They drained the excess water back into the Nile and planted their winter crops in the muddy fields. They were able to harvest one crop a year with this method. In limited areas, they could use **perennial irrigation** devices to obtain a second or even a third crop.

Achieving perennial irrigation on a vast scale—and thus boosting crop production enormously—required the construction of barrages and dams, which were innovations of the 19th and 20th centuries in this part of the world. Late in the 19th century, French and British colonial occupiers of Egypt, anxious to raise Egypt's exports of cotton (a summer crop demanding perennial irrigation), began Egypt's conversion from basin to perennial irrigation by constructing a number of **barrages**, or low barriers designed to raise the level of the river high enough that the water flows by gravity into irrigation canals. Barrages are not designed to store large amounts of water, a function now performed by two dams in Upper Egypt (the old Aswan Dam, completed in 1902, and the huge Aswan High Dam, completed in 1970; see Figure 6.24) and several others along the Nile and its tributaries in Sudan and Uganda.

The enormous Aswan High Dam (•Figure 6.1.10) is located on a stretch of the Nile known as the First Cataract (the Nile's many **cataracts** are areas where the valley narrows and rapids form in the river). Its reservoir, Lake Nasser, stretches for more than 300 miles (480 km) and reaches into northern Sudan. Like all giant hydrological projects, the Aswan High Dam has generated benefits and liabilities, and its construction was controversial. On the plus side, by storing water in years of high flood for use in years of low volume, the High Dam provides a perennial water supply to the Nile Valley's 11,000 square miles (28,500 sq km) of irrigated fields, thus extending the cultivation period. The dam also made possible the "reclamation" of

Compliments of NASA/Earth Sciences and Image Analysis Laboratory at Johnson Space Center.

• **Figure 6.1.10** Egypt's Aswan High Dam, on the First Cataract of the Nile River.

previously uncultivated land on the desert fringe adjoining the western Nile Delta. Navigation of the Nile downstream from (north of) Aswan is much easier, and the country's industries rely heavily on the hydroelectricity produced at the dam. Egypt weathered a two-year drought in the late 1980s thanks to the excess waters stored in Lake Nasser. The dam also eliminates the prospect of catastrophic flooding, which sometimes occurred before it was built.

The Aswan High Dam has also had drawbacks. The dam has caused the water table to rise, making it harder for irrigated soil to drain properly. When farmers use too much water, standing water evaporates and leaves a deposit of mineral salts—a problem known as **salinization**—that causes the once-fertile soil to lose its productivity. In addition, perennially available canal waters are ideal breeding grounds for the snail that hosts the parasite that causes schistosomiasis, or bilharziasis, a debilitating disease affecting a large proportion of Egypt's rural population. Mediterranean sardine populations, now deprived of the rich silt that nurtured their feeding ground, have plummeted off the Nile Delta, and the sardine industry has faltered. Without the free fertilizer the silt offered, Egypt's bill for artificial fertilizers has increased. Generally, however, Egyptians are very proud of the High Dam. It has increased domestic food and crop export production dramatically.

Egypt's most serious problem is that its population has grown even more dramatically. The country was once a great exporter of crops such as corn, wheat, barley, rice, millet, fruit, vegetables, and sugarcane, but its population has grown so explosively that food now comprises over 30 percent of Egypt's import expenditures. Cotton has long been the main commercial export crop for the hard currency Egypt desperately needs. Rural population densities are already among the highest in the world, and the crowding is growing worse despite a steady migration to Egypt's primate
111 city, Cairo (population 15.9 million; •Figure 6.1.11)—the largest metropolis in Africa and the Middle East—and other cities. So far, the government's efforts to lure urban dwellers and factories out of the Nile Valley to desert "satellite cities" have borne little fruit. Egypt is much more successful in exporting its brains and brawn abroad: Egyptians are prominent among the teachers, doctors, lawyers, engineers, and laborers working in the wealthy Persian-Arabian Gulf countries, and the remittances they send home pump much-needed cash into Egypt's economy.

Joe Hobbs

• **Figure 6.1.11** Cairo, on the banks of the Nile River.

Egypt's industries are concentrated in Cairo and in the Mediterranean port of Alexandria. They include cotton textiles, food processing, clothing manufacture, chemical fertilizers, and cement. Automobiles and other consumer durables are assembled primarily from imported components. The country is self-sufficient in some vital minerals, including oil, and exports some oil. But Egypt lacks the mineral wealth to support extensive industrialization, and for its development, the country depends heavily on foreign loans and grants. Income from tourism, particularly to the outstanding temples and tombs of ancient Egypt and to the world-famous scuba-diving sites on the Red Sea and Gulf of Aqaba, helps redress the unfavorable trade balance, as do transit fees from ships using the Suez Canal. However, tourism is a very vulnerable resource, as the number of visitors plummets each time violence—including al-Qa'ida attacks on resorts in the Egypt's Sinai Peninsula—rocks the 235
Middle East.

Egypt is the giant of the Arab world, its largest country in terms of population, and its most influential in regional and international politics. The Western powers recognize its key role in mediation between Palestinians and Israelis. The West is anxious to see that Egypt does not become poorer than it already is and that the already huge gap between the rich and poor does not widen, perhaps throwing its population into the embrace of militant Islam. And Egypt's strategic location astride the Suez Canal is also a major concern for the West.

The United States is Egypt's chief non-Arab ally, lavishing huge sums of civilian and military aid in support of the autocratic regime led by President Hosni Mubarak. To quell real and perceived threats by Islamists, Mubarak has maintained a state of emergency for more than two decades and has effectively stifled most opposition. The ranks of outlawed and grassroots Islamist movements have grown at the same time. Egyptian society has generally reembraced traditional Islamic values in dress, education, relations between the sexes, and other social and cultural practices. A radical fringe has periodically attacked the interests of Egypt's 6 million Coptic Christian minority and foreign targets such as the tourism industry.

6.1.3 Sudan: Bridge between the Middle East and Africa

Egypt is bordered on the south by Africa's largest country, the vast, tropical, and sparsely populated republic of Sudan (population 38.6 million) (•Figure 6.1.12). Formerly controlled by Britain and Egypt, Sudan gained its independence

Figure 6.1.12 Principal features of Sudan.

in 1956. This southern neighbor is vitally important to Egypt, for it is from Sudan that the Nile River brings the water that sustains the Egyptian people. The Nile's major tributaries join in Sudan, and storage reservoirs exist there on the Blue Nile, the White Nile, and the Atbara River, all of which benefit Egypt as well as Sudan itself (see Figure 6.24, page 233).

The Nile's main branches—the White Nile and Blue Nile—originate at the outlet of Lake Victoria in Uganda and of Lake Tana in Ethiopia, respectively, and flow into
233 Sudan. In southern Sudan, the river passes through a vast wetland called the Sudd, where much water is lost to evaporation. The Jonglei Canal, an artificial channel to straighten the river's course and increase the volume and velocity of its flow through the Sudd, was under construction for a long time. Civil war in Sudan halted its progress in 1983. Predicting the death of the Sudd wetlands if the canal is completed, many environmentalists would be pleased if work did not resume.

Sudan is overwhelmingly rural. The largest urban district, formed by the capital, Khartoum (population 8.3 million, including the adjoining cities of Omdurman and Khartoum North), is located strategically at the junction of the Blue and White Niles. Transportation in Sudan as a whole is poorly developed. Several cataracts prohibit navigation on the Nile north of Khartoum, but the White Nile is continuously navigable at all seasons from Khartoum to Juba in the deep south. Most of the country lacks railways or good highways. The main route for rail freight connects Khartoum with Port Sudan, a modern, well-equipped port that handles most of Sudan's seaborne trade.

About 7 percent of Sudan is cultivated, but agricultural productivity is low. The largest block of irrigated land is found in the Gezira region between the Blue and White Niles. The country's main cash crop is irrigated cotton. Many Sudanese support themselves by raising cattle, camels, sheep, and goats. The country's industries are meager. The most important resource is oil, which Sudan began exporting in modest quantities in 1999. Sudan's reserves are not large, but a sudden inflow of oil money ever since then has made Sudan one of the fastest-growing economies in the world. The government has used substantial resources to improve the country's infrastructure, but deep poverty—and a growing gap between the newly wealthy minority and the vast poor majority—remains widespread.

Most of Sudan is an immense plain, broken frequently by hills and mountains. Climatically and culturally, the area is transitional between the Middle East and sub-Saharan Africa. A contrast exists between Sudan's arid Saharan north, peopled mainly by Arabic-speaking Muslims, and the seasonally rainy equatorial south, with its grassy savannas, papyrus wetlands, and non–Arabic-speaking peoples of many tribal and ethnic affiliations who practice both Christian and animistic faiths. About three-fourths of Sudan's total
population is Muslim. 222

Much political friction exists between north and south. A civil war raged between the two sections from 1983 to 2003, with southern factions of the **Sudanese People's Liberation Army (SPLA)** engaging government army troops from the north. As many as 2 million people, mostly civilians, died, and another 4.5 million were driven from their homes. Among the reasons for this conflict were historical antagonisms and economic disparities between the more developed north and the less developed and poorer south, along with efforts by the Arab-dominated government to impose Islamic law (*sharia*) and the Arabic language on the south and to exploit the south economically. The conflict helped keep Sudan mired in poverty. SPLA rebels repeatedly attacked the country's single export pipeline leading from the oil-producing area around Bentui to the coast at Port Sudan (see Figure 6.C, page 230). Bentui is close to the southern SPLA stronghold, and the county's largest oil reserves are within SPLA territory.

Negotiations between the government and SPLA rebels led in 2005 to a peace agreement and the tentative end to Sudan's long civil war. The government in Khartoum now recognizes the autonomy (but not independence) of southern Sudan, also known as New Sudan, and allows oil revenues to benefit both southerners and northerners. They parties agreed to hold a referendum in 2010 in which southerners would decide if they want independence separate from the north. Meanwhile, some of the 3 to 5 million refugees displaced from southern Sudan by two decades of civil conflict

began going home. Most are internally displaced persons
58 (IDPs, meaning they sought refuge within their own borders), of which Sudan has more than any other country in the world. They are rebuilding a homeland in which virtually all infrastructure was destroyed.

Drought has regularly visited southern Sudan and adjacent Ethiopia during recent years, and Sudan has periodically hovered near catastrophic famine. Huge tides of refugees have flowed back and forth between Sudan and Ethiopia. International relief agencies have been allowed to provide food and medical aid in only a small portion of the stricken zone. Sudan's military regime has meanwhile been accused of gross human rights violations, including ethnic cleansing of the Nuba, Nuer, Dinka, and other tribespeople
147 in the south. Human rights groups have accused the government of promoting the abduction and enslavement of non-Muslim (mostly Dinka) southerners, who usually end up serving Arab masters in the north.

A new humanitarian crisis emerged in 2004 when government-backed militias called **Janjawiid** and Sudanese army regulars began to carry out ethic cleansing in Sudan's westernmost province, Darfur. The victims (as many as 400,000 dead by 2007) were blacks of the Zaghawa, Massaliet, and Fur tribes. They were targeted because of competition with Arabs over access to land and water and initially because of government fears that these groups might link up with antigovernment rebels in the south. Early on, the crisis in Darfur went relatively unnoticed by the international community, but it gradually acquired visibility as the situation worsened. The United States and other countries accused Sudan's government of genocide. African Union (AU) forces were called in as peacekeepers to quell the violence, but
480 their effectiveness proved limited.

Sudan's foreign relations have changed dramatically twice since 1989, when an Islamic military regime overthrew an elected government. The new rulers allied themselves with militant Iran and also briefly provided refuge for Osama bin Laden. In 1993, after obtaining information linking Sudanese officials with the first bombing of New York City's World Trade Center, the United States branded Sudan a terrorist nation and halted economic aid. Following the analysis of intelligence information about bombings of U.S. embassies in Kenya and Tanzania, in August 1998 American cruise missiles obliterated a pharmaceutical factory, alleged to be a chemical weapons factory, in Sudan's capital. After that, and especially after 9/11, Sudanese relations with the West improved considerably, and the United States came to regard Sudan as an ally in its war on terrorism.

6.1.4 Libya: Deserts, Oil, and a Reformed Survivor

Egypt's neighbor to the west, Libya—formerly an Italian colony—became an independent kingdom in 1951. In 1969, Libya became a republic after Colonel Muammar al-Qaddafi and other army officers led a coup against the monarchy. Some 97 percent of the country's people are Muslim Arabs and Berbers.

Libya has three major geographic divisions: Tripolitania in the northwest, Cyrenaica in the east, and the Fezzan in the southwest (•Figure 6.1.13). The principal cities are the capital, Tripoli (population 1.9 million), in Tripolitania and Benghazi in Cyrenaica.

Most of Libya lies in the Sahara and is too dry to support cultivation except in scattered oases watered by wells and springs. The country is sparsely inhabited; its 6.2 million people are heavily concentrated in coastal lowlands and low highlands along the Mediterranean, where limited winter rains and irrigation with groundwater allow small-scale cultivation of typical Mediterranean crops. Thanks to **center-pivot irrigation**—a method in which the irrigation equipment rotates around a pivot, creating a circular crop pattern—wheat is grown in southern Libya. Deep wells there tap nonreplenished **fossil waters** that accumulated in deep limestone aquifers in ancient times when rains fell abundantly in what is now the Sahara. In an ongoing project called the **Great Manmade River** (•Figure 6.1.13), these same fossil waters are diverted from wells 1,155 feet (347 m) deep around Tazirbu in east central Libya and Hasquina in west central Libya by pipelines 13 feet (4 m) in diameter stretching hundreds of miles to the Mediterranean (see Figure 6.14). The goal of this effort, which Colonel Qaddafi calls "the eighth wonder of the world" and critics abroad decry as "Libya's pipe dream," is agricultural self-sufficiency through the irrigation of fruits, vegetables, and wheat in northern Libya.

Crude oil, petroleum products, and natural gas represent nearly all of Libya's exports (primarily to western Europe) and have brought prosperity to the country since the 1960s. The government spends much of the oil revenue on public works, education, housing, aid to agriculture, and other projects for social and economic development. The country has also spent large sums on weapons.

Libya's revolutionary government was long involved in an exchange of hostile rhetoric and sometimes hostile action with the United States, its Western allies, and Israel. In 1986, U.S. warplanes bombed Tripoli and Benghazi in reprisal for alleged terrorist acts against American citizens instigated by Libya. After Western intelligence authorities linked two Libyans to the 1988 bombing of Pan American Airlines' Flight 103 over Lockerbie, Scotland, Libya was subjected to an international air embargo.

Apparently conceding that his country's isolation was devastating the economy, Colonel Qaddafi decided to make amends with the West. Libya admitted to its role in the Flight 103 bombing and agreed to pay billions of dollars in restitution to the victims' families. Libya also announced that it was giving up development of weapons of mass destruction, including nuclear weapons. That proclamation led to much praise from Western governments and to the progressive renewal of diplomatic and economic relations.

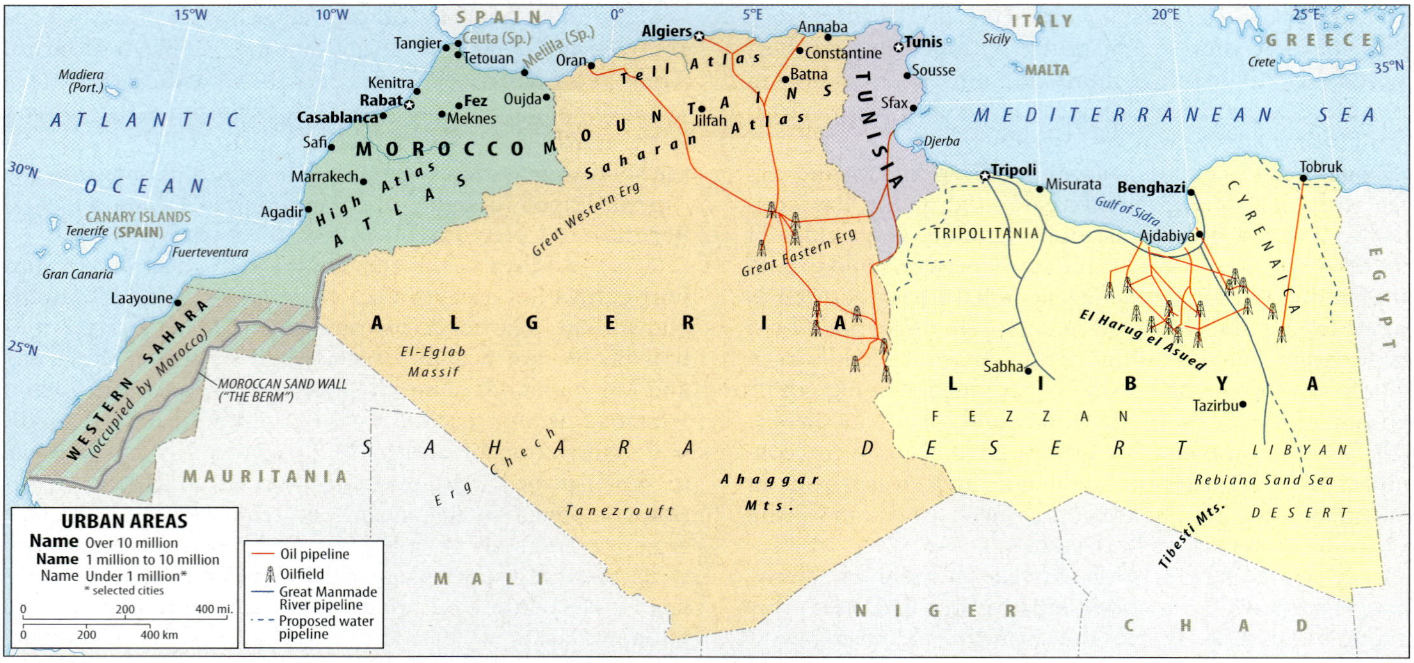

• **Figure 6.1.13** Principal features of Libya, the Maghreb countries, and Western Sahara.

• **Figure 6.1.14** Laying pipe for Libya's Great Manmade River.

The U.S. removed Libya from its list of "terrorist states" in 2006, and American oil companies are getting back to work in this oil-blessed country. Growing numbers of international tourists are visiting Libya's outstanding Roman sites, pristine beaches, and desert wilderness.

6.1.5 Northwestern Africa: The Maghreb

Arabs know the northwestern fringes of Africa as the Maghreb (meaning "western land"). Most of this area's people are Muslim Arabs. The ethnically distinct Berbers, who are most numerous in Morocco and Algeria, converted to Islam after the Arabs brought the new faith into the region in the 7th century. Many Berbers now speak Arabic, and most have adopted Arab customs. Majority Arabs discriminate against them in some countries. Berbers in Algeria, who make up 10 percent of the population, have demanded that the government recognize the Berber language and help the economically disadvantaged Berbers achieve better working and social conditions.

The Maghreb includes four main political units: Morocco, Algeria, Tunisia, and the disputed Western Sahara (see Figure 6.1.13. Morocco is geographically and culturally the closest to Europe; only the Strait of Gibraltar, a mere 11 miles (18 km) wide, separates it from Spain (•Figure 6.1.15). Islamic influences crossed the strait beginning early in the 8th century and prevailed in southern Spain until 1492, when Spain's monarchs ordered both Jews and the Muslim Moors out of the country. France became the domi-

Compliments of NASA/Visible Earth.

• **Figure 6.1.15** The Strait of Gibraltar (center) from space, with the southern portion of the Iberian Peninsula at the top of the image and extreme northwestern Africa at the bottom.

nant power in the Maghreb during the 19th and early 20th centuries. French investment developed mines, industries, irrigation works, power stations, railroads, highways, and port facilities, but French rule was very unpopular. In 1956, after long agitation by local nationalists, Tunisia and Morocco secured independence. In Algeria, independence in 1962 came only after a bitter civil war lasting eight years.

Most people in the Maghreb live in the coastal belt of Mediterranean climate (called "the Tell" in Algeria), with its cool-season rains and hot, dry summers. Here, in a landscape of hills, low mountains, small plains, and scrubby Mediterranean vegetation, people grow the familiar products associated with the Mediterranean climate (•Figure 6.1.16). The region's major cities are on the coasts: on the Mediterranean, Tunisia's capital, Tunis (population 2 million), and Algeria's capital, Algiers (population 5.9 million), and on the Atlantic, Morocco's fabled Casablanca (population 3.2 million), and the capital, Rabat (population 1.7 million).

Inland, the Atlas Mountains extend in almost continuous chains from southern Morocco to northwestern Tunisia, with the highest peaks in the High Atlas Mountains of Morocco. They block the path of moisture-bearing winds from the Atlantic and the Mediterranean, and some mountain areas receive 40 to 50 inches (c. 100 to 130 cm) of precipitation annually, sometimes more. The precipitation nourishes forests of cork oak or cedar in some places. In Algeria, the mountains form two east-west chains: the Tell Atlas nearer the coast and the Saharan Atlas farther south.

Nicholas Devore/Stone/Getty Images

• **Figure 6.1.16** The western Maghreb has breathtaking physical landscapes, from the Mediterranean regions in the north to the Atlas Mountains and Sahara in the south. Traditional village and urban architecture lends a unique character to the human environment. This is Vallée du Dades in Morocco, with the Atlas Mountains in the background.

All the countries reach the Sahara to the south. Some pastoral nomadic tribespeople, including the legendary Tuareg, still migrate through the Sahara with their camels,
sheep, and goats. Clusters of oases exist in a few places, 217
such as in the dramatic sandstone Ahaggar Mountains of southern Algeria, which rise high enough to catch moisture from the passing winds. A line of oases fed by springs, wells, mountain streams, and ***foggaras*** (tunnels from mountain water sources, known as ***qanats*** in Iran) lies along or near the southern base of the Atlas Mountains. Several large areas of sandy desert (***ergs***) and a barren gravel plain, the Tanezrouft, occupy portions of the Algerian Sahara.

The Maghreb economies benefit from a valuable mineral endowment. Algeria's oil and natural gas, located in the Sahara, are the region's greatest assets (see Figure 6.1.13). The oil flows through pipelines to shipping points on the Mediterranean in Algeria and Tunisia, and two pipelines carry Algerian natural gas to Europe. Petroleum products, crude oil, and natural gas make up almost all of Algeria's exports by value. Morocco's exports also include some minerals, notably phosphate, plus agricultural exports and clothing. Consumer industries, including plants that assemble foreign-made components, are the second-leading industries in the Maghreb countries.

Algeria's oil boom went bust in the 1980s, but population growth continued, fueling an economic crisis and shortages of food and consumer goods that contributed to widespread unrest in the 1990s. Algeria built up a staggering foreign debt that exacerbated the instability. Discord in Algeria reverberated through the entire Maghreb and across the Mediterranean to the region's former colonial power, France. During almost eight years after 1992, when Algeria's military government nullified the results of parliamentary elections that would have given a majority of

234 the seats to the Islamic Salvation Front (FIS), civil strife claimed an estimated 150,000 Algerian lives. Violent sympathizers of the FIS also carried their insurgency to France in the 1990s, bombing government and civilian targets to protest France's support of the military government in Algiers. The 2000s have seen Algeria's 34 million people reconciling with the West and enjoying the strong economic growth prompted by high energy prices.

Morocco and Tunisia viewed events in neighboring Algeria with anxiety. Citing the violence in Algeria, Tunisia banned Islamist political parties in 1994. The government of Tunisia (population 10 million) has promoted family planning and women's rights and is often cited as an "oasis" of openness and progressive change in the region—but it also closely monitors its citizens' e-mails and keeps a tight lid on Internet access. Tunisia also has the Maghreb's strongest economy, with active agricultural, mining, energy, tourism, and manufacturing sectors.

In Morocco, King Muhammad VI rules over a constitutional monarchy that promises political accommodation of Islamic groups, in contrast with Algeria's experience, but human rights monitors are increasingly alarmed by the young king's muzzling of opposition voices. Morocco's close alliance with the United States in the war on terrorism made it a target of an al-Qa'ida attack in 2003. Morocco is the poorest of the Maghreb countries. But there has been some progress in this nation of 31 million in recent years. The country has shown exceptional success in bringing down birth rates; whereas the average Moroccan woman had seven children in 1980, she now has two or three.

Some significant international problems exist in this region. Morocco would like Spain to return to Moroccan control the two Spanish enclaves of Ceuta and Melilla on Morocco's Mediterranean coast. Many African refugees find their way into these enclaves and eventually into Spain itself and onward into other destinations in Europe. About 30,000 Moroccans emigrate to Spain each year. The two countries have contesting claims to a number of small islands near the Moroccan coast.

Both Morocco and Algeria have long been at odds over the future of Spain's former dependency known as Western Sahara. This sparsely populated coastal desert (estimated population 500,000), which contains rich phosphate deposits, adjoins Morocco on the south and is claimed by that country. When Spain relinquished control in 1975, the territory was partitioned between Morocco and Mauritania, but in 1979, Mauritania withdrew its presence. Algeria objected to Moroccan control and, together with other outside parties, gave assistance to an armed resistance movement in Western Sahara against Morocco called the **Polisario Front.** In 1989, Morocco and the Polisario Front agreed to a referendum on independence, and a peace treaty was signed in 1991. To influence the outcome of the referendum in its favor, Morocco relocated 40,000 Moroccans into the region in 1991; they remain there today under Moroccan guard. The Moroccan government has subsequently used tax breaks and other subsidies to attract new Moroccan settlers there.

Such questionable actions led the United Nations to repeatedly suspend the referendum and finally call it off in 2000. Morocco continued to assert its claim to Western Sahara. It sold Western Sahara's offshore fishing rights to European countries. To stave off Algerian interests, Moroccan technicians also erected a 1,500-mile (2,400-km) sand barrier parallel to Western Sahara's border with Algeria, reinforcing it with land mines and 150,000 troops (see Figure 6.1.13). West of this berm, in Morocco-held territory, is the world's largest reserve of phosphate, used in fertilizers, toothpastes, and detergents. The United Nations wants to help to establish limited Western Saharan autonomy under Moroccan rule, but 75 countries recognize the Polisario-proclaimed Saharawi Arab Democratic Republic as a country with a government in exile. While the status of Western Sahara remains unresolved, more than 165,000 **Saharawis** (as the indigenous people of Western Sahara are known) live in refugee camps in the Algerian desert, as they have for decades. Despite their bleak conditions, the Saharawis have achieved a 95 percent literacy rate and have a representative government with a constitution that insists on religious tolerance and gender equality.

6.1.6 The Gulf Oil Region

As discussed in Chapter 6, the Persian Gulf (Arabian Gulf) region has by far the world's largest oil reserves. Foreign com- 231
panies, mainly British and American, originally developed the oil industry in this region. These firms made huge profits from their oil concessions, but with a few exceptions, after 1972 the companies saw their properties nationalized by the local governments. However, many outside companies still work closely with national oil companies in the Gulf countries, providing them with technical and managerial expertise, skilled workers, equipment, marketing, and varied services.

Gulf oil is shipped out by a combination of pipelines and ships. Most oil produced in the Gulf region is transported to market by oceangoing supertankers. These ships load oil at terminals spaced along the Gulf. Some crude oil moves by pipelines from oilfields in the Gulf countries to tanker terminals on the Mediterranean and Red Sea coasts (see Figure 6.C, page 230). A growing proportion of extracted crude oil is refined in the Gulf countries for local use and export. Petrochemical industries exist in most of the producing countries, with the largest development in Saudi Arabia. The industries use local natural gas as their primary source of energy. Huge gas reserves are associated with oil deposits, and other major reserves exist apart from the oilfields, in "nonassociated" deposits. Growing amounts of gas are exported in liquefied form; Iran and Qatar have the largest reserves of natural gas in the Middle East.

Saudi Arabia: The Oil Colossus

With an area of roughly 830,000 square miles (2.1 million sq km), Saudi Arabia occupies the greater part of the Arabian Peninsula, the homeland of Arab civilization and the faith of Islam (•Figure 6.1.17). The interior of the country is an arid plateau fronted on the west by mountains that rise steeply from a narrow coastal plain bordering the Red Sea. Most of the plateau lies at an elevation of 2,000 to 4,000 feet (c. 600 to 1,200 m), with the lowest elevations near the Gulf and in the eastern part of the gigantic sand sea called the Empty Quarter (the Rub' al-Khali desert). Mountains (Arabic: *jebels*) rise locally from the plateau, and the plateau surface is cut by many valleys (Arabic: *wadis*) that carry water whenever it happens to rain; Saudi Arabia has no permanent rivers. Most interior sections average less than 4 inches (10 cm) of precipitation annually, but years may pass at a given locale with no precipitation at all. Rainfall is greatest along the mountainous western margin of the peninsula, especially in the southwestern Asir region, near the border with Yemen, which benefits from summer monsoon rains. About 95 percent of Saudi Arabia's crop production depends on irrigation. Cultivated land makes up only 1 percent of the country's area, yet surprisingly, Saudi Arabia is self-sufficient in many agricultural staples. It is an exporter of wheat, irrigated mostly in the central part of the country by gigantic center-pivot sprinkler systems tapping fossil waters.

Saudi Arabia's oil reserves make up more than one-fourth of the world's total. Most of the country's oilfields, including the world's largest (the Ghawar field, with enough reserves to supply U.S. oil needs for 9 years at current consumption levels), are located in the eastern margins of the country, including its adjacent seabed. Some Saudi Arabian oil is processed in refineries within the country and in nearby Bahrain, but most of the production leaves Saudi Arabia in crude form on tankers, using loading terminals that jut into both the Gulf and the Red Sea. To meet anticipated future demand, huge refining and petrochemical complexes have been developed at Jubail on the Gulf and Yanbu on the Red Sea. These cities sprang up almost overnight in the 1980s. Saudi Arabia's best-known city is ancient Mecca, Islam's holiest city. The interior city of Riyadh 227 (population 4.4 million) is Saudi Arabia's capital. The Red Sea port of Jidda, some 40 miles (64 km) from Mecca, is the main point of entry for foreigners who make the pilgrimage to Mecca.

Saudi Arabia's population of 27 million is young and growing; half of all Saudis are younger than 21, and the annual growth rate is 2.7 percent. About one-fifth of the population consists of foreign workers. The country was long concerned that its native population was too small to run the country's economy and stave off cultural domination by foreign workers. The government promoted population

• **Figure 6.1.17** Principal features of the Arabian Peninsula.

growth and passed **Saudization** laws to put Saudis in jobs previously held by foreigners. Now the fear is that the indigenous Saudi population is growing much faster than job opportunities will, perhaps creating a young, unemployed, and disaffected population that might agitate for change. Since the height of the oil boom in 1981, the population has more than doubled, and per capita income, despite the recent surge in oil revenues, has fallen.

There are striking income discrepancies in this desert kingdom; although the sharing of oil wealth brought once undreamed-of prosperity to most of the population, the Shiite Muslim minority in the kingdom's Eastern Province remains poor (•Figure 6.1.18). The country's transition to a standard of living comparable to the more developed countries is not complete; the infant mortality rate, for example, remains high. The oil will of course run out one day, so the biggest challenge to Saudi Arabia's development is to diversify away from its near complete dependence on that commodity. The government is working to strengthen Saudi Arabia's global roles in finance, business services, trade, and manufacturing.

Despite the Western-style luxuries that oil wealth has brought to Saudi Arabia, the conservative Saudi family that rules this absolutist monarchy insists on maintaining strict local Muslim traditions. The legal and religious systems are based on interpretations of Islam by the Wahhabist sect, followers of the 18th-century Islamic reformer Muhammad bin Abd al-Wahhab, who joined forces with the founder of the Saudi dynasty. **Wahhabism** emphasizes fastidious piety, rejection of non-Muslim beliefs and cultures, corporal and capital punishment, and complete segregation of men and women in public life. Women are subject to many restrictions; for example, they are prohibited from driving automobiles, and although almost 60 percent of university graduates are women, they make up only 5 percent of the workforce (•Figure 6.1.19).

The vast Saudi royal family (which has more than 7,000 princes) keeps a wary eye on many internal and external threats. Al-Qa'ida has bombed Saudi civilian compounds and U.S. military facilities in an effort to purge the region of secular and "imperialist" Western interests and to destabilize the monarchy in order to replace it with an Islamic government. U.S. forces did withdraw from Saudi Arabia 236
after the 2003 Iraq War began. Countercurrents are now buffeting the desert kingdom. While struggling to prevent outside influences from changing Saudi ways, the government is finally allowing tourists to visit, encouraging foreign investment, and signing international trade and human rights treaties.

Bahrain, Qatar, and the United Arab Emirates

The sheikdom of Bahrain (ruled by a monarch called a *sheikh* or *emir;* population 800,000) occupies a small archipelago in the Gulf between the Qatar Peninsula and Saudi Arabia (see Figure 6.1.17), to which its main island is connected by a modern causeway. In addition to being an oil producer, Bahrain is a center of oil refining, seaborne trade, ship repair, air transportation, business administration, international banking, and aluminum manufacturing using imported alumina and local natural gas. Like Saudi Arabia, Bahrain is an oil-financed welfare state. However, in contrast to its giant neighbor, Bahrain's petroleum reserves are expected to be exhausted within two decades. Bahrain is home to the U.S. Navy's Fifth Fleet, making this tiny country a vital strategic ally of the United States in a critical region. Within the region, largely Shiite Bahrain is notable for having granted women the rights to vote and run for office; in Qatar, women can vote but not be elected, and in the other Gulf countries, they can do neither.

Joe Hobbs

• **Figure 6.1.18** Saudi Arabia's Shiite Muslim minority is significantly poorer than the Sunni majority. These are Shiites in the Eastern Province town of Qatif.

Joe Hobbs

• **Figure 6.1.19** Education of women in the Gulf countries has increased dramatically in recent decades, but strong traditions keep most women out of the workforce. This is a geographic information systems (GIS) class in the Department of Geography, United Arab Emirates University.

East of Bahrain along the Gulf lie the small emirates of Qatar (population 900,000) and the United Arab Emirates (UAE) (population 4 million). Qatar is atop the world's third-largest reserve of natural gas. Only about one-quarter of the country's inhabitants are indigenous Qataris; the rest are guest workers, mainly from South and Southeast Asia, brought in to turn that mineral wealth into tangible assets. Qatar established itself as a vital U.S. ally in the 2003 war against Saddam Hussein's Iraq, permitting American warplanes and warships to be based there. Qatar is also the home of al-Jazeera, the satellite television network that upsets many Gulf autocracies and the United States with its frank criticisms and that serves as the media outlet for al-Qa'ida videotapes and audiotapes. Qatar is also the only Gulf country to maintain an open dialogue with Israel.

The United Arab Emirates is a loose confederation of seven sheikhdoms once known as the Trucial States that until 1971 were British protectorates: Abu Dhabi, Ajman, Dubai, Fujeira, Sharjah, Ras al Khaima, and Umm al Quwain. Only three of these produce oil, and only Abu Dhabi pumps it in major quantities. Qatar and the UAE are extremely arid and prior to the oil era supported only a meager subsistence economy based on fishing, pearl diving, seafaring, herding, and a little agriculture, augmented by smuggling and piracy. Today, the UAE is capitalizing on its central location between Asia, Africa, and Europe with the construction of sophisticated airport and seaport facilities. The residential and commercial developments of Dubai and Abu Dhabi have dramatically changed the landscape (•Figure 6.1.20). With the aid of an overwhelmingly foreign work-
210 force, the UAE is successfully diversifying its economies beyond the oil industry. Dubai is a well-established international trade and banking center, featuring the world's largest shopping mall, the world's tallest building, and a massive port that U.S. naval ships are permitted to use. Abu Dhabi, the largest and richest unit of the seven emirates, is hoping to establish regional supremacy as a financial center. It also aspires to become one of the great cultural centers of the Middle East, with a host of new museums, galleries, and performing arts centers. Unlike its more conservative neighbors, the UAE has promoted itself as a destination for sun-worshiping international tourists and shoppers. Dubai is the most liberal of the emirates and, in marked contrast to its teetotaling neighbors, tolerates bars and brothels.

Compliments of NASA/Visible Earth.

• **Figure 6.1.20** Perhaps nowhere on earth is humanity's transformation of nature so visible from space. The United Arab Emirates' city state of Dubai boasts the world's tallest building, the artificial deepwater part of Jebel Ali, and an artificial island complex shaped like a palm tree. The Palm Island development offers very high priced homes and apartments.

The Southern Margins of Arabia

Known to the ancient Romans as *Arabia Felix*, or "Happy Arabia," because of their verdant landscapes, Yemen (population 22 million) and Oman (population 2.7 million) occupy, respectively, the mountainous southwestern and southeastern corners of the Arabian Peninsula (see Figure 6.1.17). The present Republic of Yemen was created in 1990 by the union of the former North Yemen and South Yemen. They remain united despite a brief civil war in 1994. Its political capital and largest city is San'a (population 2 million), famed for its distinctive indigenous architecture (•Figure 6.1.21). Until the 1960s, North Yemen was an autocratic monarchy, extremely isolated and little known to the outside world. In the mid-1960s, a republican government supported by Egypt and opposed by Saudi Arabia took control of large sections at a cost of much bloodshed.

The former South Yemen, known until 1967 as Aden, a British colony and protectorate, lies along the Indian Ocean between the former North Yemen and Oman. The main city, Aden, is located on a fine harbor near the straits. Aden is perfectly located for refueling vessels that transit the Suez Canal and is close to the strategic chokepoint of Bab el-Mandeb, connecting the Red Sea with the Indian Ocean. Possession or use of this port has therefore long been prized by outside powers. After British colonial times, South Yemen became a Marxist state supported by the Soviet Union and other Communist bloc nations. Since the two Yemens were unified, Aden has been an important refueling station for the United States and its Western allies. Security at the port of Aden has been stepped up since the bombing there of the U.S.S. *Cole* in 2000, and Western governments have pressured Yemen's government to crack down on Islamist groups that advocate struggle against Western interests in the region. Yemen gave the United States permission to use a remote-controlled predator drone to kill an important al-Qa'ida leader on a remote Yemeni road in 2002. Some of Yemen's tribal peoples live in areas well beyond the effective control of the government in San'a and are known to be strongly sympathetic to al-Qa'ida. The mountainous north-central region of Yemen, known as the Hadramaut, is the ancestral homeland of Osama bin Laden's family.

Yemen receives the heaviest rainfall in the Arabian Peninsula, enjoying a summer drenching derived from the same monsoonal wind system that brings summer rain to the Horn of Africa and the Indian subcontinent. Its main subsistence crops are millet and sorghum. Most of the country's best agricultural land consists of terraces and is dedicated to the cultivation of the valuable cash crop *qat (Cathya edulis),* a stimulant shrub chewed by virtually all adults in the country on a daily basis. Most of the country's people live in the rainier highlands, and the coastal plain is extremely arid and sparsely populated.

Yemen's meager agricultural output and lack of industrial production have been compensated for during recent decades by remittances, the earnings sent home by Yemeni workers in the oilfields and services of the wealthy Gulf countries. The support of Yemen's government for Saddam Hussein's regime in Iraq during the 1991 Gulf War prompted Saudi Arabia to expel millions of these expatriate Yemeni workers, leading to a precipitous decline in Yemen's economy.

Joe Hobbs

• **Figure 6.1.21** The Palace of the Imam Yahya in Wadi Dhar, near San'a, is typical of the architectural gems that have made old San'a a World Heritage Site.

Oil production began in Yemen itself in the 1980s. The fields are located in a contested border area between Yemen and Saudi Arabia and subjected periodically to kidnappings and killings of foreign oil workers and tourists by groups opposed to Yemen's government, so production has been slow to increase.

Most of the Sultanate of Oman lies along the Indian Ocean and the southern shore of the Arabian Sea, with a small, detached part bordering the Strait of Hormuz and the Gulf. As in Yemen, the coastal region is arid, and most people live in interior highlands where mountains induce a fair amount of rain. For about 2,000 years, beginning around 1000 B.C.E., Oman was the source of most of the world's frankincense, a prized aromatic commodity that made the region a hub of important oceanic and continental trade routes. The capital, Muscat (population 900,000), has long been a vital port in trade between the Middle 229
East and South Asia. Crude oil now furnishes almost all of Oman's exports, but with the easiest-to-reach oil already tapped, production has been declining since 1997. The country's strategic position on the Strait of Hormuz, through which much of the oil produced by the Gulf countries is transported, has made it an object of concern in industrial nations, particularly Britain, with which it has historic ties, and the United States. These Western powers have rights of access to base facilities in Oman in the event of any military emergency that threatens the continued flow of oil from the Gulf region. Allied forces used the bases after Oman joined the United Nations coalition against Iraq in the Gulf War of 1991 and again in the U.S.-led invasion of Iraq in 2003.

Kuwait and the Gulf War of 1991

With 8 percent of the world's proven oil reserves within its borders, tiny Kuwait (6,900 sq mi/17,900 sq km—a little smaller than the state of New Jersey; see Figure 6.1.17) has the world's fifth-largest store of proven petroleum resources (after Saudi Arabia, Canada, Iran, and Iraq). Associated with the oil is much natural gas, which Kuwait both uses and exports. Oil revenues have changed Kuwait (population 2.8 million) from an impoverished country of boat builders, sailors, small traders, and pearl divers to an extraordinarily prosperous, completely urbanized welfare state. Generous benefits have flowed to Kuwaiti citizens, but not to the Asian and other non-Kuwaiti expatriates drawn by employment opportunities to this emirate. About half of the people residing in Kuwait are foreigners. To reduce possible security threats and become more self-reliant, the government is trying to reduce dependence on foreign workers. A huge challenge will be to provide jobs for its young population (half of its people are under 25) in non-oil-related industries and services. As Kuwait is almost completely dependent on fossil fuel commodities, it is vulnerable to developments such as alternative energy and, of course, to the eventual depletion of its own reserves.

Kuwait took center stage in world affairs in August 1990 when Iraqi military forces invaded the country. Claiming that it was historically part of Iraq and calling it Iraq's "19th province," Saddam Hussein's regime attempted to annex Kuwait. Saudi Arabia and other Gulf oil states seemed
230 in danger of invasion. Worldwide opposition to Iraq's aggression resulted in a sweeping embargo on Iraqi foreign trade imposed by the Security Council of the United Nations. Led by the United States, a coalition of 28 UN members (including several Arab states) staged a rapid buildup of armed forces in the Gulf region.

Beginning in January 1991, the coalition's crushing air war drove the Iraqi air force from the skies, massively damaged Iraq's infrastructure, and hammered the Iraqi forces in Kuwait and adjacent southern Iraq with intensive bombardment. In the ensuing ground war, the Iraqis were driven from Kuwait within 100 hours. Overall, an estimated 110,000 Iraqi soldiers and tens of thousands of Iraqi civilians were killed, while the technologically superior coalition forces suffered 340 combat deaths, of which 148 were Americans. During the conflict, which came to
be known as the Gulf War, the Iraqis also made war on the 232
environment, setting fire to hundreds of oil wells in Kuwait and allowing damaged wells to discharge huge amounts of crude oil into the Gulf.

Iraq: Vortex of Violence

Iraq (population 29 million, about 67 percent urban) occupies a broad, irrigated plain drained by the Tigris and Euphrates Rivers, together with fringing highlands in the north and deserts in the west (Figure 6.1.22). This riverine land once known as Mesopotamia has a complex geopolitical history. It has been a seat of empires, a target of conquerors, and in the 20th and early 21st centuries, a focus of oil development and political and military contention involving many other nations as well as groups within Iraq itself.

A critical (and now unfortunate) feature of Iraq's geography is its ethnic and religious complexity, with three major groups present: the Shiite Arabs, mainly in the south,

Figure 6.1.22 Principal features of Iraq.

who make up about 60 percent of the country's population; the Sunni Arabs, about 35 percent of the total and living mainly in the center (especially in the so-called Sunni Triangle); and the Kurds, mainly in the north, most of whom are Sunni Muslims and who represent 15 to 20 percent of the total (•Figure 6.1.23; see also Ethnic Geography, page 265). There are numerous smaller minorities, including Yazidi and Sunni Turkoman.

Internal sectarianism, foreign intervention, and costly conflict are not new to Iraq, and the modern problems there may be better appreciated through some understanding of the past. Iraq was a province of the Ottoman Empire (based in what is now Turkey) for almost 400 years. British forces defeated the Turks in Iraq in World War I, and Britain had a League of Nations mandate over the country until it gained its independence in 1932. Iraq's government was a pro-Western monarchy from 1923 to 1958, when an army coup made the country a republic. By the mid-1970s, Iraq had become an authoritarian one-party state ruled by President Saddam Hussein in the name of the Arab Baath Socialist Party. After 1973, increasing oil revenues from the country's huge reserves (about 9 percent of the world's total) underwrote new social programs, large construction projects, and the creation of new businesses. New high-rise buildings, busy expressways, and an active commercial and social life gave metropolitan luster to ancient Baghdad. Subsequent wars caused massive destruction in the city and in Iraq's economy. This country should have become rich from its oil, but it squandered that wealth on hapless military adventures.

• **Figure 6.1.23** Ethnicity in Iraq and neighboring regions.

Saddam Hussein's misadventures were based to some extent on the country's geography. Baghdad (population 8 million), Iraq's political capital, largest city, and main industrial center, is located strategically on the Tigris River near its closest approach to the Euphrates River. It is not a port city; Basra is Iraq's main seaport, although it is located upriver on the Shatt al-Arab, the waterway formed by the confluence of the Tigris and Euphrates. Iraq suffers from the fundamental geographic disadvantage of not having natural deep-water ports in its tiny 35-mile (55-km) window on the Gulf. One of the main reasons Iraq took on its immediate neighbors in two separate wars was to widen this window. In 1980, perceiving internal weakness in the neighboring fledgling Islamic Republic of Iran and a means of securing land and mineral resources on the Iranian side of the Shatt al-Arab waterway, Saddam Hussein launched an invasion of his more populous neighbor. Thus began an eight-year war that proved enormously costly to both sides, with an estimated 500,000 people killed or wounded. Iran mounted a far better defense than Iraq anticipated, expending its greater military personnel in bloody "human wave" assaults to counter Iraq's greater firepower from tanks, artillery, machine guns, and warplanes. A complex web of geopolitical relationships made the **Iran-Iraq War** a storm center in the political world, particularly as the superpowers and other countries sold sophisticated arms to both sides. The war ended in stalemate in 1988, with neither side gaining significant territory or other assets.

Just two years later, Saddam Hussein's military invaded Kuwait. A far more internationalized conflict, that war proved disastrous for Iraq. A formal cease-fire to the so-called **Gulf War** came in 1991, with harsh conditions imposed on Iraq by the UN Security Council. UN-sponsored 263
economic sanctions created shortages of food and medical supplies, bringing great hardship to the country's population. Iraq agreed to pay reparations for the damage its forces had caused, and it renounced its claims to Kuwait. Saddam Hussein's surviving forces then mercilessly put down rebellions by Shiite Arabs in southern Iraq and by Kurds in the north. To deter Iraq's military and provide a safe haven for the Kurds in the north and the Shiites in the south, the United Nations established "no-fly zones," where Iraqi aircraft were prohibited from operating. But there was widespread condemnation of the Western coalition in the Arab world for seemingly encouraging the Kurds and Shiites to rebel and then failing to help them.

Iraq's Shiites fared especially badly. One of the great environmental and cultural tragedies of the late 20th century was Iraq's systematic destruction after 1991 of the marshes adjoining the southern floodplains of the Tigris and Eu-

ETHNIC GEOGRAPHY

The Kurds

A mostly Sunni Muslim people of Indo-European origin, the Kurds are by far the largest of several non-Arab minorities in Iraq. They have revolted against Iraqi central authority on numerous occasions, including during the Gulf War of 1991. The Kurdish question has international implications because the area occupied by an estimated 30 million Kurds also extends into Iran, Turkey, Syria, and the southern Caucasus region (see Figure 6.1.23); an estimated 2 million more live outside the Middle East. European powers promised an independent state of Kurdistan at the end of World War I, but the governments concerned never took steps to create it. Kurds today remain the world's largest ethnic group without a country.

A book about the Kurds is appropriately titled *No Friends but the Mountains.** Turkish officials have long downplayed the identity of Kurds (who make up about 20 percent of Turkey's population), often referring to them as "Mountain Turks," and even banned or restricted Kurdish-language media in Turkey until 2002.

A Kurdish rebellion against Turkish rule, rising in part from the great poverty of the Kurds relative to the rest of Turkey's population, has smoldered and sometimes flamed since 1984. The main Kurdish resistance comes from the **Kurdistan Workers Party,** known by its Kurdish acronym **PKK.** Under Saddam Hussein, Iraq tried to "Arabize" its Kurds by forcing them to renounce their ethnicity and sign forms saying they were Arabs. However, after a no-fly zone and "safe haven" established after the Gulf War put them beyond Saddam Hussein's reach, Iraq's Kurds came to enjoy relative autonomy and prosperity. Even in the tumultuous years since the U.S.-led invasion in 2003, the Kurdish region of the north has escaped most of the bloodshed and economic chaos characteristic of the south, although recent resurgence of PKK activities have raised the prospect of an incursion by Turkey.

*John Bullock, *No Friends but the Mountains* (London: Oxford University Press, 1993).

phrates Rivers. To quell any opposition by the ancient Shiite culture of the **Ma'adan,** or **"Marsh Arabs,"** the Iraqi military built massive embankments and canals that drained and destroyed the marshes, forcing the Ma'adan to abandon their historical homeland. After 2003, the embankments were demolished, and restoration of the marshlands began. Under Saddam Hussein, Iraq's government had expanded irrigation, flood control, and hydropower through the construction of dams on the Tigris and Euphrates and their tributaries. Agricultural output in Iraq remains low, however, due mainly to the problem of overwatering, which causes the
253 salinization of once-fertile soils.

Throughout the 1990s, the regime of Saddam Hussein played a cat-and-mouse game with UN inspectors whose duty was to eliminate Iraq's ability to produce weapons of mass destruction. United States, British, and other forces repeatedly struck Iraq as the country refused to allow full access by the UN inspectors and, in 1998, attacked scores of military targets when access was denied. The Baghdad regime flouted the no-fly zones and restated its wartime position that Kuwait was an illegitimate country. International support for the sanctions against Iraq weakened greatly, particularly in the Arab world. President Saddam Hussein endured, much to the chagrin of his foes, and might have been able to hang on to power indefinitely if not for the 9/11 al-Qa'ida attacks on the United States.

Immediately after the Gulf War and during the Clinton administration of 1993–2001, the reasoning in Washington was that a unified Iraq under Saddam Hussein was better than the alternative: its possible fragmentation into the three entities of a Shiite south, a Sunni center, and a Kurdish north, some or all of which might be hostile to U.S. interests in the region. Although unsavory, Saddam Hussein could be contained. Under the new Bush administration, and especially after 9/11, the downfall of Saddam Hussein became an urgent priority, ranking alongside the war on terrorism. The Bush administration attempted to link Saddam Hussein with al-Qa'ida, particularly to garner public American support for a new war on Iraq, but such connections have never been decisively established.

Publicly, Washington built the case that Iraq was harboring and continuing to develop weapons of mass destruction that might be used against the United States and its allies. The United States was able to prompt further United Nations pressure on Iraq, which finally allowed UN weapons inspectors to resume their work. When these inspections failed to turn up weapons, the United States insisted that Iraq was hiding them and that only the use of force could remove their threat. It was necessary to effect "regime change" in Baghdad. To take on Saddam Hussein, President Bush assembled what he called a "coalition of the willing" consisting mainly of Britain, Australia, Denmark, and Poland. It fell far short of the coalition assembled during the Gulf War. France, Germany, and other traditional allies of the
United States opposed the impending war against Iraq, and 93
even within the United States there was considerable dissent against what was perceived as a war of choice rather than one of necessity.

The U.S.-led ground and air assault on Iraq (dubbed **Operation Iraqi Freedom**) that began in March 2003 led quickly and with relatively few casualties to the downfall of the Iraqi regime. With about 130,000 troops, concentrated mainly in the Sunni-dominated center of the country around Baghdad, the United States settled in for a long and troubled occupation of the country. The United States handpicked what it called the Governing Council of Iraqis, broadly representative of the country's ethnic composition, and ran many of Iraq's security and other affairs with its own Coalition Provisional Authority (CPA) until June 2004, when it turned over sovereignty to an interim Iraqi government.

The United Nations, with strong U.S. support, organized a January 2005 election for a 275-member transitional national assembly. In this, Iraq's first-ever free election, Iraq's largest ethnic population unsurprisingly won the lion's share of the seats: the Shiites, who turned out in huge numbers to vote, won 48 percent for their main political party. Anticipating the Shiite success and condemning the election as an illegitimate U.S. action, most Sunnis boycotted the vote. That made it possible for the main political party of the Kurds, a smaller ethnic group, to capture another large bloc of the assembly votes—26 percent. Iraq's Sunnis ended up with just 2 percent of the assembly seats. Having been the dominant force in Iraq for decades under Saddam Hussein (who was convicted of war crimes by an Iraqi court and hanged in 2006), the Sunnis now found themselves deprived of power and increasingly angry about the American occupation. The pro-U.S. government meanwhile faced a perplexing array of challenges in cementing a constitution for the fragile country: how much autonomy the Kurds and Shiites might receive within the country, how oil revenues would be split among the geographic and ethnic regions, and what the role of Islam should be in national affairs.

Sunni resistance against the United States grew (•Figure 6.1.24). Using roadside bombs, car bombs, small arms, and rocket-propelled grenades, Sunni militants were able to inflict American casualties on an almost daily basis. Predictably, al-Qa'ida used the U.S. occupation as a
235 rallying cry for intensified resistance against the United States. A well-organized insurgency under the leadership of Jordanian-born Abu Mus'ab al-Zarqawi announced that it was officially linked with al-Qa'ida. His **al-Qa'ida in Iraq** organization kept up a steady stream of bloody attacks on U.S. targets even after al-Zarqawi was killed in 2006. Also in the crosshairs were the Iraqis perceived to be collaborating with the Americans, particularly those who joined the new Iraqi army. The U.S. presence became increasingly confined to its fortified "Green Zone" in Baghdad. While the American death toll mounted—to more than 3,700 by late 2007, so did the Iraqi toll, with at least 75,000 dead to that date. Sunni insurgents relentlessly targeted Shiites in particular, even twice bombing one of the their holiest sites, the golden-domed Askariya Mosque in Samarra.

These Sunni assaults on Shiite sacred space helped push Iraq into **civil war,** defined as an armed conflict "between the government of a sovereign state and domestic political groups mounting effective resistance in relatively continuous fighting that causes high numbers of deaths."[1] Increasingly, this internecine violence took the form of ethnic cleansing, in which one group systematically murdered
147, members of the other. Moderate Shiite leaders such as Aya-
255 tollah Ali al-Sistani were no longer able to restrain a Shiite backlash. The more militant anti-Sunni and anti-U.S. Shiite factions, rallying chiefly behind a cleric named Moqtada al-Sadr and his Mahdi Army (Mahdi Militia), ascended. Shiites took retaliation against Sunnis, and a ceaseless cycle of revenge killings ensued.

At home, the Bush administration found itself in an increasingly difficult position. Costly in lives and dollars—with thousands of soldiers and supporting civilians killed and a price tag of over $100 billion per year—the war grew increasingly unpopular among Americans, and the president's approval rating slid. Many opposition politicians, academics, and others spoke increasingly of Iraq as "Vietnam," referring to a previous war that proved untenable and unwinnable for the United States. The Bush ad- 349
ministration faced a dilemma: "stay the course," as President Bush insisted with nearly 170,000 U.S. soldiers still on the ground in Iraq, or pull out (as the United States did in Vietnam) and risk Iraq's plunge into further chaos. Fearing anarchy, the Bush administration in 2007 opted for a "surge" in U.S. troops to maintain order, even as its main ally, Britain, drew down its military presence (primarily in Shiite southern Iraq).

That anarchy could take several forms. A U.S. withdrawal in the absence of strong Iraqi security might allow

AP Photo/Khalid Mohammed

• **Figure 6.1.24** U.S soldiers inspect the scene of a massive suicide bomb targeting a U.S. military convoy in April 2005.

al-Qa'ida and other militant Islamists to operate with impunity as they did in Afghanistan prior to 9/11, perhaps even using the country as a training ground for more attacks on the West. Al-Qa'ida and its allies would thus accomplish what they promised to do in Iraq—a very unsavory outcome for the West. The country could fragment along ethnoreligious lines into sectarian ministates, potentially including a militant Sunni state and an Iranian-leaning Shiite state. The country's fragmentation could be carefully calculated to reduce the potential fallout. The broad geographic distributions of Sunnis, Shiites, and Kurds seen in Figure 6.1.23 suggest what the boundaries might look like. The biggest problems would come in the big cities like Baghdad and Mosul, where neighborhoods of contentious groups live cheek by jowl.

Oil and Upheaval in Iran

Iran, formerly known as Persia (•Figure 6.1.25), was the earliest Middle Eastern country to produce oil in large quantities and has the world's third-largest reserves. The main oilfields form a line along the foothills of the Zagros Mountains in southwestern Iran. British-built pipelines linked the large refinery at Abadan (on the Shatt-al Arab waterway) with the inland oilfields. Now nearly all exports of crude oil from Iran pass through a newer terminal at Kharg Island (see Figure 6.C, page 230), which allows supertankers to load in deep water. Iran is also endowed with vast natural gas reserves, second only to those of Russia.

Oil is the leading source of revenue for Iran, a country where a dry and rugged habitat makes it hard to provide an agriculture-based livelihood for its population of 71 million. Arid plateaus and basins, bordered by high, rugged mountains, are Iran's characteristic landforms (•Figure 6.1.26). Encircling mountain ranges prevent moisture-bearing air from reaching the interior, creating a classic rain-shadow desert. Only about 10 percent of Iran is farmed. Enough 25
rain falls in the northwest during the winter to permit unirrigated cropping (dry farming), but rains sufficient to support intensive agriculture fall mainly on a densely populated lowland strip between the high, volcanic Elburz Mountains and the Caspian Sea. The western half of this lowland receives more than 40 inches (c. 100 cm) annually and produces a variety of subtropical crops such as wet rice, cotton, oranges, tobacco, silk, and tea. Agriculture elsewhere in Iran depends mainly on irrigation.

Many irrigated lands are on gently sloping alluvial fans (triangular deposits of stream-laid sediments) at the foot of the Elburz and Zagros Mountains; the capital, Tehran (population 12.5 million), is in one of these areas south of the Elburz. *Qanats* (underground sloping tunnels) supply water to a large share of the country's irrigated acreage. However, they are expensive to build and maintain, and an increasing share of water comes from stream diversions and from deep wells. Wheat is the most important irrigated crop.

The leading revenue earner, and Iran's third most important export, after oil and its famous carpets, is the pistachio nut. Cattle are raised in areas where there is enough moisture to grow forage for them, but drier areas support sheep and goats. Many livestock belong to seminomadic mountain peoples—the Qashqai, Baktiari, Lurs, Kurds, and others—whose independent ways have long been a source of friction between these tribespeople and the central government. Before the Islamic revolution in Iran in 1979, it was government policy to sedentarize Iran's nomads so that they did not threaten the state. 218

Before World War I, Iran was a poor and undeveloped country, in marked contrast to its imperial grandeur when Persian kings ruled a great empire from Persepolis and to its architectural florescence under earlier Muslim leaders. In the 1920s, a military officer of peasant origin, Reza Khan, seized control of the government and began a program to modernize Iran and free it from foreign domination. He had himself crowned as Reza Shah Pahlavi, the founder of the new Pahlavi dynasty. Influenced by the modernizing efforts of Mustafa Kemal Atatürk in neighboring Turkey, the new shah (king) introduced social and economic reforms. In 1941, his young son took the throne as Mohammed Reza Shah Pahlavi and set out to expand the program of modernization and Westernization begun by his father. In the 1970s, Iran played a leading role in raising world oil prices. Mounting oil revenues underwrote explosive industrial, urban, and social development.

In the countryside, the shah's government tried to upgrade agricultural productivity and rural standards of living through land reform and better technologies. But agricultural reforms generally failed, and poverty-stricken

• **Figure 6.1.25** Principal features of Iran.

Joe Hobbs

• **Figure 6.1.26** Aerial view of farms and settlements on an alluvial fan pouring from the western Zagros Mountains, Iran.

families from the countryside poured into Tehran and other cities. From this devoutly Muslim group of new urbanites came much of the support for the revolution that ousted the shah in 1978–1979. Dissent also grew because the regime spent vast sums on its military, mainly for U.S.-made weapons, thus undercutting benefits to the people.

The revolution that overthrew the shah took Iran abruptly out of the American orbit. The central figure among the revolutionaries was an **ayatollah** ("sign of God") named Ruhollah Khomeini, an elderly critic of the regime whom the shah had forced into exile in 1964. From Paris, Khomeini sent repeated messages to Iran's Shiites that helped spark the revolution. A furious tide of Shiite religious sentiment rose against corruption, police heavy-handedness, modernization, Westernization, Western imperialism, the exploitation of underprivileged people, and the monarchy, which Shiite beliefs regard as illegitimate. Growing numbers of people staged strikes and demonstrations in 1978, forcing the shah to flee the country and abdicate his throne. After hospitalization in the United States and life in exile in Panama and Egypt, the shah died in Cairo in 1981 and is buried there. The Ayatollah Khomeini returned in triumph to Iran in 1979. His followers seized and held 52 U.S. diplomatic personnel as hostages in Tehran for 444 days. U.S. President Jimmy Carter proved unable to resolve this crisis, which contributed to his defeat in his 1980 bid for reelection.

Under Khomeini (who died in 1989), the country became an Islamic republic governed by Shiite clerics, who included a handful of revered and powerful ayatollahs at the head of the religious establishment, and an estimated 180,000 priests called **mullahs.** Khomeini served as the Guardian Theologian, an infallible supreme leader enshrined by the principle of **divine rule by clerics** (*velayet-e-faqih*). These religious leaders continue to supervise all aspects of Iranian life and perform many functions allotted to civil servants in most countries. They base their authority on Shiite interpretations of Islam (about 89 percent of Iran's population is Shiite).

In contrast to their status during the shah's reign, when they were encouraged to take on larger public roles, women in post-revolutionary Iran have lost many freedoms.
Under the new regime, "polluting" influences of Western 227
thought and media were purged. However, recent years have seen a general softening of restrictions on personal freedom of expression, and the Internet and other media have introduced broader worldviews to Iran's youth, a very important cohort, as about 60 percent of the population is under the age of 30 (•Figure 6.1.27). Young people led an unprecedented series of prodemocracy protests in 2002. Five years later, Iranians of different generations took to the streets to protest government hikes on gasoline prices; ironically, this oil-rich country has a low oil refining capacity and needs to import one-third of its gasoline. Iran imports more gasoline than any country in the world except for the United States.

Prior to 2005, Iran watchers saw a tug-of-war between the moderate and reform-minded president Muhammad Khatami and the less tolerant supreme religious leader, Ayatollah Ali Khamanei (Khomeini's successor), who has lifetime control over the military and the judicial branch of government. There was speculation that the country's younger and more Western-oriented sectors might win the day and that relations with the wider world—and the United States in particular—would improve. But the winner of Iran's 2005 presidential election (for a five-year term

• **Figure 6.1.27** Iranian women demanding the release by the Iranian government of Jamil Bassam and Abrahim Khodabandeh

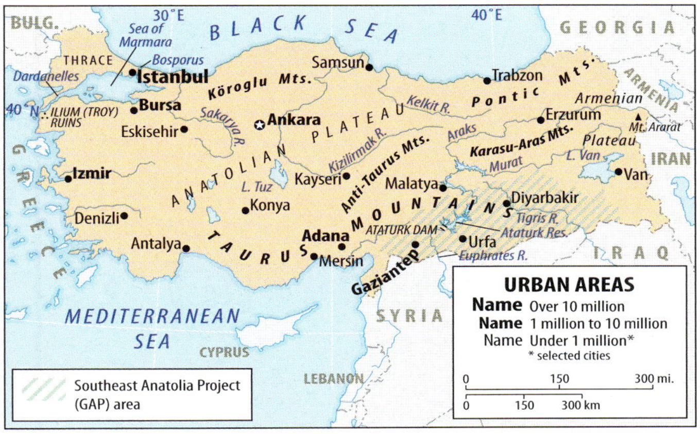

• **Figure 6.1.28** Principal features of Turkey.

of office) was Mahmoud Ahmadinejad, who outspokenly opposed American policies in the region, goaded Israel by insisting that the Holocaust never happened and calling for the country's destruction, and accelerated Iran's nuclear
234 power program. The level of rhetoric and tension between Iran and the United States escalated. American officials accused Iran of supplying weapons used to kill U.S. soldiers in Iraq, and Iran began gearing up for the possible imposition of economic sanctions by the United States and its European allies.

6.1.7 Turkey: Where East Meets West

Like Iran, Turkey is a demographic heavyweight (population 74 million people) and has long had an enormously important presence in the region (•Figure 6.1.28). The Turks who organized the Ottoman Empire beginning in the 14th century originated as pastoral nomads from central Asia, where Turkic cultures still prevail. From the 16th century to the 19th, their empire, based in what is now Turkey, was an important power. Modern Turkey was created from the wreckage of the old empire after World War I. Its founder, Mustafa Kemal Atatürk, was determined to Westernize the country, raise its standard of living, and make it a strong and respected national state. He inaugurated social and political reforms designed to break the hold of traditional Islam and open the way for modernization and Turkish nationalism.

Islam had been the state religion under the Ottoman Empire, but Atatürk divorced church from state, and to this day, Turkey is the only Muslim Middle Eastern country to
225 officially separate them. The wearing of the red cap called the *fez,* an important symbolic act under the Ottoman caliphs, was prohibited, and state-supported secular schools replaced the all-pervasive religious schools. To facilitate public education and remove further traces of Muslim culture, Latin characters replaced the Arabic script of the Qur'an. Slavery and polygamy were outlawed, and women were given full citizenship. Legal codes based on those of Western nations replaced Islamic law, and forms of democratic representative government were instituted, although Atatürk ruled in a dictatorial fashion.

Turkey has continued to have trouble establishing a fully democratic system, and there have been frequent periods of military rule (•Figure 6.1.29). Turkey's constitution invests the military with the responsibility of protecting the country's secular government, and the army has used its powers to overthrow four elected governments since 1960. In 2007, Turkish voters elected a president, Abdullah Gul, with a strong background in Islamic politics. Questions immediately arose over how long the army might tolerate his rule. There was also great controversy about the fact that President Gul's wife wore an Islamic headscarf in public: the wearing of the headscarf in public offices and schools is officially banned, and Turkey's secularists see the scarf as an affront to the country's identity.

Turkey is composed physically of two units: the Anatolian Plateau and associated mountains occupying the interior of the country and the coastal regions of hills, mountains, and small plains bordering the Black, Aegean, and Mediterranean Seas. Turkey in general has a physical presence quite unlike the popular image of the Middle East and North Africa as a desert region; it is in fact the only country in the region that has no desert, and cultivated land occupies about a third of its land area. The Anatolian Plateau, an expanse of wheat and barley fields and grazing lands, is bordered on the north by the lofty Pontic (Kuzey) Mountains and on the south by the Taurus Mountains. The plateau ranges from 2,000 to 6,000 feet (c. 600 to 1,800 m) in elevation and is highest in the east where it adjoins the mountains of the Armenian Knot. Its surface is rolling and windswept, hot and dry in summer and cold and snowy in winter, with a natural vegetation of short steppe grasses and shrubs. Production of cereals (especially barley, wheat,

Joe Hobbs

• **Figure 6.1.29** The military has a strong presence is secular Turkey. These soldiers are guarding an archaeological site in eastern Turkey, a region of Kurdish rebellion against government.

Joe Hobbs

• **Figure 6.1.30** Where there had been a bridge across the Euphrates River, there is now a ferry across the Ataturk Reservoir, upstream from the huge Ataturk Dam.

and corn) and livestock (especially sheep, goats, and cattle), employing both traditional and mechanized means, prevails in Anatolia. Coastal plains and valleys along and near the Aegean Sea, Sea of Marmara, and Black Sea are generally Turkey's most densely populated and productive areas. Here are grown cash crops such as hazelnuts, tobacco, grapes for sultana raisins, and figs. The slopes of the Pontic Mountains facing the Black Sea have copious rains in both summer and winter and support prolific tea harvests (see Figure 6.9, page 216).

Relative poverty compared with the nearby European countries is a striking characteristic of present-day Turkey. A strong rural and agricultural component in Turkish life persists, and about one-third of the population is still classified as rural. However, the country has embarked on a course of change that promises to modernize its agriculture, expand industry, and raise general standards of living. The value of output from manufacturing (mainly textiles, agricultural processing, cement manufacturing, simple metal industries, and assembly of vehicles from imported components) is already greater than that from agriculture. Some of Europe's most recognizable brand names, like Bosch, Renault, and Fiat, have factories in Turkey. Turkey does produce some oil.

Early in the 1990s, Turkey began an impressive $32 billion agricultural effort called the **Southeast Anatolia Project** (known by its Turkish acronym **GAP**). Its aim is to convert the semiarid southeastern quarter of the country into the "breadbasket of the Middle East." Surpluses of cotton, wheat, and vegetables are exported, along with value-added manufactures such as cotton garments and textiles. The centerpiece of this massive irrigation project is the Atatürk Dam on the Euphrates River. About 20 other dams on the Turkish Euphrates and Tigris are also part of the ongoing project to provide hydroelectricity as well as water (•Figure 6.1.30). The GAP is controversial. It is inundating historic settlements and priceless archaeological treasures and is compelling people to move. It is being developed in the part of the country where a restive Kurdish population is seeking recognition, autonomy, and among some factions, independence from Turkey. In addition, Turkey's Tigris and Euphrates waters flow downstream into neighboring states, raising serious questions about downstream water allocation and quality. 232

Turkey is an "in-between" country. Economically, it is well below the level of the MDCs but above most of the world's LDCs. Culturally, it is between traditional Islamic and secular European ways of living. Part of Turkey—the section called Thrace—is actually in Europe, and Turkey as a whole aspires to become more European. It has been a member of the North Atlantic Treaty Organization (NATO) since 1952. It is an associate member of the European 88
Union and is a candidate for full membership. The European Union, however, would first like to see Turkey become more European politically, for example, by diminishing the army's influence in government and by improving its human rights record. Turkey insists it has made huge strides 91
toward satisfying the minimal democratic and human rights norms for EU member countries. It has abolished the death penalty and eased restrictions it imposed on the cultural expressions of its minority Kurdish population.

At expansion talks in 2002 and 2006, EU leaders nevertheless rejected Turkey's bid and would not promise a date for membership negotiations. They argued that Turkey has still not gone far enough with democratic and human rights reforms. They probably feared the costs of absorbing Turkey's mainly low-income population into the union. Many Turks feel that their membership in the European Union was denied because they are a Muslim people. It is possible that the main reason Turkey was rejected was because EU leaders feared that Islamist terrorists would make their way from an EU Turkey into the heart of continental Europe.

The EU's rejection of Turkey may fan the growing flames of sentiment in the Islamic world that the "crusader West" is waging a systematic campaign against Muslims. Turkish public opinion has become decidedly more anti-European; a 2006 poll showed only 35 percent support for the EU, half that of 2004. A Turkey shunned by Europe may seek alliances eastward in Russia, India, and China.

Istanbul (population 12.2 million), founded as Constantinople, is Turkey's main metropolis, industrial center, and port (•Figure 6.1.31). One of the world's most historic and cosmopolitan cities, Istanbul was for many centuries the
82, capital of the Eastern Roman (Byzantine) Empire. It became
162 the capital of the Ottoman Empire when it fell to the Turks in 1453. However, the capital of the Turkish Republic was established in 1923 at the more centrally located and more purely Turkish city of Ankara (population 3.6 million) on the Anatolian Plateau. Turkey's third-largest city is the seaport of Izmir on the Aegean Sea.

Joe Hobbs

• **Figure 6.1.31** The people of Istanbul live amidst architectural treasures with roots in many civilizations.

Istanbul is located at the southern entrance to the Bosporus Strait, the northernmost of the Turkish Straits that connect the Mediterranean and Black Seas. The straits have long been a focus of contention between Turkey and Russia, who fought several wars in the past; their relations are much better now. Turkey's military posture has also been shaped by fragile relations with its NATO neighbor, Greece. 229
The main questions recently at issue have been the status of Cyprus and conflicting claims to hundreds of islands in the Aegean Sea and that sea's oil and gas reserves. Already an 136
EU member, Greece pressured Turkey to revise its policy toward Cyprus and to settle the Aegean Sea dispute as preconditions for Turkey's joining the organization.

6.1.8 Rugged, Strategic Afghanistan

High and rugged mountains dominate Afghanistan (population 32 million). This is the only landlocked nation of the Middle East and North Africa, and like Sudan and Turkey, it is a country that is transitional between regions; some geographers place it in central Asia or South Asia (•Figure 6.1.32). Historically, it has occupied an important strategic location between India and the Middle East. Major caravan routes crossed it, and a string of empire builders sought control of its passes. Today, it is a land of limited resources, poor internal transportation, little foreign trade, and much conflict; Afghanistan is one of the poorest of the world's poor countries.

Most Afghans live in irrigated valleys around the fringes of the high mountains that occupy a large part of the country. The country's second most populous area is the northern side of the central mountains, where people live in oases in foothills and steppes. Northern Afghanistan borders three of the five central Asian countries, and millions of people on the Afghan side are related to peoples of those countries. The most populous area is the northeast, particularly the fertile valley of the Kabul River, where the capital and largest city, Kabul (population 3.2 million; elevation 6,200 ft/1,890 m), is located. Most of the inhabitants of the southeast are Pashtuns (also known as Pushtuns or Pathans). Their language, Pashto, is related to Persian. 221,
The Pashtuns are the largest and most influential of the numerous ethnic groups that make up the Afghan state; the next largest are the Tajik and Hazara, respectively. 222

The independent-minded tribal Pashtun people have always been loath to recognize the authority of central governments. In the days of Britain's Indian Empire, the area saw warfare among tribes, tribal raids on British-controlled areas, and British punitive expeditions against the tribes. In those days, Peshawar—on the Indian (now Pakistani) side 327
of the Khyber Pass into Afghanistan—became the hotbed for British, Russian, and other agents playing what came to be known as the **Great Game** of vying (and spying) for strategic influence in this part of the world.

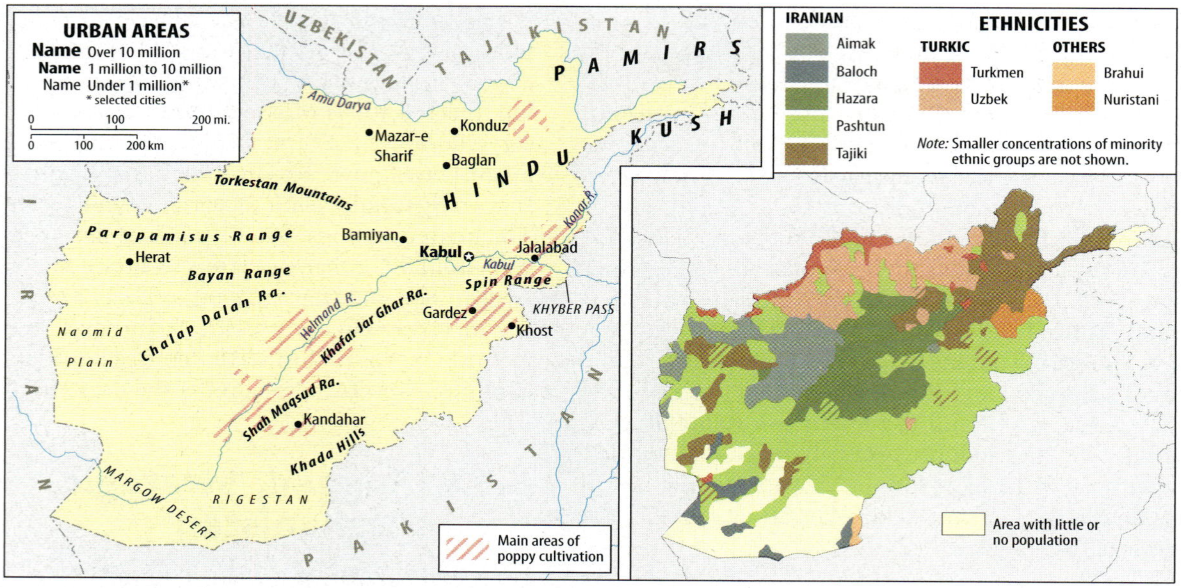

• **Figure 6.1.32** Principal features and ethnic groups of Afghanistan.

Afghanistan is overwhelmingly a rural agricultural and pastoral country, and its agriculture bears many of the customary Middle Eastern earmarks: traditional methods, simple tools, limited fertilizer, and low yields. It is so mountainous and arid that only about 12 percent is cultivated. Enough rain falls in the main populated areas during the winter for dry farming of grains (•Figure 6.1.33). A variety of cultivated crops, fruits, and nuts are important locally. Many people raise livestock on a seminomadic or nomadic basis.

346 The most successful crop has been the opium poppy, which routinely accounts for between one-third and two-thirds of the country's gross domestic product. A succession of rebel armies and then Afghanistan's government under the Taliban (see next page) used revenues from opium and heroin to obtain their arms. Afghanistan was by far the world's largest opium producer when in 2001, the Taliban reversed course and imposed a ban on cultivation. There was virtually no opium harvest that year. After the United States ousted the Taliban from power, Afghanistan regained its former status, producing fully 92 percent of the world's opium in 2007. Most of the production takes place in the southern province of Helmand, where the Taliban have retrenched themselves most successfully. Once again, the Taliban are gaining funds and strength from the drug economy. Partly in order to restrain the Taliban, Britain and the United States have stepped up efforts to eradicate opium. A policy split has developed among foreign forces: the NATO mission in Afghanistan worries that in the absence of crop alternatives, elimination of the poppy crop could turn desti-
548 tute farmers into vengeful fighters. One option rarely called for but perhaps most appropriate in Afghanistan's future, would be to allow the country to produce opium legally for pharmaceutical needs, as a handful of countries, including Turkey and India, are allowed to do.

Through most of the 20th century, this highland country was remote from the main currents of world affairs. After the Islamic revolution in Iran in 1979, however, Afghanistan's location next to that oil-rich country made it once again the target of foreign interests. The Soviet military intervention of 1979 and the ensuing devastation catapulted the country into world prominence. The USSR's motives for its invasion of Afghanistan may have included a desire to prevent by force the spread of Iranian-style Islamic fundamentalism into Afghanistan (which is 80 percent Sunni and 19 percent Shiite) or into the central Asian Soviet republics deemed vital to the superpower's security. For its part, the United States warned the Soviet Union that it would not tolerate further Soviet expansionism.

The ensuing war entailed the widespread killing and maiming of civilians, destruction of villages, burning of crops, killing of livestock, destruction of irrigation systems, and sowing of land mines over vast areas (see Insights, page 274). Soviet ground and air forces caused several million Afghan refugees to flee into neighboring Pakistan and Iran. Arms from various foreign sources filtered into the hands of the ***mujahidiin***, the anti-Soviet rebel bands that kept up resistance in the face of heavy odds. The United States was one of the powers supporting the rebels and in that sense waged a proxy war against the Soviet Union in Afghanistan. In its waning years, the USSR recognized that it could not win its "Vietnam War," and its troops withdrew from Afghanistan by 1989. It is estimated that of the 15.5 million people who lived in Afghanistan when it was invaded in 1979, at least 1 million died, 2 million were displaced from their homes to other places within the country, and 6 million fled as refu-

THOMAS J. ABERCROMBIE/National Geographic Image Collection.

Figure 6.1.33 The Bamiyan Valley and the Koh i Baba Range of mountains Afghanistan.

gees into Pakistan and Iran. By 2007, about two-thirds of those refugees had returned.

Support that the United States and moderate Arab states such as Egypt and Saudi Arabia lent to the *mujahidiin* and sympathetic Arab fighters soon came back to haunt those countries in a phenomenon that came to be known as **blowback.** Emboldened by their victory against the Soviets, the most militant Islamists among the fighters turned their attention to the United States and its Middle Eastern allies. Chief among them was Osama bin Laden, who developed "the base" (*al-Qa'ida* in Arabic) of thousands of Afghans, Arabs, and other anti-Soviet war veterans he could call on
235 to wage a wider *jihad*. Bin Laden's organization trained an estimated 10,000 fighters in al-Qa'ida camps in Afghanistan and from this unlikely, remote setting devised spectacular acts of terrorism: an assassination attempt on Egyptian President Hosni Mubarak, the bombing of U.S. military barracks in Saudi Arabia, the bombing of U.S. embassies in Kenya and Tanzania, the attack on the U.S.S. *Cole,* and
501 the 9/11 attacks. The East African bombings in 1998 led within days to a cruise missile strike by the United States on six al-Qa'ida camps near Khost, Afghanistan. Many al-Qa'ida personnel were killed, but not its top leaders.

The *mujahidiin* succeeded in overthrowing the Communist government of Afghanistan in 1992, but after that, rival factions among the formerly united rebels engaged in civil warfare. During this period, most other countries utterly neglected Afghanistan, inadvertently promoting the growth of militant movements (which tend to thrive in remote, underdeveloped, and ungoverned regions). By 1996, one of the rebel factions, the **Taliban** (backed by Saudi Arabia and Pakistan), gained control of most of the country, including the capital of Kabul. Proclaiming itself the sole legitimate government of Afghanistan, the Taliban imposed a strict code of Islamic law in the regions under its control and gained international notoriety for its austere administration. The Taliban removed almost all women from the country's workforce, forbade public education of girls, and outlawed "un-Islamic" practices such as dancing, flying kites, watching television, keeping birds, and trimming beards.

The Taliban continued to make advances against its opponents inside Afghanistan (particularly the **Northern Alliance,** whose leader, Ahmed Shah Massoud, was assassinated just days before and in apparent preparation for the 9/11 attacks) and, by 2001, controlled 95 percent of the country's territory. The neighboring central Asian countries, Russia, and even Iran grew increasingly fearful of the spread of the Taliban's extreme interpretation of Islam into their nations. Russia ended up supporting some of the rebels it had fought because they were now fighting the Taliban. Russia's once unlikely ally in supporting those rebels was the United States, which successfully lobbied the United Nations to levy economic sanctions against Afghanistan in 1999. The United States hoped that economic losses would pressure the Taliban into turning over bin Laden for prosecution and also announced a $5 million reward for information that would lead to bin Laden's capture or death.

The bin Laden bounty rose into the tens of millions of dollars after September 11, 2001. Named as the mastermind of the attacks against New York and Washington,

INSIGHTS

Land Mines

The scars of war will last a long time in Afghanistan. One of the country's urgent reconstruction tasks is to rid the landscape of millions of land mines sown during decades of war. These weapons have already killed or maimed more than 200,000 Afghans since 1979, and they continue to kill or maim about 60 Afghans, the great majority of them civilians, every month.

Land mines are antitank and antipersonnel explosives, usually hidden in the ground, designed to kill or severely wound the enemy. There are many kinds: Soviet troops planted at least nine varieties in Afghanistan, including seismic mines, which are triggered by vibrations created by passing horses or people; mines with devices that pop up and explode ("Bouncing Betties") when a person approaches; "butterfly" mines meant to maim rather than kill; and explosives disguised as toys, cigarette packs, and pens (• Figure 6.1.C).

There may be as many as half a million land mines in Afghanistan, or 2 per square mile (5/sq km). Reconstruction of this agricultural country cannot begin in earnest until the mines are cleared. The obstacles are enormous in rugged Afghanistan. Mines typically remain active for decades after they are planted and tend to last longer in arid places. The mountainous terrain prevents systematic clearance of the explosives. Many mines were scattered indiscriminately from the air and will be difficult to locate. Heavy rains move the mines downhill and away from known locations. And it is expensive to clear mines; Kuwait spent $800 million to rid its landscape of them after the Gulf War. So far, Western countries have donated more than $100 million toward the expense of clearing Afghanistan's land mines, and over 200,000 been cleared; Russia has contributed nothing.

Joe Hobbs

• **Figure 6.1.C** Inexpensive to manufacture and deploy, land mines are a scourge that outlast conflict and bring suffering to civilians.

Land mines are a global problem. Up to 70 million mines contaminate the land in over 80 countries, and they kill or maim more than 15,000 civilians each year. The numbers and concentrations in some countries are staggering. The leader in numbers is Egypt, with 22 million mines; in density of mines, Cambodia leads with about 100 per square mile (40/sq km). The United Nations estimates that with current technology, it would take over $30 billion and 1,100 years to clear the world's land mines.

Several developments may portend an eventual end to the land mine scourge. Paul McCartney and other celebrities have led high-profile campaigns against these weapons (see http//:landmines .org). The world's most visible critic of land mines, Princess Diana of Wales, was killed in an automobile accident in Paris in 1997. In her wake, there were renewed calls for a moratorium on the manufacture and use of antipersonnel land mines. Two months after Diana's death, anti–land mine activist Jody Williams won the Nobel Peace Prize. Soon after, representatives from 125 nations met in Ottawa, Canada, to sign a treaty strictly banning the use, production, storage, and transfer of antipersonnel land mines. Many of the largest producers, users, and exporters of these explosives, including the United States, Russia, China, India, Pakistan, Iran, Iraq, and Israel, refused to sign. The United States argued that land mines were necessary to protect South Korea against an invasion from North Korea and failed to win an exemption on this count
that would have allowed it to sign the treaty. 304

Too many countries regard these as legitimate weapons to make the prospect of a mine-free world likely. Costing as little as $2 each, they are very affordable weapons for cash-strapped warring LDCs (and so have been dubbed the "Saturday night specials of warfare"). For now, these hidden horrors continue to be planted and to be cleared, as one demining specialist put it, "one arm and one leg at a time."

bin Laden became the "world's most wanted" as U.S. President George W. Bush evoked a Wild West vow to have him captured "dead or alive." In crafting its pronounced war against terrorism (which it promised to carry anywhere in the world necessary), the U.S. administration identified the Taliban as al-Qa'ida's mentor and targeted both organizations for elimination.

Within a month of the attacks on the United States, American warplanes and special forces struck in **Operation Enduring Freedom** against Taliban and al-Qa'ida facilities and personnel around Afghanistan. The British military assisted in the air campaign, and most of the ground fighting was carried on by the Northern Alliance, the Taliban's rival and nemesis. The United States successfully persuaded Pakistan to drop its support of the Taliban and join the effort, and much of the military campaign was based on U.S.-Pakistani cooperation. One after another of the Taliban's
urban strongholds fell, including Kabul and the Taliban 303

spiritual capital, Kandahar. Withering assaults from U.S. warplanes crushed the tunnels and other hideouts used by al-Qa'ida guerrillas, and many al-Qa'ida prisoners were taken to the U.S. naval base in Guantánamo Bay, Cuba. The elusive bin Laden, however, along with the Taliban leader Mullah Omar, slipped away, perhaps seeking safe harbor in the ethnic Pashtun region of western Pakistan.

The reconstruction of Afghanistan began almost as soon as the Taliban were driven from power. From the billions of dollars in aid pledged, a new highway has been built connecting Kabul with Kandahar. Hospitals are being built in a country that has some of the highest infant and maternal mortality rates in the world. The numbers of Afghans enrolled in primary schools rose from 25 percent in 1999 to over 90 percent in 2007, and girls and women are again being educated. There are plans to replant landscapes that have been almost completely deforested. Afghanistan has resources of its own. Further prospecting may reveal that it has the world's largest copper deposits, and it is known to have large stores of high-grade iron ore, along with some gas, oil, coal, and precious stones.

Progress is slow because the country is so devastated and because of the tenuous security situation. American forces and the Afghan government headed by Hamid Karzai effectively took control of only a small area around Kabul, and NATO troops were stationed in the troubled south of the country. Much of the country remains in the hands of warlords whose allegiance is up for grabs. Al-Qa'ida and Taliban fighters operate in the rugged countryside, sometimes carrying fighting into Kabul and other cities. There are fears of a resurgent Taliban engaging Western forces in a conflict that would parallel the long war against the Soviets in Afghanistan. But if peace could prevail, it would be possible to rebuild this shattered land and bring hope to a country that is all too symptomatic of the troubled Middle East and North Africa.

SUMMARY

- → The United Nations Partition Plan of 1947 attempted a two-state solution to the dilemma Britain had created by promising land to both Arabs and Jews in Palestine. It envisioned geographically fragmented states, making each side feel vulnerable to the other. War prevented the plan's implementation. The Arab-Israeli conflict has continued from that time, with Israel gaining more territory and the indigenous Palestinians failing to acquire a country of their own. Principal obstacles to peace between Israelis and Palestinians are the status of Jerusalem, the potential return of Palestinian refugees, the future of Jewish settlements in the Occupied Territories, and Israeli construction of a security barrier that penetrates into portions of the Palestinian West Bank.
- → Israel's eastern neighbor, Jordan, has a majority Palestinian population.
- → In the mid-1970s, Israel's northern neighbor, Lebanon, was plunged into a devastating civil war that sparked when minority Christians failed to yield power to majority Muslims and was protracted by an Israeli invasion to crush the Palestine Liberation Organization. Lebanon's "Cedar Revolution" of 2005 led to Syrian troop withdrawal from the country, but the following year saw Israel attempt to crush the Shiite organization Hizbullah. That war destroyed much of Lebanon's rebuilt infrastructure and apparently strengthened Hizbullah politically.
- → Egypt is a populous country with a limited inhabitable area along the Nile. It has transformed the Nile Valley with thousands of years of habitation and recent large engineering works, but it remains a poor country dependent on outside assistance.
- → Sudan is a transition zone between the Middle East and sub-Saharan Africa. Its Muslim Arab government has tried to impose its will forcefully on the non-Arab south, promoting civil war. The conflict may have ended, but Sudanese Arabs have carried out ethnic cleansing against non-Arabs in the Darfur region of western Sudan.
- → Most of Libya lies in the Sahara, and its primary natural resource is oil. A massive irrigation project is tapping ancient underground waters. Colonel Muammar Qaddafi has given up support for international terrorism and weapons of mass destruction programs, leading to much improved relations with the West.
- → The Maghreb ("western land") is composed of Morocco, Algeria, Tunisia, and the disputed Western Sahara. Morocco is the region's poorest country, Algeria has the greatest mineral wealth, and Tunisia is the most progressive. Morocco controls Western Sahara, whose Saharawi people want independence.
- → Two-thirds of the world's proven petroleum reserves are concentrated in a few countries that ring the Persian-Arabian Gulf. Saudi Arabia controls more than one-quarter of the world's oil. Growing populations and the prospect of the oil running out present some critical challenges. Some of the countries, notably Saudi Arabia, are very traditional, whereas the United Arab Emirates and Bahrain have tolerated social change. On the southern end of the Arabian Peninsula, Oman has enjoyed oil wealth, but Yemen remains poor.
- → Under Saddam Hussein, Iraq squandered its oil wealth on military misadventures, including invasions of Iran and Kuwait. The Kuwait invasion led to an enormous U.S.-led counterattack that devastated the country's infrastructure and subjected it to years of economic sanctions. The U.S.-led invasion of Iraq in 2003 resulted in Saddam Hussein's downfall and death, and an

uncertain future for this ethnically diverse country (with large Shiite Arab, Sunni Arab, and Kurdish populations) engaged in civil war.

- Iran's oil revenues were used to modernize and Westernize the nation during the Pahlavi dynasty. Rapid social change and uneven economic benefits precipitated a revolutionary Islamic movement, which forced the shah to abdicate and flee in 1979. This theocratic state has had difficult relations with the United States, most recently over its alleged nuclear weapons program. Iran's young population is agitating for change, but the ruling clerics are resisting.
- Turkey was formerly the seat of power for the Islamic Ottoman Empire. It is a secular nation with membership in NATO, and hopes to join the European Union. It is blessed with freshwater resources that it has mustered with the hope of creating the "breadbasket of the Middle East."
- Afghanistan is one of the world's poorest nations. It is landlocked and has limited resources, poor internal transportation, and little foreign trade. Opium is its main export. Its location along the ancient caravan routes and adjacent to large oil reserves have made it the target of stronger powers. Backed by the United States, rebels succeeded in driving out a Soviet occupation force in the 1980s. That costly conflict was followed by a period of civil war during which outside countries neglected Afghanistan. The militant Islamist Taliban came to power and protected the presence and training of al-Qa'ida militants. Osama bin Laden and others planned anti-Western attacks, including 9/11, from Afghanistan. Al-Qa'ida was driven into hiding from there in U.S.-led attacks following 9/11, and the Taliban were deposed from power. Reconstruction of the country is progressing, but a Taliban insurgency continues.

KEY TERMS + CONCEPTS

1948–1949 Arab-Israeli war (p. 241)
1973 Arab-Israeli war (p. 242)
1982 Lebanese war (p. 250)
2006 Israeli-Hizbullah war (p. 251)
al-Aqsa *Intifada* (p. 246)
Alawite (p. 251)
al-Qa'ida in Iraq (p. 266)
Ashkenazi Jews (p. 243)
ayatollah (p. 268)
barrage (p. 252)
basin irrigation (p. 252)
Battle of Beirut (p. 250)
blowback (p. 273)
Camp David Accords (p. 243)
cataract (p. 252)
Cedar Revolution (p. 251)
center-pivot irrigation (p. 255)
civil war (p. 266)
divine rule by clerics (p. 268)
Druze (p. 243)
erg (p. 257)
"facts on the ground" (p. 244)
Falashas (p. 243)
final status issues (p. 246)
foggara (p. 257)
fossil waters (p. 255)
Gaza-Jericho Accord (p. 245)
ghurba (p. 244)
"gift of the Nile" (p. 252)
Great Game (p. 271)
Great Manmade River (p. 255)
Green Line (p. 246)
Gulf War (p. 264)
Intifada (p. 247)
Iran-Iraq War (p. 264)
Janjawiid (p. 255)
Jewish settlements (p. 244)
Judaea and Samaria (p. 244)
kibbutzim (p. 247)
Kurdistan Workers Party (PKK) (p. 265)
Labor Party (p. 247)
land mines (p. 274)
Law of Return (p. 243)
Lebanese civil war (p. 249)
Likud Party (p. 244)
Ma'adan ("Marsh Arabs") (p. 265)
moshavim (p. 247)
mujahidiin (p. 272)
mullah (p. 268)
Northern Alliance (p. 273)
Occupied Territories (p. 242)
Operation Enduring Freedom (p. 274)
Operation Iraqi Freedom (p. 265)
Oslo I Accord (p. 245)
Oslo II Accord (p. 245)
Palestine Liberation Organization (PLO) (p. 245)
Palestinian Authority (PA) (p. 245)
Palestinians (p. 241)
perennial irrigation (p. 252)
Polisario Front (p. 258)
pre-1967 borders (p. 241)
qanat (p. 257)
right of return (p. xxx)
"road map for peace" (p. 246)
Saharawis (p. 258)
salinization (p. 253)
Saudization (p. 260)
security fence (p. 246)
Sephardic Jews (p. 243)
Six-Day War (p. 242)
Southeast Anatolia Project (GAP) (p. 270)
Sudanese People's Liberation Army (SPLA) (p. 254)
"Switzerland of the Middle East" (p. 249)
Taliban (p. 273)
two-state solution (p. 241)
United Nations Resolutions 242 and 338 (p. 247)
Wahhabism (p. 260)

REVIEW QUESTIONS

1. What precipitated the various conflicts between Israel and its Arab neighbors? How did each of these conflicts rearrange the political map?
2. What lands did the peace process of the Oslo Accords yield to Palestinian control? What is the current situation in those areas? Who controls the Gaza Strip and the West Bank?
3. What are the major ethnic groups and territories of Lebanon?
4. How did Egypt transform the Nile Valley from a system of basin irrigation to one of perennial irrigation?
5. In what ways is Sudan's north different from its south? What significant conflicts have affected the country?
6. What is the role of the Great Manmade River?
7. What are the main resources of the Maghreb countries? What issues do they have with countries outside the region?
8. Which Persian-Arabian Gulf countries have the most oil and natural gas? Have Oman and Yemen benefited from the oil boom?
9. What happened when Saddam Hussein's troops invaded Kuwait? What happened when they invaded Iran? What critical decisions must the United States make in Iraq?
10. What is the ethnic composition of Iraq? How did that consideration cause the United States to avoid invading the country in 1991?
11. Who are the Kurds, and where do they live?
12. What were some of the grievances against the Pahlavi dynasty that led to revolution in Iran? What changes has the Islamic regime brought to the country?
13. In what ways is Turkey an "in-between" country?
14. What were Afghanistan's shifting fortunes between 1979 and the present?

NOTES

1. Quoted in Nicholas Sambanis, "It's Official: There Is Now a Civil War in Iraq." *New York Times,* July 23, 2006, p. WK-13.

Joe Hobbs

View across the Swat Valley in northern Pakistan, with the Hindu Kush Mountains rising along the Afghan border in the distance.

CHAPTER 7
A GEOGRAPHIC PROFILE OF MONSOON ASIA

Monsoon Asia is a great triangle that runs from Pakistan in the west to Japan in the northeast and the island of New Guinea in the southeast (•Figure 7.1). In this land area of less than 8 million square miles (20,761,000 sq km)—about twice the size of the United States (•Figure 7.2), or less than one quarter of the earth's landmass—lives more than half of the world's population (•Figure 7.3). The cultural makeup of these 3.6 billion people is as diverse and expressive as the physical environments they inhabit. The world's highest mountain peaks, some of the longest rivers, and the earth's most highly transformed and densely settled river plains are the settings for some of the oldest civilizations and the most modern economies. Just as this region has played a major role in human development over many millennia, it is a major force in world affairs in the 21st century. Some observers say this will be known as the **Asian Century** for the ways in which the region will reshape the globe.

chapter outline

chapter objectives

This chapter should enable you to

- → Recognize the role of seasonal monsoon winds and rains in people's livelihoods and perceptions
- → Appreciate China and India as the demographic giants and surging economies of early-21st-century Asia
- → Learn how unique spatial and spiritual considerations have influenced the traditional layout of Asian settlements
- → Know the basic beliefs of Hinduism, Buddhism, Confucianism, and Daoism
- → Become familiar with the pros and cons of the Green Revolution
- → Understand the geopolitical dimensions of the tensions between India and Pakistan, North Korea and the West, and Islamists and governments in Pakistan and Indonesia

---

The countries of eastern Eurasia were at one time referred to as "the Orient," from a Latin word meaning "facing the east" or "facing the sunrise." (The West was then known as "the Occident," derived from a word meaning "facing the sunset.") But this terminology reflected the view from Europe and became associated with Western stereotypes and misconceptions about Asia. Consequently, those terms have fallen out of favor. This text refers to the region shown in Figure 7.1 as Monsoon Asia because it connotes a major physical attribute that affects most of this region and also in acknowledgment of the achievements of the geographer George Cressey, who coined the name and wrote so well of the region.

7.1 Area and Population

Monsoon Asia encompasses the following subregions, which in some geography books are classified as separate regions:

- East Asia, which includes Japan, North and South Korea, China, Mongolia, Taiwan, and many near-shore islands
- South Asia, which includes Pakistan, India, Sri Lanka, Bangladesh, the mountain nations of Bhutan and Nepal, and the island country of the Maldives
- Southeast Asia, which includes both the peninsula jutting out from the southeast corner of the Asian continent, on which are located Myanmar (Burma), Thailand, Laos, Cambodia, Vietnam, Malaysia, and Singapore, and the island world that rings this peninsula, which includes the countries of Indonesia, the Philippines, Brunei, and Timor-Leste (Table 7.1; see also Figure 7.1).

Demographic Heavyweights of Monsoon Asia

Perhaps the most critical fact of Monsoon Asia's geography is that this region is home to about 54 percent of the world's population (•Figure 7.3; see also Figures 3.14 and 3.15, pages 57 and 58). Two countries alone, India and China, together have 2.4 billion people, or 37 percent of the world total. Their extraordinary demographic weight reflects thousands of years of human occupation of productive agricultural landscapes, combined with 20th-century advances in health technology that reduced death rates. 50
The highest population densities are found in the areas of most abundant precipitation or surface water and with the most productive soils; rivers and coastal plains are especially densely settled. Rugged mountains, almost waterless deserts, and until recently, tropical rain forest vegetation have limited populations elsewhere.

Some of the world's highest urban densities are found in the cities of Hong Kong, Macao, and Singapore, but in Mongolia and Nepal, population densities are extremely low. Singapore is one of the few nations to claim a 100 percent urban population; in contrast, only 8 percent of Timor-Leste's population is urban. A mere 3 percent of Bhutan is arable, whereas farmers in India and Bangladesh cultivate more than 50 percent of their lands.

Population Growth Patterns

It is not possible to generalize about population growth in Monsoon Asia. Growth rates range from 0 percent per year in Japan to a very high 3.3 percent per year in Timor-Leste. The region does reflect the general worldwide trends of poorer countries having higher growth rates and wealthier countries having lower growth rates. The region is demographically and otherwise recognizable as one of mainly less developed countries (LDCs), with some very noticeable exceptions. Japan may be seen as a classic postindustrial country and, like Italy, must worry about a declining, aging population in which young people will have to shoulder an increasing economic burden to support their elders. As 72, 297
discussed later in this chapter, there are other strong economies in the region.

One of the outstanding anomalies in Asia's population equation is China. With a per capita GNI PPP of just $7,730, China might be expected to have a rather high population growth rate like India, which has a similar economic ranking. But in the 20th century, China's population was so alarmingly large and rapidly growing that its Communist leaders decided to restrain its growth drastically. China adopted a coercive "one-child policy" that has brought its growth rate to just 0.5 percent per year, representing

• **Figure 7.1** Monsoon Asia.

a doubling time of 140 years (see Geographic Spotlight, pages 286–287). That is a remarkable turnabout from the situation in 1964, when China had a growth rate of 3.2 percent and a doubling time of 22 years.

To reduce its poverty, India has made great strides in reducing its population growth, but not in China's coercive fashion. India is projected to overtake China as the world's most populous country in 2040, when it will have an estimated 1.5 billion people. A major wild card in India's population deck is HIV/AIDS. The scale of the problem there is not certain; 2007 estimates put the number of Indians infected with HIV at between 2 and 5 million, with the U.S. Central Intelligence Agency projecting up to 25 million cases by 2010. The potential for a pandemic in the world's second-largest country is strong, and much will depend on how India's government chooses to fight the disease. About 80 percent of India's HIV infections are spread by heterosexual contact, and public health authorities have so far been shy to mount a strong campaign about sexual ethics and practices. China is thus far not in the peril that India is; 2007 estimates put the number of people infected at under 1 million, and plans were announced to provide free HIV tests for all Chinese citizens.

7.2 Physical Geography and Human Adaptations

Monsoon Asia's physical geography may be described broadly as consisting of three concentric arcs, or crescents, of land: an inner western arc of high mountains, plateaus, and basins; a middle arc of lower mountains, hill lands, river plains, and basins; and an outer eastern arc of islands and seas (•Figure 7.4, page 284).

TABLE 7.1 Monsoon Asia: Basic Data

Political Unit	Area (thousands; sq mi)	Area (thousands; sq km)	Estimated Population (millions)	Estimated Population Density (sq mi)	Estimated Population Density (sq km)	Annual rate of Natural Increase (%)	Human Development Index	Urban Population (%)	Arable Land (%)	Per Capita GNI PPP ($US)
South Asia	**1,732.6**	**4,487.4**	**1,499.3**	**865**	**334**	**1.7**	**0.595**	**28**	**47**	**3,480**
Bangladesh	55.6	144.0	149.0	2,680	1,035	1.9	0.530	23	62	2,340
Bhutan	18.1	46.9	0.9	50	19	1.3	0.538	21	3	N/A
India	1,269.3	3,287.5	1,131.9	892	344	1.6	0.611	28	54	3,800
Maldives	0.1	0.3	0.3	3,000	1,158	1.6	0.739	27	13	N/A
Nepal	56.8	147.1	27.8	489	189	1.9	0.527	14	21	1,630
Pakistan	307.4	796.2	169.3	551	213	2.3	0.539	34	28	2,500
Sri Lanka	25.3	65.5	20.1	794	307	1.2	0.755	30	14	5,000
Southeast Asia	**1,735.3**	**4,494.4**	**574.4**	**331**	**128**	**1.4**	**0.716**	**38**	**14**	**5,420**
Brunei	2.2	5.7	0.4	182	70	1.6	0.871	74	1	N/A
Cambodia	69.9	181.0	14.4	206	80	1.7	0.583	16	21	2,920
Indonesia	735.4	1,904.7	231.6	315	122	1.4	0.711	42	11	3,950
Laos	91.4	236.7	5.9	65	25	2.4	0.553	19	3	2,050
Malaysia	127.3	329.7	27.2	214	82	1.8	0.805	62	5	11,300
Myanmar	261.2	676.5	49.8	191	74	0.9	0.581	28	15	N/A
Philippines	115.8	299.9	88.7	766	296	2.1	0.763	48	19	5,980
Singapore	0.2	0.5	4.6	23,000	8,880	0.6	0.916	100	1	31,700
Thailand	198.1	513.1	65.7	332	128	0.7	0.784	31	29	9,140
Timor-Leste	5.7	14.8	1.0	175	68	3.3	0.512	8	4	N/A
Vietnam	128.1	331.8	85.1	664	256	1.3	0.709	25	20	3,300
East Asia	**4,545.6**	**11,773.1**	**1,543.0**	**339**	**131**	**0.5**	**0.770**	**48**	**13**	**10,320**
China	3,696.1	9,572.9	1,318.0	357	138	0.5	0.768	44	15	7,730
Japan	145.9	377.9	127.7	875	338	0.0	0.949	79	12	33,730
Korea, North	46.5	120.4	23.3	501	193	0.9	N/A	60	20	N/A
Korea, South	38.3	99.2	48.5	1,266	489	0.4	0.912	82	17	23,800
Mongolia	604.8	1,566.4	2.6	4	2	1.2	0.691	59	1	2,280
Taiwan	14.0	36.3	22.9	1,636	632	0.3	N/A	78	24	N/A
Summary Total	**8,013.5**	**20,754.9**	**3,616.7**	**451**	**174**	**1.1**	**0.688**	**38**	**20**	**6,710**

Sources: World Population Data Sheet, Population Reference Bureau, 2007; Human Development Report, United Nations, 2006; World Factbook, CIA, 2007.

The inner arc includes the world's highest mountain ranges, interspersed with plateaus and basins. In the south, the great wall of the Himalaya, Karakoram, and Hindu Kush Mountains overlooks the north of the Indian subcontinent. In the north, the Altai, Tien Shan, Pamir, and other towering ranges separate this region of Asia from the countries of central Asia. Between these mountain walls lie the sparsely inhabited Tibetan Plateau, at over 15,000 feet (c. 4,500 m) in average elevation, and the dry, thinly populated basins and plateaus of Xinjiang (Sinkiang) and Mongolia.

The middle arc, between the western inner highland and the sea, is occupied by river floodplains and deltas bordered and separated by hills and low mountains. The major features of this area are the immense alluvial plain of northern India, built up through ages of meandering and deposition by the Indus, Ganges, and Brahmaputra Rivers; the

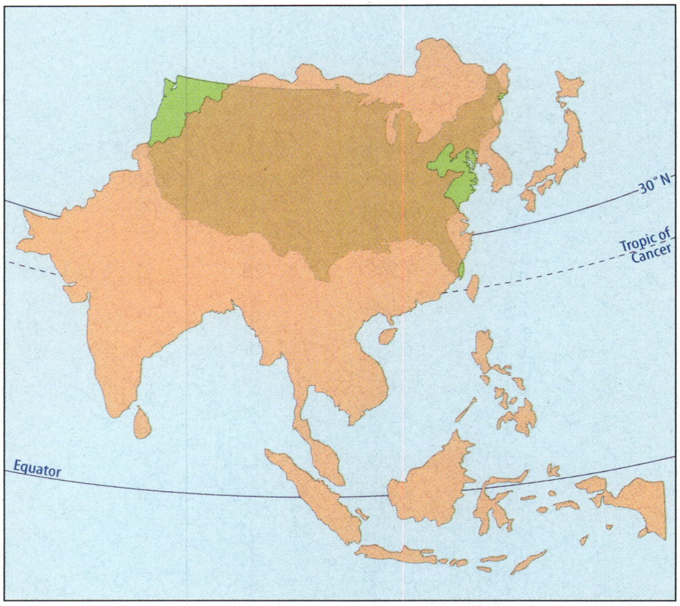

• **Figure 7.2** Monsoon Asia compared in latitude and area with the conterminous United States.

hilly uplands of peninsular India, geologically an ancient plateau; the plains of the Irrawaddy, Chao Praya (Menam), Mekong, and Red Rivers in the Indochinese Peninsula of Southeast Asia, together with bordering hills and mountains; the uplands and densely settled small alluvial plains of southern China; the broad alluvial plains along the middle and lower Chang Jiang (Yangtze River) in central China and the mountain-girded Red Basin on the upper Chang Jiang; the large delta plain of the Huang He (Yellow River) and its tributaries in northern China, backed by loess-covered hilly uplands; and the broad central plain of northeastern China (Manchuria), almost enclosed by mountains.

The outer arc is an offshore fringe of thousands of islands, mostly grouped in great **archipelagoes** (clusters of islands) bordering the mainland. On these islands, high interior mountains (including volcanoes) are flanked by coastal plains where most of the people live. Most of the islands are in three major archipelagoes: the East Indies, the Philippines, and Japan. Sri Lanka, Taiwan, and Hainan are large, densely populated islands outside these archipelagoes. Between the archipelagoes and the mainland lie the East and South China Seas and, to the north, the Sea of Japan (known in Korea as the East Sea). At the southwest, the Indian peninsula projects southward between two immense arms of the Indian Ocean: the Bay of Bengal and the Arabian Sea.

Climate and Vegetation

Monsoon Asia is characterized generally by a warm, well-watered climate, but there are nine distinct types of climate in the region: tropical rain forest, tropical savanna, humid subtropical, warm humid continental, cold humid continental, desert, steppe, subarctic, and undifferentiated highland (•Figure 7.5, left). These are generally associated with predictable patterns of vegetation, which have been heavily modified by millennia of human use (•Figure 7.5, right).

The tropical rain forest climate zone and vegetation types are typically found within 5 to 10 degrees of the equator and are characteristic of most parts of the East Indies, the Philippines, and the Malay Peninsula. Precipitation is spread throughout the year so that each month has considerable rain. The amount of precipitation is at least 30 to 40 inches (c. 75 to 100 cm) and often 100 inches (c. 250 cm). Average temperatures vary only slightly from month to month; Singapore, for example, has a difference of only 3°F between the warmest and coolest months. High temperatures prevail year-round, although some relief is afforded by a drop of 10° to 25°F (6° to 14°C) at night, and sea breezes refresh coastal areas. Winter in climates such as this, it can be said, "comes only at night."

The tropical savanna climate, like the tropical rain forest climate, is characterized by high temperatures year-round but is generally found in areas farther from the equator, and the average temperatures vary more from month to month. The savanna has a better-defined dry season, lasting in some areas as long as six or eight months each year, creating a problem for agriculture. Tropical savanna climates occur in southern and central India, in most of the Indochinese Peninsula, and on eastern Java and the smaller islands to the east.

In Asia, the characteristic natural vegetation associated with this climate is a deciduous forest of trees smaller than those of the tropical rain forest. Tall, coarse grasses such as those found on African and Latin American savannas grow only in limited areas.

The humid subtropical climate zone prevails in southern China, the southern half of Japan, much of northern India, and a number of other countries. It is characterized by warm to hot summers, mild or cool winters with some frost, and a frost-free season lasting 200 days or longer. The annual precipitation of 30 to 50 inches (c. 75 to 125 cm) or more is well distributed throughout the year, although monsoonal trends produce a dry season in some areas. The natural vegetation is a mixture of evergreen hardwoods, deciduous hardwoods, and conifers.

The humid continental climates characterize the northern part of eastern China, most of Korea, and northern Japan. They are marked by warm to hot summers, cold winters with snow, a frost-free period of 100 to 200 days, and less precipitation than the humid subtropical climate. Most areas experience a dry season in winter. The main natural vegetation is a mixture of broadleaf deciduous trees and conifers, although prairie grasses probably formed the original cover in parts of northern China. Extreme northeastern China and northern Mongolia have a subarctic climate associated with coniferous trees.

Steppe and desert climates are found in Xinjiang and Mongolia and in parts of western India and Pakistan. Due

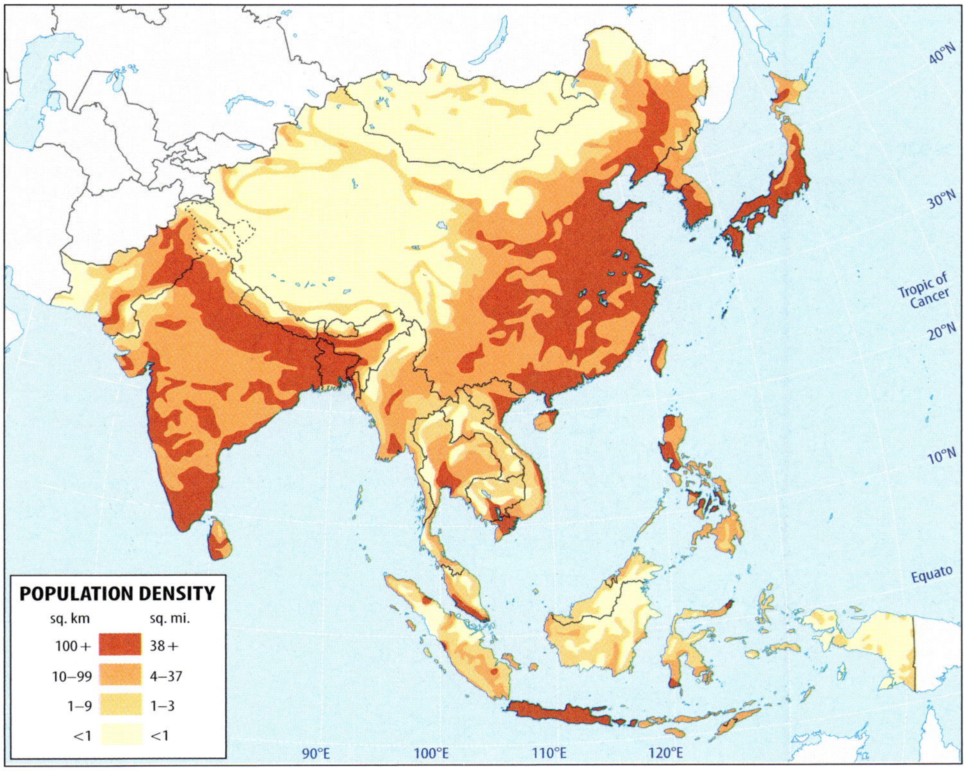

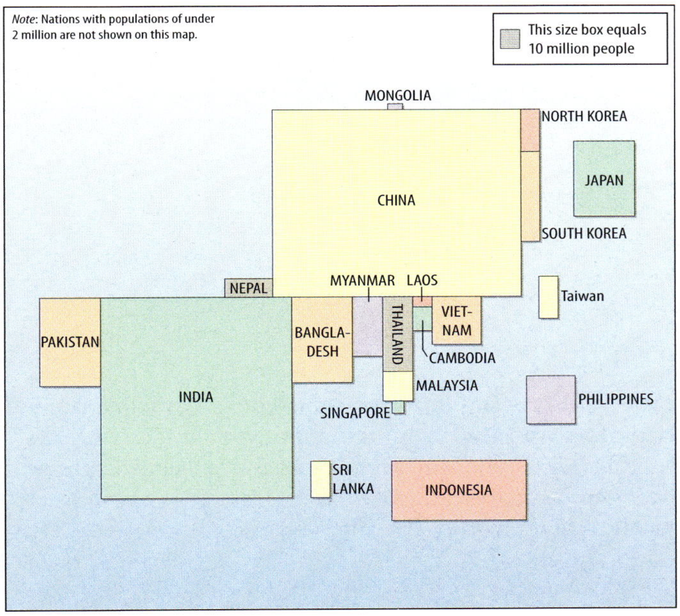

Figure 7.3 Population distribution (top) and population cartogram (bottom) of Monsoon Asia.

to limitations on agriculture, along with very high mountains, they are the region's least inhabited climate zones. The undifferentiated highland climate is most extensive in the Tibetan Highlands and adjoining mountain areas. It is made up of a broad range of montane microclimates that limit plant growth and vary from place to place according to elevation, aspect, and latitude.

The Monsoons

The **monsoons** are the prevailing sea-to-land and land-to sea winds that are the dominant climatic concern for people living in this world region (Figure 7.6). They play a significant role in both wet and dry environments and are especially influential in the coastal plains and lowlands of South Asia, the peninsula and islands of Southeast Asia, and the eastern third of China. The ways in which the sea and land absorb **insolation** (heat from the sun) are different, causing the instability in air masses that creates monsoonal winds. If coastal waters and the adjacent coastline receive about equal amounts of warmth from the sun, the land becomes warm or hot much more quickly than the seawater. As a result, air over the land begins to become unstable and to rise. This ascending air creates a low-pressure attraction for the air masses over the water, and marine winds begin to blow toward the land. In Asia, this is the **summer monsoon,** characterized by high humidity, moist air, and generally predictable rains. Because of the moisture carried by such air masses, these wind shifts are the sources of major rainfall in the late spring, summer, and early fall seasons. Agriculture and patterns of human activity are keyed to these incoming rains.

If there is a little bit of elevation in the landscape—hills or mountain flanks—the moist air is driven higher, where it cools and releases even more rain in the pattern of orographic precipitation.

In the **winter monsoon,** the land loses its relative warmth while the sea and coastal waters stay warm longer. As a

• **Figure 7.4** Physical geography of Monsoon Asia.

result, the wind shifts and air masses begin to flow from the inland areas of Asia toward the sea. However, because there is little moisture in the source areas over land for the winter monsoon, little moisture is picked up, and much less rain results. A monsoon climate, overall, is characterized by spring and summer precipitation and a long dry season in the low-sun (winter) cycle. Japan, however, experiences an unusual pattern of winter precipitation related to the monsoon wind shifts. The winter winds blow out of China and East Asia, sweep across the Sea of Japan (East Sea), and pick up moisture, dropping heavy, wet snow on the west coast of the Japanese islands.

The wet summer monsoon is of critical importance to the livelihoods of a large portion of humanity, particularly the people of India. By early May in India, the land has typically been parched by months of drought, and the rising temperatures become almost insufferable. No thought can be given to farming until the rains come. Anticipation and discomfort rise steadily until the monsoon finally "breaks," bringing great relief in the form of much needed rain and lower temperatures (•Figure 7.7). Sometimes the onset of the rain comes with particular ferocity, as related by this Indian writer:

> That year the monsoon broke early and with an evil intensity such as none could remember before. It rained so hard, so long and so incessantly that the thought of a period of no rain provoked a mild wonder. It was as if nothing had ever been but rain, and the water pitilessly found every hole in the thatched roof to come in, dripping onto the already damp floor. If we had not built on high ground the very walls would have melted in that moisture.[1]

India and Bangladesh have experienced particularly heavy early monsoonal rains of this kind for the past three

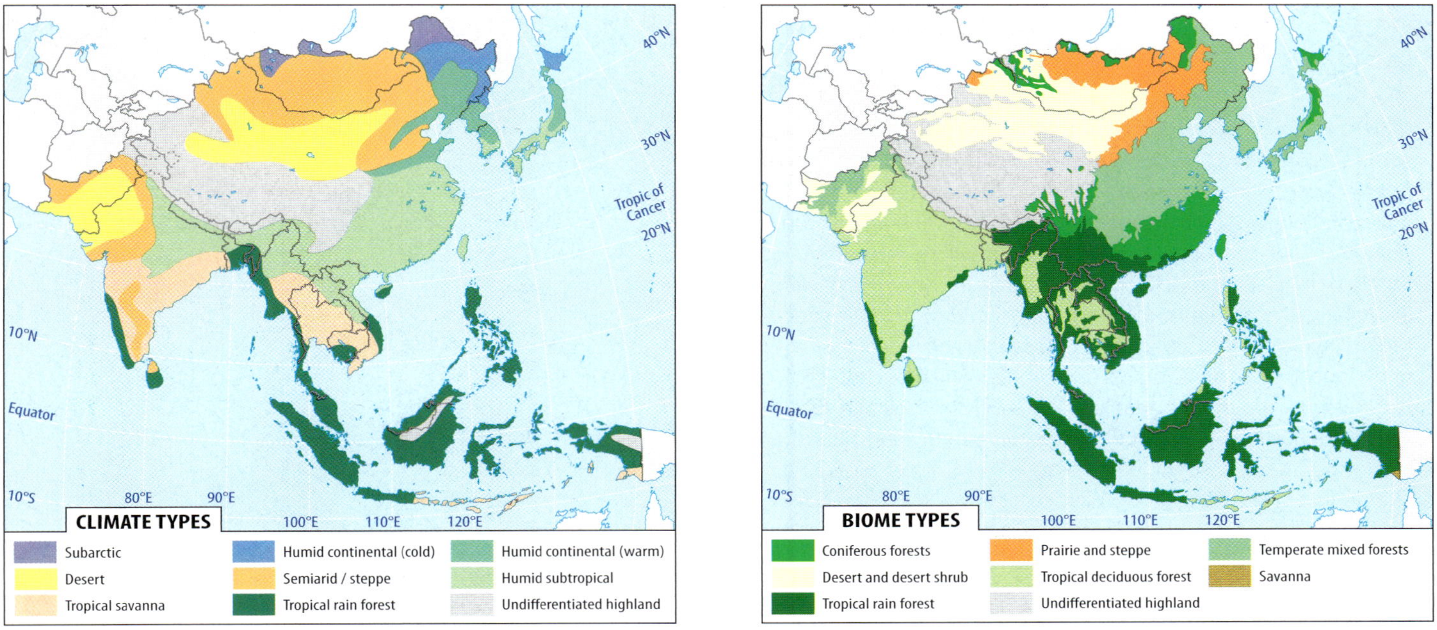

• **Figure 7.5** Climates (left) and biomes (right) of Monsoon Asia.

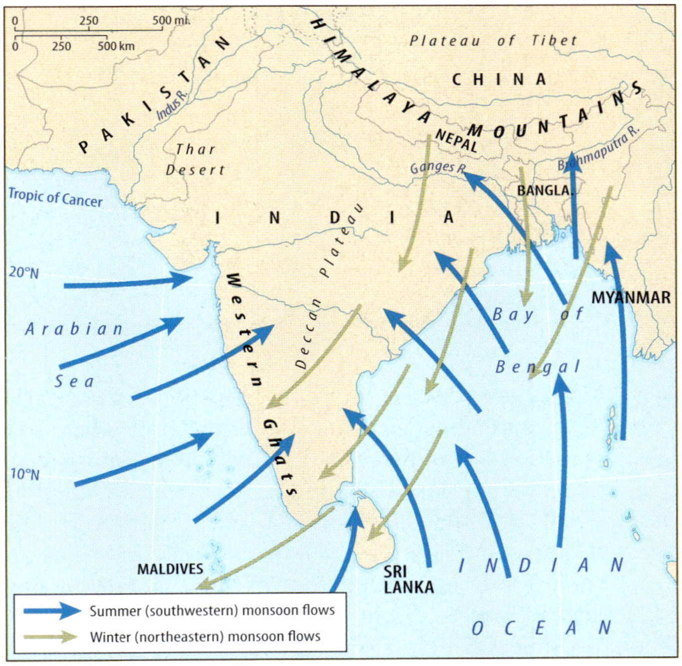

• **Figure 7.6** The monsoons at work.

• **Figure 7.7** Torrential monsoon rains are regular and usually welcome features of land and life in much of the region. This is a flooded road in south central Sri Lanka.

decades, with significant loss of life. Inevitably, some scien-
34 tific fingers point to global climate change as the culprit.

Agricultural Adaptations

The wet and warm climates associated with the monsoons might be expected to offer outstanding opportunities for agriculture (•Figure 7.8). However, the high temperatures and heavy rains promote rapid leaching of mineral nutrients and decomposition of organic matter, so many soils are infertile. The lush vegetation cover in these low-latitude areas is deceiving because most of the system's nutrient matter is in the vegetation itself rather than in the underlying soil. When Asian farmers clear the tree cover, they often find that the local soils will not support more than one or two poor harvests (see Insights, page 289).

The Importance of Rice

Agriculture in Monsoon Asia—with the principal exception of Japan—is characterized by the steady input of arduous manual labor. Known as **intensive subsistence agriculture,** it is built around the growing of cereals, especially

GEOGRAPHIC SPOTLIGHT

China's One-Child Policy

China's population of 1.3 billion represents nearly one-fifth of the world's people and is increasing by 6.6 million each year. Population is a very serious matter for a less developed country whose area of about 3.7 million square miles (9.4 million sq km) is only slightly smaller than the United States but whose inhabitants outnumber the U.S. population by about five to one. But a glimpse at China's age structure diagram, with its wide middle ages but tapering youth, reveals that China's is one of the world's great success stories in terms of population management (• Figure 7.A).

The drive for smaller families was motivated by the fact that China's population was surging while per capita food output was increasing at only a very modest rate. In 1976, China's Communist regime instituted one of the most stringent and most controversial programs of birth control ever attempted. It culminated in the **one-child campaign,** which aims to limit the number of children per married couple to one (• Figure 7.B). The government tracks individual family birthing patterns—especially in the cities—and maintains surveillance through local authorities. It dispenses free birth control devices, free sterilization operations, free hospital care in delivery, free medical care for the child, free education for the child, an extra month's salary each year for parents who comply, and other favors and preferences to induce compliance with the one-child-per-family norm.

Those who violate the norm are subjected to constant social and political pressure, denial of the privileges accorded one-child families, pay cuts, and fines. A woman who becomes pregnant after having had one child is pressured to have a free state-supplied abortion, with a paid vacation provided. China's government hopes to continue to decrease childbearing so that population will stabilize

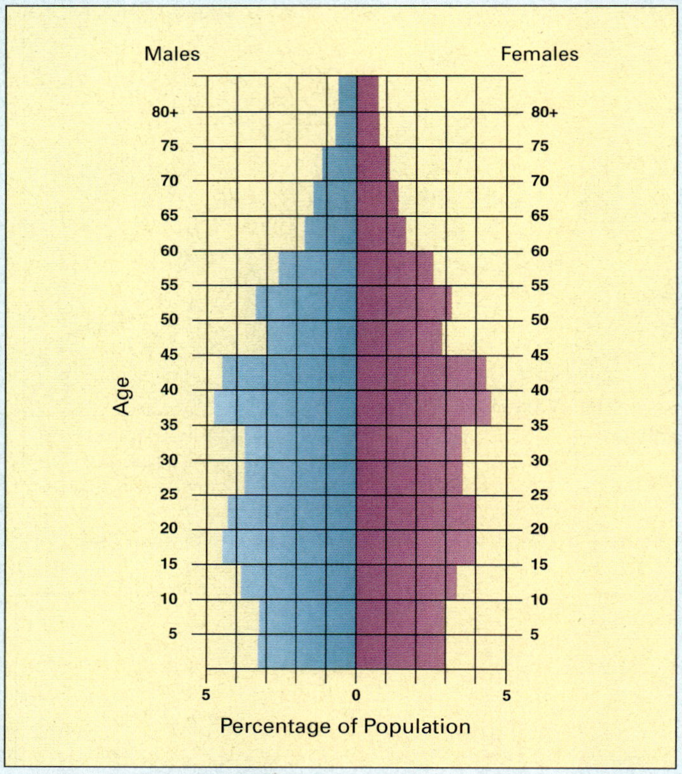

• **Figure 7.A** The age structure diagram of China (2008) reveals the country's success in slowing population growth.

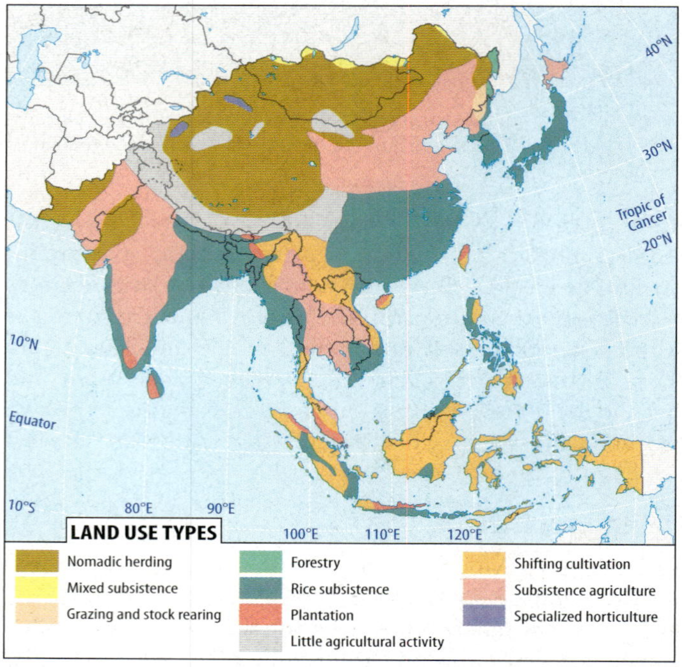

• **Figure 7.8** Land use in Monsoon Asia.

rice, which is the premier staple of Monsoon Asia and the crop of choice in areas with adequate rainfall or where irrigation waters are available (see the areas of rice subsistence in Figure 7.8). Farmers use organic fertilizers from both their animals and their own latrines and irrigate to the extent that they can or need to. Where natural conditions are not suitable for irrigated rice, grains such as wheat, barley, soybeans, millet, sorghum, or corn (maize) are raised. Where possible, farmers grow cash crops for local markets. Large-scale, commercial plantation agriculture is described in Module 7.2 on Southeast Asia, the subregion in which cash cropping has had the strongest role.

Shifting cultivation represents a precarious adaptation to local soil and climate conditions, capable of sustaining only small populations for brief periods in particular locales. At the other end of the spectrum of human adaptation to Asian lands is **wet rice cultivation,** capable of producing two or three crops per year and sustaining large populations over long periods of time. Wet rice agriculture takes place both in lowland floodplains and on carefully constructed upland **terraces** requiring enormous inputs of human labor (the term **paddy,** also spelled **padi,** may refer to the rice itself or the

GEOGRAPHIC SPOTLIGHT

China's One-Child Policy *continued*

Joe Hobbs

• **Figure 7.B** Strong disincentives mean one child is the norm for Chinese couples.

before reaching projections of 1.4 billion by 2010 or 1.5 billion by 2025. The higher number may be more likely because government officials report widespread disregard of the "one couple, one child" policy. Reasons for noncompliance vary. In some cases, parents feel wealthy enough to pay the fines of 50,000 yuan ($6,200) or more per extra child. Many rural dwellers think their extra children will escape detection by authorities, who are most effective in the cities. Anticipating their years as elders, some couples want the "safety net" of additional children—especially boys. If a young urban couple has just one child and it is a girl, it is likely that she will marry and become, by Chinese tradition, such a part of the husband's family that she may not be available to care for her parents as they get older. China has traditionally used the son's family as the accommodation of choice for senior citizens.

Some parents want to produce a son if their first child was a girl, and many girls are given to orphanages (China is the world's leading source country of adopted children, about 90 percent of them girls). Female feticide is increasingly associated with the desire to have a son. Couples often insist on an ultrasound test to determine the baby's gender, and about 90 percent of girls detected in this way are aborted. In some provinces where the practice is widespread, three boys are born for every two girls. As in India, this practice is beginning to have serious repercussions, reflected
in a shortage of eligible brides. Rural women have been abducted 321
and trafficked as virtual slaves to other parts of the country where they become brides. On the positive side, the relative shortage of women in China has helped enhance their status, particularly in urban areas, where they are also becoming more influential in the workplace.

field in which it is grown). Terraces have been built almost everywhere in Asia where there is a combination of sloping land, available irrigation water, and productive soil (•Figure 7.9). The dramatic stairstep terraces of the island of Luzon in the Philippines are the most famous, but equally dramatic cultural landscapes can be found in Pakistan's Punjab, in South China, and in many parts of Southeast Asia.

Terraces have the important function of preventing or dramatically slowing soil erosion, as the terrace walls allow heavy rainfall to run off without carrying topsoil away. Doing what the Chinese call "teaching water," people use canals and pipes to direct irrigation water from a source to the terrace, which acts like a kind of dam to hold the water in place. Wet rice is actually immersed in standing water for extended periods, creating an environment unfavorable to many crop pests. The productivity of wet rice agriculture is sometimes exceptionally high, supporting as many as 500 people per square mile (200/sq km) on the volcanic soils of Java and Bali in Indonesia. This extraordinary output is a contributing factor to the high densities and overall populations of some of the Asian countries and regions, including Java. One anthropologist has observed that most of Monsoon Asia's rice is eaten within walking distance of where it is grown.[2]

Is There a Correlation between Agriculture and Culture?

Anthropologists have observed some unique cultural adaptations to opportunities for terraced wet rice. Group welfare takes precedence over individual interests, for example. Individual farmers cannot cultivate at will but must work with the community to ensure that crops are staggered over time. This allows water to be shared fairly and efficiently and gives some fields the chance to dry out while others are irrigated. This agricultural system also prevents the use of tractors and other mechanical technology. Over much of Asia today, just as in ancient times, wet rice cultivation is the productive of intensive work by many human hands.

In some cases, population growth has outpaced the ability of even the most productive wet rice farming to feed the people, creating the classic scenario of "people overpopu-
lation." In Nepal, for example, high rates of population 61
growth on limited areas of moderately sloped, terraceable

Joe Hobbs

• **Figure 7.9** These meticulously crafted and maintained terraces produce a high yield of rice for the Hmong people of northern Vietnam.

57 land have been a push factor behind migration. After the 1950s, when malaria began to be eradicated in the lowland Terai region, millions of Nepalis migrated there, cleared the forests, and planted crops. For decades, the Terai served as a virtual "safety valve" for overpopulation in Nepal. Now that the Terai itself is mostly settled, there is nowhere for landless Nepalis to go but up. They clear and cultivate increasingly steep lands on which it is impossible to build terraces. Without terraces in place to hold the topsoil, heavy monsoon rains cause relentless erosion of their farmlands. Locally, the consequences are that the eroded plots cannot be cultivated again, and landslides occur downslope, often causing loss of life.

Much farther away, in the lowlands of India and Bangladesh, the increased sediment load contributed by erosion in Nepal causes rivers to swell out of their banks and bring extensive flooding—again, with losses of crops and lives. This sequence of events, beginning with overpopulation in Nepal and ending with flooding in Bangladesh, has been described as the **theory of Himalayan environmental degradation.**[3]

Where Asians Live

Monsoon Asia has some of the world's largest cities, including the greatest of all, Tokyo, with a metropolitan population of 37 million. About 1.3 billion people, or 38 percent of the region's total, live in urban settlements. But 2.3 billion, or 62 percent, do not. The main unit of Asian settlement is the village. Most of Asia's roughly 2 million villages are essentially groups of farmers' homes, although some villages are home to other occupational groups, such as miners or fisherfolk. Clusters of houses bunched tightly together are typical, and inexpensive and simple structures—often made of local clays and other building materials—are characteristic. Indoor plumbing continues to be the exception in village homes, but electricity lines have been extending steadily throughout Asia. So have cell phones and satellite dishes; it is not unusual today to see what appear to be very "traditional" Asians enjoying some of the latest innovations in global communications (•Figure 7.10). In Monsoon Asia, as elsewhere in the developing world, many people are making the jump directly from 19th- to 21st-century technology. Wireless technology is especially 47
appropriate where landlines and other infrastructure are difficult and expensive to install and where bureaucracies often slow down applications for access to utilities.

The original site selection for Asian villages was usually closely related to natural conditions and to perceived spiritual circumstances as well (see Geography of the Sacred, pages 290–291). With consideration always given to the possibility of flooding monsoon rains, lowland villages tend to be situated on natural **levees** (raised riverbanks built up by deposition of sediments during floods), dikes, or raised mounds. Early villages in Indonesia were often built in defensible mountain sites. European colonists sometimes

Joe Hobbs

• **Figure 7.10** Across Monsoon Asia, satellite dishes sprout incongruously from traditional dwellings.

INSIGHTS

Shifting Cultivation

Many Asian farmers, particularly in tropical rain forest regions, practice **shifting cultivation** (also known as **swidden cultivation** or **slash-and-burn cultivation**) or land rotation between crop and fallow years (Figure 7.C). The procedure begins with people clearing forest cover and then burning the dried vegetation during the dry season. Ash from the burn provides short-lived fertility to the soil in which farmers plant their subsistence crops. Most tropical soils lose their natural fertility quickly when they are cropped and after two or three harvests must be rested for several years (often 10 to 15 years) before they will again produce a crop. During this **fallow** period, the land reverts to wild vegetation.

The system of shifting agriculture is widespread in the tropics. Although it is not very productive, it does provide a bare living for people too poor to afford fertilizer or farm machinery. Traditionally, this system has minimized soil erosion because most of the land is not in cultivation at any given time. Unfortunately, there has been a recent widespread trend for farmers to reduce the length of the fallow period in the cycle of shifting cultivation or to abandon it altogether. This is a direct result of a surging growth in population: with more mouths to feed, in many places there is simply not enough land to provide the "luxury" of letting it lie fallow.

The tragic consequence is that short-term overuse of the land is eroding the soil. On the **lateritic soils** characteristic of many tropical regions, excessive exposure of the land to the sun's ultraviolet radiation and to repeated wetting and drying turns the ground to a bricklike substance called **laterite** or **plinthite,** which is essentially impossible to cultivate. When the best available lands are exhausted, farmers are often "marginalized" to places unsuitable for farming, such as semiarid lands or slopes that are too steep to till without causing disastrous erosion. Such dilemmas of land use are typical of the world's tropics from Indonesia to West Africa and 46, 48
Amazonia.

Joe Hobbs

Figure 7.C Shifting cultivation (swidden or slash-and-burn cultivation) is widespread in the world's tropical regions and can be sustained as long as the land is given sufficient time to recover.

rejected traditional settlement patterns to serve their own interests. Dutch colonial administrators in Indonesia, for
57, 357 example, required that villages be built along main roads and trails in the lowlands, making it easier to exercise control, collect taxes, and draft soldiers or laborers for roadwork and other projects.

The majority of people in Monsoon Asia are rural, but the region's future, in terms of demographic weight, economic power, and cultural change, is unquestionably focused on the cities. Some of the countries are already highly urbanized, notably South Korea (82 percent), Japan (79 percent urban), and Taiwan (78 percent) and entities like Singapore and Brunei that are effectively city-states. The layouts and architectures of the cities span a continuum from the ancient, walled fortress city to the modern Western-style metropolis, and many, like Beijing in China and Lahore in Pakistan, accommodate both. A striking pattern across the region is the steady stream of rural-to-urban migration, consisting of people motivated by both push and pull factors to leave the relative poverty of the countryside
57 in search of employment and even prosperity in the city. Many Asian cities are hard-pressed to provide accommodation and services to their swelling populations, and where the political system allows (notably in China), restrictions on internal migrations are imposed.

7.3 Cultural and Historical Geographies

Monsoon Asia has been home to some of the most important cultural developments of humankind in landscape transformation, settlement patterns, religion, art, and political systems. From Korea (not Gutenberg's Germany, as is widely thought) came the first movable printing type. From China came gunpowder, paper, silk, and china (porcelain). From India came the great faiths of Hinduism and Buddhism. Handmade textiles, artwares, bone and leather and paper products, and precious metals and gems have flowed from Asia into global trade for millennia. Spices, foodstuffs, and exotic plants and products from the East Indies have been major commodities in Asian-European trade and commerce for more than five centuries. Some of the world's most important domesticated plants and animals originated in Monsoon Asia, including rice, cabbage, chickens, water buffalo, zebu cattle, and pigs.

Ethnic and Linguistic Patterns

As would be expected of such a vast region, in many places fragmented by islands and mountain barriers, the ethnic and linguistic composition is rich and complex (Figure 7.11).

GEOGRAPHY OF THE SACRED

The Korean Village

In the West, people are accustomed to communities laid out on a perfect grid of streets intersecting at right angles, with their homes and businesses symmetrically fronting those streets. But in large parts of Asia, settlement and residence considerations are far more varied, especially because they are tied to spirituality, aesthetics, and topography. Villages, and the homes in particular, tend to be arranged for maximum harmony with the natural and spiritual environments—a practice known as **geomancy**—and constitute a special realm that may be considered sacred space. The traditional Korean village illustrates these unique Asian qualities.*

From ancient times, Koreans developed ways of determining the most auspicious village sites. The principles of *feng shui* and of Confucianism played strong roles in how they adapted settlements and homes to geographic conditions. ***Feng shui*** (pronounced "fung-*shway*'") is a traditional theory for selecting favorable sites for buildings, homes, and cities and for guiding many other urban and architectural planning decisions.† It is essentially an ordering device for environmental planning and for daily living. *Feng shui* (the words mean "wind" and "water") attempts to balance ***yin*** and ***yang*** to achieve harmony. *Yin* (conceived of as negative, dark, and feminine) and *yang* (representing the positive, bright, and masculine) stand for all the dualistic, opposing forces that may be reconciled harmoniously, including female and male, night and day, moon and sun, cold and hot, and right and left. *Feng shui* principles have begun to appear in American suburban home architecture, as in houses built to conceal overhead beams and sharp protrusions (which are disharmonious) and provided with windows opening to the east (a harmonious situation).

The theory of *feng shui* and the Confucian view of nature have some common characteristics that affect site, settlement, and residence designs. Confucianism reflects the sociopolitical philosophy of the Chinese philosopher known in the West as Confucius (see page 296). One of the main Confucian principles is that architecture and the lived environment should be in harmony with nature. This can be seen in the Confucian institutes for higher learning, called *seowon,* which in Korea peaked in the Joseon Dynasty, around 1400 C.E. Each *seowon* had a lecture hall where Confucian scholars gathered to learn philosophy. The buildings were laid out in such a way that from the inside of the complex, the scholars could see distant landscapes. The scholars' minds and spirits were enhanced as they gazed at the mountains and contemplated nature's grandeur. They believed that the earth's power was an invisible force that would help them become men of dignity and eminence.

The principles of *feng shui* and Confucianism are also apparent in the design of Korean villages. Each village is laid out to achieve harmony with the mountains—an important consideration in a land that is 70 percent mountainous. Rising behind many villages is a mountain, called *jinsan,* which is a spiritual focus for the villagers. *Jinsan* serves as the village symbol and guardian. Typically, a small stream flows in front of the village, and beyond it, there is a flat expanse of farmland. Beyond those fields is another mountain, called *ansan,* facing the village (•Figure 7.D).

These geographic conditions are economically advantageous to the community, providing it with productive fields and a good supply of water. But just as important, this situation is psychologically, spiritually, and symbolically beneficial to the villagers. The natural features are meant to embrace the village, as Koreans say a mother would hold a child. Korean historical records describe the traditional village as a place of comfort and serenity that safeguards against negative natural forces.

Central and South Asia gave birth to the **Indo-European language family,** which contains tongues as diverse as English and Hindi. **Hindi** and **Urdu,** which are major languages in India and Pakistan, respectively, are in the **Indo-Iranian subfamily** that includes such other major South Asia languages as **Pashto** (spoken in western Pakistan and Afghanistan), **Sanskrit** (which is extinct as a vernacular language but is still used in Hindu ritual), **Sinhalese** (the majority language of Sri Lanka), **Punjabi** (spoken on both sides of the northern frontier between Pakistan and India), Nepal's national language of **Nepali, Bengali** (the language of Ban-
81, 140 gladesh), and Romany, the tongue of the Roma (Gypsies) of Europe.

When the first of the Indo-European languages was carried from central Asia into India—apparently by the ethnic group known as the **Aryans,** between 2000 and 1000 B.C.E., languages in the ancient and indigenous **Dravidian language family** were being spoken there by peoples who had much darker skin than the newcomers. The Dravidian languages, today found almost exclusively in southern India and northern Sri Lanka (with the exception of a small pocket in Pakistan), include **Tamil, Malayalam,** and **Telegu.** Along the eastern side of the Himalayan mountain wall separating South Asia from China, people speak languages in the **Tibeto-Burman subfamily** of the **Sino-Tibetan language family. Tibetan,** spoken farther north in Tibet, and **Burmese,** spoken to the east in Myanmar (Burma), are also in the Tibeto-Burman subfamily.

The greatest numbers of people in China speak languages in the **Sinitic subfamily** of the Sino-Tibetan languages, overwhelmingly **Chinese** and its variants, including **Mandarin, Xiang, Min,** and **Yue.** The Chinese government has adopted Mandarin (Putonghua) as the country's official dialect, but only about half the population can speak it. Almost all of these more than 1 billion speakers of about 1,500 Chinese dialects belong to the **Han Chinese** ethnic

GEOGRAPHY OF THE SACRED

The Korean Village *continued*

Symbolic guardians are associated with the topography. Ideally, there is the Green Dragon to the east, the White Tiger to the west, the Black Tortoise to the south, and the Red Bird to the north. These guardians, named for figures in Chinese astronomy, could have varied manifestations. The Green Dragon, for example, could be a hill or a river to the east of the village, while the White Tiger could be a mountain or a road on the west side. One enters the village through a symbolic portal. For many villages, a pair of "spirit posts" stands at the foot of a nearby mountain. The posts mark the preliminary boundary to the outside, and they ward off evil spirits.

Not only is the village site selection process different from that in the West, but the settlement pattern is uniquely Asian too. Unlike in Western settlements, no more than two houses should be built side by side, they should not stand parallel to each other, and their gates should never face each other. The side street should curve naturally between the home lots. Overall, then, Koreans choose to emphasize the surrounding topography rather than geometric order in the construction of their communities and homes.

The traditional Korean home is adapted to the natural environment, embracing its surroundings while offering protection, comfort, and order. As in the *seowon,* the layout brings outdoor scenery into the house itself. Homes are also adapted to the environment in their construction materials. A thatched roof and earthen walls effectively block heat in the summer and insulate in winter, a perfect combination for Korea's sweltering summers and raw winters. The mulberry paper covering lattice windows diffuses sunlight, providing a pleasant natural light while absorbing sound and allowing some ventilation. The home's eaves are built so that sunlight penetrates deep into the house during the cold weather but never during summer—a rather recent innovation in the West, where it is known as "passive solar" design.

Joe Hobbs

Figure 7.D The Korean village of Yangdong is nestled between mountains that are separated by a river and lush rice fields.

*This discussion is based largely on two unpublished manuscripts, "Living Spaces of Korean Architecture" and "In Tune with Nature: Rural Villages and Houses in Korea," by the Korean geographer Sang-Hae Lee, of Sungkyunkwan University, and personal talks with him. I am grateful for his contributions.

†A useful introduction to *feng shui* may be found at FengShuiHelp.com, http://www.fengshuihelp.com/history.htm.

group, which is culturally, politically, and economically the dominant ethnic group of China. Northern and northwestern China, along with Mongolia, are home to speakers of languages in the **Altaic language family,** including **Uighur, Kazakh, Turkic, Mongolian,** and **Tungus.** Cut off from their Altaic-speaking kin are the **Korean** speakers of the Korean Peninsula and the **Japanese** speakers of Japan. Japan's indigenous **Ainu** language has no known linguistic relatives and is nearly extinct.

The people of Taiwan include Chinese-speaking migrants from China and indigenous speakers of an unrelated language. Most linguists assign this tongue to the **Austronesian subfamily** of the more widespread **Austric language family.** Extreme southern China and adjacent portions of Vietnam and Laos are linguistically complex, reflecting the presence of many minority ethnic groups. Several are kin to the peoples of peninsular Southeast Asia, including the **Hmong** of Laos and Vietnam, who speak a language in the **Hmong-Mien subfamily** of the Austric language family. Most people in Laos and neighboring Thailand speak **Lao** and **Thai** in the **Tai-Kadai subfamily** of the Austric family. The majority languages of Cambodia and Vietnam, respectively, are **Khmer** and **Vietnamese,** in the **Mon-Khmer** branch of the **Austro-Asiatic subfamily** of the Austric languages.

Southward and eastward into peninsular Malaysia, Indonesia, and the Philippines, the linguistic map is exceptionally rich and fragmented. The eight languages of the Philippines belong to the Austronesian group. Also in the Austronesian branch is a multitude of languages spoken in Indonesia, including **Malay, Dayak, Kayan-Kenyah, Barito, Javanese, Bugis, Tomini,** and **Timorese.** In many cases, a single language is spoken almost exclusively by an ethnic group inhabiting a single island. In other cases, as will be seen in Module 7.2, different ethnolinguistic groups sometimes live in conflicted proximity to one another on a single island. The large island of New Guinea and some of

• **Figure 7.11** Languages of Monsoon Asia.

its smaller island neighbors are home to the 750 different **Papuan languages,** which are not related to Austronesian tongues.

Colonialism and regional economic enterprise led to the infusion of distant tongues into many parts of Monsoon Asia. British colonists brought English to India, where it still serves as a lingua franca in this linguistically complex land. The British also introduced Tamil-speaking Dravidians from south India into what is now Malaysia, where they worked the rubber plantations and tin mines and

where their descendents remain as an important minority
354 today. Another small but economically prosperous minority
in several countries (and a majority in Singapore) is ethnic
354 Han Chinese. Some people in Indonesia speak Dutch, reflecting that country's former colonial status, and the Philippines' former status as a U.S. commonwealth is reflected in the fact that English remains one of that country's two official languages.

Religions and Philosophical Movements

Monsoon Asia is, along with the Middle East, one of the world's two great hearths of religion. Hinduism, Buddhism, and related traditions of Confucianism and Daoism—collectively practiced or observed by a whopping 1.7 billion, or 25 percent of the world's population—originated in this region. The basics of these **belief systems** are introduced here. The precepts of Islam and Christianity, also practiced in the region, are described in Chapter 6. The faiths with small numbers of adherents, like Zoroastrianism and Sikhism, are described in the contexts of particular countries in the following modules. Religion is sometimes, unfortunately, at the root of domestic or international conflict, and these cases are also discussed in the appropriate places.

The geographic distribution of the region's faiths is depicted in •Figure 7.12. As this map reveals, Hinduism is dominant in India, although many other religions are practiced there as well. Islam prevails in Pakistan, Bangladesh, most parts of the East Indies, parts of the southern Philippines, parts of western China, and among the Malays of western Malaysia. Various forms of Buddhism are dominant in Myanmar, Thailand, Cambodia, Laos, Tibet, and Mongolia. Most people in Sri Lanka and Nepal are Buddhists or Hindus. The religious patterns of the Chinese, Vietnamese, and Koreans are more difficult to describe. Among them, Buddhism, Confucianism, and Daoism have been influential, often in the same household. Growing prosperity among the Chinese has been accompanied by a renewed interest in spiritual matters. China officially recognizes and even encourages five faiths (Buddhism, Islam, Taoism, Catholicism, and Protestantism) but discourages or forbids some sects and belief systems, including Tibetan Buddhism; the quasi-Buddhist movement known as Falun Gong (Falun Dafa, a practice of calisthenics and meditation that the government calls an "evil cult"), and "house church" Protestantism. In Japan, religious affiliations often overlap, with an estimated 84 percent of the Japanese population sharing Buddhism with the strongly nationalistic religion of **Shintoism,** itself a blend of nature reverence, Japanese folklore, and Chinese ritual. The Philippines and Timor-Leste, with large Roman Catholic majorities, are the only predominantly Christian nations in Asia, although Christian groups are found in all other countries in this region. About one-third of South Koreans are Christian, and Korean missionaries work in many countries of the world.

Other customs include **ancestor veneration,** which is especially prominent among Chinese, Japanese, Vietnamese, and Koreans. Most homes in these countries contain small shrines with photographs and other memorabilia of deceased ancestors at which family members place fresh fruits and other offerings on a daily basis. It is crucial in these cultures that the ancestors never be forgotten (•Figure 7.13).

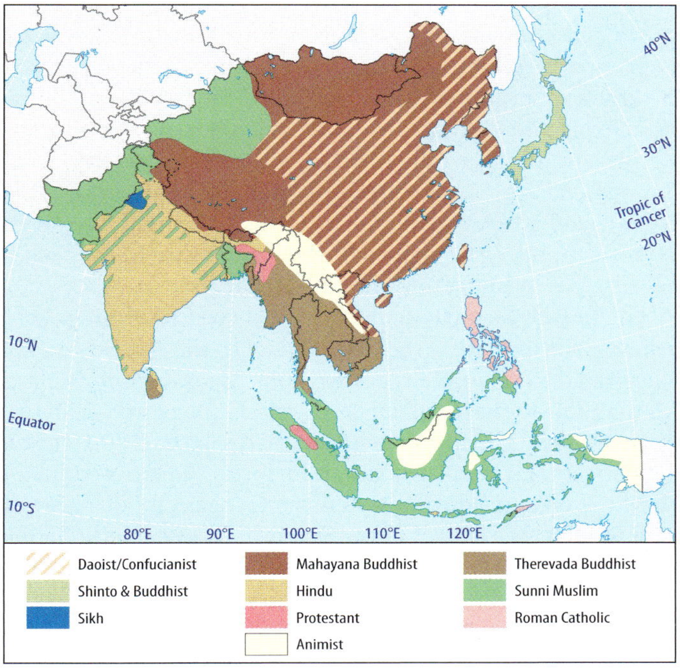

• **Figure 7.12** Religions of Monsoon Asia.

Joe Hobbs

• **Figure 7.13** Ancestor veneration is not a religion per se but an important spiritual tradition that requires the living to remember and pay tribute to their forebears. This is a shrine to the deceased parents of an elderly woman in a southern Thai village.

GEOGRAPHY OF THE SACRED

The Sacred Cow

Many attitudes, beliefs, and practices associated with cattle make up the world-famous but often poorly understood "sacred cow" concept of India. There are nearly 200 million cattle in India, representing about 15 percent of the world total and the largest concentration of domesticated animals anywhere on earth. India's dominant religion of Hinduism forbids the slaughter of cows but allows male cattle and both male and female water buffalo to be killed. Reverence for the cow is well founded. Indians favor cow's milk, ghee (clarified butter), and yogurt over dairy products from water buffalo. They value cows as producers of male offspring, which serve as India's principal draft animals. Both cows and bullocks provide dung, an almost universal fuel and fertilizer in rural India.

Reverence for the cow in particular and cattle in general pervades Hindu religion and mythology. The bull Nandi is associated with the Hindu god Shiva, the cow is linked to the goddess Lakshmi, and Krishna (an incarnation of the god Vishnu) was a cowherd. People allow cattle to roam the streets of Indian cities (•Figure 7.E). As a symbol of fertility, cows are associated with several deities. The mother of all cows, Surabhi, was one of the treasures churned from the cosmic ocean. Hindus honor cows at several special festivals. They use cow's milk in temple rituals. They believe that the "five products of the cow"—milk, curds, ghee, urine, and dung—have unique magical and medicinal properties, particularly when combined. In India, there are about 4,000 *gaushalas,* or "old folks' homes," for aged and infirm cattle.

Remarkably, however, the subcontinent countries of Pakistan, Bangladesh, and India are major exporters of leather and leather goods. In India, the leather industry is mainly in the hands of Muslims, who do not share Hindus' restrictions on killing or eating 325
cows. Muslims are forbidden to eat pork; therefore, pigs, a major food resource in many developing countries, are of little importance here. They are eaten mainly by Christians, very low-caste Hindus (some of whom are pig breeders), and tribal peoples.

Joe Hobbs

• **Figure 7.E** People and cows share street space throughout India.

Many members of hill tribes in Southeast Asia believe that natural processes and objects possess souls, a doctrine known as **animism.** Such indigenous animistic beliefs have been incorporated into a number of mainstream religions, creating vibrant and highly varied belief systems by Western standards. This blending, known as **syncretism,** has parallels elsewhere in the world, for example, among the
537 Maya of Central America.

Hinduism is a regionally varied faith and a belief system so complex that it is extremely difficult to summarize. However, it has the following characteristic elements. It lacks a definite creed or theology. It is very absorptive, encompassing an unlimited pantheon of deities (all of which, most Hindus say, are simply different aspects of the one God, Brahman) and an infinite range of types of permissible worship (•Figure 7.14). Most Hindus recognize the social hierarchy of the caste system, deferring authority to the highest (Brah-
325 min) caste. They practice rituals to honor one or another of the principal deities (Brahman the creator, Vishnu the preserver, and Shiva the destroyer) and thousands of their manifestations (*avatars*) and lesser deities, attending temples to worship them in the form of sanctified icons in which the gods' presence resides. They believe in reincarnation and the transmigration of souls and revere many living things (see Geography of the Sacred, above). They are supposed to be tolerant of other religions and ideas. They participate in folk festivals to commemorate legendary heroes and gods. To earn religious merit and to struggle toward liberation from the bondage of repeated death and rebirth, they make pilgrimages to sacred mountains and rivers.

The Ganges is a particularly sacred river to Hindus, who believe it springs from the matted hair of the god Shiva. The pilgrimage to celebrate the festival of Ardh Kumbh Mela at Allahabad, at the confluence of the Ganges and Yamuna Rivers, drew 75 million pilgrims in January 2007. It was the largest pilgrimage in human history. The most enduring Hindu pilgrimage destination is the Ganges city of Varanasi (Benares) in the state of Uttar Pradesh. Many elderly people go to die in this city and be cremated where their ashes may be strewn in the holy waters. Indian authorities have passed laws, so far ineffective, to clean up the Ganges, polluted in part by incompletely cremated corpses; many of the faithful

Joe Hobbs

• **Figure 7.14** Hinduism has a complex and very colorful pantheon. This is Vishnu, preserver god, and his consort, Lakshmi, goddess of wealth and prosperity.

poor cannot afford to buy the fuel needed for thorough immolation. Scavenging water turtles released into the river at Varanasi help dispose of the cadavers.

Buddhism is based on the life and teachings of Siddhartha Gautama, known as the Buddha or the Enlightened One (•Figure 7.15). The Buddha was born a prince in about 563 B.C.E. in northern India.[4] Although presumably born a Hindu, he came to reject most of the major precepts of Hinduism, including caste restrictions. When he was about 29 years old, he renounced his earthly possessions and became an ascetic in search of peace and enlightenment. The Buddha eventually settled for a "middle path" between self-denial and indulgence, and he meditated through a series of higher states of consciousness until he attained enlightenment (*bodhi*). From that point, he traveled, preached, and organized his disciples into monastic communities. In his sermons, he described the "Four Noble Truths" revealed in his enlightenment: life is suffering; all suffering is caused by ignorance of the nature of reality; suffering can be ended by overcoming ignorance and attachment; and the path to the suppression of suffering is the Noble Eightfold Path, made up of right views, right intention, right speech, right action, right livelihood, right effort, right mindedness, and right contemplation.

An important Buddhist concept is the doctrine of **karma,** a person's acts and their consequences. A person's actions lead to rebirth, in which good deeds of the previous life are rewarded (for example, by which one could be reborn a human) and bad deeds punished (for example, by being reborn as a resident of hell). The goal of the practicing Buddhist

Joe Hobbs

• **Figure 7.15** The Buddha has many manifestations. This temple of the Theravada faithful is in Chiang Mai, Thailand.

is to attain **nirvana**, a transcendent state in which one is able to escape the cycle of birth and rebirth and all the suffering it brings.

Buddhism evolved into two separate branches: **Theravada** ("Way of the Elders") and **Mahayana** ("Great Vehicle"). (Mahayanists sometimes apply the term Hinayana, "Lesser Vehicle," in a derogatory fashion to the Theravada.) Theravada Buddhism is strongest in Sri Lanka, Myanmar, Laos, Cambodia, and Thailand. Theravada Buddhists claim to follow the true teachings and practices of the Buddha. Mahayana Buddhism originated in India and then diffused along the Silk Road to central Asia, Tibet, and China and eventually into Vietnam, Korea, Japan, and Taiwan. Mahayana Buddhists accept a wider variety of practices than those espoused by the Theravada, have a more mythological view of the Buddha, and are interested in broader philosophical issues.

Confucianism is not as much a belief system as it is a sociopolitical philosophy that serves on its own or is blended with other religions, particularly among ethnic Han Chinese. It is based on the writings of Confucius (Kung Futzu, or "Master Kung," who lived from 551 to 479 B.C.E.), collected primarily in his *Analects*. Mencius (c. 371–288 B.C.E.) later became a major force in the widespread diffusion of this belief system throughout China. Following a period in which Communist authorities disavowed them, Confucian ideals are making a major comeback in China today.

The major tenets of Confucianism are embodied in a system of ethical precepts for the proper management of society, emphasizing honor of elders and other authorities, hierarchy, and education. Confucius never proclaimed his beliefs to constitute a religion; he was simply attempting to create a social contract between different classes central to Chinese government and society. He viewed his philosophy as secular, and its diffusion to Korea, Japan, and Vietnam has been part of the cultural baggage taken abroad by a steady stream of Chinese emigrants to those places.

Another important Chinese school of thought, second only to Confucianism and also widely blended with Buddhism among ethnic Han Chinese, is **Daoism** (sometimes spelled **Taoism**; •Figure 7.16). Its philosophy comes from a body of work known as the *Daodejing (Tao-te Ching)*, or in English, *Classic of the Way and Its Power*, ascribed to Chinese Lao-tzu (Laozi), who lived in the sixth century B.C.E. Daoism encourages the individual to reject Confucian-style social conformity and seek to conform only to the underlying pattern of the universe, the "Way" (Dao, or Tao). To follow that way, the individual should "do nothing," meaning nothing unnatural or artificial. One can become rid of all doctrines and knowledge to achieve unity with the Dao and thereby gain a mystical power. With that power, the individual can transcend everything ordinary, even life and death. Daoists hold the simple earthly life of the farmer in high esteem.

Joe Hobbs

• **Figure 7.16** Daoism is imbued with images of heroic figures associated with such attributes as good health, long life, prosperity and bountiful harvests. This is a temple in Shanghai, China.

Effects of European Colonization

Like much of the developing world, Monsoon Asia is a region where Western colonialism reshaped many traditional geographic patterns (•Figure 7.17). By the end of the 15th century, Portugal and Spain had begun to extend economic and political control over some islands and mainland coastal areas of South and Southeast Asia. In the 18th and 19th centuries, the pace of colonization and economic control quickened, and large areas came under European domination. By the end of the 19th century, Great Britain was supreme in India, Burma (modern Myanmar), Ceylon (today's Sri Lanka), and Malaya and northern Borneo (now parts of Malaysia); the Netherlands possessed most of the East Indies (Indonesia); France had acquired Indochina (the region containing Vietnam, Laos, and Cambodia) and small holdings on the coasts of India; and Portugal had Goa and

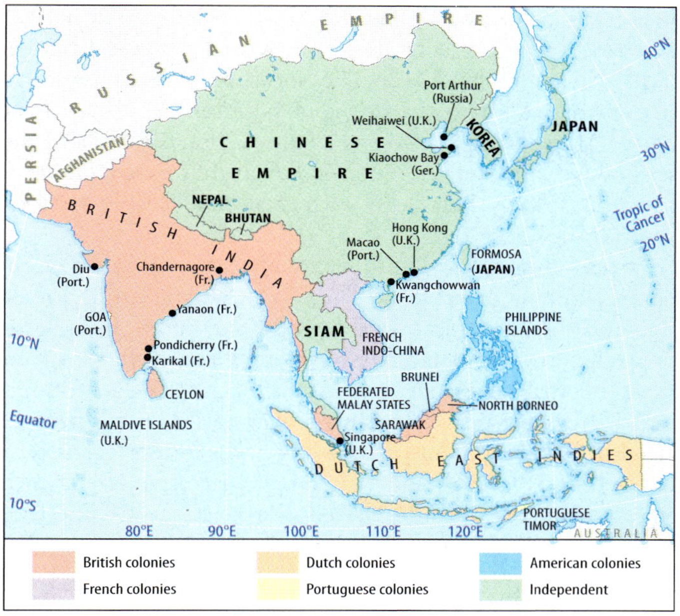

Figure 7.17 Colonial realms and independent countries in Monsoon Asia at the beginning of the 20th century.

Diu in India, Macau in China, and Timor (today's Timor-Leste) in the East Indies. Although retaining a semblance of territorial integrity, China was forced to yield possession of strategic Hong Kong to Britain in the 1842 Treaty of Nanking and to grant special trading concessions and extraterritorial rights to various European nations and the United States through the latter half of the 19th century. In 1898, the Philippines, held by Spain since the 1500s, came under the control of the United States.

A few Asian countries escaped domination by the Western powers during the colonial age. Thailand (historically known as Siam) formed a buffer between British and French colonial spheres in peninsular Southeast Asia. Japan withdrew into almost complete seclusion in the mid-17th century but emerged in the late 19th century as the first modern, industrialized Asian nation and soon acquired a colonial empire of its own. Korea also followed a policy of isolation from foreign influences until 1876, when Japan began to colonize it. It is noteworthy that Japan, South Korea, and Thailand would emerge as very or relatively prosperous countries, whereas most of the former colonies lagged in economic development. Dependency theorists point to these discrepancies to affirm their argument that coloniza-
45 tion hampered development.

Monsoon Asia was an extraordinarily profitable region for the colonizers. Western nations extracted vast quantities of tropical agricultural commodities and in turn found large markets for their manufactured goods. Westerners also invested heavily in plantations, factories, mines, transportation, communication, and electric power facilities. Some of the region's most important cities, including Shanghai in China, Kolkata (Calcutta) in India, and Singapore, were developed mainly by Western capital as seaports serving Western colonial enterprises.

Western domination of Asia ended in the 20th century for a variety of reasons. The two world wars weakened the West's ability to conduct its colonial affairs in the region, Japan rose to great power status and challenged the West early in World War II, and effective anticolonial movements arose in nearly all areas subject to European control. After World War II, all colonial possessions in Asia gained independence. The last to revert were Hong Kong (returned by Britain to China in 1997) and Macau (returned by Portugal to China in 1999). Until this era of independence, however, 20th-century Asia was marked by revolution, war, and considerable turmoil, as described in the following modules.

7.4 Economic Geography

Despite any stereotypes that might linger in the West, Monsoon Asia is not a quaint, old-timey backwater—especially in economic terms. It is a dynamic region with some of the world's strongest and fastest-growing economies. The world's second- and fourth-largest economies are there: Japan and China, respectively. The economies of China and India are surging, with both goods and services inundating the global economic system. Their clout, their accomplishments, and their ambitions are enormous. Some economists and equities marketers have taken to conflating the two countries and calling them "Chindia," a double-barreled engine of economic growth that cannot be stopped and that should attract all serious investors. "Chindia is at the intersection of emerging markets and natural resources," one mutual fund solicitation writes, "Chindia is already consuming a quarter of the world's copper, a third of the world's steel production and more than half of the world's cement. That consumption is projected to rise dramatically by 2020."[5]

It must be noted, however, that this growth is occurring against the backdrop of extraordinary poverty, and many hundreds of millions of people remain desperately poor. While Monsoon Asia as a whole has taken great strides at reducing poverty rates, the gap between rich and poor is growing in many countries. This is particularly true in India, where the overall economy has grown about 8 percent annually since 2000; yet in that same period, there has been almost no reduction in childhood malnutrition, which afflicts almost half of India's young. Poverty tends to be deepest in the rural areas, fueling the classic push factor of
rural-to-urban migration. 57

Despite being generally a region of less developed countries, Monsoon Asia is home to the strong, industrialized, export-oriented economies of South Korea, Taiwan, Hong Kong and Singapore, which late in the 20th century came to be known as the **Asian Tigers** (or "Little Dragons"). Some economists consider the rapidly industrializing Southeast Asian countries of Thailand, Malaysia, Indonesia, the Philippines, and Vietnam the **New Asian Tigers** or **Tiger Cubs**, following the path of rapid economic growth blazed by their neighbors.

Japan was the first Asian country to develop modern cities and modern types of manufacturing on a large scale. It has been a major industrial power for more than a century. China and India are much better supplied with mineral resources than Japan. With the world's largest populations, they also have cheaper labor. But Japan's labor force is, at least for the time being, generally more skilled, and it leads Asia in high **value-added manufacturing**—the process of refining and fabricating more valuable goods from raw or semiprocessed materials. Now, however, there are signs that Japan's preeminence is waning as China's economy surges. The biggest economic story in Monsoon Asia is the rise of China. The highlights of that story are told here and the rest in Module 7.3, while India's economic geography is discussed in Module 7.1.

China's Surging Economy

In conventional terms of gross domestic product (total output of goods and services), China has the world's fourth-largest economy (after the United States, Japan, and Germany). The more intriguing measure of purchasing power parity—which takes account of price differences between industrialized countries like the United States and emerging-market countries like China—places China as the second-largest economy, after the United States (India is fourth by this measure). China, incidentally, is second only to the United States in energy consumption; Russia is third, Japan is fourth, and India is fifth.

However it is weighed, China's economic growth is soaring, with an average annual rate of 10 percent between 1990 and 2007. China's exports (along with imports) increased over 900 percent during those years, growing an average of 50 percent a year. China's share of the world's goods and services also increased 250 percent between 1990 and 2007 to 7 percent (compared with 11 percent for both the European Union and the United States). Just a few years ago, China was notable for making low-value-added products—cheap toys, for example—that the United States and other richer countries really had no interest in making. There were shirts and shoes, too, most of them manufactured in more than 20,000 factories in the special economic zone (SEZ) of Shenzhen, adjoining Hong Kong.

Now China is making at least a little bit of almost everything, including about three-quarters of the toys sold in the United States (•Figure 7.18). The country is making especially rapid progress in manufactures and exports of information technology (IT) hardware. Already China is the world's third-largest manufacturer of personal computers, sold under such brand names as Dell and IBM. China is poised to displace Taiwan as the regional leader of PC manufacturing and is projected to be the world's largest producer of IT hardware by 2012. The roots of China's boom in IT may be traced to the dot-com bubble that burst in the United States in 2000. Many of the talented Chinese who worked in California's Silicon Valley at that time went back to China, taking their skills with them in a kind of "reverse brain drain."

Recent changes in China's global status and in government economic policy have fueled the boom. China joined the World Trade Organization in 2001, prompting a surge in foreign investment. China sweetens the deal for investors with incentives like subsidized loans, tax exemptions, and 50 percent discounts on land prices. The addition of Chinese brawn and brains creates a nearly perfect investment climate. Chinese labor is much less expensive than is typical of most more developed countries, and China has a rapidly growing pool of university-trained engineers and other innovators.

Joe Hobbs

• **Figure 7.18** Mechanized silk production in China.

China's Economic Impact

China's economic boom is having enormous consequences for other countries, particularly its Asian neighbors. Japanese electronics companies are cutting costs by moving increasing amounts of production to China. Surging investment in China is linked to disinvestment elsewhere, especially in Southeast Asia. That region's Tigers and Tiger Cubs—notably Malaysia, Singapore, and Thailand—dominated the 1990s with manufactures and exports of medium- and high-tech industrial products. With its growing consumer appetite, China was buying such goods from these countries. But now the manufacturing is shifting to China from those countries—some of them still smarting from an economic downturn in the late 1990s. South Korea's Samsung will soon make China the main base for production of its flat screens and computers. Japan's Toshiba makes televisions only in China. Dell recently moved some of its computer-making facilities from Malaysia to China. Between 1999 and 2004, Singapore lost more than 35,000 jobs, most of them to China. Motivated by lower costs in China, Japan's largest electronics manufacturer, Matsushita, has eliminated about 40 percent of its production and sales contractors in Southeast Asia and shifted new assets to China.

With China's increasing gravity, Southeast Asian economies may at least temporarily be relegated to supplying low-end products, especially food and raw materials, to China. In turn, they may end up buying cheap Chinese manufactured goods, harming their own fledgling industries. They may try to counter this trend by developing higher-value niche products to meet demands in China. Thailand, for example, wants to boost manufactures of health care products, while Singapore focuses on biomedical products and financial services. Poorer Indonesia will probably be able to count on helping satisfy China's growing appetite for natural gas to fuel its southern industrial cities. Meanwhile, the critical economies of Japan, Taiwan, and South Korea are increasingly vulnerable to China's ascendancy as investors are lured from those countries by China's lower labor costs and its potential 1.3 billion consumers.

Cheap labor is unlikely to make China a permanent magnet, however. As affluence rises there, so will wages, and China's less prosperous neighbors will again attract investment. China's one-child policy also means it will have fewer industrial workers in the future. Mindful of this demographic problem, farsighted multinational companies are boosting manufacturing investment in India, where the pool of young workers is likely to remain large and growing for decades to come.

China's rise is chipping away at U.S. economic dominance in the region. In one country after another (including the most critical, Japan and South Korea), China has eclipsed the United States as Asia's most essential trading partner. In the meantime, the U.S.-China relationship is growing, mainly in China's favor. Of China's exports, about 20 percent go to the United States (its largest single trading partner), but its imports from the United States lag far behind. This U.S. trade deficit with China threatens to become a sore point in relations between the countries. Textile and other industries in the United States are pressuring the U.S. government to pass punitive tariffs on Chinese imports, a move that could precipitate a costly trade war. Problems with tainted food and toy imports have already led to bans of some Chinese products in the United States.

The United States is also furious at China's reluctance to stop the counterfeiting or "piracy" of American products, particularly computer software and DVDs of Hollywood films. China is the epicenter of prolific Asian trade in pirated products (•Figure 7.19). Authorities in the region periodically crack down on sales of such goods, but the vendors are usually on a hair trigger, ready to disappear on a moment's notice with their goods in tow, only to reappear moments later when police leave. There is also growing concern in the United States over the perceived loss of jobs through outsourcing to India (see Insights, page 300).

The Green Revolution

Most Asian countries do not want to improve their economies only by enhancing their outputs of services and manufactured goods. They also want to boost agricultural self-sufficiency and crop exports. Revolutionary changes in crops are at the heart of efforts to reshape Asia's agricultural economies. One of the most significant agricultural innovations in Asia during the past century is the development of new seed types and innovations in planting, cultivating, harvesting, and marketing of crops, efforts known
collectively as the Green Revolution. These developments 30
are not exclusive to Monsoon Asia, but this is where most of the growth and investment have taken place.

Ever since 1962, when the International Rice Research Institute (IRRI) was founded in the Philippines, there have

Joe Hobbs

• **Figure 7.19** Counterfeited goods made in China and sold throughout Asia bear the brand names of U.S. and other manufacturers but are sold to happy consumers at a fraction of the cost of the genuine product. This is a CD and DVD vendor in Laos.

INSIGHTS

Outsourcing

Outsourcing (also known as **offshore outsourcing** or **offshoring**) is the flight of technology and other jobs from countries with high manufacturing and service costs to countries with low manufacturing and service costs. India is the world's largest recipient of these jobs, and the United States is the leading outsourcer.

In the United States, there is growing concern and even alarm over the trend of outsourcing jobs to India. Many of these jobs are in computer programming and a wide variety of technical support, from computer problem phone calls to X-ray diagnosis. Many are mundane, like bill processing and order taking. Increasingly, however, high-paying white-collar employment is being outsourced to India, in such fields as investment banking, aircraft engineering, and pharmaceuticals research. Top Western corporations like the United States' Cisco Systems, the world's leading maker of communications equipment, are beginning to relocate substantial numbers of their senior executives to India, where much of the firm's work is actually done.

India is an especially attractive venue for outsourcing for several reasons. It has a large population of well-educated people, most of whom speak excellent English. Those who work the telephone call centers typically receive training in vernacular American English and strive to perfect their American accents while shedding their Indian and British pronunciations. The bottom line is another major advantage: Indian skilled labor is cheap. The average computer programmer in the United States earns $75,000 per year, while his or her counterpart in India earns $9,000 per year.

Finally, the Internet and superb telephone communications, ironically combined with India's physical distance from the United States, provide some remarkable opportunities. While people in the United States sleep, the day shift is on in India. An American doctor can X-ray a patient in the afternoon and by early the next morning have the analysis—performed overnight in India. On the down side, when an American consumer places a technical service call for a computer problem, he or she may be speaking with a very tired Indian employee working the graveyard shift. There are many accounts of fatigue and burnout among Indian call center employees.

What about the loss of American jobs due to outsourcing? By conservative estimates, 3.3 million U.S. service jobs will move offshore by 2015. Although this sounds alarming, it amounts to just over 200,000 jobs per year, which is a small fraction of the approximately 3.5 million jobs that are created in a typical year in the United States. But some projections are more alarming. In 2007, a former official of the U.S. Federal Reserve described outsourcing as a "third Industrial Revolution," predicting it would take jobs from as many as 42 million American workers in the coming years.* Outsourcing companies generally argue that reducing the costs of their services reduces inflationary costs for their consumers and improves efficiency and productivity. This rationale is, of course, cold comfort to the American worker whose job has moved offshore.

*Alan S. Blinder, quoted in Anand Giridharadas, "India's Edge Goes beyond Outsourcing." *New York Times*, April 4, 2007, p. C4.

been efforts to use science to increase food yields (particularly of rice) to stave off hunger and generate export income. Notable success has been achieved in breeding the new high-yield varieties of seed stock, and there has been a large upsurge of production in certain areas where the new strains have been widely introduced. Many of the new varieties are bioengineered or genetically modified (GM)
89 crops. This means that through precise manipulation of the genetic components of a crop, it can not only produce a higher yield but also be more resistant to drought, flood, or pests or generate a higher amount of a desired nutritional component such as vitamin A. China, second only to the United States in the global biotechnology industry, has produced genetically engineered rice, corn, cotton, tobacco, sweet peppers, petunias, and poplar trees.

Because the underlying premise of the Green Revolution is economic—more crops will be produced at lower cost for higher revenue—most countries in the region are racing to increase their biotechnological output, especially in an effort to catch up with China and ensure they do not lose positions in the global marketplace. Malaysia hopes to build Biovalley, a biotechnology research park near Kuala Lumpur, and Indonesia is planning a similar Bio-island. One of Biovalley's goals is to create palm oil trees genetically modified to produce the raw material for specialized plastics used in medical devices (•Figure 7.20). South Korea, Japan, and India have large and well-funded

Joe Hobbs

• **Figure 7.20** Palm oil trees in Malaysia are being genetically modified to produce specialized plastics. The palm oil comes from the cluster of nuts at the base of the fronds.

biotechnology research programs. Japan has so far resisted production of GM food crops out of fears of possible health hazards, a concern that most Europeans and their governments share.

For Asian farmers to capitalize fully on the Green Revolution, they must overcome many obstacles. Success requires levels of capital that are often beyond the means of peasant farmers, landlords, and governments. Such expenditures are needed for water supply facilities (for example, the tubewells that now number hundreds of thousands in the Indo-Gangetic Plain of the northern Indian subcontinent), chemical fertilizers, and chemicals to control weeds, pests, and diseases. And as agriculture becomes more mechanized, considerable increases in the costs of machinery, fertilizers, and fuel have to be borne by farmers who have, in many cases, had little experience with the cash economy. Governments must improve transportation so that the large quantities of fertilizer required by the new seed varieties can be delivered in a timely fashion. Not only must there be these associated infrastructural changes to support this "revolution," but the crop calendar of the farmer becomes much less forgiving because many of the newest seed grains demand more precise water, fertilizer, and cultivation requirements than traditional grains. Grain storage facilities, now subject to plundering by rats, must be improved.

Overcoming the financial obstacles is rendered more difficult by the widespread system of **share tenancy.** Farmers who are share tenants generally have no security of tenure on the land and thus cannot be sure that money they invest in the Green Revolution will benefit them in the future. They may not wish to assume any additional risk, even if credit on reasonable terms is available. If they remain on their holdings, landlords may take up to half of the increased crop but bear little or none of the additional expense. Landlords may in turn be content to collect their customary rents without expending the additional capital necessary in this new mode of farming, or they may try to turn tenants off the land to create larger spatial units that they themselves can farm more profitably with machinery and hired labor. In fact, landowners with large holdings often become the chief beneficiaries of the new technology, with many smaller farmers becoming a class of landless workers hired for low wages on a seasonal basis or migrants in the rural-to-urban migration stream.

There are other problems associated with the Green Revolution. Economic dislocations result when rice-importing countries become more self-sufficient, causing hardships for rice exporters. Large infusions of the agricultural chemicals associated with high-yield varieties have negative repercussions on natural ecosystems. Of much wider scope is concern that the development of a limited number of high-yield crop varieties grown in vast monocultures will dramatically reduce the genetic variability of crops, in essence interfering with nature's ability to adjust to environmental changes.

7.5 Geopolitical Issues

In the Middle East and North Africa and in Russia and the Near Abroad, principal geopolitical concerns focus on the production and distribution of energy resources. In Monsoon Asia, by contrast, some of the most serious geopolitical issues are prospects for what may be done with weapons created from a particular energy source: nuclear energy. There are also concerns about traditional fossil fuels (see Geography of Energy, page 302), Islamist terrorism, and the security of shipping lanes.

The preceding discussion of the mushrooming economic clout of China and India also raises another critical geopolitical reality: Asia is emerging as center of gravity that will seriously challenge the century-long primacy of the United States in world affairs. Japan long stood alone as a pillar of strength in Asia, but with China and even India surging ahead, regional might has been reshuffled. Eyes in the region and around the world will be focused especially on China, which some analysts fear may use military power to enhance its economic clout.

Nationalism and Nuclear Weapons

As discussed in Module 7.1, after its independence from Britain in 1947, India emerged as an avowedly secular democratic state. But after the 1998 victory of the Hindu nationalist Bharatiya Janata Party (BJP), hopes for new privileges emerged among India's Hindu majority. Among BJP supporters, there was hope for a "Hindu bomb," a counterweight to a long-feared "Islamic bomb" in neighboring Pakistan.

On May 11, 1998, much to the surprise of U.S. and other Western intelligence agencies, India conducted three underground nuclear tests in the Thar Desert. With the blasts, India's government seemed to be trying to stake India's claim as a great world power, exhibit its military muscle to Pakistan and China, and garner enough political support for the BJP to form an outright parliamentary majority in the future. Initial reactions among India's vast populace were highly favorable; 91 percent of residents polled within three days of the event supported the tests. Outside India, there was alarm. India had defied an informal worldwide moratorium on nuclear testing that went into effect in 1996, when 149 nations (not including India and Pakistan) signed the **Comprehensive Test Ban Treaty** (also known as the **Nuclear Nonproliferation Treaty,** or **NPT**), which prohibits all nuclear tests.

The world's eyes quickly turned to India's neighbor to the west. Governments pleaded with Pakistan to refrain from answering India with nuclear tests of its own, arguing that Pakistan would have a public relations triumph it if exercised restraint: It would appear to be a mature and responsible power, whereas India revealed itself as a dangerous rogue state. But Pakistan's leaders felt obliged by their population's demands for a tit-for-tat response to India's

GEOGRAPHY OF ENERGY

The Spratly Islands

About 60 islands make up the Spratly Island chain, which lies in the South China Sea between Vietnam and the Philippines (see Figure 7.2.1, page 336). They are an idyllic tourist destination where divers can hire luxury boats to explore the coral reefs and palm-lined beaches of remote atolls. However, the islands are much more significant for their strategic location between the Pacific and Indian Oceans. During World War II, Japan used the islands as a base for attacking the Philippines and Southeast Asia. Still more significant, as much as $1 trillion in oil and gas may lie beneath the seabed around the Spratlys.

Not surprisingly, many nations covet control of the Spratlys. Six countries claim some or all of the islands, and international law is not equipped to sort out these conflicting claims (see the U.N. Convention on the Law of the Sea, page 407). China, Vietnam, and Taiwan claim sovereignty over all of them, while Malaysia, Brunei, and the Philippines claim some of them. During the Cold War, the competing nations felt it was too hazardous to push their claims on the islands. As the Cold War drew to a close, however, the situation became more volatile. All the contenders except Brunei placed soldiers, airstrips, and ships on the islands. In 1988, the Chinese navy invaded seven of the islands occupied by Vietnam, killing about 70 Vietnamese soldiers. In 1992, China again landed troops in the islands and began exploring for oil in a section of the seabed claimed by Vietnam. In 1995, China moved to expand its territorial designs on islands already claimed by the Philippines and has since built what it calls "shelters," which look like fortifications to the Filipinos. Late in the 1990s, Malaysia occupied two disputed Spratly Islands and began building on one of them. In 2004, Vietnam began renovations of an airport on one of the islands, saying it was necessary to boost tourism there. China condemned the construction as a violation of its territorial sovereignty, an accusation it repeated in 2007 when Vietnam negotiated natural gas development with foreign companies in the Spratlys.

Indonesia, which has no claim on the Spratlys, sponsored unofficial workshops on joint efforts in oil exploration among the six contenders in an effort to ward off a potential crisis. But so far, China, which has proclaimed not only the Spratlys but also the entire South China Sea as its own, has refused to discuss the issue at the official level. There are fears that as the countries' petroleum needs grow, each will seek more aggressively to gain control of the Spratlys (•Figure 7.F). The smaller powers fear that China could turn the islands into a kind of permanent aircraft carrier that could be used to dominate them militarily. An incident in these remote islands could trigger a much wider and more serious conflict in Asia.

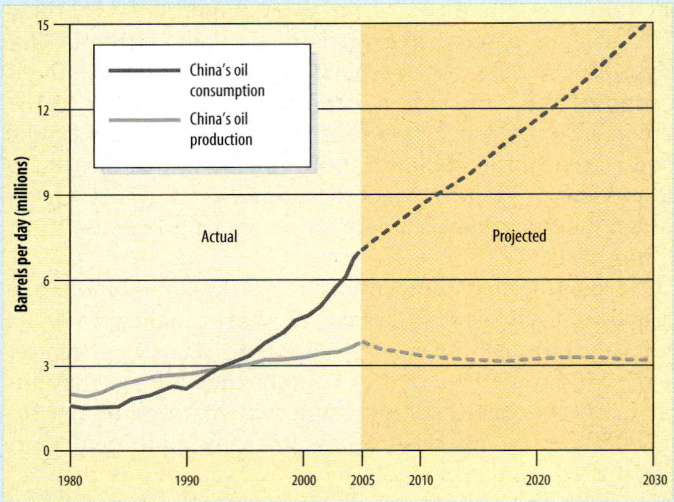

• **Figure 7.F** With China's oil appetite expected to grow enormously, every potential source of oil—like the Spratly Islands—takes on unprecedented importance.

blasts and soon followed with six nuclear tests. India and Pakistan thus joined only five other nations—the United States, Russia, China, the United Kingdom, and France—in acknowledging that they possess nuclear weapons.

There is disagreement about what this recent regional and global shift in the balance of power means. Some analysts fear an escalating nuclear arms race that, perhaps ignited by a border skirmish in Kashmir, could lead to the **mutually assured destruction (MAD)** of Pakistan and India. There are concurrent fears of nuclear war between China and India. Others, particularly in South Asia, argue that the weapons represent the best deterrent against conflict, as they did for decades between the countries of NATO and the Warsaw Pact. However, few people anywhere argue with the contention that the nuclear rivalry between India and Pakistan has taken an enormous economic toll in two nations that need to wage war on poverty.

To the United States, both India and Pakistan are **pivotal countries,** defined by several influential historians as ones whose collapse would cause international migration, war, pollution, disease epidemics, or other international security problems.[6] The disposition of their nuclear arsenals is one of the main reasons for their being considered pivotal. If the government of Pakistan were to fall, Pakistan's nuclear weapons could come into the possession of a rogue element or a new government hostile to the West. India also has much to fear. Either a right-wing or an Islamist government in Pakistan would be far more likely than the current regime to take India on in the disputed Kashmir region. Such 314
a confrontation could set the stage for a nuclear exchange

between Pakistan and India, not only decimating those countries but also sending economic shock waves around the world. India's recently surging economy has been a major force in India's recent diplomatic overtures to Pakistan; nothing discourages investment like war or the fear of it.

U.S.-Pakistan Relations since 9/11

Pakistan became even more pivotal with the events of 9/11. Up to that point, the United States had slapped tough economic sanctions against Pakistan (and less severe ones against India) for the nuclear weapons tests. The United States pursued a diplomatic courtship with India, both as a counterweight against China (India's longtime foe in the region) and as a means of expressing displeasure with Pakistan (China's ally) for helping Afghanistan's Taliban harbor Osama bin Laden. But the United States and Pakistan did an abrupt diplomatic about-face and embraced one another after 9/11. Pakistani President Pervez Musharraf instantly dropped support for the Taliban and quietly allowed the United States to use its territory to prepare for the assault on the Taliban and al-Qa'ida in Afghanistan. In return, the United States forgave much of Pakistan's debt to the United States and lifted its sanctions against Pakistan. To avoid isolating India, the United States also lifted its post–nuclear test sanctions against that country.

The United States has dramatically strengthened military ties with both countries, recognizing Pakistan in 2004 as a major non-NATO ally—one of just 11 countries to have that designation—and acknowledging India the same year as a "strategic partner." These designations opened the doors of both countries to major new shipments of American war matériel. But India would be favored: in 2007, Washington singled out India as its most critical strategic ally in the region by offering nuclear fuel and technology to New Delhi.

Under U.S. and allied attack in Afghanistan after 9/11, Taliban and al-Qa'ida fighters retreated to and regrouped in western Pakistan, particularly in the semiautonomous federally administered tribal areas (FATA; see Figure 7.1.17, page 326). Here the populace is mainly Pashtun and is sympathetic to the causes of their Taliban ethnic kin and
221 their al-Qa'ida spiritual kin (•Figure 7.21). This is an area where Pakistani government authority has long been kept at bay and where it was considered extremely difficult to insert Pakistani forces on a sustained, aggressive basis. Bowing to American pressure, Pakistan's government periodically undertakes military operations in the FATA but considers sustained effort there potentially detrimental to its survival. Even without the direct insertion of U.S. troops, assertive Pakistani intervention in the FATA could lead to a popular uprising, a military coup, or both. Not just the Pashtuns but the vast majority of Pakistan's ethnic groups are strongly against U.S. interventions in the region, and two factions in the army (one right-wing, another Islamist) stand ready to challenge the government's authority. If the government were to fall, there would be grave concerns about Pakistan's nuclear weapons.

Joe Hobbs

• **Figure 7.21** The Pashtun of western Pakistan are a proudly independent people who are opposed to American interests in the region.

What Does North Korea Want?

There are also major concerns about nuclear weapons in Northeast Asia. Ever since suffering Hiroshima and Nagasaki, Japan has had an official policy never to develop or use atomic weapons. But Japan worries about the potential nuclear threat from three adversaries, all of which possess nuclear weapons: Russia, China, and North Korea. The Japanese feel that the West dismissed potential threats from Russia too readily when the Soviet Union dissolved. Japan still has territorial disputes with Russia, particularly involving the four Kuril Islands of Kunashiri, Etorofu, Shikotan, and Habomai, which the government of Josef Stalin seized at the end of World War II and lie just off the northeastern coast of Hokkaido (see Figure 7.4.1, page 393). Japan and China also have a territorial dispute concerning the East China Sea, and if major oil reserves are discovered there, as anticipated, relations between the two countries could deteriorate. China continued to test nuclear weapons through 1996, adding to tensions in Japan. Finally, Japan fears reunification in Korea, which, as a former Japanese colony from 1910 to 1945, has a historical dislike of Japan. There is speculation that Japan may do the unthinkable 406
and the officially disavowed—develop nuclear weapons—to counter the perceived threat of North Korea's nuclear weapons program.

The world looks nervously at the troubled relations between North and South Korea and between North Korea and the West. A crisis flared in 1994 when North Korea refused to permit full inspection of its nuclear facilities by the **International Atomic Energy Agency (IAEA).** The country was suspected of separating plutonium that could be used in making nuclear bombs. Over time, the United States, South Korea, and Japan worked out an agreement in which North Korea would agree to freeze development of nuclear weapons in exchange for the others' providing fuel oil and assistance in building nuclear power plants. Those nuclear reactors would be of the "light water" variety, much less likely than North Korea's plutonium-based reactors to be "dual use" for both military and civilian purposes. The Clinton administration trumpeted the agreement as a success, but it had several flaws. One was that North Korea did not have to dispose of its existing nuclear fuel; it simply had to stow it safely away—meaning that it could quickly be reactivated.

North Korea pressed ahead with a program to build missiles capable of carrying nuclear warheads and in 1998 launched such a missile (minus the warhead) over Japan. This incident prompted the United States to renew its resolve to establish an antimissile defensive "shield" over the United States—the so-called **Strategic Defense Initiative (SDI),** which detractors soon dubbed the "Star Wars Initiative," dating to the Reagan administration of the 1980s. Some analysts believe that the United States is reluctant to see the two Koreas reunite because it would remove much of the justification on which the missile shield program is based. It might also put pressure on the United States to reduce its military presence in the western Pacific. They add that Japan, too, would not want to see Korean reunification because it would remove Japan's justification for building
383, 400 up its defenses, which are designed less for confrontation with North Korea than with China.

President George W. Bush came into office with a hard line against North Korea, effectively suspending dialogue and technical assistance for the still uncompleted nuclear power plants. In 2002, he proclaimed that North Korea was one of three countries making up an "axis of evil," along with Iraq and Iran, signaling that the United States was more interested in confronting than accommodat-
234, 265 ing North Korea. Late that year, North Korea dropped a virtual bombshell on the United States by admitting—after being confronted with evidence collected by U.S. intelligence agencies—that it did indeed have an active nuclear weapons program (developed, as it turns out, with Pakistani technical assistance). This admission followed earlier strident denials of such a program by North Korean officials. The American side, focused on developing the military campaign in Iraq, had been content to let the problem of North Korea's weapons program simmer in the background. By most measures, North Korea should also have been content to keep quiet; the United States, South Korea, and Japan had for almost a decade been providing food and fuel to the economically beleaguered country, and any belligerence by North Korea could end that aid.

What, then, was North Korea hoping to accomplish by revealing its nuclear weapons program? There are several possibilities. The least likely is that North Korea was instigating a military confrontation with the United States and its allies South Korea and Japan. In any war scenario, those allies would obliterate North Korea. The United States is obliged to defend South Korea in the event of a war with the North and would effectively wield its huge military advantage (•Figure 7.22). However, that victory would come at an enormous price for South Korea and possibly Japan. North Korea has a standing army of nearly 1 million soldiers, an impressive military arsenal including short- and medium-range missiles, a stockpile of chemical and biological weapons that it might not be shy to use, and now, presumably, at least a few nuclear weapons that it could deploy. The casualties in South Korea would be enormous, whether from a conventional or an unconventional assault from the North.

It is much more likely that by disclosing its nuclear weapons program, North Korea sought guarantees that it could avoid war and also gain even more assistance from the West. Since the early 1990s, its nuclear weapons program has been the only leverage North Korea has had to coax desperately needed supplies from abroad. North Korea had only to mention the prospect of reactivating its nuclear weapons program to get more concessions from the United States and its allies—especially food aid during a succession of droughts and floods in the 1990s. In effect, North Korea was extorting money from the United States and its allies, which have been happy to ante up as a means of containing the rogue nation. By admitting to the nuclear weapons program, North Korea may have felt it could get even more aid in technology, fuel, food, or other forms. It may have been right.

A series of on-again, off-again negotiations known as the "Six Party Talks" (held between North and South Korea, the United States, China, Japan, and Russia) culminated in 2007 with an agreement: the United States would help unfreeze North Korean funds in a Macau bank, take steps to remove North Korea from its list of countries supporting terrorism, lift trade sanctions in place since the Korean War ended in the 1950s, and with along with its negotiating partners restart the flow of fuel oil to North Korea; in return, North Korea would close its nuclear plants and allow them to be inspected by international monitors. In the deal, North Korea did not promise to dismantle its nuclear facilities at Yongbyon but only "abandon" or "disable" them, which it did do by the end of that year. North Korea is not required to destroy any existing stocks of nuclear weapons, and much uncertainty remains. The United States apparently hopes that the involvement of China and the other countries will make North Korea less likely to restart its nuclear program this time.

Joe Hobbs

• **Figure 7.22** A little humor eases the tension for U.S. troops serving along what may be the world's tensest border, the "no-man's-land" separating North and South Korea. Even a minor incident here could touch off a conflagration that might take a huge toll in human lives.

Islands, Sea Lanes, and Islamists

As related in Module 7.2, Indonesia—another pivotal country from the U.S. perspective—is a very ethnically complex nation whose integrity is threatened from a variety of secessionist movements in locales as far apart as Aceh in the west and Papua (formerly Irian Jaya) in the east (see Figure 7.2.18, page 357). A possible **domino effect** scenario for Indonesia concerns foreign powers: If one province falls or proclaims its independence, many others will, and the largest country in the region will break apart. Countries like the United States fear what would happen to vital international shipping lanes should several new and perhaps militant states emerge from Indonesia's fragmentation. There have already been scores of pirate attacks on vessels plying the strategic Strait of Malacca, a critical chokepoint through which a quarter of the world's trade passes, including two-thirds of the world's shipment of liquefied natural gas and half of all sea shipments of oil (it ranks second only to the Strait of Hor-
229 muz as an oil shipping lane). Security in the Strait of Malacca is a source of great concern to Japan, which imports 80 percent of its oil on ships that pass through the strait. Aceh's vast natural gas deposits and Papua's copper and gold are also seen as critical in the global economy, so like Indonesia, many world powers are anxious to see that they stay in the
360 "right" hands. The United States gives generous amounts of foreign aid to Indonesia, but the American presence is increasingly unwelcome because so many Indonesians, who constitute the world's largest Muslim population, perceive the post-9/11 United States as anti-Muslim.

U.S. intelligence agencies fear that Southeast Asia, and Indonesia in particular, will emerge as an Islamist terrorist hearth to replace Afghanistan. Al-Qa'ida and its affiliate organizations are known to be active in Southeast Asia. Two of the 19 men who carried out the 9/11 attacks 235
planned part of the operation in Malaysia's capital, Kuala Lumpur. Malaysia is an unlikely refuge for groups like al-Qa'ida, mainly because it sees itself as a progressive engine of high-tech economic growth and would be loath to suffer the sanctions or other fallout associated with harboring terrorists. However, Malaysia is a convenient transit and staging point for terrorist interests, and some members of al-Qa'ida and affiliated groups are Malaysian nationals. Citizens of any predominantly Muslim country may enter Malaysia without obtaining a visa (and thus escape most opportunities for a background check).

Mainly Chinese, prosperous, aggressively policed Singapore is the region's most unlikely terrorist refuge of all. But with its resident population of 17,000 Americans and its numerous Western commercial and military activities (including regular visits by U.S. naval vessels), Singapore is an attractive target. Evidence uncovered in an al-Qa'ida "safe house" in Afghanistan that was examined by U.S. intelligence experts revealed a plot that would have carried out ammonium nitrate truck bombings against U.S. and British embassies and other facilities in Singapore in 2002. An al-Qa'ida "sleeper cell" based in Singapore had been activated from Malaysia by Riduan Isamuddin (also known as Hambali), an Indonesian citizen who hosted the two 9/11 hijackers on their visits to Kuala Lumpur, was a suspect in the U.S.S. *Cole* bombing, and was arrested by intelligence authorities in 2003. These sleepers, eight of whom had been trained by al-Qa'ida in Afghanistan, were Indonesians.

Indonesia is a secular rather than a religious state, and a moderate form of Islam prevails in politics and in everyday life. What concerns Western intelligence authorities and Indonesia's moderate leadership are the more militant Islamist organizations that have emerged in the past decade and have apparently gathered strength since 9/11. One of these is **Laskar Jihad,** created in 2000 to mobilize Indonesian Muslims to fight Christians in Indonesia's Sulawesi and Muluku Islands and Americans throughout the country. Another organization with suspected terrorist links is **Jemaah Islamiah (JI),** also known as the Islamic Group, led by Abu Bakar Baasyir (Abu Bakar Bashir), whose nominal role was preacher in an Islamic school in Solo on the island of Java. (Western authorities are much concerned about such religious schools, known as *pesantrens* in Indonesia. Like the *madrasas* of Pakistan and the Arab heartland of the Middle East, these are venues for both religious and secular learning, and some of the schools' curricula are im-
327 bued with virulent anti-Western content.) Baasyir admitted to having tutored 13 of the men arrested in connection with the plot to blow up Western targets in Singapore. These included Hambali, al-Qa'ida's point man in Southeast Asia, whose regional affiliation is with JI. Intelligence authorities believe that Baasyir was also the ringleader for a series of attacks that would have been carried out against American embassies and naval vessels in several Southeast Asian countries on the first anniversary of the 9/11 attacks. Finally, after the ferocious bombings of nightclubs on the Indonesian island of Bali in 2002 in which both al-Qa'ida and Baasyir were implicated, Indonesian authorities imprisoned Baasyir. Amid furious international criticism, they released him after two years.

Indonesia has much going for it as a potential new hearth for al-Qa'ida affiliated activities. Much like Afghanistan was prior to 9/11, it has a predominantly Muslim population, including some extremist factions that would be willing to provide safe haven; it has a largely poor and otherwise disaffected population (in theory, providing recruits for a militant cause); it includes remote locales that make suitable weapons and tactics training grounds; and it has regions that the government is unwilling or unable to control. In 2003, the director of Indonesia's national in- 273
telligence agency acknowledged that al-Qa'ida had set up training camps in the country. U.S. authorities believe that Southeast Asia, with Indonesia as its centerpiece, has the world's highest concentration of al-Qa'ida operatives outside Pakistan and Iraq.

Why, then, does the United States not have troops on the ground in Indonesia or advisers there (as in the Philippines) to train Indonesian forces in antiterrorism tactics? The problem is that Indonesia is even more difficult than Pakistan for U.S. antiterrorism forces to operate in. It is more geographically complex and even less receptive to a U.S. presence. Perceived intervention by the United States could be enough to spark a now largely neutral populace into violent anti-Americanism and perhaps bring down Indonesia's government. Washington fears that many states, some of them anti-Western, could emerge in Indonesia's place and threaten American interests in the country—notably its oil, natural gas, and copper resources and its location astride vital shipping lanes.

This concludes an introduction that has set the stage for further exploration of land and life in the subregions and countries of Monsoon Asia.

SUMMARY

- Monsoon Asia includes the countries of Japan, North and South Korea, China, Taiwan, Pakistan, India, Sri Lanka, Bangladesh, Bhutan, Nepal, Maldives, Myanmar (Burma), Laos, Thailand, Cambodia, Vietnam, Malaysia, Singapore, Indonesia, the Philippines, Brunei, Timor-Leste, and numerous islands scattered along the edges of this major continental bloc. The term *monsoon* is used because of the central role this wind and precipitation pattern plays throughout this region.
- This is the most populous world region, with 54 percent of the world's people. It includes the world's most populous countries, China and India. Population growth rates in the region vary widely, from zero growth in Japan to over 3 percent per year in Timor-Leste.
- Three concentric arcs make up the broad physiography of the region: an inner arc of the high mountain ranges, the Himalaya, Karakoram, and Hindu Kush; a middle arc of major floodplains, deltas, and low mountains; and an outer arc of thousands of islands, including the archipelagoes of the East Indies, Philippines, and Japan.
- Major rivers include the Indus, Ganges, and Brahmaputra in South Asia; the Irrawaddy, Chao Praya (Menam), Mekong, and Red in Southeast Asia; and the Chiang Jiang (Yangtze) and Huang He (Yellow) in East Asia.
- The region's climate types and biomes include tropical rain forest, savanna, humid subtropical, humid continental, steppe, desert, and undifferentiated highland. Shifting cultivation and

wet rice cultivation are important forms of agriculture. Wet rice cultivation produces very high yields and is associated with dense human populations.

→ Although Monsoon Asia has some of the world's largest cities, about 62 percent of the region's population is rural. Many uniquely Asian traditions helped shape village settlement planning and home design.

→ Monsoon Asia's ethnic and linguistic compositions are diverse. Major language families are Indo-European, Dravidian, Sino-Tibetan, Altaic, Austric, and Papuan.

→ Major religions and sociopolitical philosophies of Monsoon Asia are Hinduism, Islam, Buddhism, Confucianism, Daoism, and Christianity.

→ Great Britain, the Netherlands, France, and Portugal were the most important colonial powers in this region. Most of these domains were relinquished by the middle of the 20th century, and the later British return of Hong Kong and the Portuguese return of Macao (both to China) closed the colonial period.

→ Japan has the region's strongest economy, second only to that of the United States. Recent decades have seen strong economic growth among the Tigers and Tiger Cubs of the region, including South Korea, Taiwan, Singapore, Thailand, and Malaysia. The most important emerging economic power is China, whose large and inexpensive labor force has recently been attracting investment away from other parts of the region. India's well-educated and inexpensive labor force is contributing to its strong economic growth. There are fears in the United States about the outsourcing of American jobs to India.

→ The Green Revolution is a broad effort to increase agricultural productivity in dominant crops. Biotechnology has produced crops that are more drought- and pest-resistant and capable of creating much higher yields, but genetic engineering of crops has perceived risks. Profits and other benefits from the Green Revolution are not uniformly spread.

→ There are several major geopolitical issues in Monsoon Asia. The traditional enemies Pakistan and India (both pivotal countries from the U.S. point of view) now possess nuclear weapons. There are fears that destabilization in Pakistan might allow the weapons to fall into the wrong hands. Pakistan's support of the United States in its war on terrorism is very risky because a popular backlash could bring down the government. North Korea has nuclear weapons and is apparently trying to use them as bargaining chips to get more food aid and other assistance from the West. Antigovernment and anti-Western Islamists are active in Indonesia, another country viewed as pivotal by the United States. Strife and fragmentation in Indonesia could threaten vital oceanic shipping lanes.

KEY TERMS + CONCEPTS

Ainu (p. 293)
Altaic language family (p. 293)
- Japanese (p. 293)
- Kazakh (p. 293)
- Korean (p. 293)
- Mongolian (p. 293)
- Tungus (p. 293)
- Turkic (p. 293)
- Uighur (p. 293)

ancestor veneration (p. 294)
animism (p. 294)
archipelago (p. 282)
Aryans (p. 291)
Asian Century (p. 278)
Asian Tigers (p. 298)
Austric language family (p. 293)
- Austro-Asiatic subfamily (p. 293)
  - Mon-Khmer (p. 293)
    - Khmer (p. 293)
    - Vietnamese (p. 293)
- Austronesian subfamily (p. 293)
  - Barito (p. 293)
  - Bugis (p. 293)
  - Dayak (p. 293)
  - Javanese (p. 293)
  - Kayan-Kenyah (p. 293)
  - Malay (p. 293)
  - Timorese (p. 293)
  - Tomini (p. 293)
- Hmong-Mien subfamily (p. 293)
  - Hmong (p. 293)
- Tai-Kadai subfamily (p. 293)
  - Lao (p. 293)
  - Thai (p. 293)

belief systems (p. 293)
Buddhism (p. 295)
Comprehensive Test Ban Treaty (p. 301)
Confucianism (p. 296)
Daoism (Taoism) (p. 296)
domino effect (p. 305)
Dravidian language family (p. 291)
- Malayalam (p. 292)
- Tamil (p. 292)
- Telegu (p. 292)

fallow (p. 288)
feng shui (p. 290)
geomancy (p. 290)
Han Chinese (p. 293)
Hinduism (p. 294)
Indo-European language family (p. 291)
- Indo-Iranian subfamily (p. 291)
  - Bengali (p. 291)
  - Hindi (p. 291)
  - Nepali (p. 291)
  - Pashto (p. 291)
  - Punjabi (p. 291)
  - Sanskrit (p. 291)
  - Sinhalese (p. 291)
  - Urdu (p. 291)

insolation (p. 283)
intensive subsistence agriculture (p. 286)
International Atomic Energy Agency (IAEA) (p. 304)
Jemaah Islamiah (JI, or the Islamic Group) (p. 306)
karma (p. 295)
Laskar Jihad (p. 306)
laterite (p. 288)
lateritic soils (p. 288)
levee (p. 290)
Mahayana Buddhism (p. 296)
monsoon (p. 283)
mutually assured destruction (MAD) (p. 302)
New Asian Tigers (p. 298)
nirvana (p. 296)
Nuclear Nonproliferation Treaty (NPT) (p. 301)
offshore outsourcing (offshoring) (p. 300)
one-child campaign (p. 284)
outsourcing (p. 300)

paddy (padi) (p. 287)
Papuan languages (p. 293)
pivotal countries (p. 302)
plinthite (p. 288)
shifting cultivation (p. 288)
share tenancy (p. 301)
Shintoism (p. 294)
Sino-Tibetan language family (p. 292)
  Sinitic subfamily (p. 292)
    Chinese (p. 292)
    Mandarin (p. 293)
    Min (p. 293)
    Xiang (p. 293)
    Yue (p. 293)
  Tibeto-Burman subfamily (p. 292)
    Burmese (p. 292)
    Tibetan (p. 292)
slash-and-burn cultivation (p. 288)
Strategic Defense Initiative (SDI) (p. 304)
summer monsoon (p. 284)
swidden cultivation (p. 288)
syncretism (p. 294)
terraces (p. 287)
theory of Himalayan environmental degradation (p. 289)
Theravada Buddhism (p. 296)
Tiger Cubs (p. 298)
value-added manufacturing (p. 298)
wet rice cultivation (p. 286)
winter monsoon (p. 284)
yang (p. 290)
yin (p. 290)

REVIEW QUESTIONS

1. What countries make up Monsoon Asia?
2. Use the concept of the three arcs to trace the outlines of Monsoon Asia's geography. What are the largest plains and river valleys of the region? What are the most important mountain ranges? What are the major island groups?
3. What are the region's most populous countries? Where are the highest and lowest population densities? Which countries have the highest and lowest population growth rates?
4. What are the monsoons? What roles do they play in climates and human activities?
5. What types of environments are shifting cultivation and wet rice cultivation practiced in? What are some of the basic methods and land use considerations of both types of agriculture?
6. What roles have *feng shui* and Confucianism played in the site selection and design of homes, schools, and villages in parts of Asia?
7. What are some of the innovations and products that originated in Monsoon Asia and diffused from there?
8. What are the major linguistic and ethnic groups of Monsoon Asia? What are the principal religions and sociopolitical philosophies?
9. Who were the major colonial powers in Monsoon Asia? What regions did they colonize?
10. What are the region's most significant economic trends?
11. Why are India, Pakistan, and Indonesia considered pivotal countries for U.S. interests? Has the United States been forced to take sides in the traditional enmity between India and Pakistan?

NOTES

1. Kamala Markandaya, *Nectar in a Sieve* (New York: Day, 1954), p. 57.
2. John Reader, British anthropologist and traveler, cited in Fred Pearce, "Terraces: The Other Wonders of the World," *Eurozine,* March 2001, http://www.eurozine.com/articles/2001-03-01-pearce-en.html. Accessed July 11, 2007.
3. Jack D. Ives and Bruno Messerli, *The Himalayan Dilemma: Reconciling Development and Conservation* (New York: Routledge, 1989).
4. Descriptions of Buddhism and Daoism are based on the *Microsoft Encarta Reference Library,* 2004.
5. U.S. Global Investors, "What You Need to Know about Chindia." Postcard from U.S. Global Investors, Inc., 2006. See also Frank E. Holmes, *Chindia and Its Global Impact* (San Antonio, Tex.: U.S. Global Investors, 2007), http://www.usfunds.com/docs/presentations/chindia/viewhtml.asp. Accessed October 12, 2007.
6. Robert Chase, Emily Hill, and Paul Kennedy, "Pivotal States and U.S. Strategy," *Foreign Affairs,* January-February 1996, pp. 33–51.

This Hindu saddhu is one face in the rich cultural fabric of South Asia. This ascetic is dedicated to prayer and abstention from worldly pursuits and depends on charity of the faithful. Joe Hobbs

MODULE 7.1
COMPLEX AND POPULOUS SOUTH ASIA

A triangular peninsula thrusts southward 1,000 miles (c. 1,600 km) from the Asian landmass, splitting the northern Indian Ocean into the Bay of Bengal and the Arabian Sea. The peninsula is bordered on the north by the alluvial plain of the Indus and Ganges (Ganga) Rivers, and north of these rise the highest mountains on earth. The entire unit—peninsula, plain, and fringing mountains—is known as the Indian subcontinent. It contains the five countries of India, Bangladesh, Pakistan, Nepal, and Bhutan in an area about half the size of the 48 conterminous United States (●Figure 7.1.1). Off the southern tip of India, across the narrow Palk Strait, lies the island nation of Sri Lanka, which shares many physical and cultural traits with the subcontinent. Farther to the southwest is the Maldives, a country of some 1,100 small, low-lying islands. These seven countries make up the subregion of South Asia.

Module outline

Module objectives

This module should enable you to

- → Recognize the tensions between Muslims and Hindus in South Asia that often give rise to violent conflict
- → Appreciate how the partition of India created lasting problems in political relations, resource use and allocation, and industrial development
- → See that food production has kept pace with enormous population growth in South Asia
- → Consider the role of high technology in reshaping the economies and outlooks of South Asia's peoples
- → Recognize the serious consequences of political insurgencies in Kashmir, India, Pakistan, Nepal, and Sri Lanka

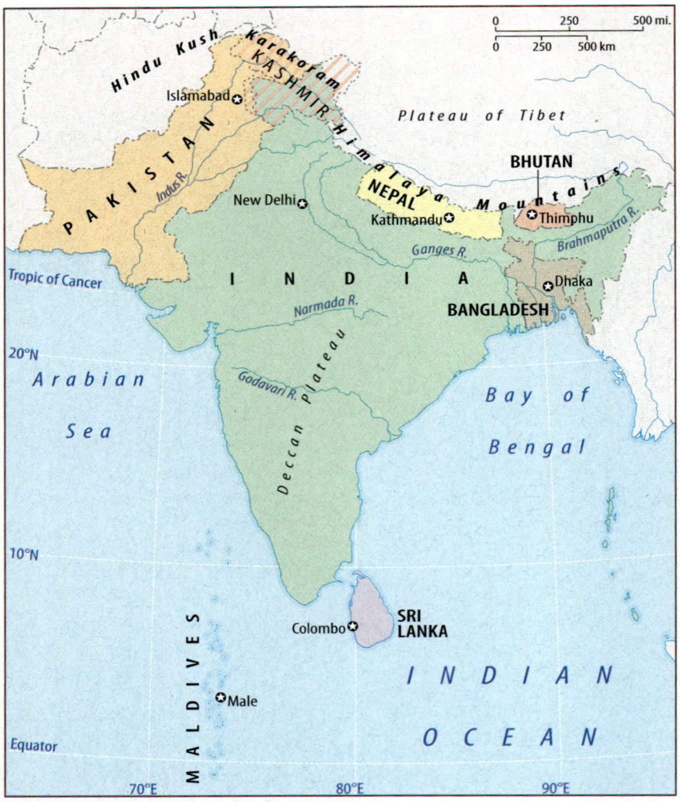

• **Figure 7.1.1** Political units of South Asia. The status of Kashmir, disputed between India and Pakistan, with some parts occupied by China, remains unresolved.

A world-class giant, India outranks the other South Asian nations in both population (1.1 billion) and area (1.2 million square miles, or 3 million square kilometers, about the combined size of the states of Alaska, Texas, California, and Colorado). Bangladesh (149 million) and Pakistan (169 million) are much smaller than India in area and population, but both are among the world's most populous countries. Nepal and Bhutan, on the southern flank of the Himalaya Mountains, are small, rugged, and remote, with fewer people (27 and 1 million, respectively). They serve as buffer states between India and China, which have engaged in sometimes violent border disputes since the early 1950s. Strife has not spared Sri Lanka, whose 20 million people have experienced a civil war for more than two decades. The 300,000 people of the Maldives enjoy relative peace and tranquillity on their isolated islands. Altogether, about 1.5 billion people, more than one of every five on earth, lived in South Asia in 2007.

South Asia is a paradox. The true South Asia is found in the villages, where 7 of every 10 persons live and where many peasant farmers live and work much as their ancestors have for millennia. But the true South Asia is also found in its great cities (which in India the government is progressively renaming): Kolkata (formerly Calcutta), Mumbai (formerly Bombay), Bengaluru (formerly Bangalore), Karachi, and Dhaka are central to the identities and, in some cases, the cutting edge of technology economies of this region's people. South Asia is thus both ancient and modern, rural and urban.

7.1.1 The Cultural Foundation

An enormous range of ethnic groups, social hierarchies, languages, and religions can be found among regions of South Asia and even within a single settlement. The subcontinent is the most culturally complex area of its size on earth, and its civilizations have roots deep in antiquity. The
Indian subcontinent is one of the world's culture hearths. 41
Among the animals first domesticated in the region were zebu cattle (Indian humped cattle), which today are important livestock in many world regions. After 3000 B.C.E., some of the world's earliest cities (notably Mohenjo-Daro) developed on the banks of the Indus River in what is now Pakistan (•Figure 7.1.2). The peoples of the **Harappan civilization** were probably the ancestors of the Dravidians, represented today mainly by the Tamils of southern India.

Between 2000 and 1000 B.C.E. came the Aryans (or **Indo-Aryans**), herders of cattle and horses from central Asia who

Joe Hobbs

• **Figure 7.1.2** Mohenjo-Daro, on the banks of the Indus River in what is now central Pakistan, was one of the world's earliest cities, with an estimated 50,000 people in 2500 C/D/F/ The written language of the Harappan civilization has not yet been deciphered.

were the source of many ethnic and cultural attributes that are dominant in the subcontinent today. These include the lighter complexions than are typical of the Dravidians, the numerically dominant Indo-European languages such as Hindi, and the faith of Hinduism, which developed mainly from the beliefs and customs of the Aryans. In the eighth century, Arabs introduced the religion of Islam to what is now Pakistan. Waves of Muslim conquerors who came through mountain passes from Afghanistan and central Asia spread Islam farther afield in later centuries. The last great Islamic invasion, by a Mongol-Turkish dynasty called the **Moguls** (**Mughals**), began in the early 16th century, soon after the Europeans first came to India by sea. Before it succumbed to rebellious local Hindus and Sikhs and to the British Empire, the Mogul empire ruled nearly the entire subcontinent from its capital at Delhi. Some of the region's most spectacular artistic and scientific achievements date to the peak of Mogul power in the 16th and 17th centuries, including the Red Fort in Delhi and the tomb of Emperor Shah Jahan and Empress Mumtaz Mahal—the astonishing Taj Mahal in Agra (• Figure 7.1.3).

There are major social and political divisions within the subcontinent that relate to religion. As will be seen later in this module, religious differences between the two largest religious groups—Hindus (whose religion is described in Chapter 7, pages 294–295) and Muslims (whose religion is described in Chapter 6, pages 225–227)—are often troublesome. Hinduism was the dominant religion at the time Islam made its appearance in this region. It continues to have the largest number of adherents, including an estimated 13 percent of the population of India and 4 percent of Nepal. Muslims are majorities in Pakistan (97 percent) and Bangladesh (83 percent). Muslims make up a seemingly small percentage of India's population, but they number nearly 150 million people, making India home to the world's third largest Muslim population, behind Indonesia and Pakistan.

Seldom in history have two large groups with such differing beliefs lived in such close association with each other as Hindus and Muslims in the subcontinent. Islam holds to an uncompromising monotheism and prescribes uniformity in religious beliefs and practices. Hinduism is monotheistic for some believers but polytheistic for others and asserts that a variety of religious observances is consistent with the differing natures and social roles of humans. The exuberant and loud celebrations of the Hindu faith are a striking contrast to the austere ceremonies of Islam. Islam has a mission to convert others to the true religion, whereas most Hindus regard proselytizing as essentially useless and wrong; one is born or reborn as a Hindu, so conversion is not an issue. Islam's concept of the essential equality of all believers is in total contrast to the inequalities of the caste system endorsed in Hinduism (see page 325). Islam's use of the cow for food is anathema to Hinduism. And then there are religion-based politics, which often turn differences into discord and conflict, as will be seen later.

In addition to Muslims, there are other significant religious minorities in India. About 20 million **Sikhs** are concentrated mainly in the prosperous state of Punjab. **Sikhism** is India's fourth-largest religion, after Hinduism, Islam, and Christianity. The faith emerged in the 15th century as a religious reform movement intent on narrowing the differences between Punjab's Hindus and Muslims. Sikhs regard men and women as equals and disavow the caste system recognized by Hindus. Sikh men are recognizable by their turbans, which contain hair that should never be shorn (although a growing number of Sikhs insist on doing so), by a bracelet called a "bangle" worn on the right arm, and by their ubiquitous surname Singh, meaning "lion."

There are about 25 million Christians in India, most of whom live in the south of the peninsula; Buddhists, Jains, Parsis, and members of a variety of tribal religions make up the numerous smaller remaining religious minorities. Most of the estimated 7 million Buddhists are recent converts from among India's lowest castes. Numbering perhaps 75,000 and concentrated mainly in Mumbai, the famously entrepreneurial **Parsis** have attained wealth and economic power far out of proportion to their number. Their religion is the pre-Islamic Persian faith of **Zoroastrianism**, a monotheistic faith known mainly for its reverence of fire. Zoroastrians follow the teachings of the prophet Zarathustra, who was born around 600 B.C.E. in Persia (modern Iran). He preached that people should have good thoughts, speak good words, do good deeds, and believe in a single god, Ahura Mazda ("Wise Lord"). Persia's Zarathustras, as the Zoroastrians are also known, fled Persia in two waves, first during the conquest by Alexander the Great around 300 B.C.E. and then after Islam swept into Persia in the seventh century. They settled in China, Russia, and Europe, where they became assimilated with other cultures. In India, they

Joe Hobbs

• **Figure 7.1.3** For Indians and foreigners alike, the Taj Mahal in Agra symbolizes the cultural achievements of the Indian subcontinent.

maintained a separate existence until quite recently; now about one in three takes a spouse of another religion and abandons Zoroastrianism. Many scholars consider theirs an endangered religion, with only 124,000 to 190,000 followers worldwide.

India's 6 million **Jains** also have influence beyond what their numbers suggest, as they control a significant share of India's business. They live mainly in the western state of Gujarat. **Jainism,** founded on the teachings of Vardhamana Mahavira (a contemporary of the Buddha, c. 540–468 B.C.E.), is renowned for its respect for geographic features and animal life. Jains believe that souls are in people, plants, animals, and nonliving natural entities such as rocks and rivers. Jainism has taken the principle of nonviolence (*ahimsa,* also present in Hinduism) to mean they should not even harm microbes. Therefore, Jain worshipers often wear masks to prevent inhalation of microscopic organisms, and Jains cannot be farmers because they would have to destroy plant life and living organisms in the soil (•Figure 7.1.4).

Joe Hobbs

• **Figure 7.1.4** A priest in a Jain temple in Mumbai wears a mask so as not to breathe in and thereby harm microbes.

Sri Lanka has two major ethnic groups, distinguished from each other by religion and language: the predominantly Buddhist **Sinhalese,** making up about 74 percent of the population, and the Hindu **Tamils,** about 9 percent (•Figure 7.1.5). The light-skinned Sinhalese are an Indo-Aryan people who settled in Sri Lanka about 2,000 years ago. The dark-skinned Tamils, whose main area of settlement is the Jaffna Peninsula and adjoining areas in the north, are descendants of early invaders and more recent imported tea plantation workers from southern India. The
328–329 migrations of Tamil tea workers from India to Sri Lanka, and their subsequent travails there, are part of the legacy of British colonial rule.

7.1.2 Geographic Consequences of Colonialism and Partition

On a physical map, the Indian subcontinent looks like a discrete unit, marked off from the rest of the world by its mountain borders and seacoasts. But the social complexity of this area is so great that it has never been unified politi-

Joe Hobbs

• **Figure 7.1.5** Buddhism is the majority faith in Sri Lanka. This is the Vatadage Temple at Polonnaruwa, where the relic of the tooth of the Buddha was said to have been kept in the twelfth century.

cally, except when Britain ruled. For more than a century, until 1947, India was the most important possession of the British Empire—the **"jewel in the crown."** India then consisted of the entire area of the present India, Pakistan, and Bangladesh. It had a classic colonial relationship with Britain, supplying the motherland with an almost endless pool of low-cost labor and valuable raw commodities like cotton and opium while providing Britain with a huge market for manufactured goods.

British colonialism left indelible marks on the subcontinent's cultural and economic landscapes, some favorable and others decidedly not. To serve the needs of their empire, the British developed the great port cities of Karachi, Bombay (now known as Mumbai), Madras (now Chennai), and Calcutta (now Kolkata) and India's current inland capital of New Delhi. The British created a lasting positive legacy in developing India's infrastructure, particularly the superior rail transportation networks that still assist industrial development in India, Pakistan, and Bangladesh.

Parvinder Sethi

• **Figure 7.1.6** The central rail station in New Delhi, India, is one of the hubs in a vast network dating to the British colonial period.

The vast rail network created by the British has been expanded and improved by the independent nations. In addition to transporting goods to internal markets and to ports for export, this system carries millions of passengers daily (•Figure 7.1.6).

Around the world, the colonizing power frequently favored one ethnic or socioeconomic group over another, sowing the seeds for eventual discord. This was the case in British India, where the formerly subordinate Hindus came to dominate the civil service and most businesses. Many Muslims feared the results of being incorporated into a state with a Hindu majority, and their demands for political separation led in 1947 to the creation of two independent countries: the secular but predominantly Hindu nation of India and the Muslim nation of Pakistan. The respective secular and religious identities of both countries remain critical issues today. Pakistan had two parts, West Pakistan and East Pakistan, separated by 1,000 miles (c. 1,600 km) of Indian territory. In 1971, an Indian-supported revolt in East Pakistan led to the birth there of the new independent country of Bangladesh.

The partition of colonial India precipitated all kinds of problems for the new countries, some of them still prominent today. The largest problem is the disputed Kashmir region (see Problem Landscape, pages 314–315). Immediately preceding and following partition in 1947, violence broke out between Hindus and Muslims, and hundreds of thousands of lives were lost in wholesale massacres. More than 15 million people migrated between the two countries. Pakistan today has a large population known as **Mohajirs,** or "migrants," the Muslim immigrants and their descendants who poured into the region from India when partition occurred. Particularly in Karachi and elsewhere in the southern province of Sind, violent conflict frequently occurs between rival factions among the Mohajirs and between the Mohajirs and other ethnic groups, including Pathans (Pushtuns) and Biharis.

PROBLEM LANDSCAPE

Kashmir

Since 1947, India and Pakistan have been in conflict over the status of Kashmir (called Jammu and Kashmir in India), a disputed province straddling their northern border (• Figure 7.1.A and 7.1.1, page 310). Before independence, Kashmir was a princely state administered by a Hindu maharaja. About three-fourths of Kashmir's estimated population of 15 million is Muslim, which is the basis of Pakistan's claim to the territory. But under the partition arrangements, the ruler of each princely state was to have the right to join either India or Pakistan, as he chose. Kashmir's Hindu ruler chose India, which is the legal basis of India's claim.

After partition, fighting between India and Pakistan led to the demarcation of a cease-fire line—the "line of control" that still separates the forces—leaving eastern Kashmir, with most of the state's population, in India and the more rugged western Kashmir in Pakistan. Beginning in the mid-1950s, China pressed its own claims to remote northern mountain areas of Kashmir and occupied some of the border territories. Pakistan ceded part of the occupied territory to China and established friendly relations with its giant northern neighbor, but India rejected China's claims (relations between these countries remain chilly). India's subsequent small-scale military actions failed to dislodge Chinese forces. India now holds about 55 percent of the old state of Kashmir, Pakistan 30 percent, and China 15 percent. India has another dispute with China over China's control of 34,750 square miles (90,000 sq km) in India's northeastern state of Arunachal Pradesh.

Three wars between India and Pakistan have effected little change in Kashmir. In 1965, conflict began in Kashmir, spread to the Punjab, and escalated to a brief but indecisive full-scale war involving tanks, airborne forces, and widespread air raids. Renewed hostilities in Kashmir in 1971, as part of the war in which India supported the revolt of Bangladesh against Pakistan, again did not alter the political landscape of Kashmir. During that conflict, the Siachen Glacier in the Karakoram Mountains, at about 20,000 feet (2,800 m), earned the title of "world's highest battlefield."

Kashmir's Muslim majority escalated the campaign for secession from India in 1989, and the strife has continued ever since. India and Pakistan came perilously close to the brink of war over Kashmir in 2002. Late in 2001, armed insurgents attacked India's parliament in New Delhi in a failed attempt to assassinate Indian politicians. India's government identified the assailants as members of two Muslim rebel movements (Lashkar-e-Taiba and Jaish-e-Muhammad) based in Kashmir. India further accused Pakistan's

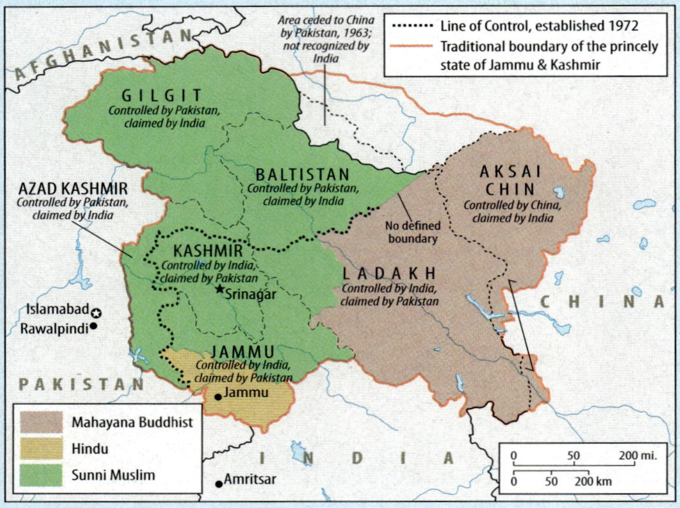

• **Figure 7.1.A** Political boundaries of Kashmir. See Figure 7.1.1 on page 310 for its context.

As a consequence of the partition, India received almost all of the subcontinent's modern industries and most of its mineral wealth and energy supplies. Pakistan, including the present Bangladesh, was left with a small industrial infrastructure and fewer natural resources. Pakistan and Bangladesh are therefore still very far behind India in both high-tech and traditional industrial output. The burlap industry is a good example of how partition frustrated development in the region. The province of Bengal, in colonial India's eastern Ganges-Brahmaputra Delta region, is the world's greatest production area for jute, the raw material for burlap. With independence, most of the jute-growing areas were isolated on the Pakistani (now Bangladeshi) side, while the jute mills, which had developed in the Calcutta area, are in India. After independence, India struggled to promote jute cultivation and Bangladesh to promote jute processing. There were also problems in how to divide up fresh water among the newly independent countries (see The World's Great Rivers, page 316).

7.1.3 Natural Regions and Resources

Physical conditions are very important in the lives of the rural majorities of South Asia. People's well-being depends largely on local agricultural resources, and industrialization is based mainly on other locally available natural resources. Fortunately, South Asia has some generous endowments of natural resources and a huge and talented labor force to work with them. It is also prone to colossal natural hazards, including cyclones, floods, droughts, landslides, and earthquakes. The subcontinent may be roughly divided into three natural areas: the outer mountain wall, the northern plain, and peninsular India (see Figure 7.4, page 286).

PROBLEM LANDSCAPE

Kashmir *continued*

government of supporting these rebel groups and therefore being ultimately responsible for the attack on India's parliament. Indian military forces began amassing on the line of control in Kashmir, and Pakistani troops responded in kind. Tremendous international diplomatic pressure was applied to both sides in an effort to prevent all-out war.

The pressure was particularly intense for Pakistan. Pakistani President Pervez Musharraf responded by outlawing and beginning to crack down on five militant groups, including the two blamed for the December attack. It was a risky move, especially because many people in Pakistan believed their leader was forsaking a legitimate struggle in Kashmir and "selling out" to the United States and other Western powers in the wake of the 9/11 attacks. While Musharraf's actions did at least temporarily defuse the explosive situation in Kashmir, many observers remained concerned that Pakistan would prove unable to rein in Muslim attacks against Indian targets in Kashmir. There were and remain many questions about how much control Pakistani leaders have over their own intelligence service, the ISI (Inter-Service Intelligence Agency), which may contain rogue elements intent on continued support of the militants fighting India.

Joe Hobbs

• **Figure 7.1.B** Kashmir, "Paradise on Earth" but for the bloodshed there. This is Mt. Rakapashi, 22,500 ft. (7,788 m) high.

The Kashmir conflict has been very costly to both sides, with more than 65,000 people killed since 1965, about half of those since 1989. India's international reputation suffered because of the country's unyielding position on the region. Pakistan has steadily spent about a third of its budget on defense, mostly focused on a potential engagement with India over Kashmir. There would be economic benefits for both India and Pakistan if lasting peace were achieved. "Kashmir is a Paradise on Earth," wrote Samsar Kour.* The stunning snow-crowned peaks and flower-laden valleys of this western Himalayan region once lured millions of Indian tourists and ranks of Western trekkers each year, and the tourism development potential is enormous (•Figure 7.1.B). In 2003, Indian and Pakistani leaders finally met to speak about resolving the Kashmir problem and declared a formal cease-fire in the region for the first time in 14 years. Both countries, and the wider world, await more moves toward a permanent peace agreement.

*Samsar Chand Kour, *Beautiful Valleys of Kashmir and Ladakh* (Mysore, India: Wesley Press, 1942), p. 2.

The Outer Mountain Wall

Between 40 and 50 million years ago, a triangular piece of the earth's crust completed its tectonic journey across the Indian Ocean and collided with southern Asia (•Figure 7.1.7). The collision resulted in the thrusting up of a mas-
21 sive and almost solid wall of mountains. That wall, made up mainly of the Himalaya Mountains and the Karakoram Range, forms an inverted U to define the northern bound-
284 ary of the subcontinent. From each end of this massive wall, lower mountain ranges trend southward to the sea.

Pakistan's borders with Iran and Afghanistan traverse the rugged western section of the mountain wall. Desert and steppe climates reach upward to high elevations. These mountains provide only a few crossing places, including the storied Khyber Pass through the Hindu Kush Mountains on the border between Pakistan and Afghanistan (see Figure 7.1.17). These passes have long been gateways into India for conquerors from the outer world, including the Aryans and the Arabs.

The towering northern segment of the mountain wall has been much less passable, but a trickle of trade and cultural exchange has crossed these mountains for millennia. The Himalayas extend about 1,500 miles (c. 2,400 km) from Kashmir in the west through the countries of Nepal and Bhutan to the northeastern corner of India (•Figure 7.1.8). Paralleling the Himalayas on the northwest and separated from them by the deep gorge of the upper Indus River in Kashmir lies another lofty range called the Karakoram. The two ranges contain 92 of the approximately 100 world peaks rising higher than 24,000 feet (7,315 m), including the two tallest: Mount Everest (29,035 ft, or 8,850 m), in the Himalayas on the border between Nepal and China, and K2, or

THE WORLD'S GREAT RIVERS

Indus, Ganges, Brahmaputra

South Asia's Indus, Ganges, and Brahmaputra Rivers are the waterways on which some of the earth's most significant civilizations developed, and they are still vital to the livelihoods and spirituality of a large portion of humanity. The allocation of these precious and meaningful waters since India's independence from Britain has been problematic.

Since 1947, the region of the subcontinent called the Punjab, meaning "Five Waters" (referring to the five Indus River tributaries), has been split between two countries, with its rivers flowing through Indian territory before entering Pakistan (•Figure 7.1.C; see also Figures 7.1.13 and 7.1.17). In geographic and geopolitical terms, India is an upstream country and Pakistan is a downstream country where Punjab waters are concerned. There are no internationally accepted laws regarding water sharing between upstream and downstream states, and the upstream country often exercises its geographic advantage to siphon off what the down-
233, 353 stream country views as too much water. **Riparian** (river-owning or river-bordering) countries often come close to, or actually go to, war over the problem. In many cases, formal treaties have averted what could have become major international conflicts.

After 1947, relations between India and newly independent Pakistan deteriorated in part because of the disputed waters of the Indus and the five Punjab rivers. For many years, the parties could not agree on how much of the water India could divert and how much should remain for Pakistan. The countries finally reached a satisfactory resolution in the 1960 **Indus Waters Treaty.** The agreement allocates the water of the three eastern rivers of the Punjab to India, which in return allows unrestricted and undiminished flow of the two others, and the Indus River itself, into Pakistan. The two countries have abided by the agreement ever since, despite their periodic wars and constant state of tension. Today, the biggest problem over Indus waters is within Pakistan, with downstream Sind province complaining that upstream Punjab province is committing "water robbery."

Another landmark water-sharing agreement was signed between India and Bangladesh in 1996. For two decades, these nations had a bitter disagreement about the allocation of water from the Ganges River. In the 1996 accord, which is meant to be in effect for 30 years, upstream India guarantees a larger share of water to downstream Bangladesh. This achievement may open the way to the integrated development of what is known as the Ganges-Brahmaputra-Barak (GBB) river basin, a watershed that is home to more than 500 million people in Bangladesh, Bhutan, China, India, and Nepal. Cooperation to settle the vital issue of water allocation among these countries could promote peace and reduce poverty in the region.

Joe Hobbs

• **Figure 7.1.C** The confluence of the Gilgit (left) and Indus (right) Rivers, in Pakistan's section of the disputed Kashmir region.

Mount Godwin Austen (28,268 ft, or 8,616 m), in the Karakoram within the Pakistani-controlled part of Kashmir.

Geographic surveys recently quieted a brewing debate over whether K2 might actually be higher than Everest: The Himalayan peak is the highest. Both great ranges, however, are young, dynamic, and still rising, so K2 could conceivably outgrow Everest. While the peaks are climbing, their glaciers are retreating. There is more ice here than anywhere else on earth outside Greenland and the po-
34 lar regions, but it is melting quickly. Some fingers point to human-made greenhouse gases as the culprit, and by one estimate, at current rates of melting, these glaciers would be gone by 2035.

The mountains along and near the subcontinent's border with Myanmar (Burma) are much lower than the Himalayan ranges, but cloaked with dense vegetation and soaked with rain, they are also nearly impenetrable. Cherrapunji in the Khasi Hills of India has an average annual rainfall of 432 inches (11 m), more than nine-tenths of it falling in the summer. It ranks with a station on Kauai in the Hawaiian Islands as one of the two spots with the greatest annual rainfall ever recorded. Cherrapunji reportedly has the highest alcoholism rate in South Asia because so many people stay indoors and drink on rainy days!

The Plains of India, Pakistan, and Bangladesh

Just inside the outer mountain wall lies the subcontinent's
northern plain. This large alluvial expanse contains the 322, 324
core areas of the three major countries. In the west, the plain is split between Pakistan and India. The Indus River and its tributaries and **distributaries** cross the portion lying in Pakistan and water this country's two most populous provinces, Punjab and Sind, and its major cities of Kara-

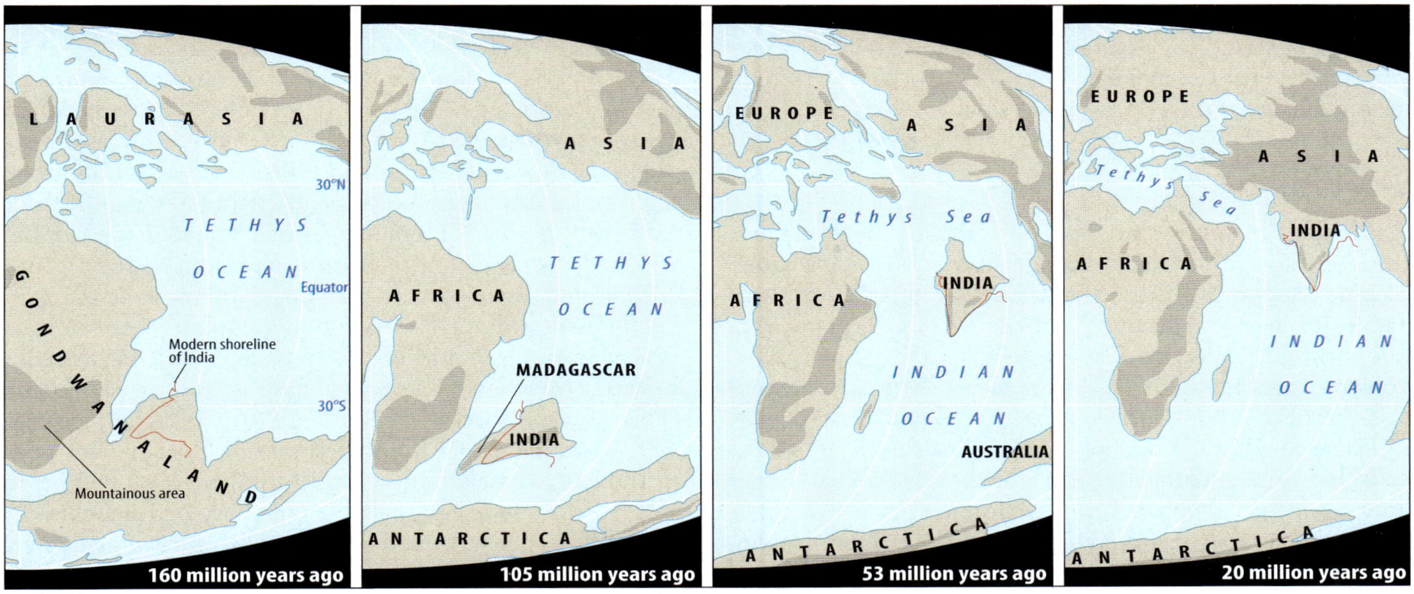

• **Figure 7.1.7** Most of India was once an island. Tectonic movement transported it across the Indian Ocean. When it collided with the Eurasian landmass, the world's highest mountains were formed.

• **Figure 7.1.8** The roof of the world, near Mount Everest in eastern Nepal. Generations of Sherpa farmers and herders have cleared all the forest that once cloaked the hillsides in the foreground.

chi and Lahore. The climate of the western plains is desert and steppe; the dry region straddling the two countries is known as the Thar Desert, or the Great Indian Desert. Irrigation water from the rivers sustains highly productive agriculture and millions of people here.

East of the Punjab lies the part of the northern plain traversed by the Ganges (Ganga) River and its tributaries, especially the Jumna (Yamuna). The western portion of this region is known as the Doab, "the land between the two rivers." The northern plain is India's core region, containing more than two-fifths of the country's population as well as its capital at New Delhi, many other major cities, and sacred places in the religions of Hinduism, Buddhism, and Jainism. This region of India is relatively well watered, especially toward the east, where agriculture based on irrigated rice supports high population densities. The southeastern part of the northern plain lies in the region of Bengal, which is essentially the huge delta formed by the combined flow of the Ganges and Brahmaputra Rivers. Bengal is divided between India and Bangladesh and has an extremely dense population, with the major cities of Kolkata on the Indian side and Dhaka on the Bangladeshi side.

Peninsular India

The southern peninsular portion of the subcontinent, which is entirely within India, consists mainly of a large volcanic plateau called the Deccan. It is generally low, less than 2,000 feet (c. 600 m) above sea level. Its rivers run in valleys cut well below the plateau surface. This topographic problem, together with the rivers' seasonal flow pattern, makes it difficult to use their waters for irrigation of the upland surface without large inputs of capital and technology. Wells and rain catchment ponds called tanks are used, but water is in short supply and population density is low compared with the better-watered parts of the northern plain (• Figure 7.1.9).

Hills and mountains outline the edges of the Deccan. In the north, belts of hills (the Central Indian Hills) separate the plateau from the northern plain. On both east and west, the plateau is edged by ranges of low mountains called *ghats* ("steps") overlooking alluvial plains along the coasts of the Bay of Bengal and the Arabian Sea. The low and discontinuous eastern mountains are known as the Eastern Ghats. The higher and almost continuous western mountains are the Western Ghats. The two ranges merge near the southern tip of the peninsula in clusters known locally as the Nilgiri, Palni, and Cardamom Hills. They are really rather high mountains, with some peaks rising above 8,000 feet (c. 2,400 m). The colonizing British founded several "hill stations" in them for recreation and administration during summer months, when the climate on the subcontinent's lowlands is stifling. They are now popular tourist destinations for Indians during that season. The coastal plains between the ghats and the sea support very dense populations, especially in the far south along the Malabar Coast of western India and on the corresponding Coromandel Coast in the east. India's largest city, Mumbai, is on the peninsula's northwest coast, and the large city of Chennai is on its southeast coast. Inland are the major cities of Hyderabad and Bengaluru.

Joe Hobbs

• **Figure 7.1.9** It is usually the task of women to draw water from communal supplies such as this one in a small settlement near Mumbai.

Climate and Agricultural Conditions

Climatic conditions in South Asia range from Himalayan ice to the tropical heat of peninsular India and from some of the world's driest climates to some of the wettest. Climatic and biotic types in the subcontinent include undifferentiated highland climates in the northern mountains; desert and steppe in Pakistan and adjacent India; humid subtropical climate in the northern plains; tropical savanna in the peninsula; and rain forest along the seaward slopes of the Western Ghats and in parts of the Ganges-Brahmaputra Delta and the eastern mountains near Myanmar (see Figure 7.5, page 287). Most of the region's original vegetation cover is gone, but there are numerous large national parks and protected areas, especially in India, designed to ensure that important remnants will remain. The region's wildlife still includes tigers, elephants, rhinoceroses, and in western India, the sole surviving population of the once widespread Asiatic lion (see Figure 2.12g, p. 29).

Heat is nearly constant except in the mountains. The tropical peninsula is hot all year. The subtropical north has stifling heat before the arrival of the wet monsoon season, and there are warm conditions even in the winter. Some of the warmest temperatures on the globe occur in the plains of Pakistan and northwestern India. Delhi averages 94°F (34°C) in both May and June, the months when the highest temperatures occur over most of the subcontinent. High humidities accompanying the impending monsoon rains combine with the heat to create almost intolerable conditions. The monsoon rains bring cooling relief to the subcontinent—but sometimes also catastrophic flooding.

India's monsoons include both the **wet southwest monsoon,** which brings the short, wet summer between June and September, and the **dry northeast monsoon,** which creates mostly arid conditions the rest of the year; these are the typical monsoon patterns of Asia described on pages 283–285. There are two main arms of the wet, or summer,

Joe Hobbs

• **Figure 7.1.10** Aridity and deforestation are among the urgent environmental problems faced by Pakistan. Foraging by goats and cutting trees for fuel in an already dry environment have created a lunar landscape in the country's Karakoram Mountain region.

monsoon. One, approaching from the west off the Arabian Sea, strikes the Western Ghats and produces heavy rainfall on these mountains and the coastal plain. The amount of rain diminishes in the interior Deccan rain shadow region to the east of the mountains, where barely enough rain falls to support dry farming, and drought sometimes brings crop failures. The second arm of the summer monsoon approaches from the Bay of Bengal, bringing moderate rainfall to the eastern coastal areas of the peninsula and heavy rains to the Ganges-Brahmaputra Delta region and northeastern India. This is the monsoon that feeds cyclones (hur-
328 ricanes). Wet monsoon winds pass up the Ganges Valley to drop moisture that diminishes in quantity from east to west.

Both arms of the monsoon bring some rain to Pakistan, but it is so little that semiarid or desert conditions prevail in most areas (•Figure 7.1.10). Blowing mainly from land to sea, the dry monsoon brings dry and cooler weather to most parts of the subcontinent. An exception is in the far south of the peninsula, where the heaviest rainfall of the year falls along the eastern coast and in adjacent uplands from October through January.

With precipitation concentrated in a short part of the year, people must extend the crop season by irrigating. They go to great lengths to get additional water onto the land, and groundwater is in increasingly short supply. Spurred on by cheap electricity and the absence of any legal constraints, farmers pump out as much water as they can afford the technology to reach. India is considering an overhaul of its agricultural policies, especially by revising water laws and switching to more drought-tolerant crops. Drought conditions in a country often visited by flooding monsoon rains may seem ironic, but the problem is that India's water storage capacity remains low. The conventional solution—building more and larger dams—is problematic because of environmental and social concerns.

Large dams and associated canal systems have been built in the steppe and desert areas of the Punjab and the lower Indus Valley (Sind province) in Pakistan, and more are planned. A massive irrigation and hydroelectric power project is under way in the north central part of peninsular India, where the Sardar Sarovar Dam was recently constructed on the Narmada River (•Figure 7.1.11). Largely because the dam project would displace 320,000 tribal and rural people (known as "oustees" in the project language) and inundate about 44 square miles (114 sq km) of cropland and 50 square miles (130 sq km) of forest, environmentalists and human rights advocates both in India and abroad fiercely opposed the dam. The World Bank withdrew its financial support in response to their protests, but asserting that its benefits will greatly outweigh its costs, India pressed ahead with the dam's construction. It is the centerpiece of the massive Narmada Valley Development Project,

• **Figure 7.1.11** The controversial Sardar Sarovar Dam, on India's Narmada River.

where 30 dams will displace about 1 million people by the projected completion date of 2025.

Correlations between people and food supplies, both spatially and in raw numbers, are always critical in South Asia (see Regional Perspective, page 321). The spatial distributions of food crops and population densities correlate
282 with the amount of water available in the subcontinent. Rice is the basis of life in the wetter areas and, with its high caloric yields per acre, is associated with the highest population densities. Rice dominates in the delta area of Bengal (in both India and Bangladesh), in the adjacent lower Ganges Valley, and in the coastal plains of the peninsula. India's Punjab (which produces just over half of India's food crops) and Pakistan's Sind and Punjab provinces produce surplus rice, including the basmati variety, which is an important export. Irrigated wheat is the staple crop and food in the drier upper Ganges Valley of India and in the dry Punjab of India and Pakistan. In these areas, crop yields and population densities are intermediate. Unirrigated plantings of sorghum and millet prevail over most of the Deccan plateau and in other areas where it is difficult to supplement the low rainfall through irrigation. Yields from these crops are low, and so are population densities.

Maize (corn) is an important grain in the lower regions of the Himalayas. But rice is the preferred food almost everywhere, and where enough water is available, there are patches of irrigated rice. Rice, cotton, jute, and tea are the main crops grown in the subcontinent for industrial use and for export. Of the four, only tea is exported on a sizable scale, and only tea is a plantation crop. The plantations were developed, and are still largely owned, by British interests. Production is greatest in the northeastern state of Assam, with another center in the mountainous far south of peninsular India (•Figure 7.1.12). Cotton is grown mainly in the interior Deccan. India's large textile industry absorbs most of the production. Coffee, rubber, and coconuts are minor plantation crops, and India exports products made from coir (coconut fiber).

Joe Hobbs

• **Figure 7.1.12** A tea plantation in the Western Ghats of eastern Kerala state. These mountains have been widely cleared of native forest by logging and agricultural expansion.

The remainder of this module deals with economic and other important geographic characteristics and problems of each of the South Asian countries.

7.1.4 India: Power, Courage, and Confidence

Despite its overwhelmingly rural population (72 percent, living in about 600,000 villages), India boasts some of the world's **megacities** (extremely large cities), notably Mumbai and Kolkata (•Figure 7.1.13). These and many smaller cities are at the heart of India's current economic boom, which is based more on information technology than on old-fashioned heavy industry.

It is difficult to make generalizations about India in economic or most other matters because there is such a vast continuum of Indian experience. Visitors to India may be astounded to see both the latest computer technologies and the labors of human hands performing, on a large scale, the tasks routinely done by machines in more developed countries—for example, people making gravel by hammering boulders into pebbles. And always looming over India are questions of poverty, caste, and religious strife. Poverty is arguably the most prominent issue: nearly one in three Indians is among the official ranks of the poor, and the country is ranked 126th of the 177 countries on the Human Development Index. 44

Urban and Economic Geography

India's recent high rate of economic growth (an average of 8 percent per year between 2001 and 2007) owes largely to important reforms and gives new meaning to its ancient "three lions" symbol, standing for power, courage, and confidence.

Until the 1990s, India was intent on keeping out most foreign manufacturing in favor of industrial self-sufficiency. Most of the economy was socialized, with many state-owned industries. With these policies, India's industries expanded and diversified, but profitable exports were limited. After that, India opened up to more imports, privatized many state-run industries, and boosted exports (but in notable contrast with China, India still imports more goods than it exports). India's vast pool of well-educated people responded in creative ways to new opportunities. The result was remarkably strong economic growth—so strong, in fact, that there is nervousness about a "bubble" that could burst; some observers fear that skyrocketing real estate prices, stock indexes, inflation, and borrowing all point in that direction. 295

India's diverse and rapidly expanding manufactures include old-economy products like cars, trucks, motorcycles, bicycles, and farm machinery, along with high-tech data processing equipment, silicon chips for electronics, and computer software. India's leading factory industries are cotton textiles, jute products, food processing, and iron and steel. Unlike China, India has abundant reserves of iron ore

REGIONAL PERSPECTIVE

Keeping Malthus at Bay

Poverty and human health are serious problems in South Asia, and the prospect of continued high population growth raises the question of whether growth in food supplies can avert the proverbial Malthusian crisis (see Chapter 3, page 60).

India's population has surged since independence, from 352 million in 1947 to 1.1 billion in 2007. In recent decades, the overall trend has between toward lower population growth rates. Fifty years ago, the average Indian woman had six children; now she has three. That rate is still well above the natural replacement rate, however, that would keep the population steady. The population base is already so vast that even modest growth will add huge numbers. India is predicted to overtake China as the world's most populous country by 2040.

For many people, the name "India" invokes an image of grinding poverty, and with good reason: it is estimated that fully one-third of the population is "abjectly poor," defined as living on less than $1 per day, and almost half of India's children are malnourished. But perhaps in defiance of Malthus, an increased population has so far succeeded in producing more economic resources overall. A few people in India are very wealthy, and there is an emerging middle class of as many as 300 million people—the world's largest middle class (•Figure 7.1.D). India's economy has grown impressively in recent years. An apparent problem is that the benefits of the growing economy are unevenly distributed. Most of the economic growth benefits those who are already better off, widening the gap between rich and poor. About 800 million Indians are either unemployed or underemployed. In its bid for reelection in 2004, India's ruling party, the BJP, promised to make the country an economic superpower by 2010. In a stunning setback viewed as a referendum by the poor, who felt that India's growth would continue to pass them by, the BJP lost the election to the opposition Congress Party. Population growth in the long run may further distort India's economic imbalances.

Joe Hobbs

• **Figure 7.1.D** This Indian family visiting the Red Fort in Agra is among the growing ranks of the subcontinent's middle class.

The issue of population growth is also critical for Pakistan, which has never mounted an effective family planning program, mainly because of opposition from Muslim religious authorities, who regard birth control as an intervention against God's will (Pakistan's birth rate in 2007 was 31 per 1,000, compared with 24 per 1,000 in India). Muslim Bangladesh, along with Nepal and Bhutan, also have high birth rates, but they are much lower in the most critical nation of Bangladesh than they once were. In the 1970s, the average woman had seven children, while now she has three, and contraception use nationwide has increased from 4 percent then to more than 50 percent now. South Asia's success story in bringing population growth under control is Sri Lanka, with a birth rate of just 18 per 1,000 and annual population growth of 1.2 percent.

Agricultural output in South Asia has increased since independence. Most notably, despite its huge and growing population, India has achieved self-sufficiency in grain production, with a fourfold increase in output since 1950. Almost half a million "ration shops" sell subsidized food staples to the country's poorest people. The successes of agriculture in the subcontinent have been due mainly to the increased use of artificial fertilizers, the introduction of new high-yield varieties of wheat and rice associated with the Green Revolution, more labor provided by the growing rural population, the spread of education, the development of government extension institutions to aid farmers, and better irrigation. 299

One of the challenges confronting India is to improve the status of women. Despite a 1961 ban on dowries (money and gifts given by a bride's parents to the groom), the practice is continuing. So is the rate of killing brides who do not provide enough dowry. The burden placed on the bride's family is one reason that many prospective parents choose to abort female fetuses, which can be detected with ultrasound technology. India banned this use of ultrasound in 1996, but the practice continues. Female feticide—particularly prevalent among India's more educated and affluent, who can afford the ultrasound screening—and infanticide have taken tens of millions of young and unborn girls. In some of India's states, especially Haryana, Punjab, Delhi and Gujarat, one result is a serious shortage of local brides; for every 1,000 men, there are 900 women. The gap is being filled by marriages arranged with women from distant Indian states and Bangladesh.

to draw on for its steel production, which has surged in recent years. India is both a major user and major exporter of cotton textiles and clothing. Favoring cotton manufacturing are a huge home market, inexpensive labor, and prolific domestic production of cotton.

India makes a wide range of chemical products, from vitamins to fertilizers. The pharmaceutical industry is flourishing. Much to the chagrin of Western drug companies, their Indian counterparts routinely replicated foreign drug formulas and produced copycat medications for sale at a

• **Figure 7.1.13** Principal features of India.

fraction of their cost in the West. India finally stopped argu-
ing that it was a poor country needing low-cost medicines
and in 2006 bowed to the pressure to join global patent
465 standards. India's success in providing lifesaving medica-
tions to its population cannot be disputed; life expectancy
has risen by 20 years since the country became independent.
One of the big questions is if such improvements in the qual-
ity of life can be sustained for India's growing population.

While manufacturing is growing quickly, the economic
sector with the greatest current growth is information tech-
nology, which employed 1.3 million of India's 400 million
workers in 2007. India's software exports, two-thirds of
which go to the United States, have been growing in recent
years at rates of up to 50 percent annually, and technology
and service jobs have been increasingly outsourced to India 300
from the United States.

Information technology is helping bridge the digital di-
47 vide separating the rich and poor in India. Internet cafés are turning up even in remote villages, where their operators use diesel generators to keep computer terminals from failing during frequent power outages. The Internet is debuting in many rural areas with Wi-Fi access, thanks to high-range wireless transmitters. Increasing Internet usage (now only 16 per 1,000 in India, versus 57 per 1,000 in all of Asia and 96 per 1,000 worldwide) may bring significant social and economic improvements to the vast numbers of India's rural poor. For example, villagers might be able to use this technology to gain unbridled access to land records, rather than having to bribe officials to get them. Those seeking work abroad could download passport applications instantly rather than spend time and money working through intermediaries. Farmers could get information on crop prices in different towns to improve their bargaining positions during harvests. These kinds of advances are already happening in parts of rural India, where a private company called ITC has set up Internet access so that farmers can check the price of soybean futures on the Chicago Board of Trade. Armed with this knowledge, the farmer can decide whether to sell his soybeans immediately or wait a while longer until prices rise. In this arrangement, known as the **e-Choupal initiative** (a *choupal* is the leading farmer in a district), ITC is encouraging farmers to sell their better soybeans directly to the company rather than to the village traders or government markets from which ITC previously had to buy the crops. Mobile phones are also aiding entrepreneurship in all sectors of India's population, including farmers. At less than 1 U.S. cent per minute, India's cell phone rates are the lowest in the world, and half of the country's population is expected to own cell phones by 2010.

Some of India's intellectual productivity is lost to other countries, perhaps slowing down potential economic growth. India suffers from a brain drain, the emigration
59 of highly trained and highly educated individuals abroad. Chances are that the American reader of this text has been or will be treated in a hospital or clinic by a physician who emigrated from India, attracted by higher U.S. wages. The United States' "brain gain," in the form of highly qualified medical personnel, is inevitably India's loss because better development requires better human resources. Indian computer scientists, engineers, nurses, and other professionals have in recent years followed the trend in "exports" of doctors to the United States and other countries. In part reflecting the fact that Indian universities are not yet up to world-class standards, American universities currently train more students from India (about 80,000) than from any other foreign country. Most return to India, and others are eventually wooed into returning. Altogether, 20 to 30 million Indian citizens live abroad, with about 2 million in the United States and about 3 million in the oil-exporting countries of the Middle East. Another 2 million Bangladeshis and 1 million Pakistanis work in the Middle East,
many in semiskilled or unskilled jobs. The **remittances** they 146
send home are vital to South Asia's economies.

India faces several challenges to growth. One is upgrading energy sources and distribution. India has some hydropower, coal deposits, and a modest oil output from fields in Assam, Gujarat, and offshore near Mumbai. Nuclear power produces about 3 percent of its electricity. Like the other South Asian countries, however, India is heavily dependent on imported energy. Infrastructure needs improvement. Long neglected, the country's decrepit road system is getting a major makeover, comparable to the interstate highway system built in the United States in the mid-20th cen-
tury. At the heart of this effort is the **Golden Quadrilateral,** 626
a four- and six-lane highway system running 3,625 miles (5,833 km) through 13 Indian states and connecting 19 of the 35 Indian cities with more than a million inhabitants (see Figure 7.1.13). The highway's purpose is to speed up the pace of development and integration in this vast and complex country. It will certainly facilitate social and economic exchange, and health experts warn that the road will also be a major transmission route for HIV/AIDS. An estimated 11 percent of India's truckers are HIV-positive. India's labyrinthine bureaucracy, by far the world's largest and a legacy of the country's past socialistic policies, also challenges the country's development. In an effort to lure foreign investors who might otherwise be scared off by India's red tape, the country has set up numerous Special
Economic Zones (SEZs) patterned after those of China. 385
Located near the larger cities, these industrial sites promise tax breaks, speedy paperwork, and reliable electricity supplies to companies that agree to locate here.

India's industries are concentrated in its cities. On the peninsula's northwest coast is the largest, Mumbai (population 20.4 million). The country's cotton mills are concentrated mainly in the metropolitan area of Mumbai and neighboring areas. Mumbai is known as **India's Wall Street** and is also its Hollywood (it calls itself **Bollywood**), the center of a prolific and accomplished film industry (the world's largest) that produces movies of international stature as well as the regular fare that a huge domestic market consumes passionately (•Figure 7.1.14). In the south central peninsula is Bengaluru, known as **India's Silicon Valley,** where bustling operations of Microsoft, Motorola, Intel, Texas Instruments, and a host of other high-tech firms present a startling contrast to the largely rural and unchanging image that many outsiders have of India. The largest metropolis of the far south, Chennai, is a seaport on the Coromandel Coast.

At the western end of the Ganges-Jumna section of the subcontinent's northern plain, the old fortified capital of Delhi arose as a bastion against invaders from beyond the western mountains (•Figure 7.1.15). The present adjoining capital of New Delhi was founded after 1912 as the capital of British-controlled India. Today, the population of Delhi and New Delhi is an estimated 18 million, making it India's second-largest metropolis. Its main industries are

Joe Hobbs

• **Figure 7.1.14** Mumbai's Bollywood produces the romance and action films that are the staples of Indian cinema. The advertising is misleading; Indian films tend to show little flesh.

Joe Hobbs

• **Figure 7.1.15** The Mogul Emperor Shah Jahan established Delhi as India's capital early in the 17th century. His monumental Red Fort (top left) still dominates this old city. Adjacent New Delhi is India's capital today.

manufacturing, textiles, electronics, and chemicals. Far to the southeast is the English-founded port of Kolkata in the state of West Bengal. A city of 14.9 million, Kolkata has a worldwide reputation for teeming slums and dire poverty—Rudyard Kipling called it "The City of Dreadful Night"—and it was this Indian city to which the extraordinary nun Mother Teresa dedicated her life's work of aiding the poor. Within India, Kolkata is recognized as a great center for culture and the arts. The surrounding Bengal region is an important industrial area where the country's jute, iron, and steel industries are concentrated. The most valuable mineral resources in India, including iron ore, coal, manganese, chromium, and tungsten, are found in this region. These resources have led to the rise of small to medium-sized industrial and mining cities (notably Jamshedpur) in an area that otherwise is one of the least developed parts of the country.

Social and Political Challenges

Partition following independence did not solve India's problems. Serious issues have persisted among the country's ethnic and religious groups. Some of the most problematic involve language, nationalism, and caste (see Insights, page 325). Whatever its travails, India is admired for its tireless and record-setting democracy. Every time a national election is held there, it is the largest in human history.

India has 17 major languages and more than 22,000 dialects. Many people speak two or even three languages, but overall literacy rates are low (about 61 percent for the country as a whole), and communication is often a problem. Since British colonial times, English has been the lingua franca and the de facto official language of the subcontinent. However, only a small percentage of the population

INSIGHTS

The Caste System

One ancient Hindu belief is that every individual is born into a particular **caste,** a social subgroup that determines the individual's rank and role in society. Castes form a hierarchy, with the Brahmin caste at the top (comprising about 5 percent of all Hindus) and three others (Kshatriya, Vaisya, and Sudra) below. Caste membership is inherited and cannot be changed. Particular castes are associated with certain religious obligations, and their members are expected to follow traditional caste occupations. Marriage outside the caste is generally forbidden, and meals are usually taken only with fellow caste members.

At the bottom of the social ladder are the **Dalits,** or *scheduled castes,* once known as **untouchables,** accounting for about 20 percent of all Hindus. They are not part of the caste system but are "outcastes," and according to Hindu belief, they are not reincarnated. The untouchables were so called because they traditionally performed the worst jobs, such as the handling of corpses and garbage; therefore, their touch would defile caste Hindus.

This ancient, rigid social system is changing. India's constitution outlawed the caste system in 1950, but a single decree could not undo thousands of years of tradition. Brahmin privileges are now being increasingly challenged and restricted by more laws. Indian law explicitly forbids recognition of Dalits as a separate social group. Confronting the fact that upper castes account for less than one-fifth of India's population but command more than half of the best government jobs, the Indian government has instituted an affirmative action program to allocate more jobs to members of lower castes. Indian leaders of high caste have championed the Dalits' cause for both moral and practical reasons. Politicians use caste divisions to their advantage to bring out voters, promising if elected to act in the interests of their respective large social blocs. In 1997, a Dalit was elected president of India for the first time in the country's history. Since then, Dalits have won more regional and local offices.

The Dalits are voting in greater numbers than ever before with the conviction they can use India's democratic system to overcome ancient discrimination against them. They are making progress. Disintegration of the caste system has been especially rapid in the cities, where people of different castes must mingle in factories, in public eating places, and on public transportation. In India's vast rural areas, the caste system is more entrenched.

is literate in English. This fact, along with nationalist sentiments, has prompted calls for indigenous languages to be adopted as official languages. After independence and partition, disputes arose in each country over which language should be chosen. Pakistan's leadership insisted that Urdu be the sole official language of the country, even in East Pakistan, where Bengali prevailed and Urdu was virtually unknown. The linguistic divide would become one element
291 in the eventual strife between the two Pakistans. India decided on Hindi, an Indo-European language spoken by the largest linguistic group, and made it the country's national language in 1965. This government decision resulted in strong protests, especially in the Dravidian south. The result has been a proliferation of official languages to satisfy huge populations; India now has 15, all of which appear on the country's paper currency. Rioting between speakers of different languages has occurred repeatedly, and demands by language groups have reshaped the boundaries of several Indian states.

Hindu nationalism is a serious issue in a country with such large religious minorities. The country is supposed to be secular, but in recent years, Hindu nationalists
301 have heavily influenced its politics. The explicitly sectarian Bharatiya Janata Party (BJP) held a majority of parliamentary seats, and the position of prime minister, between 1998 and 2004. This Hindu nationalist party would like India's Hindu majority to have the strongest say in the country's affairs. Until recently, the BJP platform included the adoption of a code of civil law that would strip Muslims of separate laws in matters of marriage, divorce, and property rights.

Sectarian violence (known in the region as **communal violence**) between Hindus and Muslims is a constant threat to India's social and political fabric. Contention over places sacred to both faiths sometimes ignites widespread violence. The 16th century Babri Mosque in the city of Ayodhya in Uttar Pradesh was a place of prayer for Muslims on a spot revered by the Hindus as the birthplace of the god-king Ram, an incarnation of the god Vishnu. Backed by Hindu nationalists in the provincial government, a mob of about 250,000 Hindu fundamentalists demolished the mosque in December 1992 with the intention of building a Hindu temple on the site. The ensuing communal violence between Hindus and Muslims throughout India, even among the cosmopolitan and usually tolerant urbanites of Mumbai, left thousands dead.

Ever since this incident, as a means of garnering votes from Hindu Indians, Hindu politicians have made veiled promises to proceed with construction of the temple. In 1998, the BJP platform included a pledge to build a Hindu temple atop the ruins of the mosque at Ayodhya. His apparent support for the temple helped sweep the BJP leader, Atal Vajpayee, into power as prime minister that year. The conflict over this multireligious sacred site in India has parallels with the Temple Mount dispute between Israel and
the Palestinians. Politicians in India today are increasingly 245
aware of the threat that Hindu-Muslim discord may pose to the country's development. Acknowledging the widespread

• **Figure 7.1.16** The Golden Temple (established in 1577 and rebuilt in 1764), in Amritsar, is the holiest shrine of the Sikh religion.

discrimination against Muslims in Indian society, they are considering affirmative action programs to improve the stature of the country's Muslims.

In India's Punjab, where Sikhs make up 60 percent of the population and Hindus 36 percent, the 1980s and early 1990s were violent years during which about 20,000 people died in armed clashes. Sikh factions were intent on establishing their own homeland of **Khalistan,** prompted to do so in part by New Delhi's plans to divert water from the Punjab. They challenged Indian authority and turned Amritsar's Golden Temple, the holiest site in the Sikh faith,
311 into their military stronghold (•Figure 7.1.16). In a controversial move to quell the revolt, Prime Minister Indira Gandhi ordered troops to storm the Golden Temple in June 1984. The resulting deaths and desecration led directly to Gandhi's assassination by her own Sikh bodyguards later that year. Indian authorities accused Pakistan of arming the Sikh militants.

Other trouble spots for India are the northeastern states of Assam, Manipur, Nagaland, and Arunachal Pradesh, an extraordinarily diverse region with more than 200 ethnic groups and followers of many different religions. Several rebel groups have been active in this region for years, agitating for autonomy and independence from India. The government has begun investing more resources in the area, especially infrastructure to boost trade ties with Myanmar. There are also troubles in half of India's states from Maoist insurgents, known as **Naxalites,** who claim to champion the causes of the rural poor and promise to carry violent rebellion into India's cities and ultimately expand their "compact revolutionary zone" to include the entire country.

7.1.5 Pakistan: Faith, Unity, and Discipline

Pakistan (•Figure 7.1.17) is not the demographic giant that India is, but its 169 million people face many challenges, especially owing to a small resource base relative to the population, a general shortage of industrial output, and local and geopolitical tensions that are destabilizing forces. Unlike India, Pakistan is not a secular country but an Islamic one whose motto is "faith, unity, and discipline."

• **Figure 7.1.17** Principal features of Pakistan.

Urban and Economic Geography

As mentioned earlier, the 1947 partition left Pakistan with 313
far fewer industries and less mineral wealth than India received. The country has yet to make up for that setback. Many of the country's bills are paid by foreign aid donors, with a large share of those contributions tied directly to Pakistan's perceived importance in the U.S.-led war on terrorism. 301

Pakistan does have some assets to build on. Its cotton-textile industries are well established, and irrigated plantations of cotton in the Punjab and lower Indus Valley make Pakistan the world's fourth-largest cotton producer (after China, the United States, and India). Textiles, yarn, ready-made garments, and raw cotton constitute about two-thirds of Pakistan's exports. Rice is another important export. Substantial deposits of natural gas may further enhance the nation's development.

Opium was once a major illegal export crop of Pakistan, 272
but stepped-up law enforcement has dramatically cut production. However, processing of the raw opium into heroin is still a significant industry in the rugged mountain region of the Northwest Frontier Province and nearby Federally Administered Tribal Areas, situated along the frontier with Afghanistan and inhabited mainly by Pashtun peoples. Domestic heroin addiction is a growing problem, with an estimated 3 million Pakistanis dependent on the drug.

Pakistan's major cities, which are also the main industrial centers, are on plains near the Indus River: the seaport of Karachi (population 12.3 million), located at the western edge of the Indus River Delta in Sind province, and the cultural and textile manufacturing center of Lahore (population 6.6 million; •Figure 7.1.18) in the Punjab. Farther north, at the foot of the Himalayas, is the capital city of Islamabad ("City of Islam"), a much smaller, very modern administrative metropolis.

Social and Political Challenges

Pakistan has internal problems centering on religious divisions, mainly between the majority Sunni Muslims (77 percent of the country's population) and Shiite Muslims (20 percent). Shiite Muslims opposed the government's efforts in the 1980s to establish an avowedly Sunni Islamic state, and there was violent conflict between members of the two sects. Pakistani factions who preferred a Western-style secular democracy also rioted in protest against this **Islamization** of the government. Democratic elections in the 1980s and 1990s twice installed Benazir Bhutto, a liberal, Western-educated woman, as the country's prime minister. Her regimes were so riddled with corruption that she was twice voted out of office to be replaced by Nawaz Sharif, who promised to cleanse the country of corruption and embrace the traditional Islamic system of law known as ***sharia.*** Sharif was in turn ousted in a military coup in 1999.

Sharif's successor, General Pervez Musharraf, took the helm of a country torn between the secular wishes of its Western-oriented middle class and popular calls for greater Islamic influences and fewer ties with the West. The growing Islamist sentiment is focused both on ousting India from disputed Kashmir and on resisting the U.S. and other Western intervention in the region in the wake of the 9/11 terrorist attacks. Much of the dissent is based in the Federally Administrated Tribal Areas (FATAs) and in the Northwest Frontier Province, especially in and around its principal city, Peshawar (•Figure 7.1.19). At the time of the attacks on the United States, there were already 3 million mainly ethnic Pashtun refugees from Afghanistan living in camps outside the city. Hundreds of thousands more 272
poured in from Afghanistan in the weeks following 9/11, but a reverse flow has taken place since then. Within these impoverished, desolate encampments, much anti-U.S. and generally anti-Western sentiment still simmers.

An important breeding ground for anti-Western (and anti-Indian) thought and action in Pakistan is the institution known as the ***madrasa,*** an Islamic school or seminary (•Figure 7.1.20). There are as many as 8,000 *madrasas* in 306
Pakistan, where approximately 700,000 children (mainly boys) are educated. Their curriculum is based largely on study of the Qur'an (Koran) and other Islamic texts, and some instructors promote an interpretation of the Islamic concept of *jihad* ("struggle"), in which Muslims are urged 235
to fight India, the United States, and other entities branded as "the infidel." Many of the *jihadis* ("holy warriors") who join groups like al-Qa'ida, Lashkar-e-Taiba, and Jaish-e-Muhammad have been educated in these schools. But *madrasas* also offer much legitimate instruction and help compensate for Pakistan's poor public educational system. Pakistan's leaders are in a difficult position, at once called on by the outside world to cleanse the *madrasas* of militant content and by their own people to uphold the long-cherished

Joe Hobbs

• **Figure 7.1.18** Lahore in the Punjab region of Pakistan, is the country's cultural center. It boasts many fine Islamic monuments, including this seventeenth-century Badshahi Mosque, built by the Mogul emperor Aurangzeb.

Joe Hobbs

• **Figure 7.1.19** Pakistan's Northwest Frontier Province city of Peshawar has long been a haven for spies, smugglers, and insurgents.

Joe Hobbs

• **Figure 7.1.20** Islamic education in Pakistan's *madrasas* begins at an early age. These boys are memorizing the Qur'an.

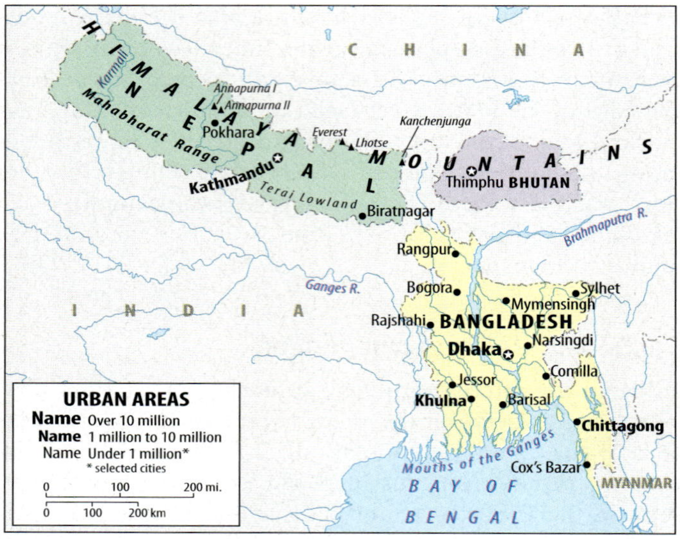

• **Figure 7.1.21** Principal features of Bangladesh, Nepal, and Bhutan.

institution of the religious school. Public opinion against President Musharraf surged after July 2007, when he ordered a bloody armed invasion of a *madrasa* held by militants in the nation's capital. His position became even more untenable following the assasination of Benazir Bhutto in December 2007, during her final bid to be elected the country's prime minister. A Pakistani group affiliated with al-Qa'ida claimed responsibility for her killing, but many Bhutto supporters held Musharraf responsible.

7.1.6 Vulnerable Bangladesh

Formerly East Pakistan, Bangladesh is about the size of the state of Iowa or the nation of Greece—not an important fact until one considers that it is home to 149 million people (•Figure 7.1.21). With its limited resources, people-overpopulated Bangladesh is South Asia's third-poorest country on a per capita GNI PPP basis, after Bhutan and Nepal. Poverty is so deep that many Bangladeshi see neighboring India as a land of opportunity. About 20 million Bangladeshis are illegal immigrants living mainly in India's West Bengal province, where authorities fear they will form permanent Muslim-majority enclaves and perhaps threaten national security. Legal migrant labor is one of Bangladesh's major "exports"; more than 6 million Bangladeshis work abroad, sending home an estimated $6 billion in remittances each year. But India has had enough; in 2007, it completed construction of an iron fence 8 feet (2.5 m) high along its 2,500-mile (4,100-km) border with Bangladesh.

The country's most notable success has been the development of cotton textile industries. If you are American, you almost certainly own clothing that was made in Bangladesh. Textile, jute, leather, and seafood export industries are concentrated in Dhaka (formerly Dacca; population 12.3 million), the capital and largest city, and Chittagong, the main seaport.

The Bangladesh portion of the Ganges-Brahmaputra Delta ranks, along with the Indonesian island of Java, as one of the two most crowded agricultural areas on earth. This area is subject to catastrophic flooding (•Figure 7.1.22). Massive tropical cyclones (hurricanes) ravage 289
Bangladesh regularly in September and October. Deforestation upstream on the steep slopes of the Himalayas has increased water runoff and sediment load in the Ganges, also contributing to Bangladesh's flood problems.

Bangladesh and adjacent portions of India must also keep a wary eye on the potential for the sea level to rise if the world's temperatures increase, as predicted by many climate change models. The country's environment minister 34
has insisted that 20 percent of Bangladesh will be under water by 2015 if nothing is done to control global warming. The Intergovernmental Panel on Climate Change recognizes this as one the world's most endangered areas, predicting a rise in sea level there of 23 inches (c. 58 cm) by 2100. On the Indian side of the delta, scientists are documenting the shrinkage of the river delta islands known as the Sundarbans. Renowned as habitat for the Bengal tiger, the Sundarbans are succumbing to a combination of sea level rise and deforestation that makes them very prone to flooding and erosion.

7.1.7 Nepal and Bhutan: Mountain Kingdoms

The kingdoms of Nepal (population 27 million), with its capital at Kathmandu (•Figure 7.1.23), and Bhutan (population 1 million), with its capital at Thimphu, are perched on South Asia's mountain wall (see Figure 7.1.21). Their landlocked locations and difficult topographies have contributed

JAMES P. BLAIR/NGS Image Collection

• **Figure 7.1.22** Bangladesh is both a low-lying coastal land and a downstream country of several large watersheds. As a result of storms blowing in from the Bay of Bengal and of accelerated runoff from rainfall and snowmelt on deforested mountains upstream, it is subjected to devastating floods.

Joe Hobbs

• **Figure 7.1.23** Nepal's capital, Kathmandu, sprawls across a plain in the country's "middle" mountain range. In the foreground is the Royal Palace.

to their slow economic development. Both countries have productive agricultural regions in the "middle ranges," the bands of foothills between the vast lowland of the subcontinent and the towering Himalayan peaks. In Nepal, when the capacity of this productive area was saturated, many
61 poor peasants moved to the lowland region known as the Terai. The shortage of quality farmland for Nepal's mainly rural (86 percent) population of 24 million is a critical problem. Nepal's mostly deforested landscape bears witness to relentless pressure to provide new land for agriculture and wood for fuel and construction (•Figure 7.1.24).

The conservative Buddhist kingdom of Bhutan has been more careful in opening its doors to tourism, placing restrictions on the numbers permitted in and on the routes they may follow. Tourism and hydroelectric power (mainly for Indian needs) are the most promising areas for future development of this poor country. This is the only country in the world whose politicians claim to disdain conventional economic growth; their mantra is **"gross national happiness"** for the people of Bhutan. They measure happiness using criteria such as equality, good government, and harmony with nature.

But the Bhutanese "Shangri-la" is not without its religious and political problems. Bhutan is ruled from the capital by a Buddhist king of the Drukpa tribe, to which most of the government ministers also belong. The king wants the monarch's power to become more symbolic, with real authority turned over to a multiparty parliament of elected officials. Critics complain that the king will not tolerate any real opposition, however, particularly from the ethnic Nepalis, who are Hindus and who for decades have been migrating out of their own overcrowded Himalayan kingdom to work the productive soils of Bhutan. Ostensibly as a means of promoting national unity, Bhutan's king in the 1990s outlawed the use of the Nepali language in Bhutan's schools and insisted that all inhabitants of Bhutan wear the national dress and hairstyle of the Drukpa tribe. Violent clashes ensued between ethnic Nepalis and Drukpas, and more than 100,000 ethnic Nepalis fled back to Nepal, where they languish in refugee camps.

Joe Hobbs

• **Figure 7.1.24** People are constantly collecting, hauling, storing, and using wood in Nepal, creating tremendous pressure on the country's remaining forests.

Nepal itself has suffered violent consequences of one-party rule and of palace intrigue. The country was an absolute monarchy until 1991, when prodemocracy riots and demonstrations forced King Birendra to accept a new role as constitutional monarch and permit elections for a national parliament. King Birendra and many other royal family members were shot to death at a dinner gathering in 2001, killed by Nepal's crown prince in a drunken rage over his mother's refusal to allow him to take the bride of his choice. Birendra's successor, King Gyanendra, soon declared a state of emergency and dismissed the democratically elected parliament. He claimed that these steps would quell a Maoist insurgency that began in 1996 with promises to alleviate rural poverty and rid the country of its caste system, which funnels wealth, power, and land ownership into relatively few hands. By 2004, conflict between government troops and the Maoist rebels of the **Communist Party of Nepal** had taken more than 9,000 lives, and the rebels claimed to control 80 percent of the country. The violence snuffed out the country's vital tourism industry.

As his rule stood at the brink of collapse in 2006, Nepal's king made concessions. His government signed a peace agreement with the Maoist rebels, and a temporary constitution and parliament (which included Maoist members) were instituted to address demands for more democracy. Despite the Maoist concessions, other violent opposition has emerged, particularly among groups wanting to end the long-standing political and economic supremacy of the top two Hindu castes in Nepal. The Hindu majority is angry that the government's deal with the Maoists included a declaration that Nepal was henceforth a secular rather than a Hindu nation. Betting on peace, however, tour agencies are clamoring to get their clients back on Nepal's mountain trails, especially in the Everest region of the east and the Annapurna area of the west.

7.1.8 Sri Lanka: Resplendent and Troubled

Sri Lanka (population 20 million), formerly known as Ceylon, is a tropical island country with affinities to the Indian subcontinent. Its name conveys its physical beauty: Sri Lanka means "resplendent isle" in the native Sinhalese language, and medieval Arabs knew it as Serendip, source of the concept of serendipity (•Figure 7.1.25). The island consists of a coastal plain surrounding a knot of mountains and hill lands (•Figure 7.1.26). Most people live in the wetter southwestern portion of the plain, in the south central hilly areas, and in the drier Jaffna Peninsula of the north. Coconuts and rice are the major crops of the low southwestern coast and Jaffna Peninsula, while tea and rubber plantations dominate the economy of the uplands. Colombo (population 2.5 million), in the southwest, is the capital, chief port, and only large city.

Sri Lanka's economy is highly commercialized. Three cash crops—rubber, coconuts, and the world-famous Ceylon tea—supply about 11 percent of export earnings. Sri Lanka is the world's second-largest exporter of tea. Textiles and clothing manufactured in the Colombo area make up about two-thirds of the country's exports by value. Sri Lanka is nearly self-sufficient in rice production, and with declining rates of population growth, its development prospects are good, particularly if its civil conflict can be ended.

External influences have diversified the island's population and sown seeds of unrest. Centuries of recurrent invasion from India were followed by domination by the Portuguese in the early 16th century, the Dutch in the 17th,
and the British from 1795 until the country was granted 296
independence in 1948 as a member of the British Commonwealth. This eventful history, plus the long-standing commercial importance of the sea route around southern Asia, has given Sri Lanka a polyglot population that includes Burghers (descendants of Portuguese and Dutch settlers) and Arabs.

Sri Lanka's civil conflict reflects antagonisms between ethnic groups and discontent with economic and political conditions, especially among the minority Tamils (9 percent of the population). Since 1983, more than 70,000 people have been killed and many more made homeless as the Tamils of the country's north have fought for autonomy or independence from Sri Lanka's majority Sinhalese gov-

Joe Hobbs

• **Figure 7.1.25** Landscapes made verdant by abundant rains are typical of all but northernmost Sri Lanka. This is Sigiriya, in the central part of the island.

• **Figure 7.1.26** Principal features of Sri Lanka.

ernment. Tamils complain that the Sinhalese treat them as second-class citizens and deprive them of many basic rights. Tamil fighters want to establish their own homeland, **Tamil Eelam,** along the east, north, and west of the country. In 1987, guerrilla attacks by the organization of Tamil separatists called the **Tamil Tigers,** officially the **Liberation Tigers of Tamil Eelam (LTTE),** led to military action by Indian troops invited to help restore order. India withdrew in 1990, but the troubles continued, and India worried that Tamil war refugees from Sri Lanka could flood southern India or that Sri Lankan infiltrators could inspire Tamil separatism within India itself.

The 1991 assassination in southern India of Indian Prime Minister Rajiv Gandhi, son of the slain Indira Gandhi, was linked to Indian sympathizers with the Tamil separatist movement in Sri Lanka. In 1995, Sri Lankan government troops succeeded in retaking the Tamils' geographic stronghold on the Jaffna Peninsula. The Tamil guerrillas retreated to the wild forests just to the south and continued their insurgency with the sinking of Sri Lankan naval ships, a devastating car bombing of downtown Colombo, an attack on the Sinhalese Buddhists' holiest site (the Temple of the Tooth in Kandy), an assassination attempt on the Sri Lankan prime minister, and a spectacular assault on Colombo's airport that destroyed most of the national airline's planes. Many of these were suicide bombings, a favored tactic of the Tigers.

The country's leadership is being variously advised to crush the revolt militarily or to pursue a course of devolution that would turn over federal powers to outlying states—a possible track for granting autonomy in the dis-
83 puted territories to the Tamils. A cease-fire was declared by both sides in 2002, but the Tamil Tigers promised more violence if their basic demands for self-rule were not met. They have delivered, including another bombing of Colombo's airfield in 2007, and intensified fighting on both sides makes it clear that the cease-fire has expired. Until a formal agreement is reached, the rebels will govern their de facto state of Tamil Eelam; they have their own tax, legal, health, and educational systems and operate banks, newspapers, and radio stations. Unrest in rebel-held areas has made it particularly difficult to rebuild northern sections of the area hit hard by the December 2004 tsunami (see pages 340–341). Early hopes that this catastrophe would have led the warring parties to resolve their differences have been dashed.

7.1.9 The Laid-Back, Low-Lying Maldives

Aurora/Getty Images

• **Figure 7.1.27** This aerial view of Male (pronounced *mah*-lay), the Maldives' capital, reveals how vulnerable the site is to storms and sea level rise. This is the world's most densely populated city.

The roughly 1,100 islands that make up the Maldives (population 300,000; capital, Male, population 90,000), an independent country since 1965, appear to be the very essence of tropical paradise (•Figure 7.1.27; for location, see Figure 7.1.1). So prized are its palm-blessed and coral-fringed beaches that more than half a million tourists, mainly Europeans, visit the country each year. More than 60 percent of the country's foreign-currency earnings come from tourism, with fishing and clothing providing most of the rest. Some sources list the Maldives as South Asia's wealthiest country on a per capita basis. The greatest wealth is in the tourist sector, where it is concentrated in so few hands that a potentially dangerous rich-poor divide has developed.

There may be other troubles in paradise. More than 80 percent of this country's very limited land area consists of limestone atolls less than 3 feet (90 cm) above the sea. If sea levels were to rise, as predicted by many common climate change scenarios, the entire country could rapidly be submerged. This prospect led to the Maldives' president's famous cry at a global warming conference: "We are an endangered nation!"

This concludes a detailed look at the geographies of South Asia. Despite linguistic, social, religious, and political difficulties, South Asia's accomplishments in recent decades have been remarkable. India has lowered its birth rate, and its economy has blossomed. Despite its internal divisions and frictions, India has retained a representative democracy. But sectarian violence between Hindus and Muslims, a growing Hindu nationalist movement that threatens India's remarkable diversity, Sikhs' desire for autonomy, and the problem of Kashmir all pose long-term difficulties for this South Asian giant. Pakistan must walk the tightrope between support for its own majority Muslim interests and support for the secular West in its campaign against terrorism while also avoiding war with India. Both Pakistan and Nepal must confront the problems of rapid population growth, deforestation, and land degradation where the balance between people and resources is already precarious. Bangladesh faces the huge challenge of dealing with the environmental consequences of upstream deforestation and repeated flooding assaults from the Bay of Bengal. The Maldivians hope and pray that sea levels remain constant but also lobby industrial powers to cut greenhouse gas emissions. Sri Lanka must accommodate its Tamil minority. All of these nations struggle with the chronic problem of poverty, which is ultimately the source of many of the region's most severe environmental and political dilemmas.

SUMMARY

- Most of South Asia is a triangular peninsula splitting the northern Indian Ocean into the Bay of Bengal and the Arabian Sea. The entire unit is often called the Indian subcontinent.
- The Indian subcontinent is one of the world's culture hearths. Some of the world's earliest cities developed on the banks of the Indus. The Aryans arrived between 2000 and 3000 B.C.E. and were the source of many of the region's dominant cultural attributes. The Tamils and other Dravidians predated the arrival of the Aryans.
- South Asia encompasses the countries of India, Bangladesh, Pakistan, Nepal, Bhutan, Sri Lanka, and the Maldives in an area about half the size of the conterminous United States.
- The subcontinent is home to many different faiths, including Hinduism, Sikhism, Christianity, Islam, Buddhism, Zoroastrianism, and Jainism. Religious, ethnic, and other differences underlie serious and often violent conflicts. The main ones have been Hindus versus Muslims in India, Sikhs versus the government of India, and Tamils versus Sinhalese in Sri Lanka.
- Until 1947, what are now Pakistan, Bangladesh, and India formed the single country of India. For over a century, it was the most important unit in the British colonial empire—the "jewel in the crown."
- With independence, India was partitioned between avowedly Muslim Pakistan (which later divided into two countries, Pakistan and Bangladesh) and secular, primarily Hindu India. The partition had many consequences, including violence between Hindus and Muslims and their large-scale migrations, disputes over the sharing of Indus River waters, and resources and industries stranded on either side of the new partition lines.
- The most severe product of the partition has been lasting conflict in Kashmir. Several wars have been sparked or fought in the still-volatile frontier province, which, despite its Muslim majority, was joined within India in 1947.
- The Himalaya and Karakoram ranges, the world's highest mountains, enclose the subcontinent on its northern borders. They were produced by the tectonic collision of the former island of India with Asia.
- The subcontinent's northern plain contains the core areas of the three major countries of India, Bangladesh, and Pakistan. Three great rivers water this plain: the Indus River, the Ganges River, and the Brahmaputra River. Although mostly desert and steppe in Pakistan, the plain is relatively well watered in India, where it supports unusually high population densities.
- The southern portion of the subcontinent is mainly a large volcanic plateau called the Deccan. Generally low in elevation, it is rimmed by highlands: the Western Ghats, the northern Central Indian Hills, and the Eastern Ghats.
- Climates vary, with undifferentiated highland in the northern mountains to desert and steppe in Pakistan and western India. A humid subtropical climate occurs in the northern plain, tropical savanna is in the peninsula with the exception of the rain shadow in the lee of the Western Ghats, and there are rain forests in the Western Ghats' seaward slopes, the coastal plain to the south, and part of Bangladesh.
- The wet southwest monsoon is at its height from June to September, blowing in from the Arabian Sea and Bay of Bengal. The dry northeast monsoon occurs in the winter, with dry, cooler winds from the northeast coming from high-pressure systems over the continent.
- Although it is an overwhelmingly poor and rural country, India has many modern industries that, along with privatization of former state-run firms, have contributed to its substantial rate of economic growth in recent years. These include the software and computer industries of Bengaluru and the film industry of Mumbai. Pakistan and Bangladesh are much less industrialized and rely heavily on textile exports. India and Bangladesh have been successful in lowering birth rates, but population growth is a serious issue for both countries. India has managed to feed its huge population.
- South Asia provides large numbers of workers to other countries, ranging from professional to semiskilled and unskilled. Remittances sent home by these workers are important to the South Asian economies. Many of the professionals do not return, contributing to the region's brain drain.
- Some of India's large minorities have had serious difficulties with the majority Hindu population. Hindu nationalism is a problem in India, fanning the flames of communal violence between Hindus and Muslims. In the 1980s, some of the Punjab's Sikhs agitated for an independent country.
- Pakistan's government faces challenges from its Pashtun population in the country's west and northwest and from strong Islamist sentiment throughout the country.
- Bangladesh and the Maldives are threatened by serious natural hazards from the sea. Strong typhoons (hurricanes) regularly devastate large parts of low-lying Bangladesh, and most of the Maldives would be inundated if sea levels were to rise by 3 feet (90 cm).

- Nepal and Bhutan are picturesque but poor mountain kingdoms. Both have strong tourism potential, but Bhutan restricts tourists, and for a time a violent Maoist insurgency in Nepal scared many tourists away.
- Sri Lanka, the world's largest tea exporter, has a Buddhist Sinhalese majority and a Hindu Tamil minority. For more than 20 years, Tamil factions have led a violent struggle to establish an autonomous homeland on the island's east, north, and west periphery. The violence has been punctuated by peace talks that have not yet arrived at a permanent solution.

KEY TERMS + CONCEPTS

Bollywood (p. 323)
caste (p. 325)
communal violence (p. 325)
Communist Party of Nepal (p. 330)
Dalits (p. 325)
distributaries (p. 316)
dry northeast monsoon (p. 318
e-Choupal initiative (p. 323)
Golden Quadrilateral (p. 323)
"gross national happiness" (p. 329)
Harappan civilization (p. 310)
India's Silicon Valley (p. 323)
India's Wall Street (p. 323)
Indo-Aryans (p. 310)
Indus Waters Treaty (p. 316)
Islamization (p. 327)
Jainism (p. 312)
Jains (p. 312)
"jewel in the crown" (p. 313)
Khalistan (p. 326)
Liberation Tigers of Tamil Eelam (LTTE) (p. 331)
madrasa (p. 327)
megacity (p. 320)
Moguls (p. 311)
Mohajirs (p. 313)
Mughals (p. 311)
Naxalites (p. 326)
Parsis (p. 311)
remittances (p. 323)
riparian (p. 316)
sectarian violence (p. 325)
sharia (p. 327)
Sikhism (p. 311)
Sikhs (p. 311)
Sinhalese (p. 312)
Tamils (p. 312)
Tamil Eelam (p. 331)
Tamil Tigers (p. 331)
untouchables (p. 325)
wet southwest monsoon (p. 318)
Zoroastrianism (p. 311)

REVIEW QUESTIONS

1. What groups of peoples have been most significant in forming the cultural foundation of South Asia?
2. Using maps and the text, locate the outer mountain wall, the northern plain, peninsular India, and Sri Lanka. What are the major climates, resources, and other natural attributes of these areas? What and where are the three major rivers of the northern plain?
3. How were questions of water allocation between the region's countries resolved after independence?
4. Where are the main production areas for the export crops rice, cotton, jute, and tea?
5. What are the influences of the wet and dry monsoons in South Asia? What particular problems do Bangladesh and the Maldives have in terms of climate and climate change?
6. Why are there far fewer women than men in some Indian provinces? What problems are associated with this phenomenon?
7. What are the major nonagricultural industries of South Asia? Where is India's Silicon Valley? Its Hollywood? Its Wall Street? Where are the other major industries located? Which country is the most industrialized? Which lag behind, and why?
8. What is the caste system? How does it affect employment and social relations in India? What evidence of change in this ancient system is there?
9. Where is Ayodhya? What site there is critical in relations between Hindus and Muslims in India?
10. What are Pakistan's *madrasas*? Why is their curriculum of so much concern to the government of Pakistan and to the West?
11. What ethnic and political differences have given rise to strife in Sri Lanka?
12. What issues are particularly important in the mountain countries of Nepal and Bhutan and in the lowland country of Bangladesh?
13. Why is climate change such a serious issue for the Maldives?

Digital Wisdom

Joe Hobbs

Woman and child of the Flower Hmong, Bac Ha, northern Vietnam. Southeast Asia has a wealth of cultures. The Hmong are one of 54 distinct ethnic groups in Vietnam alone.

MODULE 7.2
SOUTHEAST ASIA: FROM SUBSISTENCE FARMING TO SEMICONDUCTORS

Southeast Asia, from Myanmar (known until 1989 as Burma) in the west to the Philippines in the east, is a fragmented region of peninsulas, islands, and seas. East of India and south of China, between the Bay of Bengal and the South China Sea, the large Indochinese Peninsula projects southward from the continental mass of Asia (•Figure 7.2.1). From it, the long, narrow Malay Peninsula extends another 900 miles (c. 1,450 km) toward the equator. To the south and east are thousands of islands, the largest of which are Sumatra, Java, Borneo, Sulawesi (Celebes), Mindanao, and Luzon. Another large island, New Guinea, is culturally a part of the Melanesian archipelagoes of the Pacific World, but its western half, held by Indonesia, is politically part of Southeast Asia.

Southeast Asia contributed domesticated plants (including rice), animals (including chickens), and the achievements of civilizations to a wider world. Innovations and accomplishments diffused from several centers, which themselves were influenced by foreign cultures. Indian and Chinese traits have been especially strong in shaping the

module outline

module objectives

This module should enable you to

- Recognize the correlation between environmental opportunity, environmental constraint, and population densities
- Appreciate the catastrophic reach of the December 2004 tsunami and some of the region's other natural hazards
- See how the interconnectedness of the global economic system led to "economic meltdown" in Southeast Asia in 1997
- Appreciate the economic strength of minority Chinese populations in many of the countries and how that prosperity provokes resentment in times of difficulty
- Understand some of the major political obstacles and environmental consequences associated with large dams, in this case on the Mekong River
- Consider how China's economic ascendancy has forced some of the countries to retool their industries
- Understand how demands for self-rule threaten the cohesion of Indonesia and raise alarm there and abroad

---

region's cultural geography. Hinduism came from India, and Theravada Buddhism came from India through Sri
294 Lanka (Ceylon). The Chinese cultural influences of Confucianism, Daoism, and Mahayana Buddhism were particularly strong in Vietnam. China periodically demanded and received tribute from states in Burma, Sumatra, and Java. The Islamic way of life came with Arab traders from the west from the 13th through 15th centuries and took permanent root. Southeast Asia is the eastern margin of the world of Islam, and Indonesia has more Muslims than any other country in the world today.

The historical and contemporary monumental architectures of Southeast Asia reflect these diverse origins. In the 9th century, the advanced Sailendra culture of the Indonesian island of Java built Borobudur, the world's largest Buddhist *stupa* (shrine) and still the largest monument in the Southern Hemisphere. In the 12th century, a Hindu king built the extraordinary complex of Angkor Wat as the capital of the Khmer empire based in Cambodia (•Figure 7.2.2). The first independent Thai kingdom emerged in the 13th century, and in the 14th century, Thai kings ruling from the great city of Ayuthaya (near modern Bangkok) extinguished the Khmer empire and extended their own control into the Malay Peninsula and Burma. The world's largest mosque is in modern Brunei, a tiny oil-rich country near the northern tip of the island of Borneo, and for many years Malaysia boasted the world's tallest buildings, the Petronas Towers in Kuala Lumpur (•Figure 7.2.3).

Southeast Asia today is composed politically of 11 countries: Myanmar (formerly Burma), Thailand, Laos, Cambodia, Vietnam, Malaysia, Singapore, Indonesia, Timor-Leste (formerly East Timor), Brunei, and the Philippines. With the exception of Thailand (which was never colonized) and Timor-Leste (which gained independence first from Portugal and then from Indonesia), all of these states became independent from colonial powers after 1946. These 11 nations are fragmented both politically and topographically. Many of the countries are also culturally fragmented and have experienced strife between different ethnic groups. Outside intervention has sometimes complicated and worsened local discord, producing enormous suffering in the region and lasting trauma among many of the foreign soldiers who fought the determined inhabitants of Southeast Asia.

7.2.1 Area, Population, and Environment

It is about 4,000 miles (c. 6,400 km) from western Myanmar to central New Guinea and 2,500 miles (c. 4,000 km) from northern Myanmar to southern Indonesia. Despite these great distances, the total land area of Southeast Asia is only about 1.7 million square miles (4.4 million sq km), or about half the size of the United States (including Alaska) or China (see Figure 7.2, page 282). The estimated population of the region was 574 million in 2007, giving it an overall density of 331 people per square mile (128/sq km) (see Figure 7.3, page 282). To put this in perspective, Southeast Asia has nearly four times the overall population density of the United States but well under half that of the Indian subcontinent or the densely populated humid eastern part of China, south of the Great Wall.

Southeast Asia has been less populous than India and China throughout history. Only about 10 million people lived in the region in the year 1800, and there were no widespread dense populations. Historically, geographic and environmental obstacles probably helped keep population growth low in Southeast Asia. By land, the region is rather isolated, cut off to some extent by the rugged mountains of the northern peninsula of Indochina. Over many centuries, the ancestors of most of the present inhabitants entered the area from the north as recurrent thin trickles of population, crossing this mountain barrier as refugees. Southeast Asia is tropical, with continuous heat in the lowlands, torrential rains, prolific vegetation that is difficult to clear and

• **Figure 7.2.1** Principal features of Myanmar, Thailand, Cambodia, Laos, Vietnam, Malaysia, Brunei, Singapore, and the Philippines. (See Figure 7.2.18, page 357 for Indonesia and Timor-Leste.)

keep cleared, soils that are generally leached and poor. Historically, people in the region suffered a high incidence of disease. A four- to six-month dry season on the Indochinese Peninsula and on some of the islands presents further challenges.

Violent storms known locally as **typhoons** (elsewhere
called cyclones or hurricanes) strike often, particularly in
194 Vietnam and the Philippines (•Figure 7.2.4). More spec-
tacular environmental difficulties result from the location
396 of part of the region on the Pacific Ring of Fire. subjecting
Indonesia and the Philippines especially to earthquakes, the
20 tsunamis that earthquakes unleash, and volcanic eruptions
(see Natural Hazards, pages 340–341). Two of earth his-
52 tory's most cataclysmic volcanic eruptions took place here:
the 1815 eruption of Mount Tambora on the Indonesian island of Sumbawa killed 117,000 people, and its ash lingering in the atmosphere caused the worldwide "year without a summer" in 1816. The 1883 eruption and explosion of nearby Krakatoa (which scientists believe caused the loudest sound ever heard) also killed tens of thousands and affected the global climate for half a decade.

Typical of less developed countries (LDCs), the introduction of modern medical, food-producing, and other technologies into the region softened the blows of environmental adversity and lowered death rates, resulting in a tremendous population increase in the last century and a half. This increase, which accelerated after World War II, is now slowing but still rapid. Between 1973 and 2007, population grew from an estimated 316 million to 574 million, a 34-year increase of 258 million people, or 81 percent. As will be seen later in this module, the combination of Dutch colonial practices and localized, very productive soils gave rise to an extraordinarily dense population on
the island of Java. This Indonesian island's 125 million 355

people represent over one-fifth of the entire population of Southeast Asia. In recent times, the race between population growth and food supply has seen food supply winning in some countries but falling behind in others. Since the 1970s, increases in food production outpaced population growth in Malaysia and Indonesia. At the other end of the scale were Cambodia and Laos, in which war, repression, and inefficient communalization of agriculture slowed progress. These discrepancies are typical of Southeast Asia, home to a remarkable range of economies, from quite poor to very prosperous.

7.2.2 Livelihood Patterns

Southeast Asia as a whole is among the world's poorer regions, yet some of its countries are rather prosperous. The countries' mean per capita GNI PPP of $5,420 in 2007 was about average for the world's less developed countries. In contrast, Singapore, Malaysia, Brunei, and Thailand were well above average for LDCs. Because of their high rates of industrial productivity and economic growth, Singapore is generally included in the ranks of the economically dynamic "Asian Tigers," while Malaysia, Thailand, and sometimes Indonesia, the Philippines, and Vietnam are de-
298 scribed as either "New Asian Tigers" or "Tiger Cubs." The commercial-industrial city state of Singapore has about the same per capita GNI PPP as Germany. High-tech industries brought spectacular economic growth to these countries through much of the 1990s. In 1997, a series of events precipitated an economic crisis that roared through the region and temporarily declawed these emerging economies, but conditions have since improved (see page 344).

This region's economic activities span the entire continuum from Stone Age hunting and gathering (in Papua on the island of New Guinea) to slash-and-burn farming—one of humankind's earliest agricultural techniques—to the commercial agriculture typical of colonial times and from the export of raw materials like rubber to manufactured goods like clothing, shoes, and computer products. As Southeast Asians see it, their economic future will depend not so much on rubber, coffee, or even oil, of which some countries have abundant supplies, but on high-tech exports like semiconductors and software. But despite pockets of technological innovation, rapid urbanization, and industrialization, the majority of the people in Southeast Asia are still farmers, with 62 percent of the people classified as rural.

Jerry Alexander/Stone/Getty Images

• **Figure 7.2.2** Angkor Wat, the 12th-century capital of the Khmer empire, is a symbol of the rich cultural history of Southeast Asia.

Joe Hobbs

• **Figure 7.2.3** Modernity and prosperity characterize some nations in Southeast Asia today. At 1,462 feet (446 m), the twin Petronas Towers in Kuala Lumpur were until 2005 the tallest buildings in the world.

Shifting and Sedentary Agriculture

Permanent agriculture has been difficult to establish in many parts of this largely mountainous, tropical realm.
288 Much land is used only for shifting cultivation, in which a migratory farmer clears and uses fields for a few years and then allows them to revert to secondary growth vegetation while he moves on to clear a new patch. Since an abandoned field needs about 15 years to restore itself to forest that can again be cleared for farming, only a small portion of the land can be cultivated at any one time. This extensive form of land use can support only sparse populations. Unirrigated rice is the principal crop of shifting agriculture, with corn, beans, and root crops such as yams and cassava (manioc) also grown. In some forested areas, people supplement their farming by hunting and gathering. These migratory subsistence farmers are often of different ethnic groups from adjacent settled populations, having historically sought refuge in the backcountry from stronger invad-
260 ers (•Figure 7.2.5).

Most of Southeast Asia's people live in widely scattered, dense clusters of permanent agricultural settlements (see Figure 7.3 top, page 282). These core regions stand in striking contrast to the relatively empty spaces of adjoining districts. Superior soil fertility is the main locational factor in these areas. However, there are a few districts of dense populations without high-quality soils to support them. Most of these rely on plantation agriculture of rubber and other less demanding cash crops.

The typical inhabitant of the areas of permanent sedentary agriculture is a subsistence farmer whose main crop is wet rice, grown with the aid of natural flooding or irrigation. In some drier areas, unirrigated millet or corn replaces rice. Secondary crops that can be grown on land unsuitable for rice and other grains supplement the major crop. These secondary food crops include coconuts, yams, cassava, beans, and garden vegetables. Some farmers grow a secondary cash crop such as tobacco, coffee, or rubber. Fishing is important both on the coasts and along inland streams and lakes. Because of their importance in the food supply, many fish are harvested from artificial enclosures and flooded rice fields, in which they are raised as a supplementary "crop" (•Figure 7.2.6).

Subsistence agriculture in Southeast Asia exists alongside tropical cash crops grown for export. The region is one of the world's major supply areas for such crops, which include palm oil, rubber, and cloves. This highly commer-
45 cial type of production is a legacy of Western colonialism, which began in the 16th century. Until the 19th century, Westerners were interested mainly in the region's location on the route to China. They were content to control patches of land along the coasts and to trade for some goods produced by local people. During the 19th century, however, the Industrial Revolution in the West greatly expanded the demand for tropical products. Europeans consequently extended their control over almost all of Southeast Asia. European colonists used their capital and knowledge, along with indigenous and imported labor, to increase export production rapidly. They focused on exporting some local produce such as copra (dried coconut meat, from which oil is pressed) and spices. They also introduced some new commercial crops.

The usual method of introducing commercial production was to establish large estates, or **plantations**, managed by Europeans but worked by indigenous labor or labor imported from other parts of Monsoon Asia. There were often problems finding available local hands, so Europeans imported contract labor from India and China. Large migrations from these countries complicated an already complex ethnic mixture, setting the stage for today's unusual distribution of ethnic groups and their sometimes discordant relations.

Joe Hobbs

• **Figure 7.2.4** Typhoon Xangsane slammed ashore in central Vietnam on October 1, 2006, causing extensive wind and flood damage. This is the city of Hoi An on the following morning. *Joe Hobbs*

• **Figure 7.2.5** This couple of the ethnic group known as the Black Tai are farmers near the village of Mai Chau in northern Vietnam. They have just served tea and mangoes to guests in their single room stilt home. They are not wearing their traditional clothing, which they don only for ceremonial occasions.

Many export cash crops, some raised by small landowners, are still produced in Southeast Asia. The major ones are rubber, palm products, and tea. Nearly three-fourths of the world's natural rubber comes from the region, mainly Malaysia, Indonesia, and Thailand; Thailand is in fact the world's largest producer and exporter (see Insights, page 342).

Palm products include palm oil, coconut oil, and copra. Malaysia and Indonesia are the world leaders in producing palm oil, which now accounts for about 20 percent of edible oils worldwide and is used in a variety of detergents, soaps and cosmetics. The Netherlands and other European countries had promoted the use of palm oil for biofuels to replace fossil fuels. It has recently become apparent, however, that to meet this European demand, Malaysia and Indonesia are accelerating the clearing of natural vegetation to replace them with palm plantations. In Indonesia, peat-covered lands are being replaced with palm oil farms. The draining and burning of peat there releases large quantities of carbon dioxide into the atmosphere. By this unusual means, Indonesia became the world's third-largest emitter of carbon dioxide (after China and the United States) in 2007. Alarmed energy planners in the Netherlands and elsewhere have revised their views of palm oil as a "green" source of energy. 103, 589

Indonesia dominates the world output of coconut palm products. It is also the region's leading producer of tea and is among the world's top five producers. Other export crops grown in Southeast Asia include coffee from Indonesia, cane sugar from the Philippines and Indonesia, and pineapples (of which the Philippines and Thailand have become the world's leading producers and exporters). However, the countries of Southeast Asia are generally coming to depend less on agricultural exports and more on mining and manufacturing.

One lasting impact of Western colonialism on the economies of Southeast Asia was the stimulation of commercial rice farming in formerly unproductive areas. Three areas became prominent in commercial rice growing: the deltas of the Irrawaddy River in Myanmar, the Chao Praya in Thailand, and the Mekong in Vietnam and Cambodia. Almost impenetrable wetlands and uncontrolled floods had

• **Figure 7.2.6** A marine fish farm off the southern coast of Cambodia. This type of aquaculture is an increasingly important in subsistence and commerce in Southeast Asia and elsewhere.

NATURAL HAZARDS

The Great Tsunami of 2004

On December 26, 2004, the most cataclysmic natural event of modern times afflicted a vast portion of the earth. Beneath the Indian Ocean, just off the northwestern coast of the Indonesian island of Sumatra, a gargantuan "megathrust" earthquake measuring magnitude 9.3—with the force of more than 32 billion tons of TNT, or the equivalent of as many as a million nuclear explosions—struck at 6:58 B/N / local time. In the Java (Sunda) Trench, as much as 750 miles (1,200 km) of fault line slipped 60 feet (over 18 m) along the subduction zone where the Indian Plate is sliding beneath the Burma Plate (part of the greater Eurasian Plate; see Figure 2.1, page 21).

As the seabed of the Burma Plate instantly rose several meters vertically above the Indian Plate, the massive displacement of seawater created a series of tsunamis (*tsunami* is Japanese for "harbor wave"; the phenomenon is sometimes popularly but incorrectly referred to as a "tidal wave") that pulsed across the Andaman Sea and the Indian Ocean at up to 500 miles (800 km) per hour. On the open ocean, such waves are barely perceptible, but 20 they break on coastlines with the force of giant storm surges. The tsunami roared ashore, at heights up to 45 feet (14 m), in 14 countries: Indonesia, Malaysia, Thailand, Myanmar, Bangladesh, India, Sri Lanka, the Maldives, Seychelles, Somalia, Kenya, Tanzania, Yemen, and South Africa (•Figure 7.2.A).

These were the deadliest tsunamis in history. The total number of dead exceeded 200,000. By far the greatest number of deaths (over 130,000) were in Indonesia, where Aceh's principal city of

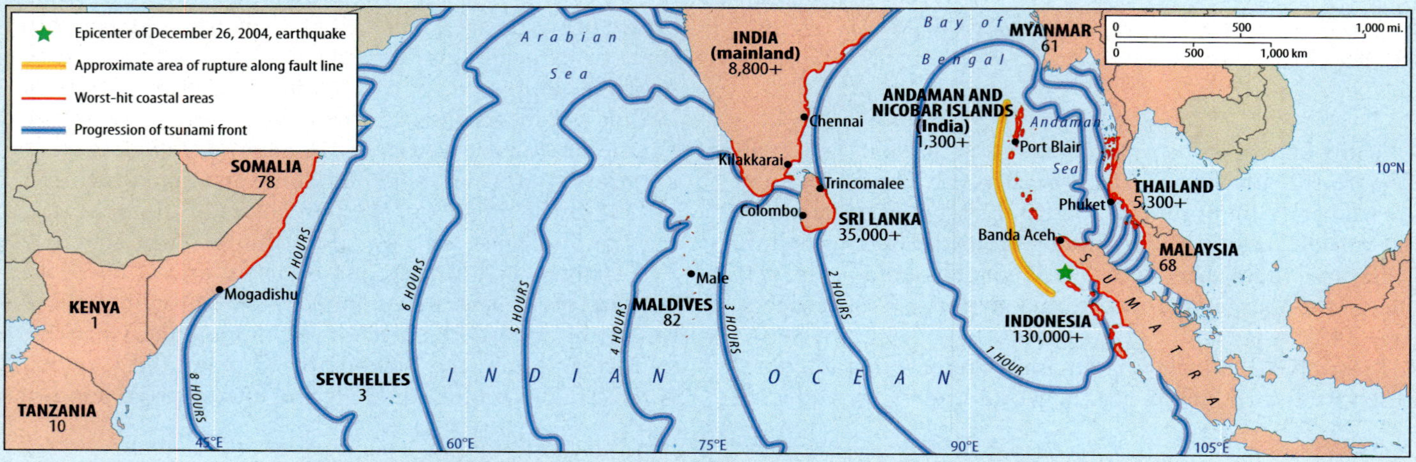

• **Figure 7.2.A** The epicenter of the earthquake, which registered magnitude 9.3, was located off the coast of Sumatra, Indonesia. It created a tsunami of tremendous force and scope.

kept these deltas thinly settled, but incentives and methods for settlement have turned them into densely populated areas within the past century. Thailand, Vietnam, and Myanmar are the region's leading rice exporters. Other Southeast Asian nations import much of this produce.

Deforestation

Near areas of dense settlement in Southeast Asia, there are still some large areas covered in primary forest, but they are shrinking as population expands and development progresses. Many environmentalists view the destruction of the region's tropical rain forest as an international environmental problem (•Figure 7.2.7). In 2007, only 28 percent of Thailand's original forest cover remained. Poorer neighbor Myanmar had a 49 percent forest cover. Vietnam lost more than one-third of its forest between 1985 and 2007, by which time only 37 percent of the original forest cover remained. Deforestation is advancing most rapidly in Malaysia, where 63 percent of the land area is still forested (official statistics often include palm oil plantations in forest estimates, so these figures should be evaluated carefully), and in Indonesia, where 49 percent is forested. Malaysia is now the world's largest exporter of tropical hardwoods. Virtually all of the primary forests in Peninsular Malaysia have already been cleared, and it is projected that commercial logging will decimate those remaining outside protected areas in Malaysian Borneo soon after 2010 (•Figure 7.2.8).

Although Southeast Asia's tropical forests are smaller in total area than those of central Africa and the Amazon Basin, they are being destroyed at a much faster rate, most significantly by commercial logging for Japanese and Chinese markets rather than by subsistence farmers. Environmentalists fear the irretrievable loss of plant and animal species and the potential contribution to global warming that deforestation in this region may cause. Malaysia and

NATURAL HAZARDS

The Great Tsunami of 2004 *continued*

Banda Aceh was obliterated (• Figure 7.2.B). Sri Lanka lost 35,000; India, 12,000; and Thailand, over 5,000. Almost all the deaths occurred within a mile of the shoreline and were caused by blunt force from debris and by drowning. An estimated one-third of the fatalities were children. There were reports of children running out to investigate fish stranded by retreating seas—a characteristic precursor to a tsunami—only to be struck seconds later by a wall of water.

As many as 2 million persons were made homeless by the disaster. An unprecedented international relief effort, designed to feed and house these refugees and to contain the spread of epidemic diseases, began within hours.

Tsunami experts concur that countless thousands of lives would have been saved had the affected Indian Ocean region been equipped with the same kind of tsunami early warning system now in place in the Pacific Ocean.[1] That investment had not been made because historically about 90 percent of the world's tsunamis have occurred in the Pacific and because many of the nations afflicted lacked the financial resources for the wave sensors and other infrastructure required by the system. Now, however, installing an early warning system in the Indian Ocean basin is an urgent priority. In this event, Indonesia would have had just a few minutes' warning, but Thailand would have had two hours and Somalia seven hours to prepare for the disaster.

Courtesy of DigitalGlobe

Courtesy of DigitalGlobe

• **Figure 7.2.B** Banda Aceh before and after the tsunami.

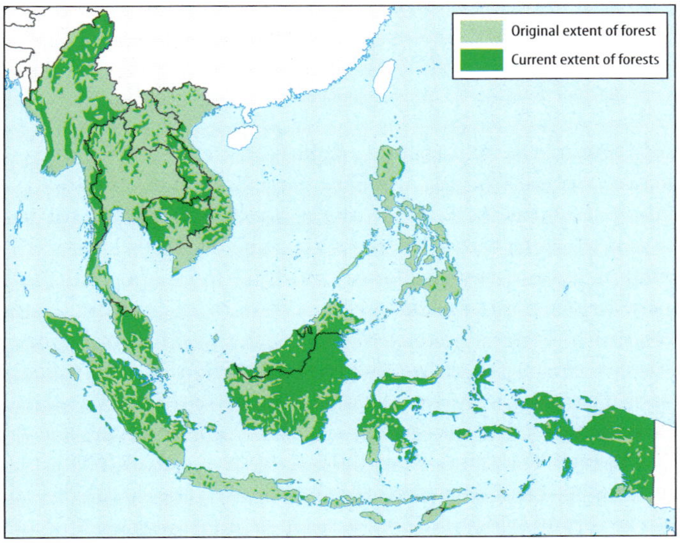

• **Figure 7.2.7** In Southeast Asia, natural forest cover has been rapidly reduced by many human uses, especially commercial logging and farming.

Indonesia have passed laws to slow the destruction, with Indonesia taking the aggressive step of banning the use of fire to clear forests. However, enforcing such legislation has so far proved impossible. Hundreds of Indonesian and Malaysian companies, most of them large agricultural concerns (for example, palm oil plantations and pulp and paper companies) with close ties to the government, continue to use fire as a cheap and illegal method of clearing forests. The World Wildlife Fund estimates that 70 percent of Indonesia's exported trees are illegal, many of them removed by "timber barons" in the country's national parks while bribed park and police officials look the other way. The urgent need for construction materials after the devastating tsunami of 2004 further increased the pressure on the country's forests.

There is a growing movement in the consuming countries to boycott the tainted timber by selling only certified "good wood." The U.S. company Home Depot, for example, sells only imported woods bearing the Forest Stewardship Council (FSC) logo.

INSIGHTS

Rubber

The earliest major cash crop introduced into Southeast Asia was rubber, first cultivated there in the 1870s. An Englishman, Sir Joseph Priestley, gave the name "rubber" to the latex of the *Hevea brasiliensis* tree in 1770, when he discovered he could use it to rub errors off the written page. Extensive exploitation of this tree in its native Brazil gave rise to the "rubber boom" of the late 19th century, when Brazilian rubber traders were so wealthy that they used banknotes to light cigars and sent their shirts to be laundered in Europe.

In 1876, the Englishman Henry Wickham smuggled 70,000 rubber seeds out of Brazil. Delighted botanists at London's Kew Gardens cultivated them and shipped them to Ceylon (Sri Lanka) and Southeast Asia for experimental commercial planting. Rubber trees proved exceptionally well adapted to the climate and soils of South Asia and then Southeast Asia (•Figure 7.2.D). Back in Brazil, however, a fungus known as leaf blight wiped out the rubber plantations, and by 1910, bats and lizards inhabited the mansions of Brazil's rubber barons. Fortunately, the Kew Gardens stock was free of the fungus; by 1940, fully 90 percent of the world's natural rubber came from Asian plantations. The desire to secure this vital resource was Japan's principal reason for occupying much of
398 Southeast Asia at the outset of World War II.

Perhaps surprisingly, real rubber is more important today than ever. Synthetic rubber, most of which is made from petroleum, now meets the greater part of the world's demand for products made originally of natural rubber. However, natural rubber has superior elasticity, and aircraft and truck tires are made almost exclusively from it. China's boom is helping ensure a strong future for future rubber producers, and in an unusual turnabout, many Southeast Asian workers are leaving the factory for the rubber farm.

Joe Hobbs

• **Figure 7.2.D** Rubber latex bleeding down a fresh incision and into the collection cup on a Malaysian plantation.

Deforestation in Southeast Asia has many actual and potential transboundary consequences. In 1997, a widespread drought attributed to the warming of Pacific Ocean waters
532 by the phenomenon known as El Niño turned the annual July-to-October burning season into a holocaust of human origin. Deliberately set fires raced out of control over large areas of Sumatra and Borneo, resulting in a choking haze that shut down airports, closed schools, deterred tourists, and caused respiratory distress to millions of people in Indonesia, Malaysia, Singapore, the Philippines, Brunei, and Thailand. Less widespread but still severe smog crises struck Malaysia and Indonesia, particularly the island of Sumatra, in subsequent years, and Indonesian authorities have fought back by revoking the licenses of plantation companies.

Many of the plants and animals in these forests are en-
demic species (found nowhere else on earth). Indonesia, 30
which contains 10 percent of the world's tropical rain forests, is known, after Brazil, as the world's second most important **megadiversity country,** with about 11 percent of all the world's plant species, 12 percent of all mammal species, and 17 percent of all bird species within its borders. The Southeast Asian region is also particularly significant in biogeographic terms because of the so-called **Wallace's Line**—named for its discoverer, the English naturalist Al- 358–
fred Russel Wallace—that divides it (see Figure 7.2.18). No- 359
where else on earth is there such a striking local change in the composition of plant and animal species in such a small area on either side of this divide. East of the line (separating the Indonesian islands of Bali and Lombok, for

Joe Hobbs

• **Figure 7.2.8** Commercial logging has removed all but small portions of the tropical rain forest (as in the lower left) in East Malaysia's states of Sarawak, seen here, and Sabah. The land is then used for oil palm plantations. The same process is underway in neighboring Kalimantan on Indonesia's side of the island of Borneo.

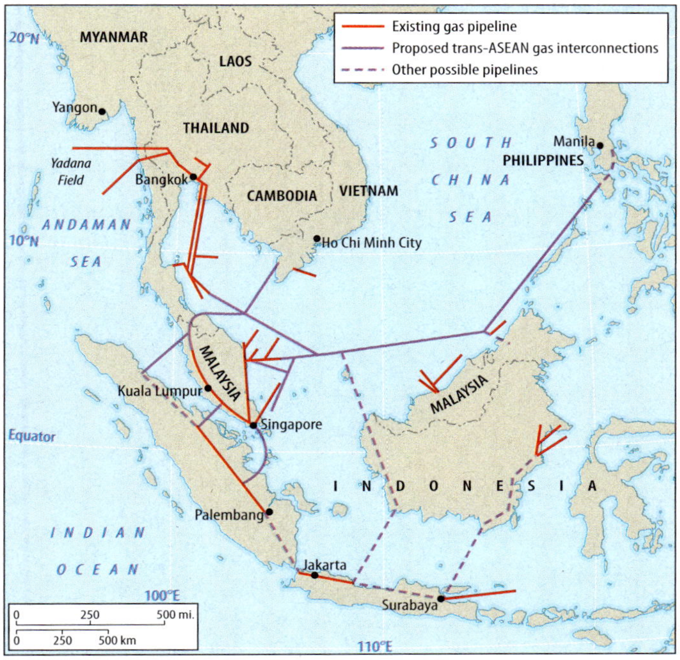

• **Figure 7.2.9** Existing and proposed natural gas pipelines for Southeast Asia.

example), marsupials are the predominant mammals, and placental mammals prevail west of the line. Similarly, bird populations are remarkably different on either side of the line.

Mineral Production and Reserves

Petroleum is the most important mineral resource in Southeast Asia. Seven of the region's countries—Indonesia, Malaysia, Brunei, Myanmar, Thailand, Vietnam, and the Philippines—produce oil. Collectively, their proven oil reserves amount to about 1.1 percent of the world total. Their combined output is about 3 percent of world oil production, with Indonesia the leading producer. Indonesia's main oilfields are in Sumatra and Indonesian Borneo (known as Kalimantan).

Although it does not loom large on the world scene, Southeast Asian oil is very important to the countries that own and produce it. Oil and associated natural gas provide nearly all the exports of Brunei and make that tiny Muslim country one of the world's wealthiest nations. That wealth is precarious, as Brunei has little to offer world markets except its finite fuels. Oil, gas, and refined products supply almost one-fourth of Indonesia's exports and about one-tenth of Malaysia's. Singapore profits as a processor of oil. The main customer for Southeast Asian oil and natural gas is Japan, the leading trade partner of several Southeast Asian countries. Natural gas is so important domestically for electricity production in several Southeast Asian countries that efforts are under way to link them in a network of pipelines, tapping mainly Indonesian and Malaysian gas reserves (•Figure 7.2.9). About 70 percent of Thailand's power production comes from natural gas from Myanmar and its own offshore fields. This relatively clean source of energy has helped clear Thailand's air of its long-standing smog problem.

There is other mineral wealth in Southeast Asia. Tin is the most important ore mined, with about one-fifth of world output from Indonesia. The Philippines produces a variety of metals—copper, chromite, nickel, silver, and gold—in significant amounts. The most serious mineral deficiency in the region is the near absence of high-grade coal. However, more energy could come from the region's additional potential for hydroelectric power. Gemstones are important exports from Myanmar and Thailand (•Figure 7.2.10).

Joe Hobbs

• **Figure 7.2.10** Star sapphires from Myanmar are among Southeast Asia's most prized gemstones.

Economic Meltdown and Its Aftermath

Following years of spectacular growth, in which economies across the region grew at rates up to 8 percent annually, 1997 saw a stunning reversal of fortunes in Southeast Asia. Until then, the region's economic tigers and cubs were praised for high rates of savings, balanced budgets, and low inflation. Export-driven growth rates were climbing, particularly with the increased manufacturing of computers and other electronics. Gradually, however, the appetite of the main consumer of Southeast Asia's products, the United States, began to wane, and trade deficits developed in the producing countries. At the same time, the region's emerging economies pursued expensive public works projects, such as Malaysia's new administrative capital of Putrajaya (see Figure 7.2.15, page 355).

In some countries, corruption added to the potential for economic crisis as insolvent companies were propped up with infusions of cash from sympathetic politicians (the phenomenon is known as **crony capitalism**). Banks extended risky loans for both private and public businesses and construction projects. The loans were generally made in U.S. dollars. When local currencies rapidly lost value against the dollar, the borrowers were unable to repay these loans. Nervous foreign investors in the region's stock markets withdrew their investments; theirs was the "hot money" that can send stock markets soaring but can also be with-
47 drawn quickly, with devastating results.

Economies melted down. Stock markets, which had seen explosive growth, imploded; Malaysia's market, for example, lost 80 percent of its value between September 1996 and September 1998. Among the region's stronger economies (excluding Vietnam, Cambodia, Laos, and Myanmar, which had far less to lose), Malaysia, Thailand, and Indonesia fell into **economic recession,** characterized by a relatively short fall in economic output and a moderate rise in unemployment. As the crisis dragged on, it threatened to tip into a full-scale **economic depression,** a devastating breakdown of the economy. Only the Philippines and Singapore escaped the bloodletting.

Economies gradually began to improve, but the economic meltdown had lasting impacts. Largely unchallenged in prosperous times, Southeast Asian ways of running governments and businesses have been under strong scrutiny ever since. Indications of this shift run the gamut from large-scale revolts against or impeachments of national leaders to parents fighting favoritism in school admissions, farmers protesting dam-building projects, peasants protesting industrial pollution, and patients suing doctors for malpractice. Examples of these events are described in the discussions of the individual countries. Despite their internal problems and differences, the countries hoped that their common membership in the Association of Southeast Asian Nations (ASEAN) might unite them against future economic calamities (see the Regional Perspective, page 345).

Although the countries of Southeast Asia have many similarities, each has its own distinctive qualities, with unique environmental features, indigenous and immigrant peoples, culture traits, and economic activities. Many of the characteristics and problems of these countries are the outcome of European imperialism, but in no two countries have the results of colonialism been the same. Here are portraits of the countries.

7.2.3 Myanmar (Burma)

Myanmar (population 50 million; see Figure 7.2.1, page 336) is centered in the basin of the Irrawaddy River and includes surrounding uplands and mountains. Within the basin are two areas of dense population: the so-called Dry Zone, around and south of Mandalay (population 1.2 million), and the Irrawaddy Delta, which includes the former capital and major seaport of Yangon (Rangoon; population 4.6 million). The Dry Zone is the historical nucleus of the country. The annual rainfall of this area—34 inches (87 cm) at Mandalay—is very low for Southeast Asia. The people are supported by mixed subsistence and commercial farming, with millet, rice, and cotton the major crops. During the past century, the Dry Zone was surpassed in population by the delta, where a commercial rice-farming economy now provides one of the country's two leading exports (forest products is the other). The Irrawaddy forms a major artery of transportation uniting the two core areas.

The indigenous Burman people, most of whom live in the Irrawaddy Basin, make up an estimated 68 percent of

REGIONAL PERSPECTIVE

The Association of Southeast Asian Nations (ASEAN)

The countries of Southeast Asia have sought to strengthen their economic and security interests by forming a regional organization. Known as the **Association of Southeast Asian Nations (ASEAN),** the coalition was formed in 1967 to promote regional trade and minimize the need for alignment with either of the two world superpowers, the United States and the Soviet Union. Singapore, Malaysia, Thailand, Indonesia, Brunei, and the Philippines were founding members. Vietnam joined in 1995 with the promise of strengthening regional ties and adopting the more open economic system that the Vietnamese call ***doi moi,*** or renovation. Myanmar and Laos joined in 1997 and Cambodia in 1999.

After U.S. troops withdrew from bases in the Philippines in 1991, ASEAN members grew concerned that they lacked a security "umbrella" and that China or possibly even Japan might become more aggressive in the region. Several ASEAN members, notably Indonesia, Malaysia, and Singapore, therefore strengthened ties with the U.S. military to deter other powers from attempts to dominate the region. The organization thus aspired to become both a large trading bloc and a substantial security counterweight to China.

By the end of the century, however, ASEAN members moved closer to China, increasingly seen as a benign and economically advantageous ally, and created an informal economic cooperation alliance with China, Japan, and South Korea known as **ASEAN+3.** China even proposed that a China-ASEAN Free Trade Area should be established by 2015. ASEAN's main goal is to adopt policies that would avert the type of economic meltdown that shook all the group's members in 1997 and 1998. ASEAN nations aspire to have, by 2015, a European-style common market that would eliminate member country barriers to trade and investment. The 90
countries are also trying to reduce dependence on the International Monetary Fund and other Western monetary agencies and countries in the event of future economic crises.

Solidarity among the member countries will demand that they first resolve quarrels among themselves. Many tensions focus on the most prosperous member, Singapore, which other members 355
view as an arrogant economic giant that grows at the expense of the other members. Indonesia notably has cut back exports of some commodities to Singapore to register its discontent.

---

the country's population. About 90 percent of Myanmar's citizens are Theravada Buddhists. Distinct ethnic groups live in the Arakan Mountains of the west and the northern highlands. Ethnic Shans live in the northern part of the Shan Plateau of eastern Myanmar, and ethnic Karens live in the southern Shan Plateau and in the delta of the Salween River. These ethnic groups account, respectively, for about 9 and 7 percent of the country's population. The Shan Plateau in northeastern Myanmar is part of the **Golden Triangle**—the frontier region of Myanmar, Thailand, and Laos—which was the world's leading source area for opium until Afghanistan pushed ahead of it in the 1990s (see the Regional Perspective, pages 346–347).

Britain conquered Burma in three wars between 1824 and 1885. It became an independent republic outside the British Commonwealth in 1948. Civil war has been a constant since independence. Communism, targeted persecution, and ethnic separatism have motivated rebellions by ethnic Karens and Shans, a number of less numerous hill peoples, and Muslims of the Rohingya ethnic group who fear the establishment of a pure Buddhist state. Beginning around 1999, however, Myanmar's ruling military government reached cease-fire agreements with most of these ethnic groups (except the Karens) and gave these groups what it called **contingent sovereignty,** offering them more civil rights and economic opportunities than previously. The government also encouraged profits made by some of these groups in the drug business to be "invested" in national development.

The economy has been badly damaged by fighting and for years was mismanaged under a form of rigid state control promoted as the **"Burmese Way to Socialism."** Myanmar became the poorest non-Communist country in Southeast Asia, a sad contrast with its reputation in early independence years for having the best health care, highest literacy rate, and most efficient civil service in Southeast Asia. It used to be the world's largest rice exporter; it now exports less than 1 percent of the world total. Myanmar's government spends comparatively less on health care for its citizens than all but three countries in the world.

In the 1990s, Myanmar began to abandon socialism in favor of a free-market economy and enjoyed considerable economic growth, with China its main trading partner and military ally. Most of the fruits of this economic growth, however, lined the pockets of the ruling military. The **State Law and Order Restoration Council (SLORC),** the military government that seized power in Burma in 1988 (and changed the country's name to Myanmar), has yielded little to popular pressure for democratization. Myanmar remains one of the world's most repressive places to live. Access to the Internet was prohibited until 1999 and is still strictly regulated. Foreign journalists are banned, and citizens may not allow foreigners into their homes. It is illegal to gather outside in groups of more than five. Giant green billboards across the country read: "Crush all internal and external destructive elements as the common enemy."

A much publicized contest, followed around the world, has been raging between the government and one of its citizens, the Nobel Peace Prize laureate Aung San Suu Kyi, who had been placed under house arrest after leading her antimilitary political party to an overwhelming victory in

REGIONAL PERSPECTIVE

Sex, Drugs, and Health in Southeast Asia

Southeast Asia has long been one of the world's main source areas for opium and its highly addictive semisynthetic derivative, heroin. For decades, global concern focused on combating the drug trade originating in this region and on dealing with the problem of heroin addiction in the consuming nations. Now, owing to the emergence of HIV/AIDS, the other traditional Southeast Asian industries of prostitution and tourism have woven drugs into a leviathan problem of global concern.

The region's drug production is centered in the Golden Triangle, where Laos, Thailand, and Myanmar historically exercised little control over their territories. Thailand has recently increased its authority substantially in the area. As in Afghanistan and Pakistan, the absence of a strong central government presence and ideal growing conditions facilitated explosive growth in drug production
272 (•Figure 7.2.E). But as Thailand and Laos exercised increasing authority in the area, drug production plummeted. Today, Myanmar is the only one of the three countries with a strong opium poppy economy, but even there, production is beginning to show signs of decline. The reason is that neighboring China, with an eye to more trade with Myanmar, is pressuring Myanmar authorities to crack down near their common border.

• **Figure 7.2.E** Raw opium oozes from slices in a poppy pod. Heroin, which is derived from opium, has a hearth in Southeast Asia's Golden Triangle region.

In recent years, with new heroin-producing areas cropping up in Mexico and elsewhere, the price of heroin for the consumer has dropped and its quality has risen. In the United States, these trends are noticeable in the growing appeal of heroin among middle and upper income groups, by smoking and other previously rare forms of heroin consumption, and by the increasing numbers of deaths due to heroin overdose.

In this region, the U.S. Drug Enforcement Administration and other international antidrug authorities focus their concerns on Myanmar, where refining of heroin from opium is done in remote laboratories. In northeastern Myanmar, armed ethnic minority groups have produced both heroin and a potent methamphetamine known as *ya baa*. They smuggle raw opium into both China and India and amphetamines into Thailand, providing investment capital to contribute to Myanmar's "development." Laundered drug money has paid for the country's recent construction boom, and the government encourages such investment from the drug sector to make up for shortfalls in legitimate export revenues.

Not all of the associated problems are exported. A flood of cheap and potent heroin hit the market in Myanmar after the current regime came to power in 1988. Heroin addiction has soared ever since. Primary delivery is by needles shared in tea stalls. The numbers of users continue to grow; perhaps 1 percent of the adult population is addicted. Accompanying the heroin use is an AIDS

the 1990 elections (which the military government nullified). In 1995, the government released "The Lady," as Suu Kyi is popularly known. She then called for dialogue with the military junta and appealed for the parliament that was elected in 1990 to be convened. The government responded by again placing her alternatively under house arrest and in prison, triggering strong economic sanctions by the United States.

Aung San Suu Kyi has called for would-be tourists to avoid Myanmar, which has abundant sites of interest but few visitors, arguing that tourism revenue would help strengthen the repressive government. She has also been successful in gaining support abroad for boycotts of U.S. and other Western companies doing business with Myanmar. American corporations including PepsiCo and ConocoPhillips have withdrawn from Myanmar, and the U.S. government has banned new investments. The United States regards the country's government as illegitimate and therefore rejects the change of the country's name to Myanmar. There is pressure on the American firm Unocal to suspend its contract to join a French firm in developing Myanmar's considerable offshore natural gas reserves in the Yadana Field of the Andaman Sea. The two companies have already built a pipeline from the field to a port west of Bangkok, Thailand, where the gas is used to generate electricity (see Figure 7.2.9). Human rights groups complain that slave labor was used to build the pipeline and that profits from gas sales help prop up Myanmar's repressive government. The pipeline has to be well defended in view of the government's many domestic opponents who might try to destroy

REGIONAL PERSPECTIVE

Sex, Drugs, and Health in Southeast Asia *continued*

epidemic, with again about 1 percent of the adult population infected. The best available figures indicate that the HIV infection rate among heroin users in Myanmar is around 37 percent.

The epidemic that began among heroin users spread quickly to Myanmar's sex industry. Recent estimates put the HIV infection rate among prostitutes at 26 percent, three times the rate in neighboring Thailand. Many young men with AIDS are shunned by their families and go to monasteries to die. Tragically, their virus has spread among the populations of Buddhist monks (there are now 400,000 in Myanmar), who share razors to shave their heads. AIDS prevention is virtually unheard of. The government had outlawed condoms until 1993. They are now legal, and over 30 million have been distributed by the government since 2001, but this is still short of what is needed. Practically no anti-HIV drugs find their way to Myanmar.

Myanmar's prosperous, cosmopolitan neighbor Thailand also has the scourge, with a 1.4 percent HIV infection rate among adults in 2007. This is a substantial reduction from the peak infection rate the country recorded in 1994. Thailand's liberal social climate has given rise to the unique phenomenon of sex tourism, with package tours catering to an international clientele (•Figure 7.2.F). Thailand's overall annual income from prostitution is estimated to be between $2 and $4 billion, and prostitutes send as much as $300 million each year to relatives in the villages from which they migrated. However, the spread of HIV infection among intravenous drug users and among prostitutes in the red-light districts of Bangkok and the popular beach resorts of Pattaya and Phuket threatens both the country's public health and its lucrative tourist trade. The Thai government mounted an all-out war on the drug trade in 2002 that resulted in the deaths of more than 2,000 people, most of them suspected drug dealers.

There are striking differences in how this region's countries are dealing with HIV/AIDS. Thailand, regarded globally as a leader in efforts to combat the problem, has run an aggressive and in-

Taimay Jones

• **Figure 7.2.F** Bangkok's red-light district was a breeding ground for HIV/AIDS until the government and nongovernmental agencies aggressively promoted a safe-sex campaign.

creasingly successful anti-AIDS public awareness campaign, resulting in a dramatic decrease in the number of new HIV infections. Thailand distributes antiretroviral drugs to all AIDS patients at
extremely low cost. Other Southeast Asian countries lag far be- 464
hind. The government of the Philippines has been unsuccessful in pursuing an anti-AIDS campaign because of opposition from the country's powerful and anticontraception Catholic clergy. Church officials have denounced the government's anti-AIDS program as "intrinsically evil" and have set boxes of condoms on fire at antigovernment demonstrations. Similarly, in the conservative and largely Islamic nations of Indonesia and Malaysia, Muslim clerics denounce their governments' anti-AIDS campaigns as efforts to encourage promiscuity.

it as a means of weakening the regime. In an apparent move to gain strategic depth from its own population and from any outside threats, the government in 2005 established a new capital at Naypyidaw, in the country's remote central region. The move did nothing to quell popular discontent with the regime. Widespread antigovernment protests, with broad participation of Buddhist monks, broke out all over the country late in 2007. Authorities responded with deadly force and the insurgency faded—at least temporarily.

7.2.4 Thailand

Governed today as a democracy in which military influence is strong and in which a traditional monarchy has mainly symbolic functions, Thailand (historically known as Siam; population 65 million) was the only Southeast Asian country to preserve its independence throughout the colonial period. This unique status was due to Siam's position as a buffer between the British and French colonial spheres.

Thailand is located in the delta of the Chao Praya River, known to Thais as the Menam ("River"). Annual floods of the river irrigate the rice that is the most important crop in this Southeast Asian land of relative abundance and progress. Thailand is the world's largest exporter of rice and dominates the niche market in the prized, aromatic and expensive jasmine variety.

On the lower Chao Praya is Bangkok (population 10.2 million), the capital, main port, and primate city. It is notoriously congested and was until recently a smoggy city. Environmental blight prompted the Thai government to crack down on carbon emissions and also to envision construction of

Joe Hobbs

• **Figure 7.2.11** The Buddhist monastery and temple of Wat Phra Kaew ("Temple of the Emerald Buddha") is one of the architectural gems of Bangkok, which is also a city of modern skyscrapers.

"Heaven City," a satellite of Bangkok designed to have little pollution or traffic gridlock.

Bangkok is a city of stunning architectural marvels, particularly Buddhist temples (•Figure 7.2.11). Bangkok's networks of canals (*klongs*), thronged with both commercial and residential activities, give it a distinctively amphibious character; it has been dubbed the **"Venice of the East."** Its "floating market" is especially renowned. Many of Bangkok's canals have been filled in with paved roads and buildings, thereby making the city vulnerable to flooding. Due to the pumping out of groundwater beneath the city, Bangkok's soils are settling downward, and the entire capital is sinking at a rate of about 2 inches (5 cm) a year. Thai officials forecast that Bangkok will be below sea level by 2050. If nearby sea levels rise, as predicted by many climate change models, the city would be imperiled.

Areas outside the river's delta are more sparsely populated and include the mountainous territories in the west and north, inhabited mainly by Karens and other ethnic groups, and the dry Korat Plateau to the east, populated by Thais, the related Laotians, and Cambodians. To the south in the Kra Isthmus, a part of the Malay Peninsula, live more than 2 million ethnic Malays. Most Malays are Muslims, but an estimated 95 percent of the people of Thailand are Buddhists whose faith includes elements borrowed from Hinduism and local spiritualism. Spectacular Buddhist temples are a prime attraction for one of Southeast Asia's largest tourist industries, as are ceremonies celebrating Thailand's monarchy and other historical traditions. The Thai people are extremely reverent of the country's king, and criticism of the monarchy is illegal. In 2007, a drunken Swiss tourist who spray-painted on images of King Bhumibol Adulyade was sentenced to ten years in jail. The king pardoned him.

The main ethnic minority in Thailand is the Chinese, an estimated 9 million, or 14 percent of the population. They control much of the country's business, a situation that many Thais resent and that makes the loyalty of the Chinese an important issue.

Although Thailand, dubbed the "Land of Smiles," has enjoyed more internal tranquillity than most of Southeast Asia, it has had its troubles, including a 2006 coup that ousted an unpopular prime minister but also diminished foreign confidence in the Thai economy. Since the late 1960s, guerrilla insurrections motivated by communism and ethnic separatism have persisted in the northeast and south. In the late 1970s, the country dealt with a mass migration of refugees from adjoining Cambodia and still hosts about 2 million refugees. Recently, Muslim ethnic Malay guerrillas seeking independence for Thailand's southernmost region have burned schools, killed Buddhist monks, and ambushed government troops in that area. Thailand annexed this region, which was formerly a Muslim sultanate, in 1902. Thai authorities today are divided on how to deal with the unrest. Some want to build a security barrier along the Malaysian border and crack down on the insurgency north of that line, but others—including the country's prime minister—would like to accommodate some of the ethnic Malay demands, including the application of Islamic law (*sharia*) in the region. 227

In spite of conflict and refugee problems, Thailand has had one of the strongest economies in Southeast Asia. From 1993 to 1996, the economy grew a robust 8 percent annually. The country shifted away from its traditional exports of garments, textiles, cut gems, and seafood in favor of medium-tech industries such as computer assembly. Tourism was the leading source of foreign exchange. Then came 1997, when Thailand was the first "domino" to fall in the Asian economic crisis (see page 344). Economic growth had attracted tens of thousands of the rural poor to the factories of Bangkok, where they made garments for such major U.S. retailers as Gap, Nike, and London Fog. Now they were laid off and joined the swelling ranks of Thailand's poor. Even before the slump, Thailand had a huge gap between the wealthy and the poor, with more than 50 percent of the country's wealth in the hands of the richest 10 percent of its population. By the beginning of the new century,

Thailand's economy was back on its feet. In recent years, it has recorded strong growth based largely on surging demand in China for electronic parts, petrochemicals, and rubber.

7.2.5 Vietnam, Cambodia, and Laos

Vietnam, Cambodia, and Laos (see Figure 7.2.1) are the three countries that emerged from the former French colony of Indochina. Vietnam is the largest and most geographically complex, with three main regions. Tonkin, the northern region, consists of the densely populated Red River Delta and surrounding, sparsely inhabited mountains; Annam, the central region, includes the wild Annamite Cordillera and small, densely populated pockets of lowland along its seaward edge; and Cochin China, the southern region, lies mainly in the delta of the Mekong River and has a high population density.

Almost 90 percent of Vietnam's people are ethnic Vietnamese (also known as Kinh), closely related to the Chinese in many cultural respects, including language and the shared religious elements of Confucianism, Buddhism, and ancestor veneration. A Catholic minority reflects the French occupation, and there are small indigenous religions. Fifty-four minorities enrich the country's ethnic makeup. Various ethnic groups known collectively to outsiders as the **hill tribes** or **Montagnards** live in the central highlands of Vietnam, where some have recently agitated for independence and religious freedom. A large Chinese minority lives in the major cities of all three countries.

Cambodia and Laos occupy interior areas except for Cambodia's short coast on the Gulf of Thailand. They are very different countries physically, and a particular culture dominates each. Cambodia is mostly plains along and to the west of the lower Mekong River, with mountain fringes to the northeast and southwest. Except for half a million ethnic Vietnamese, almost all the 14 million people of Cambodia are ethnic Cambodians, also known as Khmer. Their language is Khmer, and their predominant religion is Buddhism.

Laos is mountainous, sparsely populated with just 6 million people, and landlocked. It has 68 ethnic groups, the largest of which is the ethnic Lao, who are linguistically and culturally related to the Thais and whose dominant religion is Buddhism (•Figure 7.2.12). Their largest numbers are in the lowlands, especially in and near the capital, Vientiane. Most of the ethnic minorities of Laos live in the mountain regions, including the country's second-largest ethnic group, the Hmong.

The Vietnam War

The Vietnam War, which is such an important chapter in the American experience, had roots that preceded U.S. interests in the region. This conflict profoundly affected Southeast Asia and has a lingering legacy there today.

Joe Hobbs

• **Figure 7.2.12** The stupa of That Louang in Vientiane is Laos' most important religious building, and its national symbol.

France conquered Indochina between 1858 and 1907. The French first extinguished the Vietnamese empire, which covered approximately the territory of today's Vietnam, and defeated the Chinese, whom the Vietnamese called on for help. Later, the French took Laos (in 1893) and western Cambodia (in 1907) from Thailand. France's administration of Indochina, centered in Hanoi and Saigon, left a big cultural imprint. Its major economic success lay in opening the lower Mekong River area to commercial rice production, converting both Vietnam and Cambodia into important exporters of rice.

Japanese forces overran French Indochina in 1941, initiating five decades of warfare in the area. During World War II, a Communist resistance movement led by a Vietnamese nationalist named Ho Chi Minh carried on guerrilla warfare against the Japanese. When French forces attempted to reoccupy the area after the end of war, these forces fought to expel them. In 1954, Ho's forces destroyed an entrenched French garrison at Dien Bien Phu, in the northwestern mountains. With this defeat, France withdrew from Indochina, ending a century of colonial occupation.

Four countries came into existence with France's departure. Laos and Cambodia became independent non-Communist states. North Vietnam, where resistance to France had centered, emerged as an independent Communist state. South Vietnam gained independence as a non-Communist state that received many anti-Communist refugees from the north. Vietnamese Catholics were prominent among those who fled southward.

The partition of Vietnam into two countries was largely the work of U.S. power and diplomacy. South Vietnam was from the outset a client state of the United States, whereas North Vietnam became allied first with Communist China and then with the Soviet Union. Warfare in Vietnam gained momentum. Between 1954 and 1965, the **Viet Cong**, an insurrectionist Communist force supported by North Vietnam, achieved increasing successes in the south in its bid to reunify the country. Increasing intervention by the United States in South Vietnam escalated in 1965 and eventually involved the commitment of half a million U.S. military personnel. North Vietnam responded by fielding regular army forces against American and South Vietnamese troops. The United States never invaded North Vietnam, although American air strikes there were devastating, with more ordnance dropped on Vietnam than on all of Europe during all of World War II. American bombs, napalm (an explosive), and defoliants such as Agent Orange (dioxin) caused enormous damage to the natural and agricultural systems of Vietnam, destroying over 8,400 square miles (nearly 22,000 sq km) of forest and farmland in an effort described in military parlance as **"denying the countryside to the enemy."** Agent Orange also left a legacy of birth deformities among Vietnam's postwar generation.

The United States avoided invading North Vietnam partly because of the risks posed by Chinese and Soviet support of the North. In limiting the theater of ground warfare, however, the United States found itself unable to expel from the South both the Viet Cong and the North Vietnamese army. Those combined forces were determined, skillfully commanded, increasingly better equipped by its allies, accomplished in guerrilla tactics, and willing to bear heavy losses. More than 3 million Vietnamese soldiers and civilians died in the conflict, and about 58,000 U.S. soldiers and support staff perished. In 1973, almost all American forces were withdrawn from the costly war, which was extremely divisive at home as growing ranks of Americans questioned its rationale and protested its conduct. There are still 1,948 Americans listed as missing in action in Indochina.

North Vietnam completed its conquest of South Vietnam in 1975. Saigon was renamed Ho Chi Minh City (although most of its people still call it Saigon), after the country's leader (who died in 1969). Vietnam entered a period of relative internal peace, but there were ongoing conflicts, including Vietnam's 1979 conquest of Cambodia to oust the brutal Khmer Rouge regime (see page 352), continued warfare against Cambodian and Laotian resistance groups, and a brief border war with China. Within its borders, the Communist government of the reunited Vietnam unleashed a repression against those who had assisted the Americans, including Montagnards and members of the Cao Dai religion (•Figure 7.2.13) This backlash caused a massive outpouring of refugees, including the "boat people" (see Ethnic Geography, page 351). Pressure on perceived opponents intensified, especially for the country's ethnic Chinese after China's brief invasion of Vietnam in 1979. The two countries gradually came to terms, however, finally normalizing relations in 1991 and solving long-standing disputes such as the demarcation of their territorial waters and corresponding exclusive economic zones in the Gulf of Tonkin. They have, however, been unable to resolve disputed claims to two island groups in the South China Sea: the Paracel Islands (with rich fishing grounds and potential for oil and gas), which China took from Vietnam by force in the early 1970s, and the Spratly Islands (see Chapter 7, page 302).

Joe Hobbs

• **Figure 7.2.13** The "Divine Eye" is the symbol of God for members of the Cao Dai faith. Founded in Vietnam in 1926, the Cao Dai religion synthesizes many religions, and its pantheon includes Jesus Christ, Louis Pasteur and Victor Hugo. Cao Dai fighters resisted the French occupation, and also supported South Vietnam in its war against the North. This is the Cao Dai Temple in Danang, Vietnam.

Vietnam Today

Vietnam (population 85 million), with its capital at Hanoi (population 2.6 million; •Figure 7.2.14) is restoring its war-torn landscape through large-scale reforestation, agricultural reclamation, and nature conservation programs, aided by international organizations such as the World Wildlife Fund. Animal species new to science have been discovered in the rugged Annamite Cordillera (Truong Son Mountains). Such natural treasures are attracting international ecotourists who provide much-needed revenue, and tourism in general is booming.

ETHNIC GEOGRAPHY

The Boat People

The desperation of many people in Vietnam and its neighboring war-torn countries created the crisis of the **boat people**—refugees who fled onto the open ocean on large rafts, with chances of survival variously estimated at 40 to 70 percent. After 1974, almost 2 million Vietnamese fled their homeland, and the United States provided permanent refuge for more than half of those who resettled. Canada, France, and Australia also took in significant numbers of immigrants (•Figure 7.2.G). Large numbers of Vietnamese and Laotian refugees occupied camps in China near its borders with those countries. Boat people also filled refugee camps in Hong Kong, Malaysia, Indonesia, Thailand, and the Philippines.

The United States and other Western governments long ago reduced the size of the welcome mat for Southeast Asian refugees, but Vietnamese continue to immigrate to the United States at a rate of about 25,000 a year; they are now the fifth-largest immigrant group in the United States. Most of these immigrants are typical of an **expanding pyramid,** in which family members in the United States bring extended family members over from Vietnam. About 2,000 are classified as political refugees who survived Vietnam's postwar "education camps." Many of these are **Amerasians,** the offspring of American soldiers who served in the Vietnam War. Vietnam is also among the leading foreign sources of children adopted by American couples. Altogether, the United States is now home to about 1.2 million ethnic Vietnamese. Approximately 100,000 Vietnamese Americans visit Vietnam each year, bringing much welcome hard currency with them into Vietnam.

In 1996, Vietnamese refugee camps around Asia closed, and about 100,000 Vietnamese were compelled to return to their homeland. Just as those countries were closing their doors to Vietnam, Vietnam was undertaking political and economic reforms to provide incentives for the expatriates to return home and for resident Vietnamese to stay home. Since 1990, about 1 million overseas Vietnamese ("Viet Kieu"), mostly from the United States and France, have returned to Vietnam to stay. Their repatriation has also pumped economic capital into Vietnam's fledgling free-market economy. Many Vietnamese who once aspired to leave their home country have chosen to remain. Growing numbers leave only to earn doctorates and other professional degrees to carry home and invest in the country's growth.

Joe Hobbs

• **Figure 7.2.G** Many ethnic Vietnamese refugees settled in Canada, introducing their cuisine and culture to the new homeland.

Since its embrace in 1986 of free-market reforms in the policy known as *doi moi,* Vietnam's economy has improved markedly. The poverty rate fell by half in the 1990s, and in recent years, Vietnam has had Asia's second-fastest-growing economy, after China's. The Communist government halted collectivized farming in 1988 and turned control of land over to small farmers under 20-year lease agreements. The results have been impressive; Vietnam since 1988 has become a major exporter of rice, second in the world after Thailand. This is a vulnerable commodity, however, especially because production is concentrated along the Mekong River, which is prone to flooding in the May-to-October rainy season (see The World's Great Rivers, page 353). Vietnam is also pinning hopes on exports of its proven reserves of 2.5 billion barrels of oil, especially to energy-hungry China.

In addition to joining ASEAN, Vietnam in 1995 restored full diplomatic relations with the United States, and U.S. businesses began scrambling for consumers in this large, promising market. After the two countries signed a trade agreement in 2000, Vietnam began exporting shoes, finished clothing, and toys to the United States. Both its imports and its exports boomed when the country joined the World Trade Organization (WTO) in 2007. The dual personalities of North and South are being perpetuated in the new economic climate, with most investment and infrastructural improvement focused in the South. The North is a more austere economic landscape, and northerners dominate the civil service. Along with a growing gap between the rich and the poor, the differences between North and South could slow the country's overall development.

Joe Hobbs

• **Figure 7.2.14** The heart of Hanoi's Old Quarter.

In an effort to better integrate the nation, Vietnam recently completed a road running the entire 1,000-mile (1,600-km) length of the country, along the historic Ho Chi Minh Trail. Environmentalists fear the highway's impacts on wildlife and point out that annual heavy rains regularly bring landslides and make this mainly mountainous route unusable. Anthropologists are worried about its effects on the traditional peoples inhabiting remote areas along the road. And many in Vietnam's growing private business sector lament the road as a costly white elephant; the country already has a more practical lowland north-south highway and railway, they complain, and resources would be better spent on "new economy" infrastructure such as the Internet.

Like that of many LDCs, Vietnam's progress has been threatened by the specter of uncontrolled population growth. Since the war with the United States ended in 1975, Vietnam's population has increased by 70 percent. In view of the country's limited size and resource base, the government has stepped in with strong economic incentives for couples to have no more than two children. Birth rates have begun to fall sharply. Nevertheless, 29 percent of the population is under the age of 15, and foreign investors are pouring into Vietnam to seek this young brainpower and muscle power.

Cambodia: Away from the Killing Fields

For a time, Cambodia escaped the ills known to Vietnam, but then it suffered even worse catastrophes. After five years of fighting, a Communist insurgency overcame Cambodia's American-backed military government in 1975. Known as the **Khmer Rouge** and led by a man with the *nom de guerre* of Pol Pot, this Communist organization ruled the country for four years with exceptional savagery. In 1973, the population had been estimated at 7.5 million, but by 1979, it was estimated at only about 5 million; a third of the population had been murdered. Some intended victims managed to escape as refugees from what became known as Cambodia's **killing fields** to adjoining Thailand. The Cambodian Communists' stated policy was to build a "new kind of socialism" by eradicating the educated and the rich, emptying the cities, and breaking up families. The population of Phnom Penh, the capital and largest city, was expelled on a death march to the countryside. The city's population dropped from more than 1 million in 1975 to an estimated 40,000 to 100,000 in 1976 (its population in 2007 was 1.7 million).

The Communist-Nationalist ambitions in Vietnam came to dominate Cambodia's political landscape in 1979. Border conflicts between the two countries had been escalating, and in 1979, Vietnamese troops swiftly conquered most of Cambodia, ousting the Khmer Rouge and installing a puppet Communist government. The Khmer Rouge, supported by China (Vietnam had by then become a close ally of China's rival, the Soviet Union), began guerrilla resistance against the Vietnamese. Two non-Communist Cambodian guerrilla forces also took the field. These three resistance forces often quarreled among themselves, but all maintained the conflict with the Cambodian-Vietnamese army of the Phnom Penh government, which periodically pursued them into Thailand and engaged Thai border military units.

THE WORLD'S GREAT RIVERS

The Mekong

The Mekong River, known as "Water of Stone" in Tibet, "Great Water" in Cambodia, "Nine Dragons" in Vietnam, and "Mother of Waters" (Mae Nam Khong, contracted to Mekong) in Laos and Thailand, is Southeast Asia's great river (•Figure 7.2.H). From its source on the Tibetan Plateau in the Chinese province of Qinghai, the Mekong snakes along a 2,600-mile (4,200-km) path through China's Sichuan and Yunnan provinces, forms the boundary between Laos and its neighbors Myanmar and Thailand, winds across Cambodia, and fans out across Vietnam's densely populated Mekong Delta region before empting into the South China Sea.

Unlike the Chao Praya in Thailand, the Red River in Vietnam, and the Irrawaddy in Myanmar, the Mekong is not a great magnet for cities and civilizations. The biggest city on its banks is Cambodia's Phnom Penh, with just over a million people. The main factor limiting more settlement along the Mekong has been the difficulty in navigating the river above Phnom Penh. Historically, very little trade has been carried along much of its length; the oceans and overland routes have provided superior opportunities.

The Mekong is not economically unimportant, however. It is one of the world's most prolific fisheries, yielding about 2 million tons of fish each year, or twice the annual North Sea catch. There are more species of fish in the river (about 1,200) than in any other rivers but the Amazon and the Congo. Unique creatures inhabit its waters, including about 70 Irrawaddy dolphins and the giant catfish that attain 10 feet (c. 3 m) in length and 660 pounds (300 kg) in weight. Overfishing and the construction of dams have all but eliminated this superlative catfish throughout the Mekong Basin.

A big push is being made to capitalize on the Mekong's potential to produce hydroelectric power, with plans to build about 100 dams on the river and its tributaries. The first dam on the river itself (the Man Wan in China) was completed in 1993. China has since built another and has plans for four more. Vietnam is building five dams on the Mekong tributary called the Hoyt, which forms part of its border with Cambodia. Thailand has dammed its main tributaries to the Mekong. Laos plans to build the most dams of all—20—and thereby become mainland Southeast Asia's biggest supplier of electricity. With funding from the World Bank—despite the criticism this organization has received for supporting large dams—Laos is proceeding with construction of the giant Nam Theun 2 dam. The World Bank insists it has gone to extraordinary lengths to minimize environmental impacts from this project.

These dams will nevertheless have enormous impacts, some of them unforeseeable. It may be reliably predicted that the fish catch will fall dramatically because the main reason for the Mekong's productive fishery is its regular flooding, which will be dramatically reduced. Fishermen are already complaining of smaller catches since the dam building began. There are concerns that dams will ruin the ecosystem of Cambodia's Tonle Sap Lake, which is drained by a river that actually changes direction when downstream Mekong floodwaters inundate its basin. More than a million people make their living by fishing this lake.

As always with shared river systems, there are serious questions about how the Mekong waters should be divided. Vietnam, Cambodia, Laos, and Thailand have joined to form the **Mekong River Commission,** 233, 316 whose biggest responsibilities are to fix the minimum amount of water each country must discharge downstream on the Mekong and its tributaries and to agree on rules to ensure water quality. Both quality and quantity of water are threatened by upstream Myanmar and China, which have refused to join the Mekong River Commission.

Joe Hobbs

• **Figure 7.2.H** Ferries on the Mekong near Can Tho, Vietnam.

At first, the Vietnamese rejected international food aid and systematically starved Cambodia in an effort to weaken resistance. Later, they promoted the country's recovery.

Peace negotiations among Cambodia's contending factions led in 1991 to an agreement resulting in the restoration of democratic government under United Nations supervision. A new constitution with new leaders took effect in 1993, and Cambodia's former King Norodom Sihanouk returned to the throne. In 1997, Hun Sen, a former Khmer Rouge leader who had defected and then opposed the group, ousted the popularly elected royalist leader Prince Norodom Ranariddh in a violent coup. Under much international pressure, Hun Sen permitted elections in 1998. Finally succeeding in forming a government legitimately and peacefully, he secured Cambodia's membership in ASEAN and regained Cambodia's seat at the UN.

Cambodia is a very poor country, but more stable conditions have helped the economy improve recently. Oil and gas reserves have recently been discovered offshore, giving cause for optimism. Garment manufacturing is the largest

industry, accounting for 90 percent of export earnings. Fears that Cambodia would prove unable to compete with growing exports of cheaper clothing from China have so far not been realized, thanks to large foreign contracts from companies like Gap and Abercrombie & Fitch that
386, 540 certify Cambodia's textiles as "sweatshop-free." Tourism is the fastest-growing industry. Few visitors miss Angkor Wat, still coveted by some in Thailand because it once stood within that country's borders. Cambodians resent Thai claims to the site and also Thai ownership of many prominent tourism, telecommunications, and forestry businesses in Cambodia. That resentment led to riots in 2003—sparked by a Thai actress's claim that Angkor Wat belongs to Thailand—that did millions of dollars in damage to Thai businesses and the Thai embassy in Cambodia.

Even in this relatively peaceful time, the people of Cambodia must deal with a persistent scourge of war: land
274 mines. Sown by hostile forces during years of warfare, the antipersonnel explosives remain active indefinitely. Mines have maimed more than 35,000 Cambodians, and about 50 more casualties occur each month. People bury new mines, imported mainly from China and Singapore, to protect their property in Cambodia.

The Lao Way

Laotians also suffer from ordnance dating to the Vietnam War. Between 1964 and 1973, U.S. warplanes dropped more than 2 million tons of bombs—more than the United States dropped on Germany in World War II—on the Laotian frontier with Vietnam in an effort to disrupt Communist supply lines on the Ho Chi Minh Trail (a network of supply lines snaking through the mountains of Vietnam and Laos) and to prevent Communist troops from entering Laotian cities. This Plain of Jars region retains the distinction of being the most heavily bombed place on earth. An estimated 30 percent of the bombs failed to detonate, and today, millions of unexploded American cluster bombs remain. Smaller than a tennis ball, each cluster bomb contains more than 100 steel ball bearings. Many curious Laotian children who play with them are killed or dismembered. About 200 people are killed or maimed yearly by these explosives. Of Laos's 17 provinces, 15 contain **unexploded ordinance (UXO)**, according to UN studies, creating a serious deterrent to farming in a country with inadequate food supplies. Efforts are under way to clear the explosives, with only minimal economic assistance from the United States. The agency in charge of the cleanup hopes to clean up enough land to produce 10,000 tons of rice yearly, which would feed 50,000 people.

The administrative capital of Laos is Vientiane (with only 200,000 people), located on the Mekong River where it borders Thailand. This sleepy city poses a remarkable contrast to the nearby boomtowns of Vietnam (•Figure 7.2.15). Vientiane replaced the royal capital of Luang Prabang ("Royal Holy Image"), located upriver on the Mekong when Laos came under Vietnamese occupation in 1975. That year, the Pathet Lao organization overthrew the Laotian monarchy and established a new "people's democratic republic." The Communist dictatorship now rules Laos in a benign fashion, encouraging a market economy and foreign investment. The government calls its conservative approach to economic reform and social change the **"Lao Way."** To slow the potential flood of foreign ideas and materials into the country, Laos has permitted the building of only two bridges to Thailand across the Mekong River. However, there is an already important and growing casino gambling industry in Laos, patronized mainly by visitors from Thailand.

Joe Hobbs

• **Figure 7.2.15** Vientiane's Pataouxi, or Victory Arch, built in the 1950s, is often likened to Paris' Arc de Triomphe.

Despite gambling revenues, landlocked Laos is now one of Asia's least developed countries, ranking 133rd of the 177 countries on the United Nations Human Development 44 Index. About 80 percent of its people are peasant farmers. Exports are limited to timber, hydroelectric power (to Thailand), and garments and motorcycles assembled by inexpensive Laotian labor using imported materials. Gold and copper mining, which began in the early 2000s, may bring sorely needed revenue to Laos.

As in Vietnam and Cambodia, the Laotian economy has suffered from a brain drain prompted by years of warfare, political turmoil, and underdevelopment—conditions that caused the best-educated and most talented of the country's 59 inhabitants to flee. After the 1975 takeover, 343,000 Laotians, mostly members of the Hmong ethnic minority who had U.S. support in fighting Laotian government troops during the war, fled to neighboring Thailand. Many Hmong refugees subsequently emigrated to the United States, where 250,000 ethnic Hmongs now live, generally as an underprivileged and misunderstood underclass in cities of the Midwest and West. Back in Laos, a low-level Hmong insurgency continues against the Vientiane government.

Given the conditions of warfare, political turmoil, oppression, and inefficient economic systems, it is not surprising that the three Indochinese countries are, with Myanmar, the poorest in Southeast Asia. However, their outlook seems to be improving as tensions with the outer world relax, privatization of Communist economies increases, the supply and trade of goods and services improves, and the number of returning refugees exceeds emigrants.

7.2.6 Malaysia and Singapore

The small island of Singapore (•Figure 7.2.16; see also Figure 7.2.1) sits just off the southern tip of the mountainous Malay Peninsula. It lies at the eastern end of the Strait of Malacca, the major passageway for sea traffic between the Indian Ocean and the South China Sea. For centuries, European sea powers competed to control this passageway. Britain gained control in 1824 and used the newly established settlement of Singapore to monitor the strait. With its central position among the islands and peninsulas of Southeast Asia, British Singapore developed into a large naval base and the region's major entrepôt. Singapore is the world's second-largest port (after Shanghai) in tonnage of goods shipped.

From Singapore, the British extended their political hold over the adjacent southern end of the Malay Peninsula. This expansion gave Britain control over the part of the peninsula that is now included, along with northwestern Borneo (except Brunei), in the independent country of Malaysia. This southern end of the peninsula is known as Malaya, Peninsular Malaysia, or West Malaysia. The Borneo section includes the states of Sarawak and Sabah, together known as East Malaysia.

During the British period, Malaya developed a commercialized and prosperous economy. Tin and rubber were the major products, with oil palms and coconuts of secondary importance. Chinese and British companies pioneered the tin-mining industry in the late 19th century. The building of railroads gave access to a line of tin mines and rubber plantations in the foothills between Malacca and the hinterland of Penang (George Town). Along this line, the inland city of Kuala Lumpur (population 6.9 million) became Malaya's leading commercial center and then Malaysia's capital (see Figure 7.2.3, page 338).

Although Kuala Lumpur will remain Malaysia's official capital, all government administration has been transferred to Putrajaya, 25 miles (40 km) to the south (•Figure 7.2.17).

Taimay Jones

• **Figure 7.2.16** Singapore, located at a crossroads where the Pacific and Indian Oceans meet, is a very modern, prosperous city.

Joe Hobbs

• **Figure 7.2.17** The Prime Minister's department complex and city mosque in Putrajaya, Malaysia's new and controversial administrative capital.

It is a high-tech capital stitched together by fiber-optic cable. Malaysian officials boast that the Internet and other forms of digital communication will replace paper in the country's new administrative center.

Nearby is Cyberjaya, a new city that, as its name suggests, hopes to lure technology companies and become Asia's Silicon Valley. These cities are symbols of how Malaysia sees itself, a country leaving behind its colonial legacy as a "rubber republic" to become an engine of technical innovation and export. Putrajaya's $5 billion construction cost drew much criticism in the wake of Malaysia's financial crisis, however.

The development of a colonial economy gave Malaya an ethnically mixed population. The native Muslim Malays played only a minor role in this development. Chinese and Indian immigrants and their descendants became the principal farmers, wage laborers, and businesspeople of the tin and rubber belt. So heavy was the immigration that 23 percent of Malaysia's population of 27 million is Chinese and 7 percent is Indian. Today, the Chinese minority is quite prosperous in business, and Indians dominate many professional ranks. The growth of Singapore brought even heavier immigration so that now over three-fourths of Singapore's population of 4.6 million is Chinese.

Ethnic antagonisms between Malays and Chinese shaped the political geography of Malaysia and Singapore. Malaya accepted independence in 1957 only on the condition that Singapore not be included. This was to ensure that ethnic Malays would be a majority in the new state. In 1963, however, the Federation of Malaysia was formed, composed of Malaya, Singapore, and the former British possessions of Sarawak and Sabah in sparsely populated northern Borneo. The non-Chinese majorities in Borneo were counted on to counterbalance the admission of Singapore's Chinese. This experiment in union lasted only until 1965, when Singapore was expelled from the federation and left to go its own way as an independent state.

Within Malaysia, the government passed legislation to ensure that at least 60 percent of the university population was ethnic Malay (who with small indigenous groups are known as **Bumiputra**), with measures to ensure that 30 percent of business ownership went to Malays. Although ethnic Chinese and Indians still complain that such policies continue to shut doors of opportunity to them, the economic gap between these minorities and the majority Malays has narrowed considerably since the 1970s. But tensions between Muslims and non-Muslims in this country bear monitoring. Among Malaysia's majority Muslims, there are growing complaints that the government is too secular, although it is nominally Islamic (•Figure 7.2.18). The leading champion of the adoption of Islamic law (*sharia*) as the law of the land is the **Parti Islam se-Malaysia (PAS)**, which previously administered state governments of Kelantan and
217 Terengganu in Peninsular Malaysia.

Malaysia and Singapore are among the more economically successful of the world's formerly colonial states that

Joe Hobbs

• **Figure 7.2.18** Most ethnic Malays, like this woman, are Muslim.

became independent after World War II. Singapore is an outstanding center of manufacturing, finance, and trade along the Pacific Rim. Its industries include electronics, chemicals, oil refining, machine building, and tourism. With its electronics sector now suffering intense competition from China, Singapore is hoping to develop a strong new niche in biomedical research and other biotechnology industries; a huge biotechnology research complex called **Biopolis** has been built there. Due to restrictions on federal spending on stem cell research in the United States, many top American scientists have relocated there. Singapore suffered less than Malaysia, Thailand, and Indonesia during the region's economic crisis of the late 1990s. This was due in part to a geographic advantage: its ability to conduct trade with a vast regional hinterland in other countries without having to attend to the struggles of a large and poor rural population of its own. Perceived as an island of stability in an unsettled region, the country has also long been a magnet for investment. Wealthy Indonesians buy up real estate there, and ethnic Chinese capital and talent from all over Southeast Asia flow into Singapore. The country's prosperity is accompanied by strict attitudes toward propriety and secularism. Chewing gum is banned, graffiti artists are flogged, headscarves and other apparent religious symbols are banned in schools, and crows—famously loud, messy, and unruly—are shot.

Malaysia still exports rubber and tin but has a diversified export portfolio that includes crude oil (from northern Borneo), timber, palm oil, automobiles, and electronic components. Malaysia is the world's largest exporter of semiconductors and air conditioners. Its annual per capita GNI PPP is only about one-third that of Singapore's but is about double the average figure for the world's LDCs (see Table 7.1).

Malaysia has a national mantra, "Wawasan 2020" (Vision 2020), referring to the year by which national economic planners intend Malaysia to be a fully developed country. The vision became clouded after 1997, however, when following a decade in which the economy grew at a robust 8 percent annually, Malaysia fell into recession. The country quickly froze its ambitious and expensive development projects and adopted economic austerity measures. Like the other emerging markets, Malaysia had been benefiting from 47, 344 huge and often speculative foreign capital investments. And like the others, Malaysia suffered in 1997 and 1998 when worried foreign investors rapidly withdrew their capital.

Malaysia effectively pulled out of the orbit of globalization in 1998 when it took drastic actions to protect itself from the whims of foreign investment. Malaysian politicians blamed the International Monetary Fund and other institutions in the West for Asia's financial troubles, and these institutions in turn condemned Malaysia for retreating from the principles of the global free market. Despite the ominous downturn, by the end of the decade, Malaysia was once again registering economic growth—particularly as exports of computer chips and other electronics resumed—and was pulled out of the abyss in which neighboring Indonesia was still mired.

7.2.7 Indonesia and Timor-Leste

Indonesia (•Figure 7.2.19) is by far the largest and most populous country of the region (population 231 million) and is the fourth most populous country on earth (after China, India, and the United States). Stretching across 3,000 miles (4,800 km), its 13,600 islands comprise a land area about three times the size of Texas. Indonesia's land represents over 40 percent of Southeast Asia's land and is home to the same percentage of its population, due primarily to the dense concentration of people on the island of Java, whose 125 million people represent about 54 percent of Indonesia's population. The island's population density is about 2,500 per square mile (965/sq km). In contrast, the remainder of Indonesia, which is about 13 times larger than Java in land area, has an average of only about 150 people per square mile (58/sq km).

This extraordinary concentration of population on this island is in part a reflection of the superior fertility of Java's volcanic soils (•Figure 7.2.20). Other Indonesian islands (with the notable exception of Borneo) have areas of fertile volcanic soil but never developed the extremely high population densities found on Java. Unfriendly coastlines, coral reefs, cliffs, and extensive wetlands have hampered the economic development of some islands.

Java's extreme population density also has cultural and historical roots in the concentration of Dutch colonial activities there. The Dutch East India Company, after lengthy hostilities with Portuguese and English rivals and with indigenous states, secured control of most of Java in the 18th century. Large sections of the other islands, however, were not brought under colonial control until the 19th and early 20th centuries. The Netherlands exploited the natural

JULIO ETCHART/Peter Arnold

• **Figure 7.2.19** Volcanic landscape in the region of Mt. Bromo, eastern Java, Indonesia.

• **Figure 7.2.20** Principal features of Indonesia and Timor-Leste.

wealth of Java with the introduction of the culture system in 1830 (see Chapter 3, page 57). Under this system, Dutch colonizers forced Javanese farmers to contribute land and labor for the production of export crops. From the Dutch point of view, this harsh system was successful. The culture system was abolished in 1870, but commercial agriculture continued to expand, and its success supported increasing numbers of people. The introduction of the culture system thus coincided with an enormous increase in Java's population, which is more than 20 times larger than it was a century and a half ago.

Java's historical leadership and the island of Sumatra's current supremacy in export production are seen in the distribution of Indonesia's larger cities. Of the 15 cities having estimated metropolitan populations of more than 1 million, 9 are on Java. These include Jakarta, the country's capital, its main seaport, and the largest metropolis in Southeast Asia (population 18.2 million), and Indonesia's second-largest city, Bandung. Of the remaining Indonesian cities with more than a million people, four are on Sumatra.

The Indonesian state has had a turbulent career. Increasing nationalism during the Japanese occupation of World War II led to a bitter struggle for independence from the Netherlands following the war. Indonesia finally won independence in 1949. Another confrontation with the Netherlands in 1962 took western New Guinea (West Irian) out of Dutch control. A United Nations plebiscite in 1965 ratified Indonesian control, and relatively friendly relations were established between Indonesia and the Netherlands.

In 1965, an attempted Communist coup against the long-standing regime of Indonesian President Sukarno was unsuccessful. The ensuing retaliation resulted in the massacre of about 300,000 Communists and their supporters by the Indonesian army and by Islamic and nationalist groups.

From 1968 to 1998, Indonesia's president was an army general named Suharto, whose regime was regularly criticized for human rights abuses. Suharto ruled Indonesia undemocratically and often repressively but argued that he was a benevolent figure acting on behalf of his people. Indonesians generally tolerated his authority because the country's economy was advancing rapidly. The economic boom of the 1980s and early 1990s centered on the oil industry (in the hands of the state firm Pertamina) and on food production. Oilfields located mainly on Sumatra and Kalimantan now produce about 1.3 percent of the world's oil output. Crude oil and liquefied natural gas account for about 25 percent of Indonesia's exports and have tied its economy very closely to that of Japan, which is the main market for these products. Indonesia is a member of OPEC, the Organization of Petroleum Exporting Countries. In ad- 228
dition to oil and gas, important contributions to the export trade include clothing, wood products, various tropical agricultural products, and increasingly, electronics.

Tourism has also brought welcome foreign exchange earnings to Indonesia. The country's principal attraction by far is the legendary island of Bali, home of Indonesia's only remnant of the India-born Hindu culture that permeated the islands from the 4th through 16th centuries. Bali

is world-famous for its unique culture and forms of art and dance, which have endured and become even richer despite a massive influx of international tourism to this legendary island (•Figure 7.2.21). Bali's popularity as a destination for Western tourists put it in the crosshairs of the al-Qa'ida-affiliated Jemaah Islamiah, which in 2002 car-
305 ried out twin bombings of the nightclub district of Kuta. More than 200 people were killed, most of them young Australians; the Bali bombings have been called **"Australia's September 11."**

Indonesia's economy collapsed under the force of the spreading Asian economic crisis of the late 1990s. The rupiah, Indonesia's currency, lost 80 percent of its value in less than a year. The Suharto government removed expensive subsidies on gasoline and kerosene. Resulting protests quickly grew into vast riots. The economic troubles were
532 compounded by a prolonged drought attributed to El Niño's warming of the Pacific's waters, which caused significant declines in the country's rice harvest. Skyrocketing prices of rice and widespread shortages of the grain helped fuel widening popular discontent, which focused on two targets. First were the ethnic Chinese, who represent 3 percent of Indonesia's population but control half the economy, and second was President Suharto. Rioters looted and destroyed hundreds of Chinese-owned shops in Indonesia's major cities, forcing thousands of ethnic Chinese to flee the country. Since the Chinese had dominated the food production business in Indonesia, food shortages intensified in the wake of the disturbances. Protesters and rioters called ever more loudly for the resignation of President Suharto, on whom they also blamed their economic troubles. Bowing to the relentless pressure and averting a potential bloodbath, Suharto resigned from office in May 1998. He was succeeded by his vice president, who in turn stepped down in 1999 following widespread protests.

Indonesia's first truly democratic elections in 40 years subsequently brought to power President Abdurrahman Wahid, a widely respected Islamic scholar. His honeymoon with the public was short-lived, however, and by 2000, an official investigation revealed that he had embezzled millions of dollars of government funds. Wahid denied the allegations and insisted he would finish his term of office in 2004. He did not. Against a backdrop of massive protests across Indonesia, the president was impeached by Indonesia's supreme court, and he fled the country in 2001. He was replaced in office by Megawati Sukarnoputri, daughter of former President Sukarno, who was in turn ousted by the election of the secularist general Susilo Yudhoyono in 2004.

All of these regime changes reveal that Indonesia faces some huge problems. There is a constant threat that this vast country might splinter along ethnic lines (see Problem Landscape, pages 360–361). Many observers believe that if Indonesia's secular leadership does not deliver economic improvement to the tens of millions of disadvantaged citizens, the country will turn to Islamic rule. Rapid population growth is a serious challenge to the prospects for development. Indonesia's population grew by 90 million, or more than 63 percent, in the three decades between 1977 and 2007.

Lacking an aggressive family planning program, the government instead compelled some of its "surplus" population to move from Java, Bali, and other densely populated islands to some of the sparsely inhabited outer islands of the vast Indonesian archipelago. Begun under Dutch colonial rule, the **Transmigration Program** was one of the word's largest, most costly, and most controversial resettlement efforts. The Indonesian government picked up the tab for moving more than 6 million people, most of them landless agricultural workers and their families, to new islands

George Hunter

• **Figure 7.2.21** The cultural arts of Bali's Hindu people are world-renowned, making this Indonesian island a popular tourist destination.

PROBLEM LANDSCAPE

The Balkanization of Indonesia?

The national credo of Indonesia, seen on banners and placards throughout the country, is "One country. One people. One language." The government's constitution, in an effort to promote national unity, officially recognizes four faiths: Islam, Christianity, Hinduism, and Buddhism. But slogans and laws belie a grim reality: Indonesia may be in a process of balkanization, fragmenting into ethnically based, contentious units as happened in 148 the Balkans of southeastern Europe. The presence of some 300 different ethnic groups, combined with religious frictions, physical fragmentation, and economic problems, has made it difficult to attain peace, order, and unity in Indonesia. There is an official national language (Malay, known locally as Bahasa Indonesian), but more than 200 languages and dialects are in use. The largest ethnic group is the Javanese, who make up about 41 percent of Indonesia's population. About 86 percent of all Indonesians are Muslim, about 9 percent are Christian, and 2 percent are Hindu, living mainly in Bali.

Various groups in the outer islands have resented the dominance of the Javanese, who have traditionally asserted their power in colonialist fashion over the larger, more resource-wealthy outer islands. Such animosities have escalated at times to armed insurrections. The Suharto regime countered these militant expressions with a state ideology called ***Pancasila.*** Aimed mainly at suppressing militant Muslim aspirations, *Pancasila* was a pan-Indonesian nationalist ideology designed to neutralize all ethnic identities.

For a time, Indonesia's government vowed to break with the repressive ways of Suharto and appeased would-be separatists with promises of **liberal autonomy,** meaning that the provinces would have greater control over local administration and take more profits from the sales of local natural resources. With growing economic problems, pressure from nationalists, and accusations of corruption leveled against him, however, the government shifted to using an iron fist against every Indonesian province that might aspire to follow the path the former enclave of East Timor took.

Now the independent country of Timor-Leste (population 800,000; capital, Dili), East Timor was a Portuguese possession that Indonesia occupied after the collapse of the Portuguese colonial empire in the 1970s. The government carried on a long and bitter struggle against an independence movement led by East Timor's Catholics. In 1998, Indonesia and Portugal reached an autonomy agreement that would give the Timorese the right to local self-government.

Indonesia's incoming Wahid administration was bolder, however, and in 1999 allowed the people of East Timor to vote on whether or not they wanted outright independence. The result was overwhelmingly in favor of independence.

Despite official Indonesian blessing of the result, progovernment militias reacted to the vote with a violent rampage that left 1,000 Timorese dead and tens of thousands homeless. The United Nations dispatched a peacekeeping force to oversee the complete transition to East Timor's full independence in 2002. The country's first president was José Alexandre Gusmao, a Timorese rebel leader whom Indonesia had imprisoned in 1992 and released in 1999.

Since its birth, this tiny young country has had many woes: a breakdown of law and order, interethnic violence, and growing unemployment. The economic future of this poor nation could be brightened by resolution of an international legal wrangle over resources. Between Timor-Leste and Australia, in the portion of the Timor Sea known as the Timor Gap, lies a huge field of petroleum and natural gas. Because the distance between Timor-Leste and Australia is less than 400 miles (640 km)—making it impossible to apply the international standard of a 200-mile (320-km) offshore territorial limit—the two countries must come to an agreement on how to divide these undersea riches. Any deal is likely to yield sig- 407 nificant wealth to Timor-Leste.

Of outstanding concern for Indonesia's unity is the province of Aceh (population 4.3 million), at the northernmost tip of Sumatra. In 1976, the inhabitants of Aceh, who are a predominantly Muslim people of Malayan ethnicity, began seeking independence from Indonesia. The central government had made them a promise of autonomy with Indonesia's 1949 independence from the Netherlands but failed to keep it. The Acehnese insisted that the Jakarta government is too secular for their Muslim tastes and that too little of the revenue from sales of Aceh's abundant natural gas reserves remained in the province (which also has a wealth of oil, gold, rubber, and timber).

---

where they received houses, farmland, and subsistence payments. The results generally fell short of expectations, however. Java's population and poverty levels remained high, and the settlers' crop yields were low, reflecting lack of experience in the new lands, shortages of family labor, and difficulties in marketing crops. The new farming was also a major contributor to Indonesia's deforestation. The World Bank supported the effort but withdrew in the 1990s when human rights groups criticized the program's compulsory nature, violence flared between the resettled Javans and the indigenous peoples of the outer islands, and Jakarta beefed up its control over the outer islands, reducing the influence of indigenous peoples. Indonesia abandoned the Transmigration Program in 2000, and a much smaller—and this time, voluntary—resettlement program has taken its place.

7.2.8 The Philippines

The Philippine archipelago includes over 7,000 generally mountainous islands. The two largest islands, Luzon and Mindanao, make up two-thirds of the total area.

The Philippines were a Spanish colonial possession governed from Mexico from the late 16th century until 1898, when the United States achieved victory in the Spanish-American War and gained control of the Philippines. With

PROBLEM LANDSCAPE

The Balkanization of Indonesia? *continued*

The secessionists' main voice was the **Free Aceh Movement** (known by the acronym **GAM**), which wanted to install a member of its own indigenous royal family as president of a free Aceh. Seeing Aceh as vital to the nation's unity and economic viability and fearful that Aceh might inspire separatism elsewhere in the country, the government refused to discuss independence for the province. Violent clashes persisted, and the death toll grew—to 12,000 between 1976 and 2004. But then an unlikely arbiter appeared. The tsunami of December 2004 concentrated its unspeak- 341 able sorrow on Aceh. The international community stepped in with a massive relief effort, and Indonesian authorities listened with unprecedented interest to Acehnese concerns. GAM dropped its demands for independence, and Indonesia agreed to allow GAM to participate in local elections. Nominally secular Indonesia allowed Aceh to become the only one of its 33 provinces to adopt 227 Islamic *sharia* law. Redefining itself as the Aceh Transitional Committee, GAM laid down its weapons, and Indonesia scaled back its military presence in Aceh. The government gave amnesty to the rebels and, critically, pledged that 70 percent of the revenues from Aceh's natural resources would return to the provincial government.

Progress in Aceh raised hopes for a similar future for Papua, at the extreme eastern end of the Indonesian archipelago. Known formerly as Irian Jaya and West Irian, Papua is home to 2.7 million people of 200 different tribes and speaking 100 different languages. Most are of Melanesian origin, and most are at least nominally Christian. Collectively known as **Papuans,** they have little in common with the Javanese Muslims who control them from 2,500 miles (4,000 km) away in Jakarta. Their homeland on the western half of the island of New Guinea is of great importance to the economies of Indonesia and the United States because it contains the world's largest copper and gold mines. The American company Freeport-McMoRan, which provides more taxes to Indonesia's treasury than any other foreign source, operates these mines. There are also large assets of oil, natural gas, and timber. Indonesia has never recognized the province's 1961 unilateral declaration of independence. It is little wonder that Jakarta refuses to let its poorest but most resource-rich province go; the Dutch, coveting this wealth, originally refused to relinquish the area when it granted independence to Indonesia. Pressure from the United States led to a UN-sponsored plebiscite on independence in 1969. The vote in favor of the region's becoming part of Indonesia was widely regarded as rigged. Recently, Indonesia has promised some autonomy and a greater share of resource revenues to Papua, but Papuans complain that these promises are unfulfilled.

West of Papua lie North Maluku and Maluku, once known as the fabled Spice Islands of the Moluccas. Islamists there, who are bent on establishing *sharia* in as many locales as possible in Indonesia, have targeted the islands' majority Christian inhabitants. Over a period of three years, until a fragile peace was achieved in 2002, these Muslim fighters—often aided by regular Indonesian government troops—ethnically cleansed Christians from the North Maluku capital. Thousands of people became refugees. Humanitarian crises have become a familiar product of Indonesia's unrest. Scores of overcrowded refugee camps have sprung up in more than half of the country's 33 provinces.

Events in the Malukus reflect a broader wave of Islamist activism across Indonesia. Advocates of *sharia* have succeeded in closing bars and discos across Java and have terrorized people they consider hostile to Islam. They have scapegoated Christians, who are generally wealthier than Indonesia's Muslims, for many of the country's economic and political problems. Jakarta keeps a nervous eye on the central part of the island of Sulawesi, where tensions between Christians and Muslims are high and where the militant Jemaah Islamiah is known to be active. 306

Will Timor-Leste's independence be followed by independence for Aceh, heralding the balkanization of Indonesia? Some argue that Timor-Leste is unique; it was the only province of Indonesia colonized by the Portuguese rather than the Dutch and is the only one forcibly brought into becoming part of Indonesia. Others point out that Indonesia itself is an artificial union inherited by Dutch colonialism and is bound for devolution, the disper- 83 sal of political power and autonomy to smaller spatial units. The geopolitical concerns of larger powers may also influence the fate of some Indonesian provinces (see that discussion on page 305 in Chapter 7).

---

the exception of the Japanese occupation of 1942 to 1944, the Philippines were governed by the United States from 1898 until 1946, when the country became independent.

Spanish missionary activity in the Philippines made it the only Christian nation in Asia. The country's Spanish legacy is still important. Ever since its founding in 1571, the Spanish-oriented capital of Manila (population 18.5 million) has been the major metropolis of the islands (•Figure 7.2.22). The society created by Spain was composed of a small upper class of Hispanicized Filipino landowners and a great mass of landless peasants. Problems created by this uneven distribution of agricultural land have persisted in the Philippines.

American economic, military, and other ties with the Philippines continued after independence. Preferential treatment in the U.S. market stimulated the growth of major Philippine export industries: coconut products, sugarcane, and abaca (a fiber used in making rope and fabrics). Education in English grew so that even today, English is spoken by about 50 percent of the people and serves as a bridge between many of the diverse linguistic groups in the nation. After independence, Tagalog, one of the most widely used indigenous Austronesian languages, became the principal base for the official national language, Filipino, spoken by 40 percent of the populace. English also remains an official language. Despite the colonial history, very few people

Ron Giling/Peter Arnold

• **Figure 7.2.22** Downtown Manila.

speak Spanish. This owes largely to the fact that Spanish missionary work in the Philippines was carried on in the local languages rather than in Spanish.

The Philippine economy has expanded considerably since independence. Until the 1970s, the most striking growth was in agricultural exports such as coconut products, sugar, bananas, and pineapples and in copper and other metal exports. More recently, there has been growth in manufacturing and food production. Labor-intensive manufacturing, such as that of electronic devices and clothing, now accounts for about 80 percent of all exports. The English-language skills of Filipinos have made the
300 country an attractive outsourcing center for business processing and Web site building and editing. Japanese and American capital, together with American agricultural science and technology, has been very important in the Philippines' economic development. The United States and Japan buy more than half of all Philippine exports. Inexpensive, skilled labor is another of the country's principal exports; about 7 million Filipino workers—one of every seven—is an **overseas Filipino worker**, or **"OFW."** About one-fourth of the world's seafarers are from the Philippines. In modern Greek, the word for "maid" is *Filipenza*. Filipinos swell the ranks of workers in Taiwan, Japan, Hong Kong, the United Arab Emirates, and Saudi Arabia. From these and other countries, where they often complain of discrimination and abuse, they pump $15 billion of remittances back into their country's economy each year. This represents about 14 percent of the country's gross domestic product.

Despite encouraging progress in recent decades, the Philippine republic remains poor and potentially explosive politically. It has not joined the ranks of the Asian Tigers, although some economists consider it a Tiger Cub. The country has taken a conservative approach to its development and has borrowed money carefully to invest in modest projects, avoiding the aircraft industry, the building of a new national capital, and other such showy and expensive enterprises undertaken in other Southeast Asian nations in the 1990s. Since the mid-1970s, the Philippines has also adhered closely to the terms of loans and financial reforms imposed by the International Monetary Fund. The country did not boom, but neither did it bust in the economic crisis that rolled across Asia in 1997 and 1998. During the regional economic recovery that followed, however, the Philippines' growth continued to lag. An estimated 30 percent of the population lived below the poverty line, and unemployment stood at 9 percent in 2007.

Economic expansion in the Philippines has been hard pressed to stay ahead of the more than 70 percent growth in population since 1975. The predominant Roman Catholic culture (Catholics make up 81 percent of the population, with 10 percent adhering to other Christian denominations) encourages large families. The country's longtime leading cleric, Cardinal Sin, responded this way to a question about the population explosion: "The more the merrier." Church officials have labeled the government's current promotion of family planning as "demographic imperialism" masterminded by the United States. Unsubsidized condoms and other forms of birth control are too expensive for many to afford in a country where 17 percent of the population makes less than $1 per day. The collective result is that the Philippines has a rather high annual population growth rate of 2.1 percent. This is a marked contrast with Thailand, with its growth rate of 0.7 percent. Both countries had about the same population size in 1975, but the Philippines now has about 23 million more people than Thailand.

Compounding the problem of population growth, land and wealth are very unevenly distributed as Philippine society continues to be dominated by a few hundred wealthy families. The gap between haves and have-nots has inspired rebellion for decades. Discontented peasants supported a Communist-led revolt after World War II. It was eventually suppressed, in large part through granting land (generally on sparsely populated Mindanao) to surrendered rebels. Revolts flared up again in the late 1960s, however, and have continued ever since. A Communist group known as the **New People's Army** leads an active insurgency in many parts of the country, frequently targeting military patrols and civilian infrastructure.

On the island of Mindanao, rebels of the **Moro Islamic Liberation Front (MILF)** and a splinter organization, **Abu Sayyaf,** have long fought the government in an effort to establish a separate Islamic state. In 2001, the MILF and the government reached a cease-fire, but Abu Sayyaf accelerated its campaign, employing terrorist tactics such as kidnapping and beheading civilians. In its declared worldwide war on terrorism, the United States in 2002 began providing military advisers and weaponry to Philippine troops
fighting Abu Sayyaf. 234

This unrest has political and religious as well as economic roots. Despite a wealth of oil, natural gas, fisheries, and forests, Mindanao is the Philippines' poorest region, but the predominantly Christian central government invests little development aid in this mainly Muslim area. Hundreds of thousands of Mindanao's Muslims have fled as illegal immigrants to more prosperous Sabah, in the Malaysian portion of the island of Borneo.

The Philippines has seen some momentous political changes in recent years. In 1986, the country's dictator, Ferdinand Marcos, was ousted in a popular uprising and replaced by a democratically elected president, Corazón Aquino. Marcos and his wife, Imelda, had looted the country of billions of dollars over a period of 20 years and headed a corrupt and abusive regime virtually ignorant of the needs of the ordinary Filipino. The Marcos path seemed to be followed by Philippine President Joseph Estrada, who allegedly enriched himself in the 1990s through bribes, kickbacks, and embezzlement of public funds. Impeachment proceedings against Estrada crawled along, but he resisted their inevitable conclusion. Only massive street protests succeeded in forcing his resignation in 2001, when Gloria Macapagal-Arroyo replaced him as president.

The United States has maintained a concerned watch on the country's struggles. During the 1980s, many Filipino legislators resisted the prospect of renewing the leases under which the United States held the two military bases of Subic Bay and Clark Field on Luzon. Then, in 1991, the eruption of volcanic Mount Pinatubo put an end to the debate over the fate of these two American outposts; volcanic ash forced their closure. The Philippines has 22 active volcanoes. With volcanic eruptions, plus an average of 20 typhoon strikes yearly and regular monsoonal flooding, the Philippines earned the distinction of having more natural disasters in the 20th century than any other country. Mount Pinatubo also ejected enough ash high into the earth's atmosphere to cool temperatures globally over the following several years, at least temporarily confounding efforts to track the trend in global warming. Meanwhile, the U.S. base at Subic Bay was converted into a thriving free-trade port that the Filipinos hope will become Asia's "New Hong Kong."

The Philippines is Southeast Asia in a nutshell: There are humanized landscapes never free of nature's whims; struggles in transforming a mercantile colonial economy to one based on high-tech, global trade; seemingly intractable odds to overcome in unifying diverse peoples under one flag; and wars on poverty that have yet to be fought.

SUMMARY

- Southeast Asia is composed politically of 11 countries: Myanmar, Thailand, Laos, Cambodia, Vietnam, Malaysia, Singapore, Indonesia, Timor-Leste, Brunei, and the Philippines. With the exception of Thailand, all of these were formerly colonies of foreign powers.
- Southeast Asia contributed several domesticated plants and animals (including rice and chickens) and the achievements of civilization to a wider world.
- Relative to other parts of Asia, especially China and India, Southeast Asia has historically not had large populations or population densities. The most notable exception is on the Indonesian island of Java, where good soils and Dutch colonial polices produced a very dense population.
- This region poses numerous environmental challenges for its inhabitants, including tropical climates, poor soils, disease, long dry seasons, deforestation, heavy rains and erosion, and tectonic processes. A megathrust earthquake off the northeastern coast of Sumatra caused the great tsunami of 2004, which killed more than 200,000 people in 14 countries.
- There are more Southeast Asian villagers than urbanites. Farmers use shifting cultivation in the more marginal areas and sedentary cultivation in more productive regions.
- Plantation cash crops such as oil palm and rubber are grown for export and are the legacy of Western colonialism.
- Southeast Asia's tropical forests are being destroyed at a faster rate than others of the world, mostly for commercial logging for Japanese markets. In biodiversity terms, Indonesia is a megadiversity country whose forests are under intense pressure.
- Southeast Asia as a whole is among the world's poorer regions. The poorest countries—Myanmar, Cambodia, Laos, and Vietnam—have suffered from warfare and, in the case of Myanmar, from unwise political management. Peace has finally come to Cambodia, Laos, and Vietnam. Singapore, Malaysia, and Thailand, because of their high rates of industrial productivity and economic growth, have often been described as Asian Tigers or Tiger Cubs.
- The region's mineral wealth includes oil, gas, and a wide variety of metal-bearing ores. Indonesia has the most oil.
- Economic problems in Thailand in 1997 led to an economic meltdown in the region due to the interconnectedness of the regional and global economies. Most of the countries have since recovered from that setback, and several have experienced strong economic growth.
- Drug production, drug abuse, and HIV/AIDS are serious problems in Myanmar. Neighboring Thailand has some of these problems but has dealt more effectively with them.
- Prodemocracy efforts in Myanmar have been thwarted by official repression.
- Thailand was never colonized and is more prosperous than its neighbors. French-colonized Cambodia, Laos, and Vietnam are much poorer. U.S. forces succeeded the French withdrawal from Vietnam in the 1950s. American efforts to win a war against communism in Vietnam failed. Normal relations now exist between the United States and Vietnam, and although Communist, Vietnam encourages private enterprise and has seen recent strong economic growth.

→ Numerous dams are being constructed on the Mekong River. These will produce hydroelectricity but will damage the productive fishery of this basin.

→ Cambodia has recovered politically from the devastating days of the killing fields and has joined or rejoined significant international organizations, including the United Nations. Landlocked Laos is very poor but hopes to benefit from hydropower and mineral exports.

→ Malaysia and Singapore are former British colonies with rather strong economies based on high-tech industries. Singapore has a large and prosperous Chinese population and is developing a new niche in biotechnology.

→ The region's demographic giant is Indonesia, which is very complex culturally and has had a turbulent political past.

→ The Philippines is a mainly Catholic country and a former colony of Spain. Population growth there is relatively high, and a gap between rich and poor threatens overall development. Large numbers of Filipinos work abroad.

→ Much unrest in the region is related to tensions between ethnic groups. Following the regional economic crisis of 1997–1998, frustrations in Indonesia were directed against the prosperous Chinese minority. There are insurgents among the Muslim, ethnic Malay minority of southern Thailand. Indonesia is in danger of fragmentation. Catholic Timor-Leste has gained impendence from Indonesia, and the provinces of Aceh and West Irian are seeking self-rule or independence. Muslim insurgents in the southern Philippines and Communist rebels throughout the country have stepped up their campaigns against the government.

KEY TERMS + CONCEPTS

REVIEW QUESTIONS

1. What are the countries of Southeast Asia? What are their major physiographic features?
2. What caused the great tsunami of 2004, and what were some of its effects?
3. What generalizations can be made about population numbers and densities in this region? What factors account for these characteristics?
4. What agricultural practices are typical of Southeast Asia? What are the main subsistence and commercial crops?
5. What countries are the largest producers and consumers of Southeast Asian tropical hardwoods?
6. What is the region's most important mineral resource? Which countries have most of it?
7. What are the major goals of ASEAN and ASEAN+3?
8. What major changes are coming to the Mekong River basin?
9. What political events have hampered development in Vietnam, Cambodia, and Laos?
10. Who are the boat people? Why did they leave their homelands? Where are they now?
11. How did Malaysia transform itself from a tin and rubber exporter to a high-tech economy?
12. What are the major political and resource-related issues faced by Indonesia and the Philippines?

NOTE

1. See UNESCO, *Assessment of Capacity Building Requirements for an Effective and Durable Tsunami Warning and Mitigation System in the Indian Ocean: Consolidated Report for the Countries Affected by the 26 December 2004 Tsunami* (Paris: UNESCO, 2005), http://unesdoc.unesco.org/images/0014/001445/144508e.pdf. Accessed July 15, 2007.

Schoolchildren in Guilin, China.

MODULE 7.3
CHINA: THE GIANT

Historically, China identified itself as the **Middle Kingdom,** positioned centrally in the known world. Chinese culture is very rich, very old, and very influential. China has been home of the dominant culture of East Asia since ancient times. Nearby countries and regions carry the profound influences of Chinese culture in their languages and writing, religion, agriculture, arts, crafts, institutions, ways of thinking, and histories. The culture hearth of China has contributed much to humanity at large. Pioneering Chinese inventions and discoveries included gunpowder, paper, the wheelbarrow, and the magnetic compass. Over thousands of years, Chinese civilizations have also accomplished great engineering feats, from the Great Wall (which is, incidentally, *not,* as one urban myth would have it, the only human-made object visible from space) to the Three Gorges Dam.

Chap. 3

module outline

module objectives

This module should enable you to

- Recognize China as a land empire in which a single ethnic group added vast peripheral areas to its eastern core
- Appreciate the constraints imposed by China's Communist system on personal, political, and economic freedoms and the recent substantial easing of those restrictions
- Balance the pros and cons of China's hugely ambitious plans to transform its natural environment through dams, canals, forest restoration, and other projects
- Consider the economic discrepancies between China's rural and urban populations and the forces behind large-scale rural-to-urban migration
- Understand the regional and geopolitical risks of Taiwan's aspirations for an identity independent of China

---

From the earliest dynasties in the second millennium B.C.E. to the present, the story of China has been one of manipulation of the landscape to make it more productive. There are ever-present challenges from nature, including floods and droughts. Today, while trying to feed its enormous population and raise fiber for its huge textile sector, China is pulling away from its past as a monumental agricultural nation and is aggressively pursuing industrial development to reach its goal of becoming a "well-off society" by 2040. With China's export-driven economy booming, its 2.5 million–soldier army backed by nuclear weapons and by military spending second only to the United States, and its political influence represented with a permanent place on the UN Security Council, even the world's sole superpower, the United States, must weigh most of its geopolitical decisions mindful of this world giant. Above all, China—officially the People's Republic of China (PRC)—must be recognized as the world's most populous country, home to 1.3 billion people, or one in every five on earth. China is arguably the most important country on the planet today.

After a brief sketch of China's history, this module examines, by subregion, the physical resources for its people and some of their culture and livelihood traits. It discusses Chinese agriculture, industries, and cities. The module ends with depictions of Taiwan and Mongolia, countries whose histories and modern interests are closely linked to China.

7.3.1 Accomplishment, Subjugation, and Revolution

China (•Figure 7.3.1) was already a major political entity by the time of the Shang Dynasty (c. 1523–1027 B.C.E.). The Chinese had by then developed an agricultural system able to support large numbers of people on small areas of intensely worked land. During the Shang Dynasty, Chinese peasants occupied the great alluvial plain of the Huang He (Yellow River) in North China. There they had a garden type of agriculture based on the hoe, irrigated crops, fertilizers, intensive use of flat alluvial lands, and the terracing of fertile hillsides. From its origins in the Huang He Basin, this system diffused by migration and conquest throughout the eastern part of the country and farther afield over a period of many centuries. The Huang He region is therefore recognized as core of the Chinese culture hearth.

In the 3rd century B.C.E., the Qin (Ch'in)* Dynasty accomplished the first unification of the Chinese people into one empire (the country's Western appellation, China, is thought to be derived from the name of this dynasty, as is the combining form *Sino-*; the Chinese call their country Zhongguo). Thus began China's development as a land empire that colonized its own vast hinterland. The 164
conquerors were, as the dominant culture in China today is, ethnic Han Chinese. Under the Han Dynasty (206 B.C.E.– 293
220 C.E.), China's influence spread as its leaders extended their authority through direct conquest and by accepting the fealty and tribute of less significant powers around them. Chinese armies pushed far westward across central Asia and nearly made contact with the Romans in the vicinity of the Caspian Sea. Chinese ideas and materials traveled west along caravan routes to the Middle East and Europe.

During times of internal disorder and weakness, China was conquered by nomadic invaders, notably the Mongols from the northwest and Manchus from the northeast. But China always rebounded. Under the Yuan (Mongol) Dynasty (1280–1367), the Chinese empire controlled Burma (modern Myanmar), Indochina, Korea, Manchuria, and Tibet. The leaders of China's last dynasty, the Manchu (Qing) of 1644–1911, were of the nomadic Jurchen tribes of northeastern China. They held large areas that eventually became part of the Russian empire.

*From the late-1800s until the 1970s, Western spellings and pronunciations of Chinese words were governed by a system known as the Wade-Giles romanization system. In the 1970s the Communist Chinese government deemed this system somewhat arbitrary and embraced the more accurate Pinyin system for representing Chinese names; since then Pinyin has been standard, but many reference materials, such as older books and maps, use the Wade-Giles spellings, which is why both are given in this text: Pinyin first, with Wade-Giles in parentheses.

China under the Manchus was increasingly subjected to unwelcome pressures from the West. European ships became frequent visitors to China in the 16th century, and Europeans began to press for more trade. The Chinese, who had high-quality cottons, silks, teas, and other products, were not much interested in what the West offered, and they felt culturally superior to these intrusive "barbarians." The Chinese government strictly controlled trade, limiting it to the single port of Guangzhou (Canton) in southern China.

As their science, technology, and industrialization advanced at home, Westerners became more powerful and
42 assertive in Asia. A British expedition in the **Opium War** of 1839–1842 defeated the faltering Manchu government, forcing China to open its doors to opium (from British-controlled India) and to other British goods. Fully aware of the drug's destructive effects, Britain had carefully encouraged and fed the appetite for highly addictive opium in China, creating an enormous and very profitable market for itself.

Other European countries, Russia, and the United States joined the search for riches. China became a kind of joint colony of many competitive foreign exploiters, who got their way by occasional military actions and frequent threats. China was thus opened to Western products and then to Western-owned factories in its ports. Under the impact of Western machine-produced goods, China's handicrafts industries began to disintegrate. Foreigners in China were exempted from the control of Chinese law in an arrangement known as **extraterritoriality**, and the Western powers gradually gained control of China's taxation system.

The foreigners also extorted territorial concessions but imposed explicit colonial status on only a small part of China (see the map of colonial Asia on page 297). Britain secured the vital coastal port of Hong Kong. Portugal had earlier obtained nearby Macau. France and Germany won naval bases. Japan took Taiwan and the Ryukyu Islands, and Russia conquered parts of central Asia, Mongolia, and Siberia. **Treaty ports** were established, including Shanghai and Tianjin (Tientsin). In these ports, the foreign powers were granted special settlement areas under their own control. Chinese people could not normally reside in these areas and in some cases could not even enter them. European and American naval vessels patrolled China's coasts and great rivers to enforce foreign privileges.

The Chinese did not accept this situation passively. Disorder mounted within the country, but Western military power kept the submissive Manchu sovereigns on the throne until they were finally overthrown by the **Chinese Revolution of 1911**, led by Sun Yat-sen (sometimes called the "George Washington of China"). However, the new Republic of China had only nominal control over much of the country. Local warlords controlled many areas until in the late 1920s, the army of the **Nationalist Party** (**Kuomintang**, or **KMT**) under Chiang Kai-shek fought northward from Guangzhou and managed to impose control on most of the country.

The Nationalist regime, which established itself at Nanjing (Nanking), abolished some foreign privileges, but its strength was sapped by almost continuing civil war against the **Chinese Communist Party (CCP)** led by Mao Zedong (Mao Tse-tung) and then by war against Japan. Chiang finally drove the Communists from their bases in the hills of southeastern China in 1934. Thus began the two-year **Long March**—a more than 6,000-mile (9,660-km) flight on foot of nearly 100,000 Communists, pursued by Nationalist soldiers, from south central China north to Yenan, in Shaanxi province, which became their wartime home base.

The Sino-Japanese War of 1937–1945 compelled Mao's Communists and the Nationalists to fight in common cause against Japan. But with the defeat of Japan in World War II, the CCP and the Nationalists resumed their civil war, in which Mao's forces triumphed. In 1949, 2 million Chinese Nationalists under Chiang Kai-shek fled to Taiwan, and Mao and the CCP claimed victory on October 1, 1949. Subsequent historical events in China are discussed throughout this module, and Taiwan has its own section beginning on page 386.

7.3.2 The Setting

China's area of about 3.7 million square miles (9.4 million sq km) is only slightly larger than that of the lower 48 United States (see Figure 7.2, page 282), but its inhabitants outnumber the U.S. population by more than 4 to 1. Two main environmental regions make up this populous country: an eastern core region, known as **Humid China** or China Proper, in which the country's population, developed resources, and productive capacity are heavily concentrated, and a western periphery, or **Arid China** (see Figure 9
7.3.1). A rough boundary between the two extends from China's southern border with Myanmar to its northeastern border with Russia. This line corresponds generally to the stretch of land with average annual rainfall of about 20 inches (c. 50 cm), with Humid China to the east and Arid China to the west (see Figure 2.3, page 23, and Figure 7.5, page 287). Land uses of the regions are depicted in Figure 7.8 on page 288.

Arid China

Much of China's unproductive and thinly settled land is in the western half of the country. Here are high mountains and plateaus, together with arid and semiarid plains, where rainfall is generally insufficient for agriculture. This arid region, sometimes called Frontier China, contrasts markedly with the better-watered, more densely settled eastern half of the country and is home to only about 6 percent of China's population (see Figure 7.3a, page 282).

The principal regions of Arid China are the Tibetan Highlands, Xinjiang (Sinkiang), and Inner Mongolia (Nei Mongol). All three are identified historically with ethnic groups that are not Han Chinese (see Figure 7.11 and the ethnic geography discussion on pages 291–293). Now, after decades of politically motivated domestic migration from the east, Han Chinese comprise the great majority in Inner Mongolia and a slight majority in Xinjiang.

The thinly inhabited Tibetan Highlands, including the province of Tibet (Xizang) itself and fringes of adjoining provinces, occupy about one-fourth of China's area. Most of this vast area is a very high, barren, and mountainous plateau averaging nearly 3 miles (c. 5 km) in elevation. To the northwest, there are many high basins with internal drainages, some containing large salt lakes. To the southeast, the plateau is cut into ridge and canyon country by the upper courses of great rivers such as the Tsangpo (which becomes the Brahmaputra of India), the Mekong, and China's own Chang Jiang (Yangtze Kiang, or Yangtze River) and Huang He (Hwang Ho, or Yellow River).

Lower elevations, warmer temperatures, and greater precipitation in parts of the southeastern plateau support some extensive grasslands and stands of conifers. Around the edges of the Tibetan Highlands are huge mountain ranges: the Himalayas on the south, the Karakoram and Kunlun Shan ranges to the north and northwest, and the Qinling Shan and others to the east (•Figure 7.3.2).

Most of the 2.9 million people of the Tibetan Highlands are sedentary farmers and animal herders. A few hardy crops, especially barley and root crops, do well in this challenging environment. In grasslands at higher elevations, some pastoral nomads graze their flocks of yaks, sheep, and goats. Yaks are particularly important not only as durable beasts of burden in these high elevations but also as providers of meat, milk and butter, leather, and hair traditionally woven into cloth.

Xinjiang (meaning "New Dominion") adjoins the Tibetan Highlands on the north. It has an area of roughly 635,000 square miles (c. 1.6 million sq km) and a population of about 20 million. It consists of two great basins: the Tarim Basin to the south and the Dzungarian Basin to the north, separated by the lofty Tien Shan range. The Tarim Basin is rimmed to the south by the mountains bordering Tibet, and the Dzungarian Basin is enclosed on the north by the Altai Shan and other ranges along the southern borders of Russia and Mongolia. Both basins are arid and semiarid. The Tarim Basin is particularly dry because it is almost completely enclosed by high mountains that block rain-bearing winds. The smaller adjoining Turpan Depression drops to 928 feet (283 m) below sea level and is the second-lowest land spot on earth, after the Dead Sea. These basins include the greater part of the desert of Taklamakan, whose name means "Place of No Return," often rendered as "Once you get in, you'll never get out."

From the Taklamakan and Mongolia's Gobi Desert, and from adjacent areas despoiled by ill-conceived agricultural reclamation projects of the past, vast quantities of sand are airborne in spring sandstorms that darken skies and choke lungs downwind in Humid China and even Korea and Japan. In an effort to stem the desert's spread, the Chinese have since 1982 planted more than 40 billion jujube and

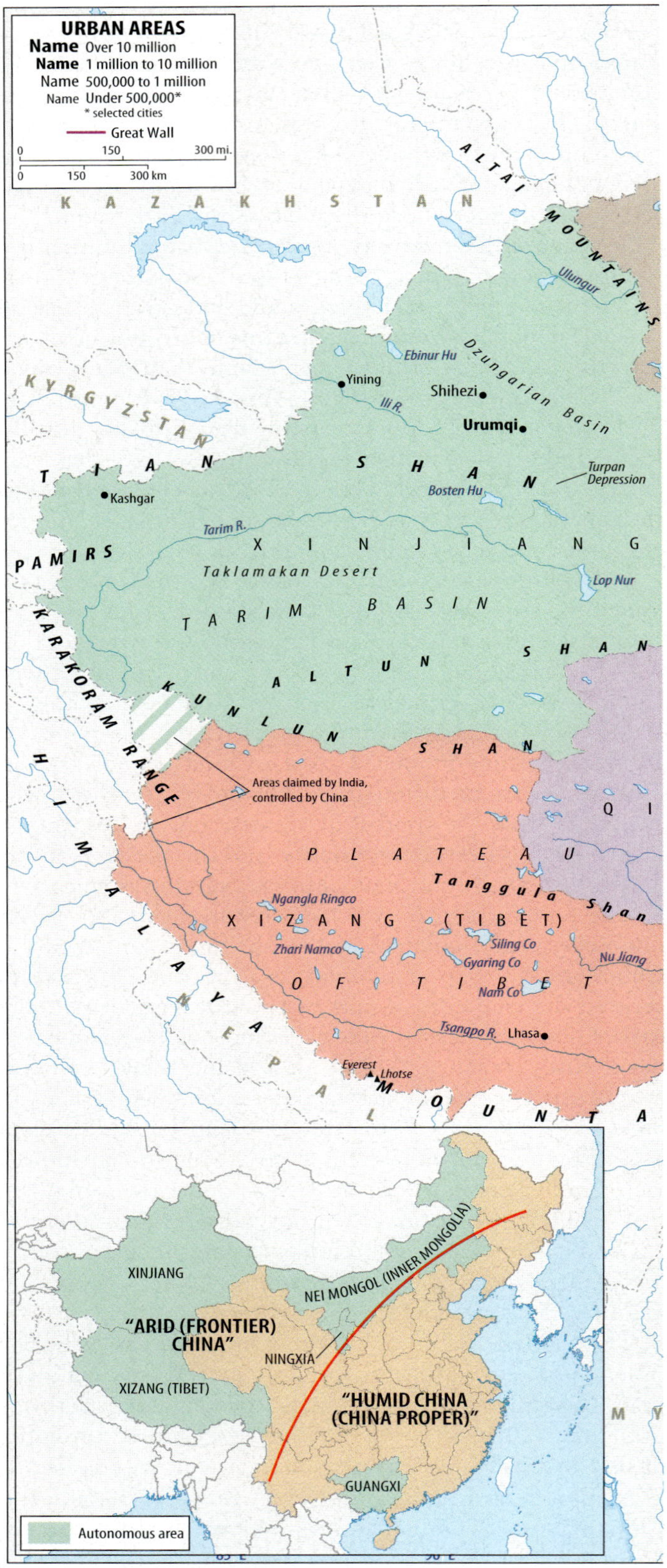

• **Figure 7.3.1** Principal features of China and Mongolia. On the inset map, the line represents the division between Humid China, or China Proper, to the east and Arid China, or Frontier China, to the west. Humid China has the great majority of the Chinese population and economic activity, and Arid China is home to many minority peoples.

Schmoekel, H./Arco Images/Peter Arnold

• **Figure 7.3.2** At 24,757 feet (7546 m), Muztagh Ata is one of the highest mountains of the Kunlun Shan, at the northern edge of the Tibetan Plateau.

Joe Hobbs

• **Figure 7.3.3** The Great Wall of China at Badaling, near Beijing. Most of the wall standing today dates to the Ming Dynasty of the 15th century. The wall proved ineffective in keeping out the Ming enemies to the north, the Qing.

other trees as "green walls" in Arid China. The government's goal for the decade 2010–2020 is to plant enough trees in this region to cover 170,000 square miles (440,000 sq km), an area about the size of California or Sweden. There are also plans use licorice plants to hold back the desert. Long valued in traditional Chinese medicines and therefore of potential economic value, licorice is drought-tolerant and helps bind sandy soils.

The great majority of Xinjiang's population is concentrated near oases located mainly around the edges of the basins where streams from the mountains enter the basin floors. For many centuries, these oases were stations on caravan routes crossing central Asia from Humid China toward the Middle East and Europe on the historic Silk Road (see Regional Perspective, page 371). The ancient routes have been superseded by modern transportation. A railroad connects Lanzhou (Lanchow; population 1.8 million), a major industrial center and supply base in northwestern China Proper, with Urumqi (Urumchi; population 1.5 million), the capital of Xinjiang. Major roads and airline routes link the oasis cities (see Figure 7.3.14).

These transport links are key elements in the latest drive to grow the economic importance of this remote part of China and bring it under firmer political control. Early in the Communist era of the mid-20th century, demobilized soldiers and urban youth from eastern China were organized to expand irrigated land on state-operated farms in this arid landscape, promoting cotton as a major crop.

Xinjiang is China's most mineral-rich region, with deposits of coal, petroleum, iron ore, and other minerals. Mining and manufacturing have expanded there. It continues to be a major focus in Chinese efforts to relocate significant numbers of ethnic Han Chinese away from the coasts (see Ethnic Geography, pages 372–373).

Northwest of the Great Wall, rolling uplands, barren mountains, and parched basins stretch into the arid interior of Asia. Here lie slightly more than 1 million square miles (2.6 million sq km) of dry, thinly grassed terrain divided almost equally between Nei Mongol (Inner Mongolia), the area nearest the Great Wall, and the country of Mongolia—known as Outer Mongolia in older literature. Most of the 24 million people of Nei Mongol, which is an autonomous region of China, are ethnic Han Chinese concentrated in irrigated areas along and near the great bend of the Huang He. Much of the region is still home to a sparse population of ethnic Mongol herders. From these austere, arid landscapes, the Mongols emerged to control China and, by the 13th century, a huge swath from the Korean Peninsula in the east to the margins of Poland in the west. The desert landscape bears many traces of its rich past: caravan routes, temples hidden in desert caves along those trails, and remnants of the Great Wall (•Figure 7.3.3). The earliest version of the Great Wall was built in the 3rd century B.C.E., but the Ming Dynasty constructed it in earnest in the 15th century C.E. in an unsuccessful bid to stave off incursions from the north.

REGIONAL PERSPECTIVE

The Silk Road

In the early Han Dynasty (c. 206 B.C.E.–8 C.E.), the Chinese began a steady exchange of goods for items from the Mediterranean trading ports. For centuries to follow, China exported silk, china, paper, and spices, and the West exported wool, gold, silver, glass, and metalware. Not produced in the West until the 6th century, silk is made from the cocoon of a caterpillar known as a silkworm and is still prized for its breathability, texture, and beauty (• Figure 7.3.A). China is still the world's leading silk producer.

The overland route for this exchange of goods—called the Silk Road or Silk Route—was approximately 4,000 miles (c. 6,400 km) long, connecting the early Chinese capital of Chang'an (now Xi'an) to ports of the eastern Mediterranean (• Figure 7.3.B). The road went alongside the Great Wall and the Taklamakan Desert, over the Pamir Mountains, and through Baghdad to the Mediterranean Sea. From there, goods were shipped to various European ports. This route also played a role in the diffusion of Buddhism into China in the 1st century C.E. and subsequently in the spread of languages, customs, and technologies. Until the Venetian explorer and trader Marco Polo and his entourage did it in the 1270s, apparently no one had ever followed the whole route. Travel was dangerous, and every caravan load had to pass numerous checkpoints and pay heavy tolls.

Joe Hobbs

• **Figure 7.3.A** Silk production is a laborious process, in which some steps must be done by human hands. Here, a Chinese worker removes each pupa (far left) from the silk cocoon (center), then stretches the silk over a frame to prepare it for drying and processing.

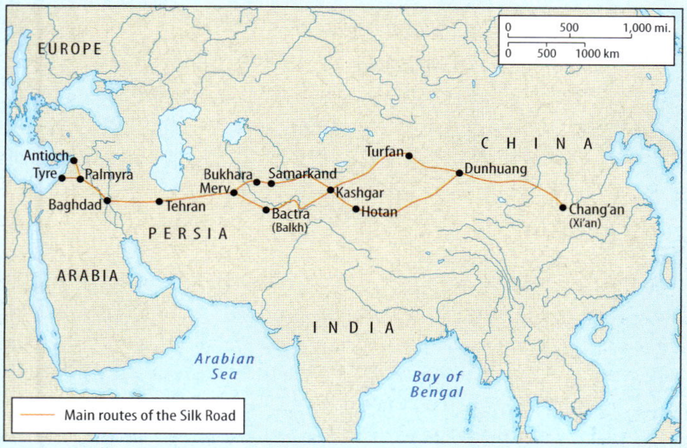

• **Figure 7.3.B** The Silk Road.

The traditional picture of nomadic Mongol tribespeople with their sheep, goats, camels, and horses is fast disappearing. The older economic core of Nei Mongol is made up of agricultural areas near the Huang He. The irrigated areas produce mainly oats and spring wheat, while unirrigated fields are used primarily for drought-resistant millet and kaoliang, a sorghum grain used for animal feed and for liquor. The main city is Baotou (Paotow; population 1.5 million), an expanding industrial center on the Huang. One of China's major iron and steel works was opened at Baotou with Soviet help in the 1950s to take advantage of nearby reserves of coal and iron ore. Nei Mongol's economy today is growing faster than that of any other Chinese province because of a boom in the production of coal, natural gas, copper, gold, and other natural resources.

Humid China

Humid China, sometimes referred to as Eastern China, Monsoon China, or China Proper—the core region—includes the most densely settled parts of the country (see Figure 7.3a, page 282). Humid China is commonly recognized in geographic literature as having two major subregions, North China and South China, which differ from each other in several physical, economic, and cultural respects. The Northeast, or Manchuria, is another subregion that straddles the line separating Humid China from Arid China, but it is important economically to the county and is therefore treated within the discussion of Humid China as the country's core region.

The Northeast

China's extreme Northeast region, better known to the West as Manchuria, lies north and east of the Great Wall. It is made up of three provinces: Liaoning, Jilin, and Heilongjiang. It has an area of about 310,000 square miles (c. 800,000 sq km) and a population of approximately 107 million. To the west are the Mongolian steppes and deserts; to the east, the area is separated from the Sea of Japan (East Sea) by the Korean Peninsula and by Russian terri-

ETHNIC GEOGRAPHY

Han Colonization of China's "Wild West"

China's growth as a land empire has involved both subjugation of people who are not ethnic Han and the colonization of those ethnic areas by ethnic Han. There are at least 56 non-Han ethnic groups in China, in total about 100 million people living mainly in the west and southwest of the country. As the Soviets did in the Soviet Union, China's Communist government granted at least token recognition of the distinctive identity and rights of five large
165, 171 minorities by creating **autonomous regions.** These are Guangxi, bordering Vietnam; Nei Mongol (Inner Mongolia) and adjacent, small Ningxia in the north; Xizang (Tibet); and Xinjiang in the far west (see Figure 7.3.1).

The Hui of Ningxia are Muslims of many different ethnicities who absorbed Chinese language and many Han customs. In contrast, the Uighurs of Xinjiang and the Tibetans have defiantly resisted becoming Chinese—and so Han colonization, or **Hanification,** focuses especially on their territories. The Chinese government aims either to assimilate these minority peoples into a broader, Han-based Chinese culture or to establish a strong enough Han demographic presence among them to dispel any hopes of autonomy or independence.

Chinese authorities have also initiated what they call the **Western Big Development Project** with the stated objective of improving locals' livelihoods enough to diminish their desire for ethnic and political separatism. Focused on the autonomous regions and four other provinces in which there are large minority populations, this project seeks to improve infrastructure like airports and roads and to boost output from factories and farms. The Uighurs, Tibetans, and other minorities generally fear that this "development" is just a cover for demographic, political, and economic domination by the Han.

Buddhism reached Tibet in the middle of the 8th century, and its arrival was followed by decades of conflict between Chinese and Indian Buddhists. The Chinese were defeated in this struggle and expelled from Tibet at the end of the 8th century. China sometimes reexerted loose control over Tibet, but Chinese authority vanished again with the overthrow of the Qing (Manchu) Dynasty in 1911. From 1912 until the Chinese Communists' violent conquest of 1951, Tibet existed as an independent state, with its capital and main religious center at Lhasa (•Figure 7.3.C).

After the Chinese completed roads to Tibet in late 1954, an increase in restrictive measures by the Communist government contributed to the rise of Tibetan guerrilla warfare. This culminated in a large-scale Tibetan revolt in 1959 and the flight of the **Dalai Lama** (the spiritual and political leader of Lamaism, or Tibetan Buddhism) and many other refugees to India. Subsequently, the Chinese drove most of the monks from their monasteries, expropriated the large monastic landholdings, implemented socialist programs, and prohibited organized religion in the rest of China. Large-scale immigration by ethnic Han picked up; about half of Lhasa's population is now Chinese.

Beijing insists it is helping Tibet with major development investment, including the $4 billion Qinghai-Tibet railway recently constructed between Beijing and Lhasa. But Tibetan resistance to Chinese rule continues. From his place of exile in Dharamsala, India, and on frequent international road trips, the Dalai Lama continues to appeal for Tibetan freedom. He asks China to grant what he calls **genuine autonomy** to Tibet, meaning that Tibetans would have authority over most matters except foreign affairs and defense, which the Chinese government would control. His cause has many prominent backers, including the American actor Richard Gere. However, China's government has so far refused to negotiate with the Dalai Lama. For many years, India had promoted the cause of Tibetan autonomy, but as its trade ties with China have grown more vital, its advocacy for Tibet has quieted.

China's government regards Xinjiang's Muslims, mainly from the 8-million-strong Uighur ethnic group, who speak a Turkic language and are kin to the Kazakhs to the west, as the most prob- 292
lematic minority groups. Many Uighurs would like to pull at least a part of Xinjiang out of China's orbit to create an independent Chinese Muslim state. Two short-lived independent Uighur republics,

tory. The Northeast consists of a broad, rolling central plain surrounded by a frame of mountains rising to about 6,000 feet (c. 1,800 m). The mountains on the east and north have much valuable hardwood and softwood timber. The central plain, oriented northeast to southwest, is about 600 miles (966 km) long by 200 to 400 miles (322 to 644 km) wide. The soils are very fertile, and there is usually enough rainfall for summer crops. Conditions are not favorable for irrigated rice, and the winters are too severe for winter grains. But soybeans, kaoliang millet, corn, and spring wheat do well during the short frost-free season of 150 to 180 days.

This was once a sparsely populated frontier region, but political instability in the early 20th century prompted the ruling Manchu Dynasty to encourage migration to the relatively open lands of the Northeast. Millions of Chinese farmers moved to these provinces, pioneered its agricultural frontier, and gave it an ethnic Han majority by the start of the Sino-Japanese War in 1937. As described on page 383, Japanese intervention initiated Manchuria's development as an important industrial region.

North China

Affected mightily by the monsoon wind patterns that characterize so much of Asia, North China has hot, humid, and rainy summers and cold, dry winters—traits typical of a continental climate. When the winter monsoon blows out of

ETHNIC GEOGRAPHY

Han Colonization of China's "Wild West" *continued*

MARTIN GRAY/NGS Image Collection

• **Figure 7.3.C** The Potala Palace in Lhasa, Tibet, is the ceremonial home of the 14th Dalai Lama, now in exile in India.

both known as the Eastern Turkestan Islamic Republic, existed in 1933 and again from 1944 to 1949, and some Uighur separatists are fighting for one again. Beyond execution of those separatists deemed to be terrorists, China's response has been to press ahead with development projects and Han immigration.

At the outset of the Communist takeover of China in 1949, the Han population in Xinjiang amounted to 6 percent of the province's population. Today, it accounts for over 40 percent, and some 250,000 more Han Chinese immigrate to the cities (especially Urumqi and Yining) annually, enticed in part by relatively high salaries and other incentives. Many of these Han migrants are in Xinjiang only a short time; they stay long enough to make some money and then return to eastern China.

Human rights monitors report systematic Han abuses of the Uighurs. Scores of mosques have been torn down, Uighur literature has been burned, Muslim clerics have been forced into "reeducation" training, Islamic schools have been closed, public prayer is now forbidden, head scarves have been banned, and prodemocracy demonstrations are brutally quashed. China's struggle to maintain a grip on its ethnically distinct regions is reminiscent of Soviet efforts in non-Russian areas, and it remains to be seen whether the land empire of China might someday crumble as the 165
Soviet empire did.

interior Asia, long, dark, cold winter days dominate North China. Many people cope with the cold with no more than the heat from the cooking stove or from the fire-heated brick bed called the *kang*.

North China receives 15 to 21 inches (c. 38–53 cm) of precipitation annually, but this is highly variable from year to year. Principal crops adapted to this climate include winter wheat, millet, and kaoliang, with some summer rice crops, corn, and a variety of vegetables.

There is a terrible history of droughts and floods. Floods in North China relate mainly to the problem of managing the Huang He (•Figure 7.3.4). The Huang drops down from its source on the Tibetan Plateau and courses for more than 1,000 miles (1,600 km) through the loess soils of North 2
China on its 2,500-mile (4,000-km) journey to the sea. The yellow loess soil blown from the Gobi Desert collapses easily into the river system, which carries it as silt and deposits it along the final 500 to 600 miles (800–960 km) of the river's flow to its delta.

This physical process is a mixed blessing. On the plus side, it has built up a broad and fertile arable area at the mouth of the river. On the negative side, this sedimentation has elevated the actual stream channel of the river above the surrounding landscape, creating a perfect setup for flooding. Before the recent construction of a series of dams, a critical system of dikes followed the river all the way from

Hiroji Kubota/Magnum Photos

• **Figure 7.3.4** The Huang He (Yellow River) passes through areas of intensively cultivated loess soils, and in this stretch has carved a deep gorge.

the Tai Hang Mountains to the delta. But when flooding breached these dikes, the river poured out of the channel and down to the lower, densely settled farmlands and cities of the North China Plain. Even after the ruptured dike was repaired and the river was contained once again, the great flow that issued forth from the elevated stream bed had nowhere to go but down into ponds, fields, and communities.

This historical pattern on the North China Plain earned the Huang He its nickname of **"China's Sorrow."** Despite these natural hazards, the Huang He and its Wei He tributary (east of Xi'an) were the historical focus for Chinese civilization. Today, many cities along the Huang He are jostling for position as newly important urban-industrial centers. The resulting building boom is putting further pressure on the water supply, now dwindling both because of increased irrigation for agriculture and because of reduced glacial cover and melting permafrost at the river's headwaters; as the ground thaws, more water saturates the soil, and less runs into the Huang He.

Much trade between North and South China has been carried along the Grand Canal, a monumental civil engineering project of the Sui Dynasty (581–618 C.E.). This 1,050-mile (1,700-km) inland waterway—the longest ever built—links Hangzhou in the Chang Jiang Delta region to the area just southwest of Beijing. It was built principally to ensure a steady and safe supply of rice from the productive, subtropical southern lands to the north and to supply the south with wheat and coal. The canal was built at a great cost in tax and human labor but this north-south artery accomplished the major feat of linking China's great west-east rivers. The canal's northern third eventually fell into disuse, but the remaining portion is heavily used and is being rehabilitated and in places even gentrified. More than 100,000 vessels use it every year, carrying hundreds of millions of tons of goods, including coal and construction materials. The Grand Canal will be even more important when the eastern portion of the Jiang Chang Water Transfer Project is complete (see Figure 7.3.E, page 377).

South China

The Qinling Mountains, reaching elevations of about 10,000 feet (c. 3,000 m), mark the traditionally recognized dividing line between North and South China. These mountains run east to west and divide stream drainages between the Huang He to the north and the Chang Jiang (Yangtze River) to the south. South China's agricultural dominance owes mainly to the 3,400-mile (c. 5,440-km) Chang Jiang, flowing from its headwaters in the Tibetan Plateau to the coastal delta just north of Shanghai. This river has played a role similar to that of the historic Missouri River in the United States, serving as a major east- 614
west transit corridor.

There are three large basins on the Chang Jiang. The two eastern ones—the Lower Chang Basin at the river's mouth and the Middle Chang Basin centered around Wuhan—are broad, densely settled areas with agricultural villages, small towns, and some expanding industrial cities (Wuhan being the most significant). Above the Middle Chang are the deep gorges recently transformed by the Three Gorges Dam (see Problem Landscape, pages 376–377). Above the gorges, and also partly inundated by the Three Gorges Dam reservoir, is the Red Basin, home to more than 100 million people. The Red Basin served as the center of the Chinese government during World War II because of its isolation and distance from Japanese bombers.

Nature has provided little level land in the basin, but rice and other crops are grown on enormous numbers of small fields in narrow ribbons of valley land and on laboriously terraced hillsides. The most prolific of China's tea-growing areas lies south of the middle and lower stretches of the Chang Jiang. China is the world's largest tea producer, and tea has long had an important place in Chinese culture (•Figure 7.3.5). A Chinese tradition holds that the seven basic necessities of daily life are fuel, rice, oil, salt, soy sauce, vinegar, and tea.[1]

Joe Hobbs

• **Figure 7.3.5** China grows some of the world's best tea and has built a rich culture around it. Chinese tea is prepared with great care and served ceremoniously.

Joe Hobbs

• **Figure 7.3.6** Tower karst along the Li River near Guilin.

South of the Chang Jiang basins is the real cultural heartland of South China: the delta region at the mouth of the Pearl River in the Guangdong lowlands. Its central features are the city of Guangzhou (Canton; population 5.7 million) and the numerous small, densely farmed, and settled lowlands on the Xi Jiang and its tributaries and distributaries. West of the Guangzhou lowlands, mostly in Yunnan and Guizhou provinces, is a deeply dissected limestone plateau 6,000 feet (1,800 m) high that features some of the world's most dramatic landscapes of **karst,** characterized by limestone pinnacles, caves, sinkholes, and "losing streams" that disappear from the surface into cavities below (•Figure 7.3.6). The area is home to many of China's small ethnic minority populations, who support themselves by hillside farming. The isolation of the region and its distance from the easy reach of national authority made it a perfect locale for opium poppy cultivation until the Communist government cracked down in the 1950s. Today the government has a development strategy to make this the world's largest source area for cut flowers, especially roses, to be exported to markets abroad.

Joe Hobbs

• **Figure 7.3.7** Rural China.

7.3.3 Issues in Chinese Agriculture

China has had productive agriculture for thousands of years, helping both to promote and to sustain the country's vast population. Today nearly 870 million Chinese (about two-thirds of China's population) live in rural settlements and are supported by agriculture (•Figure 7.3.7). These peasants must wrest from the landscape food for themselves and for the ever-expanding urban population, a burden made daunting by the fact that China has 20 percent of the world's population but less than 1 percent of its arable land.

Fully 85 percent of China is not in cultivation because of steep slopes, dryness, or short growing seasons. The country's arable area, concentrated in the plains and river valleys of the east, is about 70 percent of the arable area of the United States, but through the practice of **multiple cropping** (growing successive crops on a given field throughout the year), the sown acreage may actually be larger than the area farmed in the United States. China's territory provides, on average, about two-thirds of an acre (2,700 sq m) of arable land per person to the rural population and only one-fourth of an acre (1,000 sq m) per person to the total population. For comparative purposes, arable land in the United States provides 7 acres (over 28,000 sq m) of arable land per person to the rural population and 1.5 acres (over 6,000 sq m) per person to the total population. However, the process of desertification, caused mainly by human mis- 488
use of the landscape, is threatening China's arable land, which declined by 6 percent between 1996 and 2006.

China's agricultural history has been punctuated by extraordinary efforts to farm new lands and to raise yields. In the 1960s, China's Communist leader, Mao Zedong, proclaimed that "there are no unproductive regions; there are only unproductive people." During that decade, Mao launched a national campaign promoting self-reliance, especially through more productive agriculture. His policy of self-sufficiency in grain remains a central tenet of China's

PROBLEM LANDSCAPE

The Three Gorges Dam

The Chang Jiang ("Long River") has been central to China's identity and welfare for thousands of years. The river delivers precious water and fertile soils, creating environments that have been intensively settled and farmed, especially for rice and wheat. But this great river has also delivered tragedy in the form of devastating floods. Flooding in August 1998 affected 300 million people along the river and did an estimated $24 billion in damage, for example. Inevitably, questions arose about the advantages to be gained—especially in flood control, drought relief, and hydroelectricity production—if the river were to be dammed. The concept of a giant dam on the river dates to 1919, when it was proposed by Sun Yat-sen. His vision has been realized on a scale far grander than he could have imagined in the massive Three Gorges (Sanzxia) Dam, begun in 1994 and scheduled for completion in 2009 (• Figure 7.3.D).*

As with any large dam project, the Three Gorges may best be evaluated by considering its pros and cons. On the plus side, this largest dam ever built—1.3 miles (2.1 km) and 610 feet (186 m) high—is creating a reservoir 385 miles (620 km) long that will dramatically reduce the threat of downstream flooding. Stored waters will also help alleviate drought. As part of the larger **Chang Jiang Water Transfer Project,** the dam will also assist in the provision of water to more arid North China through three canals (• Figure 7.3.E). China has one of the world's lowest per capita water supplies and most uneven distributions of water. More than 40 percent of China's people live in the north, but less than 15 percent of the country's water is there. Chinese engineers see water transfer on a massive scale as the way to redress this imbalance.

The Three Gorges Dam will improve navigation and therefore enhance trade, especially by connecting the burgeoning inland city of Chongqing to the world abroad. Historically, the Chang Jiang was able to accommodate freighters of 10-foot (3-m) draft all the way from the East China Sea to Wuhan. Above Wuhan, the shallower river was traditionally plied by smaller freighters to Yichang, situated just below the river's spectacular gorges. Yichang was a classic **break-of-bulk point,** with freight and passengers offloaded

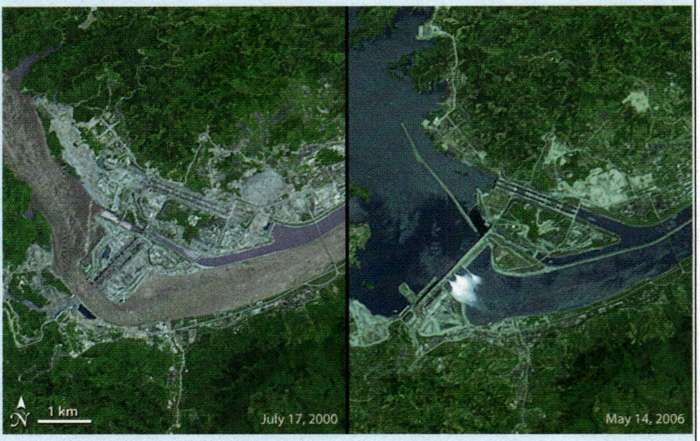

NASA/Visible Earth

• Figure 7.3.D The Three Gorges Dam, viewed from space. Construction is underway in the image dating to 2000 (left), and approaching completion in 2006 (right).

and put on flat-bottomed junks and sampans drawn upstream through the gorges to Wanxian by human trackers who pulled and rowed the vessels. With the completion of the Three Gorges Dam project, vessels with much deeper draft will be able to sail 1,500 miles (2,400 km) upstream from Shanghai all the way to Chongqing in the Sichuan Basin. Small vessels will be able to pass over the dam in a ship elevator, a vast steel box that will lift the ship the height of the dam in 42 minutes. Most ship traffic will use a system of locks with five levels, taking three hours in transit. It is projected that shipping to Chongqing will increase fivefold. Some observers say the increased commerce will transform Chongqing into a new Hong Kong. Chinese authorities view Chongqing as a new model for urban growth, as discussed on page 386.

Also on the plus side, the world's largest hydropower plant that is built into the dam will provide about 85 billion kilowatt-hours of energy annually, enough to satisfy 15 percent of China's electricity needs. Some of the power may be sold to private users, offsetting some of the dam's huge construction costs.

Most of the negatives are related to the vast reservoir forming behind the dam. When filled in 2009, the new lake, 500 feet

*Much of the information in this feature is from the film *Three Gorges: The Biggest Dam in the World,* first broadcast on the Discovery Channel on December 2, 2000.

agricultural economy today, often compelling peasants to grow grain when they could make more money by producing and selling fruits and vegetables.

Since Mao's early appeals, many efforts have been made to increase yields on already farmed land. Irrigation, water conservation, and flood control measures have been part of this effort. The Chinese have struggled both to prevent water from running off in floods and to conserve it in reservoirs for when it is needed. To achieve these aims, local communities have built large numbers of small dams, ponds, canals, and dikes. These are increasingly overshadowed by large dams and associated waterworks, such as the Three Gorges project.

A major objective of water conservancy measures has been to increase the amount of land under irrigation and thereby increase output. China has more than doubled the amount of arable land under irrigation since 1949, when Mao came to power. In arid areas, salinization is a chal-

PROBLEM LANDSCAPE

The Three Gorges Dam *continued*

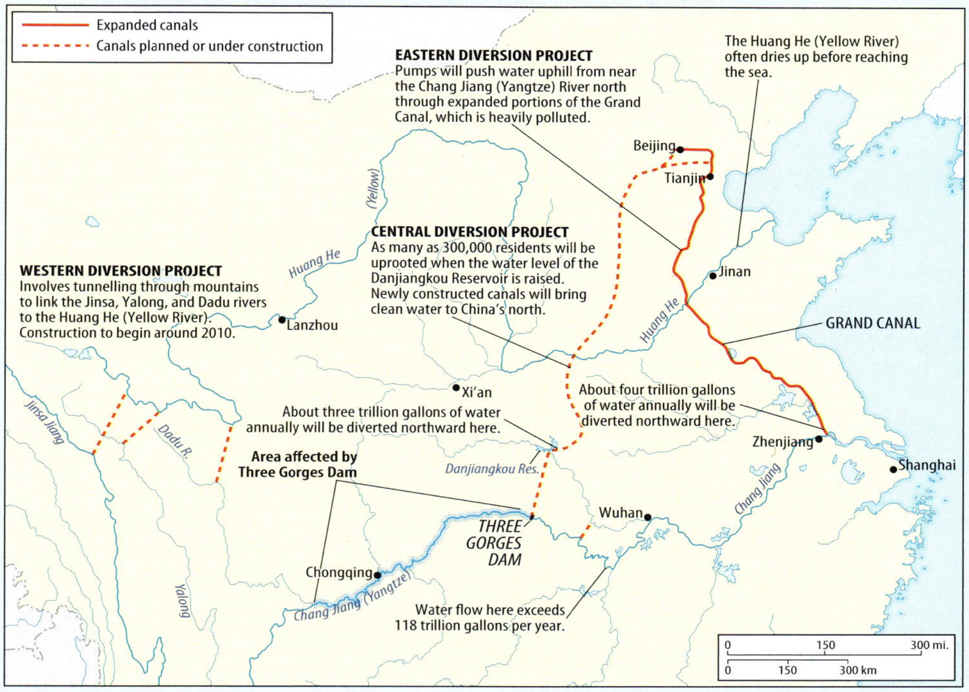

Figure 7.3.E The Three Gorges Dam and the Chang Jiang Water Transfer Project.

(150 m) deep, will inundate 4,000 villages, 140 towns, 13 cities, numerous archeological sites, and nearly 160 square miles (over 400 sq km) of farmland. Because their homes have been or will be drowned, as many as 2 million people are being resettled. Many of the settlements to be vacated are quite ancient, and their inhabitants have deep cultural roots in them. The old villages are typically situated close to fertile farmlands that are also being inundated. In the new, higher communities into which people are being located, soil conditions are generally inferior.

The reservoir will have aesthetic consequences too, replacing the world-class wonder of a wild river passing through scenic canyons. The natural marvels of the three gorges drew 17 million Chinese tourists and hundreds of thousands of foreign visitors every year. There are also practical concerns about the silt that will build up behind the dam. It is possible that this increased soil load will have the ironic effect of causing flooding upstream from the dam, even as the dam prevents flooding below. Finally, there are fears of what is almost unthinkable: that the dam could burst, causing unimaginable devastation and suffering downstream.

Although recently admitting some environmental consequences, the government of China has pursued the mighty Three Gorges Dam project without remorse and with no tolerance for dissent within China. The enormous cost of the project, nearly $30 billion, is born solely by China. Like megadams elsewhere, the 252, 319 Three Gorges Dam is a statement that its builder is a great power to be reckoned with on the world stage.

lenge; too much irrigation deposits mineral salts from the water and reduces the land's productivity. Ongoing efforts at reforestation, like those described on page 370, also relate closely to water control. Only 18 percent of China is forested. Under steady human pressure over thousands of years, most primary forests in China have been cut down. Large areas of the country are without significant vegetative cover, contributing to loss of water through rapid runoff. Planting trees is not the answer in every situation, however: in China's arid lands, too many trees would deplete the available groundwater.

Another approach to improving yields from existing lands lies in the application of scientific knowledge and technology to develop better seed, equipment, and farming techniques, together with the development of administrative means of promoting their adoption—China's version of the Green Revolution, described on pages 299–300. With more trained agricultural scientists and technicians, insecticides and herbicides are coming into wider use. These follow more rudimentary systems of pest control practiced during the 1950s and 1960s, during Mao's **Four No's Campaign** to eradicate rats, sparrows, flies, and mosquitoes. Observers were amazed at the sight of Chinese peasants fanatically hunting down the pests, motivated by government radio broadcasts over speakers placed in rural areas. A major objective was to reduce the annual losses of many millions of tons of grain. It was an unintended consequence that so many birds were killed that their natural prey, caterpillars, infested crops. Campaigns against flies, mosquitoes, and rats had additional public health benefits. They were also politically important as they demonstrated the government's capacity to marshal enormous numbers of people in a wide variety of national campaigns, many of which were focused on modifying the environment.

Increased fertilization has boosted farm productivity. The use of chemical fertilizers is growing alongside more traditional means. Chinese farmers have long fertilized their fields with pond mud, plant residues and ashes, oilseed cakes, nitrogen-rich crops raised to be plowed under for soil enrichment, and animal and human manures ("night soils"). Such traditional fertilizers, and the close contact between people and animals in rural China, play a role in the country's status as an "incubator" for disease (see Medical Geography, pages 380–381).

Beginning in the 1950s, the Communists looked for ways to increase land cultivation. Small gains in the amount of cultivated land were made through reclamation of poorly drained land and extension of irrigation into deserts. As discussed on page 375, these measures did not always succeed in the long run. Mechanization has developed slowly. The use of tractors is growing, especially in the northeastern and western regions, where more extensive farm fields are located, but human and animal power and simple hand tools are only slowly fading from the rural scene.

After 1949, when the Communist Party gained control of China, agriculture was not merely a matter for scientific and technological attention but also an engine for profound ideological change. China's leaders mandated a series of reorganizations of Chinese agriculture in successive attempts
140, 167 to solve the country's food problems using Communist ideology and control. Collectivization of agriculture began in 1955, with private farms reverting to state control in the form of agricultural cooperatives. The cooperatives were merged into even larger units in the agricultural **commune system** that began in 1958 as the cornerstone of an economic program called the **Great Leap Forward.** In the commune, 20 to 30 cooperatives consisting of over 20,000 people and 40 to 100 villages were merged into a single unit. Residents lived communally, sharing bathing, dining, and residence facilities. Any land and equipment of the former cooperative and any money or property still owned by peasants became the commune's property. An economic and administrative body controlled the labor and production and supervised all aspects of education and defense. Small-scale industries, most notably millions of "backyard steel furnaces," were also set up in the rural communes.

The legacy of this effort to bring industries to the countryside is still apparent in the Chinese economy today. There are about 25 million rural industrial enterprises with more than 125 million employees, representing about 65 percent of China's industrial labor force. Their rural-based, small-scale industries generate about half of China's annual industrial output, one-quarter of its total economic value, and more than one-third of China's exports. These successes are largely the result not of the commune movement but of more recent economic reforms.

Collectivization and communes had disappointing results, related mainly to inefficiencies inherent in trying to manage such large units and to the lack of incentives for individual farmers. An estimated 30 million Chinese died from famine attributable to the failures of the Great Leap Forward. Reforms began in the 1960s, with communes decentralized and some even divided into private farms. After the death of Mao and the 1978 ascension of Deng Xiaoping to the major leadership role in the Communist Party, free-market reforms began to take place in China. Individual households acquired long-term (30-year) "leases" on the farms, which continued to belong to the state but farmers now viewed as effectively their own. They paid fixed amounts of their production to the state and were allowed to consume or sell the rest. Chinese farmers were finally allowed to sublet land, hire labor, own machinery, and make their own land use decisions. They still lacked outright ownership of property, however, and in recent years, the government has appropriated farmland for housing and factories, creating unrest and uncertainty in the countryside.

The overall results of this experiment with a freer semi-capitalist system in agriculture have been profound. Chinese agricultural output expanded by about 50 percent during the 1980s and 1990s, and rapidly enough—when coupled with the steady decrease in birth rate—to generate an increase of 20 percent in the per capita food supply. It is remarkable that the world's largest country has generally been able to keep agricultural growth apace with population growth and provide more than 90 percent of its own supplies of rice, corn, and wheat. The challenge will be for China to continue this remarkable trend, especially as the Chinese develop new appetites for beef, beer, and other luxuries that require more inputs of energy and agriculture. Chinese agricultural plan- 62
ners are also alarmed by the accelerating loss of farmland to urban and other uses and by farmers' decisions to forsake grains in favor of more lucrative cash crops.

Chinese scientists are among the world's most fearful in forecasting the consequences of global warming. They have documented a rapid (7 percent per year) melting of glaciers on the Tibetan Plateau, a trend that threatens future water supplies to large parts of China and adjacent regions of southern Asia. According to a 2007 International Energy Agency (IEA) report, climate change could cut China's harvests of wheat, rice, and corn by 10 percent by 2030 and by 40 percent after 2050. China, which produced about one-fifth of the world's carbon dioxide emissions in 2007—slightly more than the United States—will be the industrial engine behind the largest share of global greenhouse gases for decades to come.

If nothing were done to reduce China's CO_2 footprint, the IEA warns, China's emissions would be double those of all industrialized nations by 2032. Citing its own need to develop as developed countries have, and thus refusing
35 to come on board with the existing Kyoto Protocol, China has set its own target of producing 16 percent of its energy needs from renewable resources (mainly hydropower) by 2020. The country has already taken aggressive steps to reduce energy use in factories and automobiles. The latter is a huge concern in a country that had 13 million private cars in 2007, with a projected 10-fold increase by 2020—by which time China will also be well established as a major automobile exporter. The country will also increase its output of nuclear power substantially.

7.3.4 China's Industrial Geography

China today is one of the world's great industrial powers, ranking fourth in overall economic output (•Figure 7.3.8). Remarkably, production of industrial goods did not start in China until the beginning of the 20th century. Prior to that time, imperial China's handicraft industries produced high-

• **Figure 7.3.8** Major industrial resources, industries, and industrial cities of China.

quality goods that attracted Western traders in spite of China's official disinterest in such trade. To this day, handicraft and shop-scale industries continue to supply many of China's simpler needs.

China's early factory industries were mainly foreign-owned and were attracted to China by the huge market for cheap goods, by inexpensive labor, and sometimes by certain natural resources. As in so many countries, the first factory industry to develop on a large scale was the manufacture of cotton textiles. Japanese and other foreign firms, together with some native Chinese companies, developed a major cotton industry by the 1930s. This industry, and others, developed in coastal cities such as Shanghai
and Tianjin. Meanwhile, the Japanese used the minerals 400
of the Northeast (in their puppet state of Manchukuo) to make that area the first center of heavy industry in China in the 1930s.

Communist China made major efforts to industrialize after the 1949 victory over the Chinese Nationalists. First, the industrial infrastructure, which had been badly hurt by invasion and civil war from 1937 to 1949, was brought back to prewar production levels under the new system of state ownership. China also expended considerable capital and human effort in its involvement in the Korean War 406
of 1950–1953. Aided by the Soviet Union, throughout the 1950s, the Chinese emphasized coal, oil, iron and steel, electricity, cement, certain machine-building industries, and armaments. China's resources for these industries were considerable, and the country's centrally planned command
economy—a system in which major economic activity is 167
initiated by government planners—achieved major successes in developing them, as was the case in many Communist
countries. There was also a strategic spatial component 165
to this drive, with the Soviets advising China to reduce its concentration of industrial plants along the coast. This strategy, meant to reduce the vulnerability of Chinese industry in the event of military conflict and to promote development outside the historic core, led to expansion into new industrial landscapes at Wuhan on the Chang Jiang, at Baotou in Nei Mongol, and into the far west near the Soviet border in Xinjiang.

In the 1960s, after China broke ranks with the Soviet Union, the pace of technical development, quality of output, and modernization slowed. This was a consequence both of China's near-isolation from more advanced industrial countries and of the Cultural Revolution from 1966 into the early 1970s (see Insights, page 383). This slowdown also mirrored the difficulties that non-market-based, centrally planned economies on the Soviet model typically encountered as they moved toward more consumer-oriented
industries and advanced technologies. 168

The latest phase of China's industrialization began after Mao's death in 1976, when another veteran of the Long March, Deng Xiaoping, became the country's leader. China opened its doors to the West and devoted more attention to personal incentives for agricultural and industrial success.

MEDICAL GEOGRAPHY

Deadly Viruses: Origins and Diffusion

Seemingly out of nowhere, **severe acute respiratory syndrome (SARS)** burst onto the world scene in spring 2003. But this extraordinarily virulent virus did come from somewhere: southeastern China.

As any recipient of a flu shot knows, the vaccine is typically targeted against a host of flu strains bearing Asian, usually Chinese, names such as "Shanghai" and "Beijing." SARS, like these influenza varieties, originated in a region where there is frequent close contact between people and animals. Farmers in China often cohabit with ducks, pigs, geese, and other domesticated animals and make extensive use of animal excrement as fuel and fertilizers (•Figure 7.3.F). In these close quarters, the influenza strain often starts in a duck or other animal and jumps the species barrier to affect humans. Once the sickness manifests itself in a human host, it can spread rapidly to major cities in China and thence by airliner to distant parts of the globe.

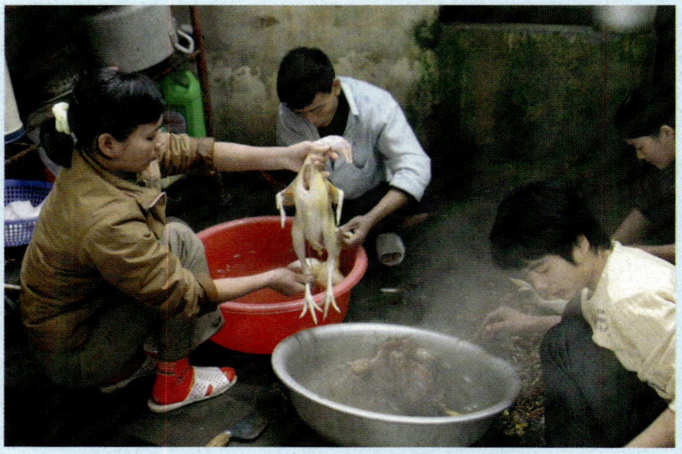

Joe Hobbs

• **Figure 7.3.F** Situations like this raise fears that the bird flu will jump the barrier between poultry and humans.

The SARS virus had similar origins, although it probably arose from close contact between animals and animal merchants in southeastern China. In Guangdong province, there are numerous large animal markets where poorly paid migrant workers carry out the sordid job of killing, skinning, plucking, gutting, and packing a wide variety of wild and domesticated creatures, including snakes, frogs, turtles, chickens, badgers, cats, rats, and civets, all bound for Chinese dinner tables.

Between November 2002 and January 2003, several of these animal merchants, along with food handlers and chefs further down the supply chain, began developing flu-like and pneumonia-like symptoms, including high fever. They had contracted SARS, an illness new to the human world and to the coronavirus family, which also includes the common cold. Some sufferers, perhaps 5 percent, died of the illness. Some sought medical attention in Guangdong hospitals. There they infected doctors, nurses, and other staff (30 percent of China's early SARS victims were hospital workers).

On February 21, 2003, a doctor from one of the hospitals managed, despite his high fever, to attend his nephew's wedding in Hong Kong. He stayed at the Metropole Hotel, where he passed the virus (probably by sneezing or coughing) on to two Canadians, an American businessman on his way to Vietnam, a man from Hong Kong, and three women from Singapore. In the days that followed, most of these sick people flew to other countries. One of the Canadian women carried the illness home to Toronto, where she died after infecting her son and several health workers. At that point, SARS was poised to become a global epidemic. By June, nearly 7,000 probable SARS cases had been reported in 29 countries, and 495 victims had died. In the United States, about 200 suspected SARS cases were reported in 38 states, with no fatalities.

The liberalization of agriculture has already been described, and related measures were introduced into trade and industry, where emphasis shifted toward light industry, consumer products, and profit-oriented enterprises. Much more private enterprise was allowed, especially after the mid-1990s. State factories were required to operate at a profit, part of which they were allowed to retain and invest in expansion. The factories introduced pay incentives for superior performance and were no longer obligated to provide lifetime jobs regardless of output. The Chinese welcomed foreign investment capital and technical expertise from Japan and the West. Great numbers of Chinese began to study abroad in graduate and professional programs; they account for the second-largest percentage of foreign students in the United States and the largest in Australia. Turning its back on the isolationist days of Mao, China sought to become a global trading power. It has succeeded handsomely in this goal (see Chapter 7, pages 298–299).

So successful has China's industrial drive been that the country is now a gigantic, insatiable consumer of domestic and imported energy and other commodities. For example, China is today the world's largest producer of steel, aluminum, magnesium, and tungsten. It is the world's leading consumer of copper, iron ore, soybeans, cotton, and wood. Heavy industry has been built on a rich domestic resource base, but it is not yet big enough to meet all the country's needs.

Because of local shortfalls, China imports vast amounts of iron ore, lead, alumina, nickel, and copper to feed its booming industries. For a time, oilfields in the Northeast, Gansu, Sichuan, and Xinjiang met China's needs, even providing some surpluses for exports. But as the economy

MEDICAL GEOGRAPHY

Deadly Viruses: Origins and Diffusion *continued*

In less than six months, SARS had become an enormous problem that posed several threats. Of most immediate concern was its impact on human health. There was no vaccine for the disease. It is especially difficult to develop a vaccine for an illness that affects both people and animals. There was no effective treatment; SARS, like all viruses, does not respond to antibiotics. In the absence of a preventive or cure, the disease follows its course: a high fever of 100°F or more is followed by chills, muscle aches and headaches, a dry cough, difficulty in breathing, and occasionally, severe pneumonia and even death.

Despite taking relatively few lives, SARS had a strong impact on human psyches and societies. The disease could have appeared anywhere and infected or killed anyone; consequently, many people—especially near the Asian epicenter—assumed the worst during the outbreak. As reports of the disease spread in spring 2003, untold numbers of people in China began wearing masks to reduce their chances of contracting the disease. They no longer shook hands. Schools, theaters, dance halls, and Internet cafés closed in many communities. Shopping malls had few shoppers. Several Asian countries stopped granting visas to Chinese visitors.

The economic costs for such a limited outbreak were also extraordinary. In the globalized world of business, many firms halted interpersonal dealings with China. Toronto lost $30 million a day in canceled conventions and other events during the brief period when the World Health Organization (WHO) posted a warning that travel there should be avoided. In spring 2003, China and South Korea lost $2 billion in the tourism, retail sales, and manufacturing sectors, and Hong Kong, Taiwan, and Singapore lost perhaps $1 billion each. Sixty percent of U.S. corporations banned travel to Asia altogether. North American airline bookings to Hong Kong fell 85 percent, and international arrivals in Singapore were down by 60 percent. Anyone diagnosed as having a fever in Singapore airport's arrival hall was unapologetically led off to two weeks of quarantine.

In summer 2003, the disease peaked and began to decline in Canada, Singapore, and Hong Kong and then subsided in its birthplace, China. SARS receded from the public imagination. So why is it the focus of this essay? Because the SARS experience may provide a preview of something of far greater consequence, also originating from the same part of the world.

Fresh on the heals of SARS came the specter of this new menace, **bird flu** (the H5N1 subtype of avian influenza A). Like SARS, this respiratory virus originated in animal species—both wild and domesticated birds—and has the potential to jump the species barrier and afflict large numbers of people. Some people working with ducks, geese, turkeys, and other domesticated fowl in East and Southeast Asia did come down with the disease, and health officials in several countries ordered hundreds of millions of birds to be killed to staunch the virus's spread.

The threat persisted. By late 2007, about 200 people, mainly in Vietnam, Indonesia, Thailand, and China, had died from bird flu. That number is small, but what most concerns health officials is that it represented 70 percent of the people who had the virus. With such a high fatality rate, health experts fear that bird flu could be the "big one": a virus like the Spanish flu that in 1918 and 1919 killed 40 to 50 million people worldwide, including 850,000 in the United States (where it apparently first manifested itself, in a soldier at Camp Funston, Kansas, on March 11, 1918). According to the WHO, even in a best-case scenario, a bird flu **pandemic** (an epidemic over a wide geographic area, affecting a large percentage of the population) would kill up to 7 million people worldwide and leave tens of millions in need of medical attention. Indonesia, China, and Vietnam have developed a flu vaccine for birds and are inoculating vast numbers, but other countries argue that the vaccine is not completely effective. Work to develop an avian flu vaccine to protect humans is ongoing.

zoomed into overdrive, China became a major oil importer, with imports increasing an average of 30 percent per year—and by 2007 was second only to the United States.

China's imported commodities come particularly from Africa and Latin America. While securing vital commodities from developing countries in these and other regions,
478, 480, 481, 538 China is also pouring economic aid into them, building roads, bridges, dams, and other infrastructure that serves the interests of those countries and, of course, China itself. China tends to attached fewer strings to its aid than the World Bank and other traditional development agencies.

Coal is king in China, and this is one commodity China does not have to seek abroad. The country has about 12 percent of the world's coal reserves (behind the United States and Russia) and is the world's largest producer and consumer of coal. The reserves are located mainly under the North China Plain and its bordering hills and in Liaoning province in the Northeast. This resource concentration has made the southern area of the Northeast the country's leading heavy-industrial region. China uses coal to meet about 80 percent of its overall energy needs and to provide 82 percent of its electricity. This is the dirtiest of all the fossil fuels, producing far more particulate matter and carbon dioxide than oil or natural gas.

What's Next for Industrial China?

There are risks for China's future industrial landscape, and it cannot be taken for granted that the country will persist as the world's economic dragon. China's extraordinary recent successes are tempered by a few concerns. One is freedom of expression. In 1989, the Chinese government cracked down brutally on prodemocracy demonstrations centered on Tiananmen Square in Beijing. Years later, there continue to be strictures on academic, media, artistic, and other freedoms that some observers believe are conducive to long-term economic well-being. China's government is one of few in the world (including those of Myanmar, Singapore, Turkmenistan, Uzbekistan, Saudi Arabia, Syria, the United Arab Emirates, Tunisia, and Iran) to censor the Internet, which is arguably one of the most important tools of economic development today. The government also monitors mobile phones, the use of which is growing 30 percent annually in China, to ensure that they too do not become instruments of protest.

Another potential problem is that China's leaders are breathlessly trying to feed what they call the "socialist market economy" with gigantic projects, thereby hoping to create jobs and stimulate economic growth. They reckon that the country needs to register at least 7 percent economic growth annually just to prevent massive unemployment and social unrest. And so one of the world's most impressive building booms ever is now under way: new subway systems, railroads, highways, bridges, sewage systems, dams, water diversion projects, a natural gas pipeline, a national electricity project, urban improvements, airports, industrial parks, and more. These efforts cost a lot of money, and the government is drawing down its national treasury to fund them. State banks are pouring billions of dollars into federal projects whose returns are uncertain.

A further hazard is that with its galloping growth, China is developing a classic "economic bubble" that will eventually burst, probably with costly results. Historically, few economies have been able to maintain torrid growth for
47 long, especially when that growth is fueled by speculative cash flows and questionable bank loans, as has been the case recently in China.

There are also environmental costs for such spectacular growth. Due to generally lax environmental quality standards, China's skies and waters are often badly polluted (•Figure 7.3.9). Twenty of the world's 30 most polluted cities are in China. The country's State Environmental Protection Administration (SEPA) reported in 2007 that 26 percent of the water in China's largest river systems was so polluted that it was dangerous for people and other living things to come into contact with or had "lost the capacity for basic ecological function." Polluted water was killing tens of thousands of people every year and reducing crop yields. SEPA warned that China's approach of "growth through industrialization" was pushing its environment "close to the breaking point."[2]

Finally, there is concern that the benefits of China's economic growth have not been distributed evenly throughout the population. Prosperity and consumption have increased dramatically, but in a remarkably unequal manner. Less than 10 percent of China's people controlled about 45 percent of the country's private wealth in 2007. Meanwhile, 17 percent of all Chinese lived on less than $1 a day—the World Bank's measure of abject poverty. These inequalities are comparable to those that existed when the revolution brought Mao's Communists to power on the promise of redistributing wealth. In 1980, China had one of the world's most even distributions of wealth, but that equation has changed dramatically. In sum, despite its phenomenal growth, China is still relatively poor (with a per capita GNI PPP of $7,730, comparable to that of Kazakhstan and Panama) and largely agricultural (see Table 7.1, page 281). The agricultural face of China is changing rapidly, though, as the cities boom.

© Lou Linwei/Alamy

• **Figure 7.3.9** Blessed with coal reserves, China is also cursed by them. Its coal-fired power plants are responsible for calamitous domestic air pollution and contribute significantly to global atmospheric warming.

INSIGHTS

Mao's Cultural Revolution

Between 1949 and his death in 1976, Mao Zedong (•Figure 7.3.G), chairman of the Chinese Communist Party, led the People's Republic of China through a demanding and sometimes counterproductive series of experiments in social reorganization. In 1957, he initiated the Great Leap Forward and the commune movement discussed earlier. The **Cultural Revolution** (1966–1972) was Mao's attempt to maintain an ongoing "permanent revolution" in China. Politically, the movement was meant to throw a newly established urban and intellectual elite off balance by promoting purges of virtually anyone who had offended Communist Party functionaries.

Local youth who donned the red armbands of the Red Guards became the agents of change in this campaign. The Red Guards were set loose on the populace to harass those who were accused of putting learning, skill, expertise, or personal matters above revolutionary enthusiasm or Marxist and Maoist goals and principles. Government oppression led to perhaps a million assassinations, executions, or suicides and to lifetime jobs for those deemed most loyal to the Maoist revolution, regardless of their performance. Reforms initiated in the late 1970s finally began removing unproductive people from their positions, but people whose lives were turned upside down by the Cultural Revolution continue to speak of themselves as a lost generation.

Despite some monumental setbacks and miscalculations, China under Mao regained a global prominence that had been denied for a century and a half under Western colonialism and interference in the nation's domestic affairs. Even though much of Mao's legacy has been dismantled, he continues to be a major icon of China's presence on the world stage.

Bettmann/Corbis

• **Figure 7.3.G** Chairman Mao greeting a crowd in 1971.

7.3.5 China's Urban and Transportation Geography

As recently as 1978, about 80 percent of China's population lived in the countryside. Although there were massive cities such as Shanghai, Beijing, and Guangzhou, most Chinese were peasants living in small villages. As a result of the greatest migration in human history, China's population is now on track to be 50 percent urban by 2010 (for comparative purposes, India is 28 percent urban, and the United States 79 percent).

As might be expected, despite the rural industries described earlier, China's industrial production is centered mainly in its cities. The country's 12 largest metropolitan cities are located in geographic clusters. This is an important spatial characteristic of China's economy. Four major clusters of cities are especially notable, in the Northeast, North China, the Chang Jiang region, and the area south of the Chang Jiang (see Figure 7.3.8, page 379).

Strong neighbors have long coveted the cities and industries of China's Northeast. After a treaty signed in 1858, tsarist Russia landlocked Manchuria by extending Russia's coastal territory all the way to northern Korea. Even today, China is eager to open a northeast access to the sea, even if only by leasing the Russian port of Zarubino. During the 20th century, the Northeast became a region of even more contention among China, Russia, and Japan. Russian interest focused on the Chinese Eastern Railway—a shortcut across the northern two provinces between Vladivostok and the Trans-Siberian Railroad east of Lake Baikal—and on the naval and port facilities of Dalian (formerly known as Darien–Port Arthur and as Lüda), which the Russians had used at various times. 184

Japan, even today, is widely resented in eastern Asia for coveting and conquering the resources of China and its neighbors. Japan was long interested especially in developing Northeast China's impressive industrial resources and its potential to produce surplus food. After the Russo-Japanese War of 1904–1905, fought mainly in Liaoning, Japan became increasingly active in the area. Finally, in 1931, Japan took over with the establishment of the puppet state of Manchukuo and in the course of about 15 years created much industrial and transportation infrastructure. In 1945, during the final days of World War II, Soviet forces wrested the area from the Japanese, stripping many of Manchuria's factories and shipping them home. Eventually, the Chinese Communists incorporated the Northeast into their own People's Republic, making this region central to their industrial development.

The cities of the Northeast include another four of China's 20 largest cities: the port of Dalian; Shenyang (Mukden), the main metropolis of a group of heavy-industry centers;

Changchun, a center of automobile and other transportation manufacturing; and Harbin, a railway center with connections to Russia and Korea. After the Communists took over in 1949 and expanded China's industrial base, these cities of the Northeast lost some of their industrial importance, but they continue to be centers of manufacturing, chemical production, and agricultural marketing and processing. The industries are centered around Shenyang. The largest center of iron and steel production is Anshan, south of Shenyang, where industry is based on deposits of iron ore and some coal in a belt crossing the southern part of the region. The Daqing oilfields, China's largest, are north of Shenyang near Harbin.

The cities of North China include the historic national capital of Beijing (Peking; population 11.8 million; •Figure 7.3.10); the port city of Tianjin (Tientsin); and far to the southwest, the very ancient city of Xi'an (Sian). Each is located at an entryway to the North China Plain: Tianjin, by the sea; Beijing, near the end of a pass leading through the uplands to the north toward Mongolia and the Great Wall; and Xi'an, in the Wei River Valley, which leads upstream to interior China and downstream to the Huang Valley and the North China Plain (•Figure 7.3.11).

The Chang Jiang Valley city group is a string of cities along and near the river. This group includes five of China's 20 largest cities, reflecting the importance of the Chang Jiang basins as a major axis of the national economy. The Chang Valley group includes a metropolis of 4 to 7 million inhabitants in each productive basin, plus a much larger overseas global connecting point as well as a transfer and connecting point for interior China. At the western end of the axis in the Red Basin is Chengdu, a provincial capital 386 and node for transport lines connecting the axis with Arid China. It lies away from the river. On the river is the Red Basin's booming metropolis of Chongqing.

Downstream lie Wuhan in the Middle Basin, an important manufacturing, transportation, and agricultural marketing center, and Nanjing in the Lower Basin. Nanjing

Joe Hobbs

• **Figure 7.3.10** The entrance to Beijing's Forbidden City, from Tiananmen Square.

Joe Hobbs

• **Figure 7.3.11** The 60,000 Muslims who worship in this Great Mosque of Xi'an (established 742 C.E.) are living evidence of the city's historic position at the eastern end of the Silk Road.

400 (a temporary capital for China during the Sino-Japanese War in the mid-20th century) is a major agricultural marketing center.

On a navigable tributary (the Huang Po Creek) near the mouth of the Chang in the Lower Basin is the port of Shanghai, the traditional center of Chinese business life and a contender with Hong Kong for the position of China's financial capital (population 15.1 million). With its location at the seaward end of this urban axis, Shanghai serves the whole Chang Valley and other areas. Built from a fishing village by foreign enterprise in the 19th century, Shanghai is the largest city in China and has the largest port in the world by tonnage. The piercing skyscrapers of its futuristic section of Pudong are just opposite the colonial waterfront known as the Bund (•Figure 7.3.12). Many Chinese and outsiders lament the loss of Shanghai's traditional and historical architecture in this city of nonstop construction.

The region south of the Chang Jiang has two of the country's 12 largest cities, including the port of Guangzhou (population 5.7 million) in the Xi (Pearl) River Delta. Guangzhou and its vicinity now form a rising center of new industries, particularly around Shenzhen, a young city that in the 1980s

Joe Hobbs

• **Figure 7.3.12** The futuristic Pudong section of Shanghai is China's modern face to the world.

was designated as one of five **Special Economic Zones (SEZs)** designated to attract investment and boost production through tax breaks and other incentives (see Figure 7.3.8). 323
The other SEZs, all with access to the sea, are nearby Zhuhai; the entire island of Hainan; farther north, Shantou and Xiamen on the Taiwan Strait; and Pudong, linked to Shanghai. Many of China's coastal cities specialize in the factory production of a single category of goods; for example, moving from Shangdong province in the north to Guangdong province in the south, one would pass through cities known for men's wear, casual wear, down-filled products, ties, socks, underwear, wedding dresses, evening gowns, and jeans.

Close to the Xi River cluster of SEZs are Hong Kong (Xianggang; population 9 million), which reverted from British control in 1997, and Macau (population 1 million), returned from Portuguese control in 1999. Each is now a **Special Administrative Region (SAR)** governed by China under a unique **"one country, two systems"** formula. In this arrangement, China's central government rules Hong Kong and Macau but grants each considerable autonomy and political liberalization. Each has the right to continue its capitalist economic development, while China proper maintains its socialist economic principles.

This formula has generally been successful, although both SARs would like to have more democratic freedoms. More than a decade after returning to Chinese control, Hong Kong continues to hold its position one of the world's busiest and most prosperous trading and financial centers. Macau's economy is based on trade with inland China, fishing, tourism, and especially gambling, which is forbidden elsewhere in China. Macau's gaming industry is more valuable than that of Las Vegas. It has helped increase Macau's overall wealth enormously, but it has also led to high prices for real estate and consumer staples. Many citizens are not receiving benefits from the new flush of cash, and thousands of them took to the streets in 2007 to protest the gaming industry.

One of the results of the clustering of cities, and the economic power they represented, was a serious problem of regional imbalance. Most of the country's industrial development and urban growth occurred along its east coast, near the traditional port and river cities that were strongly influenced by Western economic interests from the mid-19th century on. Even the rural areas of eastern China experienced much more growth relative to those in the west, especially because they were well linked to expanding markets and prospering cities. Meanwhile, people in western regions of China, mountain villages all through Humid China, and landscapes lacking in surface transportation in the southwest and parts of upland China were frustrated. They saw relatively little evidence of China's much touted economic prosperity.

Consequently and somewhat predictably, large numbers of people migrated from the less prosperous regions of the country to the promising SEZs and other urban centers of the east and south. Millions of migrant workers left their villages—now populated primarily by children and the

Joe Hobbs

• **Figure 7.3.13** China's urban workforce has a large contingent of migrants from rural areas, performing both skilled and unskilled labors. This man is making cloisonné vessels in Beijing.

elderly—to work in construction or menial labors in the coastal cities, sending much of their meager monthly earnings of $50 to $100 home as remittances (•Figure 7.3.13). People in the cities often discriminated against these migrants in education, housing, and jobs, treating them like illegal aliens. Many technically were illegal because they failed to sign work contracts that would both register them and require factories to pay fees to local cities. Police would routinely "shake down" workers without papers, seeking bribes from the frightened migrants. Some of the factories were world-class, but others were **sweatshops** where working conditions bordered on the nightmarish. Management typically suppressed strikes, banned independent trade unions, and failed to enforce minimum-wage laws for workers who toiled through 18-hour workdays and caught brief rest in prisonlike dormitories. It is little wonder that many of the eastern boomtowns saw as much as 10 percent annual turnover in the workforce and soaring crime rates.

Some of these conditions may persist for some time, but there are also signs of rapid change, reflecting the pace of China's growth. Chinese authorities are convinced that urbanization is the ticket to greater prosperity for China's huge, poor, rural population. But they want the larger cities of the interior, such as Chongqing, to absorb much of the migrant traffic that has been flowing farther east and south. Already home to 8 million people, Chongqing is poised to grow much more through its "one-hour economy circle plan." In this plan, by 2012, 2 million rural residents will be moved into newly urbanized urban areas within an hour's drive of the city center. Two million more are to be accommodated by 2017. Millions of new jobs are thus being created in the long-neglected urban centers of China's interior. At the same time, wages are rising sharply all over the country. Once pushed and pulled toward the coast, workers 298
are now finding competitive jobs closer to home. The result is a sudden and surprising shortage of wage laborers in the coastal factories. Once-confident sweatshop managers now worry that the global marketplace will shift jobs to lower-wage countries like nearby Vietnam.

• **Figure 7.3.14** One of China's most ambitious efforts in expanding its infrastructural base for continuing economic development has been its transportation system.

• **Figure 7.3.15** Principal features of Taiwan.

In another effort to redress the regional imbalances that drew so many people to the east and southeast and to integrate China's widespread resources and widely scattered producing and consuming centers into a functioning whole, the government has placed great importance on improving the transportation system (•Figure 7.3.14). The country is investing $250 billion in a national highway system to be completed by 2037 and billions more in the construction of dozens of airports. Much of the air travel will be abroad, especially as the Chinese expand their segment of the international tourism market. Estimates are that at least 50 million Chinese will travel overseas each year by 2010 and 100 million by 2020.[3]

The government has also made major efforts to improve and extend railways. Before 1949, the Northeast had an extensive rail system developed by tsarist Russia and expanded by Japan, but most of China south of the Chang Jiang floodplain and all of the western provinces were without rail lines. The postwar Communist government made South China accessible by building the first three bridges ever to cross the Chang Jiang. They pushed rail lines into South China and the Red Basin and also into Xinjiang and across Mongolia to connect with the Trans-Siberian Railroad. Since 1949, the length of China's rail network has more than tripled, and the expansion continues. However, railroad development lags behind that of roads. Except for transport over short distances, railroads are much more efficient for hauling goods than trucks are. With China's economy booming, the huge bottlenecks at ports and factories suggest that more railways need to be built.

Helen Rogers/Alamy

• **Figure 7.3.16** High-tech manufacturing is central to Taiwan's economy. This worker is assembling cameras.

7.3.6 Taiwan

Known by Westerners for centuries as Formosa (Portuguese for "beautiful"), Taiwan is separated from South China by the Taiwan, or Formosa, Strait, 100 miles (160 km) wide (•Figure 7.3.15). The island is nearly 14,000 square miles (c. 36,200 sq km) in area and is home to 23 million people, nearly all of Chinese origin. High mountains rise steeply from the sea on the eastern side of the island but slope on the west to a broad coastal plain. Taiwan lies on the northern margin of the tropics, and its temperatures are moderated by the warm, northward-flowing waters of the Kuroshio, or Japanese, Current. Irrigated rice, the main food crop, is favored by the hot, humid summers in the lowlands and by abundant irrigation water from mountain streams. Sugar and pineapples are also important. The Japanese governed the island from 1895 to 1945, developing it mainly as a sugar-producing colony.

Driven from the mainland in 1949, the Chinese Nationalist government fled to Taiwan with remnants of its armed forces and many civilian followers. Nearly 2 million Chinese were in this migration. Here, protected by American sea power, the government reestablished itself as the Republic of China with its capital at Taipei (population 8.2 million). The Nationalists, originally an authoritarian regime but with increasing elements of political democracy, were very successful in fostering industrial development. They united inexpensive Taiwanese labor with foreign capital to build one of Asia's first urban-industrial countries and one of its
Asian Tigers (•Figure 7.3.16). Its major exports are machin- 298
ery, electronics (notably computers, cell phones, modems, routers, and global positioning systems), metals, textiles, plastics, and chemicals, especially to mainland China and the United States. The population is 78 percent urban, its infant mortality rate is very low, and its birth rate is one of the lowest in Asia (see Table 7.1). The average Taiwanese citizen

is four times wealthier than the average mainland Chinese citizen. In almost all respects, Taiwan is clearly recognizable as a more developed country. A major hurdle against even stronger growth has been a lack of energy resources. Some coal and natural gas exist, and some hydropower has been harnessed, but the island depends heavily on imported oil.

Taiwan's Republic of China continues its claim as the legitimate government of China. Meanwhile, the People's Republic of China claims Taiwan as part of its own territory. The United States backed the Nationalist claim for a time, but during the 1970s, the United States and the mainland People's Republic developed closer relations. After years of opposition, in 1971, the United States supported the revocation of Taiwan's seat in the United Nations and its replacement by the People's Republic. In 1979, bowing to the wishes of the People's Republic, the United States withdrew its official recognition of Taiwan, instead recognizing China's claim of sovereignty over Taiwan. This **One China Policy** prevents the United States from having formal diplomatic relations with Taiwan. However, the United States opposes the annexation of Taiwan to China by force, vowing to defend Taiwan from attack, just as it opposed Taiwan's call for a Nationalist effort to retake the Chinese mainland during periods of internal weakness there. In the meantime, the United States is a strong backer of Taiwan, supplying it with generous packages of weapons and economic aid.

Taiwan continues to resist mainland China's overtures for reunification, which include promises of broad autonomy, and is moving increasingly away from any type of reunion. That trend reflects Taiwan's ethnic makeup. Less than 15 percent of the island's people descend from the Nationalist refugees, who with their recent ties to the mainland are known as "mainlanders." Native non-Chinese people make up about 2 percent of the population. The vast majority of the island's inhabitants are descendents of Han peoples who emigrated from China as many as four centuries ago, notably the Fujianese, who came from the southeastern coastal province of Fujian, and the Hakka, from Guangdong province. This majority is much less mainland-oriented in political outlook. Despite discrimination by the Nationalists, these "indigenous" Chinese of Taiwan have enjoyed increasing political power. In 2000 elections, their Democratic Progressive Party defeated the Nationalists for the first time in the country's history. Though a majority of Taiwanese support continuing the status quo indefinitely, President Chen Shui-bian of the Democratic Progressive Party came under pressure in 2004 by advocates of Taiwanese independence to fulfill his campaign vows to achieve formal independence for Taiwan within four years.

With Chen's election, the relationship between Taiwan and China became very tense. The two countries must confront serious choices about economic and political relations. Taiwan's economy has become very dependent on mainland China, where Taiwanese companies take advantage of low-cost mainland labor (about one-fifth the cost in Taiwan for the manufacture and assembly of products) and where they have a huge market for their products. But there is strong pressure on President Chen to sever most ties with China and assert Taiwan's independence more forcefully. These steps could have large economic consequences and invoke military intervention from Beijing—and an American military response.

7.3.7 Mongolia

Mongolia (until recently known as the Mongolian People's Republic and popularly referred to as Outer Mongolia) was once part of the Chinese empire, but it became a separate Communist country in 1924. Relations with the former Soviet Union were close, both politically and economically. Mongolia had the distinction of being the first Asian country to abandon communism (in 1990). It has an area of about 600,000 square miles (960,000 sq km) and a population of about 2.6 million. Roughly 85 percent of the population is ethnic Mongol, and the largest minority (7 percent) is Turkic. Almost all residents of the country follow Tibetan (Lamaist) Buddhism. About 59 percent of Mongolia's population is urban. Ulan Bator (population 900,000) is the capital and only major urban center.

Mongolia was the home of Genghis Khan and his grandson, Kublai Khan, who in the second half of the 13th century assembled the largest land kingdom history has ever known. From the Korean Peninsula in the east, all across the top of Asia to near the eastern border of the North European Plain, the Mongols—without benefit of a written language, cities, or most of the trappings associated with "civilization"—controlled an empire of major magnitude. 163
And although they could not maintain that expansive empire, the Mongols were able to take over China in 1279 and control it until 1368, when the Ming Dynasty ascended.

The world's largest landlocked country, Mongolia contains large desert plains in the south and east, locally termed *gobis* (hence the Gobi Desert), mountain ranges in the west, and wooded hills and mountains to the north. Most of the grassy valleys that are the archetypal steppes of Asia are also in the north (•Figure 7.3.17). These steppes are favored lands for the vast herds of livestock, mainly sheep and goats, that Mongolians graze. Much of the population still derives its livelihood from animal husbandry. The Communist government made major efforts to promote farming (especially of wheat, barley, and potatoes), but even today the total amount of cultivated land amounts to no more than 1 percent of Mongolia's area. Mongolia's export income comes mainly from mineral products (especially copper), livestock and animal products, and textiles, including cashmere wool.

Political and economic changes are progressing rapidly. In 1990, the first multiparty elections ever held resulted in a coalition government committed to changing the Soviet-style command economy to a market economy. New private businesses are being established, and many state-owned enterprises and properties are being privatized. Commercial

• **Figure 7.3.17** Many lives and much lore of the Mongolian steppe are focused on horses.

banks, business schools, and a stock market have opened, and most Russian technical and military personnel have left.

Much economic hardship has accompanied the changeover. There is tension between the Mongolian majority and the Chinese minority that so far has prospered from capitalist reforms. In 2003, the government began the privatization of what was once the exclusively government-owned land, with every Mongolian now entitled to a free plot of land. There are fears that the purchase and parceling out of the grasslands will bring an end to Mongolia's ancient nomadic way of life, which is an extensive form of land use requiring unbridled access to pasture.

Module 7.4 turns to two lands inhabited mainly by distant kin of the Mongols: Japan and Korea. Both have had complex and often tense relations with China, which cast a large shadow on the western Pacific.

SUMMARY

- China is home to approximately 20 percent of the world's population on about 7 percent of the world's land. With 3.7 million square miles (9.6 million sq km) in area, it is the world's third-largest nation and the largest in East Asia.
- China's language, literature, government, agriculture, religion, and other major cultural traits diffused from China to other parts of East Asia during the last several millennia.
- China was subjected to colonial interests in much of the 19th century, including Britain's successful bid to create large-scale and profitable opium addiction. Several European countries and Japan won territorial and trade concessions from China that did not end until after World War II with the Communist defeat of the Nationalists.
- China may be physically divided into the arid west and the humid east. An arc drawn from China's border with Myanmar northeast to Harbin is the rough dividing line between the arid west and the humid east. The majority ethnic Han are associated with the more densely populated humid east, and non-Han realms are primarily in the arid west. The Huang He (Yellow River) and the Chang Jiang (Yangtze River) descend from the Tibetan highlands.
- Humid China, also known as China Proper, is densely settled, focused on irrigated rice production, and highly dependent on the Chang Jiang and other river systems. This region is further subdivided into North China and South China. In subtropical South China, the agriculture is centered on rice. North China, which is drier than South China and has a more continental climate, is more dependent on wheat, corn, soybeans, and sorghum. The Huang He deposits beneficial silt on the North China Plain but because of regular flooding is also called "China's Sorrow."
- The Northeast, historically called Manchuria by the West, was the early industrial heartland of China. It has coal, iron ore, petroleum, and well-developed transportation systems.
- The Three Gorges Dam, under construction on the Chang Jiang, is designed to control floods, provide hydropower, and allow transfer of water to the more arid north. The project has been both celebrated and lamented because of the scale of environmental change and social dislocation associated with it.
- Revolutionary campaigns in the first three decades of government after the 1949 defeat of the Chinese Nationalists led to experiments in communal agriculture, migration control, and backyard industrialization, all of which were less significant than China's opening to the Western free-market economy after the death of Mao Zedong in 1976. Since then, China has modeled much of its most dynamic economic growth after capitalist rather than Communist economic and political models.
- Chinese agriculture in the past four decades has involved expansion of farmland, increase in irrigated land, opening new land in areas not previously farmed, and the use of more chemical fertilizers, new plant stock, and slowly increasing agricultural mechanization. Much capital has also been spent on expanding the railroad network and highway systems and upgrading the canal and river systems.
- China's urban centers have continued to grow, with China now about 44 percent urban. Inexpensive labor helped fuel the booming industries of the coastal cities. There is still poverty in China, with the cities generally wealthier than rural areas. The most prosperous areas are in the southeast, along the east coast (including the Special Administrative Regions of Hong Kong and Macau, and several Special Economic Zones), and

in farmlands near large cities, while regions in the west, southwest, uplands, and Arid China have had less economic development. Jobs and wages are now growing in more areas, and labor shortages have developed in the coastal boomtowns.

→ Taiwan (the Republic of China) continues to claim to represent the true government of China and has a tense relationship with the mainland People's Republic of China. Taiwan's current president leans toward less political accommodation with the mainland, but his country is heavily invested in mainland China's development. The United States does not have diplomatic relations with Taiwan but supports it militarily and economically.

→ One of the first countries to abandon communism in the final decade of the 20th century, Mongolia has privatized its industries and lands, perhaps threatening the pastoral nomadic livelihood for which the country is famous.

KEY TERMS + CONCEPTS

Arid China (p. 367)
autonomous regions (p. 372)
bird flu (p. 381)
break-of-bulk point (p. 376)
Chang Jiang Water Transfer Project (p. 376)
"China's Sorrow" (p. 374)
Chinese Communist Party (CCP) (p. 367)
Chinese Revolution of 1911 (p. 367)
commune system (p. 378)
Cultural Revolution (p. 383)
Dalai Lama (p. 372)
extraterritoriality (p. 367)
Four No's Campaign (p. 378)
genuine autonomy (p. 372)
Great Leap Forward (p. 378)
Humid China (p. 367)
Hanification (p. 372)
karst (p. 375)
Long March (p. 367)
Middle Kingdom (p. 365)
multiple cropping (p. 375)
Nationalist Party (Kuomintang, or KMT) (p. 367)
One China Policy (p. 388)
"one country, two systems" (p. 385)
Opium War (p. 367)
pandemic (p. 381)
severe acute respiratory syndrome (SARS) (p. 380)
Special Administrative Region (SAR) (p. 385)
Special Economic Zones (SEZs) (p. 385)
sweatshops (p. 386)
treaty ports (p. 367)
Western Big Development Project (p. 372)

REVIEW QUESTIONS

1. What are some of the innovations that began in China and spread through eastern Asia and even farther afield?
2. What were some of China's experiences with foreign powers prior to 1949?
3. What were some of the major events and other milestones of China in the 20th century? Who were the major political figures associated with these events?
4. What are some of the major differences in physical and cultural geographies of Arid and Humid China? Of North and South China? What are the distinguishing characteristics of China's Northeast?
5. For what reasons were the Great Wall and the Grand Canal built?
6. What were the Great Leap Forward, the commune movement, and the Cultural Revolution?
7. What factors may threaten the continuation of China's economic boom?
8. How can China's economy be characterized? What is its "average" standard of living based on per capita GNI PPP? How effectively is the wealth distributed? Where is it concentrated geographically, and how is the government hoping to change that pattern?
9. Where are the main clusters of China's cities? What geographic and historical factors help explain these concentrations?
10. Where are China's Special Economic Zones? What does their distribution suggest about China's economic geography? What is the history and status of each of China's Special Administrative Regions?
11. What are the major environmental, economic, and political attributes of Taiwan? Of Mongolia?

NOTES

1. Senicyang, "Chinese Tea," February 18, 2006, http://my.opera.com/chinaculture/blog/show.dml/148037. Accessed October 12, 2007.
2. Jamil Anderlini and Mure Dickie, "Taking the Waters: China's Gung-Ho Economy Exacts a Heavy Price." *Financial Times,* July 24, 2007, p. 11.
3. Howard W. French, "Next Wave of Camera-Wielding Tourists Is from China." *New York Times,* May 17, 2006, http://travel.nytimes.com/2006/05/17/world/asia/17travel.html. Accessed October 13, 2007.

Friends take a self-portrait after making a temple offering, Tokyo.

MODULE 7.4

JAPAN AND THE KOREAS: ADVERSITY AND PROSPERITY IN THE WESTERN PACIFIC

On a cluster of islands off the eastern coast of Asia, the 127 million people of Japan operate an economy smaller only than that of the United States. This powerful economic engine is on mountainous, resource-poor, crowded lands about the size of California that are home to almost half as many people as the entire United States. Japan's dramatic rise from the ashes of World War II has drastically changed the world's economic geography. This module explains that phenomenal achievement and some of the challenges associated with it.

Although the Korean Peninsula has a seemingly marginal position on the Pacific Rim and the Eurasian landmass, it, too, has a central place in geopolitical affairs, mostly because of its strategic location between Japan, China, and Russia; the prosperity of South Korea; and the sometimes dangerous backwardness of North Korea. This module examines the problems of divided Korea and its prospects for reunification.

module outline

module objectives

This module should enable you to

- → Recognize the pattern of natural hazards associated with locations where plates of the earth's crust collide
- → Appreciate how countries burdened by a lack of natural resources can become prosperous by marshaling their human resources
- → Understand the unique challenges associated with a postindustrial society that has a prosperous and aging population
- → See how a country's geographic location can make it a target of division and conquest
- → Appreciate how different political and economic systems can produce dramatically different results for almost identical peoples

7.4.1 The Japanese Homeland

There are four main islands in the Japanese archipelago (•Figure 7.4.1). On the west, the Sea of Japan—which Koreans know as the East Sea—and the Korea Strait separate the main islands from Russia and Korea. Japan faces the open Pacific on the east. Hokkaido, the northernmost island, is close to Russian-controlled islands to its north. The scenic and busy Inland Sea separates Honshu, the largest island, from Shikoku and Kyushu. South of Kyushu, the smaller Ryukyu Islands run southwestward almost to Taiwan.

All of the main islands are mountainous, with the higher peaks mostly between 5,000 and 9,000 feet (c. 1,500–2,750 m) above sea level. The islands are volcanic (•Figure 7.4.2) and are subject to frequent and sometimes severe earthquakes (see Natural Hazards, pages 396–397). Located west of Tokyo, volcanic Mount Fuji is Japan's highest mountain at 12,388 feet (3,776 m). An **active volcano** (one that has had an eruption in recorded history), in the early 2000s Mount Fuji began experiencing "swarms" of earthquakes that many geologists interpreted as signs of an approaching eruption; it would be the first since 1707. Mount Fuji was sacred to the aboriginal Ainu people of Japan as long as 2,000 years ago. It is still a holy mountain in the national Shinto faith, and about 300,000 pilgrims and tourists visit its lofty summit each July and August, when it is free of snow. The journey to the peak of this most prominent symbol of Japan is a difficult one, as suggested in the Japanese proverb "It is as foolish not to climb Fuji-san as to climb it twice in a lifetime."

Climate and Resources

Japan's location east of the Eurasian landmass at latitudes comparable to those from South Carolina to Maine strongly influences its climates (see Figure 7.2, page 282). A humid subtropical climate extends from southernmost Kyushu to north of Tokyo, a humid continental long-summer climate characterizes northern Honshu, and Hokkaido has a humid continental short-summer climate. In lowland locations, where most of the people live, the average temperatures for the coldest month range from about 20° F (–7°C) in northern Hokkaido to about 40°F (4°C) in the vicinity of Tokyo and about 44°F (7°C) in southern Kyushu. Averages for the hottest month are only about 64°F (18°C) in northern Hokkaido but rise southward to about 80°F (27°C) around Tokyo and even higher farther south. Conditions are cooler in the mountains, with snowfalls common in winter in Honshu and Hokkaido. Hokkaido has twice hosted the Winter Olympics.

Japan is a humid country. Annual precipitation of more than 60 inches (c. 150 cm) is common from central Honshu south, and 40 inches (c. 100 cm) or more fall in northern Honshu and Hokkaido. Their location off the east coast of
Eurasia puts the islands in the path of the inward-blowing 283
summer monsoon from the Pacific, and summer is the main rainy season in eastern Japan. But the islands are also exposed to the seaward-blowing winter monsoon after it has crossed the Sea of Japan (East Sea), so much precipitation also falls in winter in western Japan. Japan is subjected in summer and fall to powerful typhoons, the great low-pressure storm systems known as hurricanes in the western Atlantic
region. These add to the woes of this hazard-prone country. 338, 531, 602

Japan's natural resource base is not large enough to meet the country's growing needs. Until the middle of the 19th century, the country was nearly self-sufficient economically, although the standard of living was low. Japan was already crowded with about 30 million people, but disease, starvation, abortion, and infanticide slowed the rate of natural increase. Industrialization stimulated rapid population growth in the second half of the 19th century and later. Population growth has recently plateaued, however, as delayed marriages and the rising costs of child rearing have lowered birth rates. Meanwhile, affluence has increased dramatically. The material demands of so many people enjoying a high standard of living in a country with limited natural resources require Japan to import an astonishing 97 percent of its raw materials and energy supplies and about 60 percent of its food.

Forests, which now cover about 70 percent of the land surface, illustrate the limitations of Japan's natural resources. When it was a preindustrial and early industrial society with a much smaller population, the nation's heavily forested

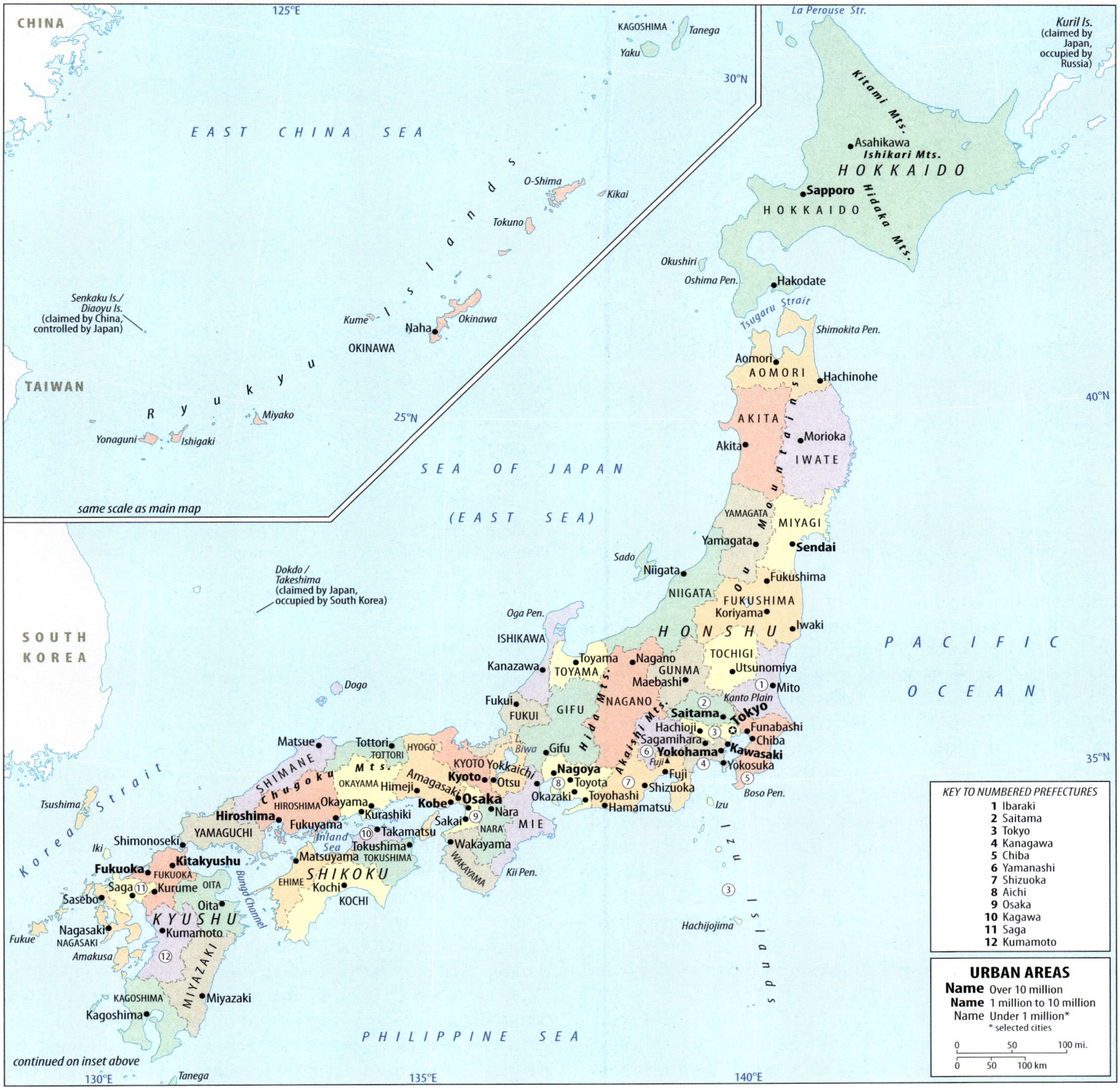

Figure 7.4.1 Principal features of Japan.

mountainsides supplied the wood it needed, but if today's demand for forest products were met with domestic supplies, the country would be rapidly deforested. A wide range of valuable trees grow in Japan, mainly broadleaf deciduous toward the south and needleleaf coniferous toward the north. Many mountainsides are quilted with rectangular plots of trees planted in rows. Most first-growth timber is long gone, but some primary forests are still protected in preserves. Forests are important in Japanese culture, and this affluent country does not wish to lose them. With its enormous demand for wood products, Japan is therefore the world's second-largest importer of wood (after China), mostly hardwoods from the tropical rain forests of Southeast Asia.

Michael S. Yamashita/Corbis

• **Figure 7.4.2** Japan has numerous volcanoes—some active and others extinct. In 1991, Mount Unzen, near Nagasaki on the island of Kyushu, erupted for the first time since 1792. Thirty-five people were killed as a pyroclastic flow inundated the village of Shimabara below.

Elizabeth Whiting & Associates/Alamy

• **Figure 7.4.3** The wood and paper characteristic of Japanese interior design contrast sharply with the drywall and metals of the West. This is a traditional Japanese guest room, or *ryokan*. The futons used for sleeping, stuffed with fragrant dried vegetation, are rolled up and put away in the daytime.

Kerosene stoves have largely replaced the charcoal formerly used for household cooking and heating, but the Japanese still use wood for many purposes. Japan's large publishing industry and other industries and handicrafts require a great deal of paper. Single-family dwellings continue to be built largely of wood, which provides more safety from earthquake shock waves than more rigid construction (•Figure 7.4.3). However, wood makes the dwellings more vulnerable to the fires that typically sweep through Japanese cities after earthquakes rupture gas lines.

Small deposits of minerals, including coal, iron ore, sulfur, silver, zinc, copper, tungsten, and manganese, supported early industrialization in Japan. These minerals are in short supply today, and like most other raw materials, they must be imported. Japan's mountain streams provided much of the power for early industrialization, but they are so swift, shallow, and rocky that they are of little use for navigation. Most of the water used for irrigation in Japan comes from these streams.

Only about one-eighth of this small country's area is arable; Japan is not one of the world's major crop-growing nations. Most cultivable land is in mountain basins and in small plains along the coast. Terraced fields on mountainsides supplement this level land. Irrigated rice, the country's basic food and most important crop, was once nearly everywhere on arable land (•Figure 7.4.4). Production of this grain is heavily subsidized by the Japanese government and promoted by a powerful farm lobby, which aims at rice self-sufficiency. The country's farmers usually produce a net export of rice. However, in the 1990s, a scarcity of rice throughout East Asia and pressure on Japan to reduce its massive trade surplus with the United States prompted Japan to open its rice market to foreign imports for the first time. Now with silos overflowing with rice and with the Japanese diet more dependent on wheat (in the form of bread and noodles) than on rice, the government is paying Japanese farmers not to grow rice. Japan's lavish subsidies of its declining, aging population of farmers are the subject of much condemnation by free-trade advocates.

Mild winters and ample moisture permit **double cropping**—the growing of two crops a year on the same field—on most irrigated land from central Honshu south. Rice is grown in summer, and wheat, barley, or another winter crop is planted after the rice harvest. Overall, about one-third of Japan's irrigated rice fields are sown to a second crop, and more than half of the unirrigated fields are double-cropped. **Intertillage**—the growing of two or more crops simultaneously in alternate rows—is also common. In addition, dry farming produces a variety of crops: sugar beets in Hokkaido; all sorts of temperate-zone fruits and vegetables; potatoes, peas, and beans, including soybeans; and in the south, tea, citrus fruits, sugarcane, tobacco, peanuts, and mulberry trees to feed silkworms, which eat their leaves.

Changes in Japan's agricultural sector have accompanied a surge in urban-industrial employment and general affluence in recent decades. Farmers have been moving to towns, and the number of agricultural workers has dropped dramatically (from about 25 percent of the labor force in 1962 to 4 percent in 2007). A huge increase in farm mechanization has made it possible to till Japan's tiny farms with far less human labor. With this technology and the heavy

Joe Hobbs

• **Figure 7.4.4** Rice fields near Tokyo. Note that people maximize the amount of land dedicated to rice by confining their settlements to the hills and that the forests on those hills have been preserved.

Joe Hobbs

• **Figure 7.4.5** Fish and seafood provide the major source of protein for the Japanese. *Joe Hobbs*

applications of fertilizers, plus improved crop varieties, Japanese agriculture has achieved the world's highest yields per unit of land.

The island people of Japan view the ocean as an important resource and a promising frontier. The country has an ambitious "inner space" program of deep-sea exploration to mine the sea's living and mineral riches. Many food fish species are in the waters around Japan, and Japanese fishermen range widely through the world's major ocean fishing
32 grounds. Japan takes more fish from the sea than any other nation—about 15 percent of the world's catch and the Japanese have the world's highest per capita consumption of fish and other marine foods, including *sashimi*, or raw seafood (•Figure 7.4.5). Some 40 to 50 percent of the protein consumed by the Japanese is in the form of fish (the figure is closer to 5 percent in the United States).

Historically, the Japanese have also enjoyed whale meat, but pressure from international environmentalists compelled Japan in 1985 to support a worldwide moratorium on
commercial whaling. However, Japan has resumed small- 123
scale whaling of minke, Bryde's, and sperm whales for what are called "scientific research" purposes. This loophole allows whale meat to be widely sold. Subjected to strong international condemnation for putting whale on the dinner table, Japanese whaling interests have accused outsiders of practicing "culinary imperialism."

In recent years, the Japanese have begun to eat less fish and rice in favor of a more Western diet. Although ocean fish continue to be the main sources of animal protein in the Japanese diet, increased consumption of meat, milk, and eggs is reflected in the growing number of cattle, pigs, and poultry on Japanese farms. The animals require heavy imports of feed, and the increasing variety in people's tastes means many more food imports. Despite the country's agricultural efficiency and in part because of its diversifying tastes, Japan is the world's largest agricultural importer country, relying on imports for 60 percent of its food needs. The United States is Japan's largest supplier of imported feeds and food, and the United States exports more food to Japan than to any other country.

Japan's Core

Most of Japan's economic activities and people are packed into a corridor about 700 miles (c. 1,100 km) long. This **megalopolis** (several large cities that make up an urban complex) extends from Tokyo, Yokohama, and the surrounding Kanto Plain on the island of Honshu in the east through northern Shikoku to northern Kyushu in the west (see Fig-
ure 7.4.1). This highly urbanized core contains more people 600
than live in the Boston-to-Washington, D.C., megalopolis on the U.S. East Coast. All of Japan's greatest cities and many less important ones have developed here, on and near harbors along the Pacific and the Inland Sea. They are sited typically on small, agriculturally productive alluvial plains between the mountains and the sea.

NATURAL HAZARDS

Living on the Ring of Fire

194, 336 The people of Japan must cope with one of the most hazard-prone regions on earth. By far the greatest threats are related to Japan's location on the Pacific **Ring of Fire**, the name given to the boundaries of the Pacific plate of the earth's crust where earthquakes and volcanic eruptions are commonplace and often catastrophic (•Figure 7.4.A). In the scheme of plate tectonics, Japan is in an area of subduction where one plate (the Pacific) is being pushed, as if by a conveyor belt, under another plate (the Eurasian), creating the country's characteristic volcanoes, earthquakes, and related features (see Figures 2.1 and 2.2, page 21).

Japan is most threatened by earthquakes on and near its islands but must also be on guard for the effects of distant earthquakes, even as far away as the shore of Nort h America. A strong quake there can produce a tsunami that may travel thousands of miles to strike and flood coastal Japan (see pages 340–341). Japan has an extensive (and expensive) system of seawalls to protect coastal communities from tsunamis. Critics argue that the walls are not high enough and promote a false sense of confidence that people can control nature.

Japan has a long history of seismic assaults. One of the most significant recent events took place on January 17, 1995, when a devastating earthquake released the tension that had built up on a transverse fault near historic Kobe, a port city of 1.5 million people. The Japanese soon labeled the 6.9-magnitude episode the Great Hanshin Earthquake, establishing its place in the annals of Japanese disasters alongside such events as the Great Kanto Earthquake (magnitude 7.9 on the Richter scale, which killed 143,000 people in and around Tokyo in 1923.

The Kobe quake was the worst disaster to hit Japan since World War II, killing 6,400 people and injuring thousands more, damaging 190,000 buildings, displacing 300,000 people, and causing at least $100 billion worth of damage. Kobe's remarkable lack of disaster preparedness and relief focused attention on a common Japanese complaint: that the government invests too much in commerce-oriented development and not enough in the welfare of its people.

Japanese authorities did not predict the Great Hanshin Earthquake. After the fact, there were reports that large schools of fish swam uncharacteristically close to the surface of the sea off Kobe in the days preceding the quake. Crows and pigeons were reportedly noisier than usual and flew erratically in the hours before the ground moved. In the weeks preceding the quake, levels of radon gas rose in a water aquifer below the city and then declined sharply after January 9. Animal behavior, water level and composition, and many other variables are often debated but still unproved barometers of an impending earthquake.

Since 1965, Japan has spent more than $1 billion in an effort to develop means of predicting earthquakes. During the 1990s, more than $100 million yearly went to this effort. But by 1998, no earthquake had been successfully predicted, and Japan suspended this expensive research program. Its critics argue that earthquakes are chaotic events and will always be impossible to predict. They also point out the hazards of issuing an earthquake warning that is not followed by a real quake. A false alarm in the Tokyo region would cost an estimated $7 billion per day in lost business and other economic disruptions. The bigger question, also applicable to such cities as San Francisco and Mexico City, is how such a great metropolis could be evacuated safely, efficiently, and with minimal panic.

With the prediction program scrubbed, Japan has redoubled its efforts to prepare for future "big ones." New structures must meet a tough safety code that aims at "earthquake-proof" construction. Much of the focus is on the Japanese capital, Tokyo, which is underlain by a complex system of active fault lines and typically suffers a big quake every 70 years (in 2007, it had been more than 80 years since the last one). Current estimates are that a big one nearing 8.0 on the Richter scale would kill 6,717 people and destroy 300,000 structures (mostly by fire) in Tokyo. These numbers do not include the effects of the tsunami an estimated 27 feet (8 m) high that would strike low-lying Tokyo within minutes of an offshore quake. Much of the devastation in Kobe resulted from the city's location on soft coastal soils and reclaimed land, which also underlie large parts of Tokyo.

With such threats in mind, the government began serious discussions about moving most national administrative functions inland to a new capital, where tsunamis are not an issue and where mountain bedrock would reduce the devastating impacts of seismic shock waves. Two sites on the island of Honshu have the most

The largest urban complex in the world—Tokyo, Yokohama, and the surrounding suburbs (population 37 million by one estimate; figures differ according to how the metropolitan area is defined)—has developed on the Kanto Plain, Japan's most extensive lowland. Tokyo grew as Japan's political capital and Yokohama as the area's main seaport. This urban complex functions today as Japan's national capital, one of its two main seaport areas (the other is Osaka-Kobe), its leading industrial center, and the commercial center for northern Japan (•Figure 7.4.6).

Beyond Tokyo, in the entire northern half of Japan, the only major city is Sapporo (population 2.5 million), on Hokkaido. About 200 miles (c. 320 km) west of Tokyo lies Japan's second-largest urban cluster, in the Kinki District at the head of the Inland Sea. Here three cities—Osaka, Kobe, and Kyoto—form a metropolis, together with their suburbs, of about 17.5 million people.

The inland city of Kyoto was the country's capital from the 8th century C.E. until 1869 (when it was moved to Tokyo) and is now preserved as a shrine city where modern

NATURAL HAZARDS

Living on the Ring of Fire *continued*

Japan is riddled with earthquakes because a huge slab of Pacific crust slides very slowly beneath the archipelago, shaking it up periodically. Quakes occur as plates fracture and rub against one another.

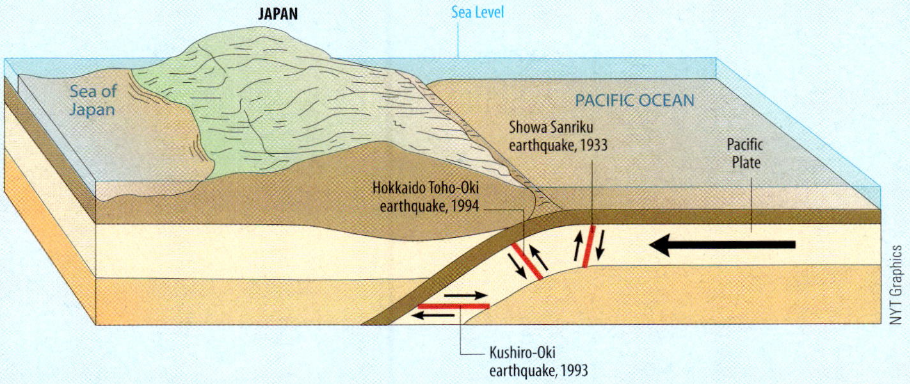

The major earthquakes that have hit Japan since the year 599. The bigger and darker the dot, the greater the energy released.

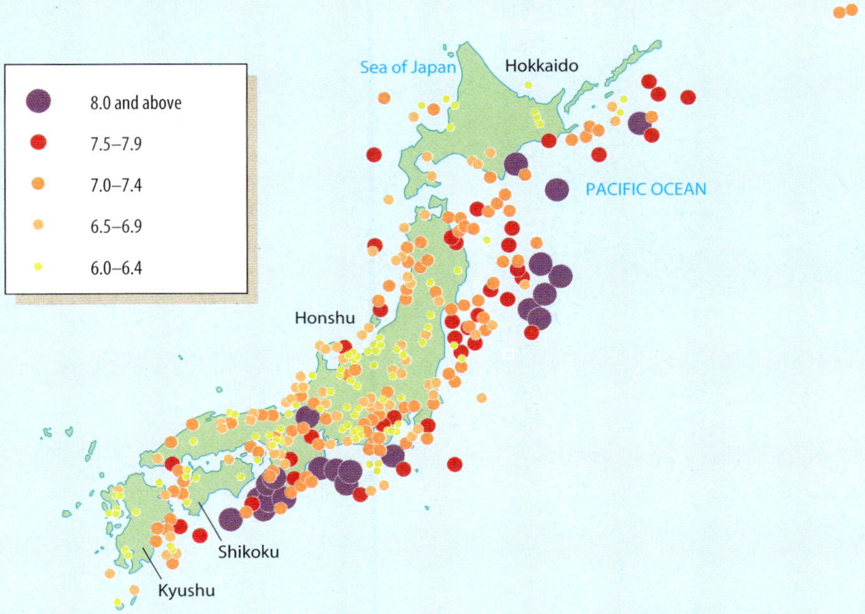

● **Figure 7.4.A** Japan's earthquakes are a consequence of the convergence of great slabs of the earth's crust.

attention as possible new capitals: Hokuto, in the northeast near Sendai, and Tokai, west of Tokyo near Nagoya. Japan's struggle against the seismically inevitable is a national priority and a symbol of the nation's determination to maintain economic success despite environmental adversity.

industrial disfigurement is prohibited (●Figure 7.4.7). Some U.S. military planners wanted to target Kyoto with a nuclear weapon in August 1945, but the decision that it should be avoided because of its historical importance spared the city. Kyoto continues to be the destination of millions of pilgrims and tourists annually.

Osaka developed as an industrial center and Kobe as the district's deepwater port. Both cities now combine port and industrial functions. Another large metropolis, Nagoya (population 8.8 million), is situated directly between Tokyo and Osaka on the Nobi Plain at the head of a bay. It is both a major industrial center and a seaport.

West of the Kinki District, smaller metropolitan cities spot the core. The largest are metropolises of more than 1.5 million people: the resurrected Hiroshima on the Honshu side of the Inland Sea, Kitakyushu (a collective name for several cities) on the Strait of Shimonoseki between Honshu and Kyushu, and Fukuoka on Kyushu. Hiroshima was destroyed on August 6, 1945, when the United States dropped the first atomic bomb ever used in warfare, detonating it

Joe Hobbs

• **Figure 7.4.6** Tokyo's Ginza District.

Bruce Burkhardt/Westlight

• **Figure 7.4.7** Kyoto has a unique status in Japanese history and religious life. This is the 14th-century Kinkakuji Temple, also known as the Golden Pavilion.

directly over the center of the city. With no topographic barriers to deter the effects of the explosion, the city was obliterated, and an estimated 140,000 residents died by the end of 1945, when radiation sickness had taken its greatest toll.

Among the several smaller cities of northern Kyushu is Nagasaki, which, on August 9, 1945, was hit by a second U.S. atomic bomb, the last used in warfare. Nagasaki was chosen during the flight mission after clouds obscured the primary target, Kokura. The bomb detonated over the suburbs rather than the city center, and hills helped diminish the explosion's impact, but even so, an estimated 70,000 people died by year's end. The bombing of Nagasaki brought Japan's surrender and the end of World War II. American decision makers insisted that the use of these weapons spared many thousands of lives, both American and Japanese, that would have been lost if the United States had instead undertaken an invasion of the Japanese homeland. Japan, which has always utterly condemned the bombing, has an official policy that it will never develop or use atomic weapons. 303

7.4.2 Historical Background

Japan has seen successive periods of isolationism and expansionism and of economic and military accomplishment and defeat. Today, it is near the top of most indicators of prosperity in the more developed countries (MDCs). The remarkable success story of Japan can best be appreciated by considering its turbulent history and the difficult home environment in which the Japanese have lived.

The Japanese, Japan's dominant ethnic group, are descended from peoples who reached Japan from other parts of eastern Asia in the distant past, especially after 7000 B.C.E. Most of these ancestors of the modern Japanese may have arrived at about the same times as the **Ainu.** Often described as the original indigenous group of Japan, the Ainu (also called **Utari**), numbering perhaps 20,000 today, are unrelated to the ethnic Japanese. Scholars dispute their ethnic and geographic origins. The majority Japanese gradually drove the Ainu into outlying areas, mainly in Hokkaido.

The oldest surviving Japanese written records date from the 8th century C.E. These depict a society strongly influenced by Chinese cultural and ethnic traits, often reaching

Japan by way of Korea. A distinct Japanese culture gradually evolved, and today, it is linguistically and in other ways
293 different from its Chinese and Korean antecedents. Traditions about Japanese origins extend back to the reign of Jimmu, the first emperor, whose accession is ascribed to the year 660 B.C.E. but who may be a mythical rather than a historical figure. Traditions of kingship are important in the Japanese worldview, which holds that the islands and their emperor are of divine origin. Jimmu was said to have descended from the Sun Goddess, and the Japanese have known their homeland as the "Land of the Gods." The Japanese concept of Japan as unique and invincible developed early, shaped by legends such as that of the ***kamikaze,*** the "divine wind" that repelled a Mongol attack on Japan in the 13th century.

The early emperors gradually extended control over their island realm. A society organized into warring clans emerged. By the 12th century, powerful military leaders called ***shoguns,*** who actually controlled the country, diminished the emperor's role. Meanwhile, nobles, or ***daimyo,*** whose power rested on the military prowess of their retainers, the ***samurai,*** ruled the provinces. This structure resembled the European feudal system of medieval times.

Adventurous Europeans reached Japan early in the mid-16th century, beginning with the Portuguese in the 1540s. Most of these early arrivals were merchants and Roman Catholic missionaries. Japanese administrators allowed the merchants to trade and the missionaries to preach freely. There were an estimated 300,000 Japanese Christians by the year 1600.

Contacts with the wider world were curtailed as Japan entered a period when military leaders imposed central authority on the disorderly feudal structure. After winning the battle of Seikigahara in 1603 with arms purchased from the Portuguese, the warlord Ieyasu Tokugawa proclaimed himself Nihon Koku Taikun ("Tycoon of All Japan") and became the first *shogun* to rule the entire country. During the era of the **Tokugawa Shogunate,** from 1600 to 1868, the Tokugawa family acquired absolute power and shaped Japan according to its will. To maintain power and stability, the early Tokugawa *shoguns* wanted to eliminate all disturbing social influences, including foreign traders and missionaries. They feared that the missionaries were the forerunners of attempted conquest by Europeans, especially the Spanish, who held the Philippines. The *shoguns* drove the traders out and nearly eliminated Christianity in persecutions during the early 17th century. After 1641, a few Dutch traders were the only Westerners allowed in Japan. Their Japanese hosts segregated them on a small island in the harbor of Nagasaki, even supplying them with a brothel so that they would not be tempted to venture out and pollute Japan's ethnic integrity. Under Tokugawa rule, Japan settled into two centuries of isolation, peace, and stagnation.

Foreigners succeeded in reopening the country to trade two centuries later, when Japan was ill prepared to resist Western firepower. Visits in 1853 and 1854 by American naval squadrons of black ships under Commodore Matthew Perry (who wanted to establish refueling stations in Japan for American whaling ships) resulted in treaties opening Japan to trade with the United States; this became known as **gunboat diplomacy.** The major European powers were soon able to obtain similar privileges. Some feudal authorities in southwestern Japan opposed accommodation with the West. United States, British, French, and Dutch ships responded in 1863 and 1864 by bombarding coastal areas under the control of these authorities. These events so weakened the Tokugawa Shogunate that a rebellion overthrew the ruling shogun in 1868. The revolutionary leaders restored the sovereignty of the emperor, who took the name Meiji, or "Enlightened Rule." The revolution of 1868 is known as the **Meiji Restoration.**

The men who came to power in 1868 aimed at a complete transformation of Japan's society and economy. They felt that if Japan were to avoid falling under the control of Western nations, its military impotence would have to be remedied. This would require a reconstruction of the Japanese economy and social order. They approached these tasks energetically and abolished feudalism, but only after a bloody revolt in 1877. The Meiji leaders forged the model of a strong central government that would modernize the country while resisting foreign encroachment. Under this system, which lasted until 1945, democracy was strictly limited, with small groups of powerful men manipulating the government's machinery and the emperor's prestige. Military leaders were prominent in the power structure.

Following their fact-finding tours of Europe and the United States, Meiji leaders pressed their people to learn and apply the knowledge and techniques that Western countries had accumulated during the centuries of Japan's isolation. Foreign scholars were brought to Japan, and Japanese students were sent abroad in large numbers. Japan adopted a constitution modeled after that of imperial Germany. The legal system was reformed to be more in line with Western systems. The government used its financial power, which it obtained from high land taxes, to foster industry. New developments included railroads, telegraph lines, a merchant marine, light and heavy industries, and banks. Wherever private interests lacked capital for economic development, the government provided subsidies to companies or built and operated plants until private concerns could acquire them. Rapid urbanization accompanied a process of rapid industrialization.

So spectacular were the results of the Meiji Restoration that within 40 years, Japan had become the first Asian nation in modern times to emerge as a world power. Outsiders often spoke of the Japanese in derogatory terms as mere imitators. The Japanese, however, knew which elements of Western technology they wanted, and they adapted them successfully to Japanese needs.

The shift from Tokugawa isolationism to Meiji openness put Japan on a pathway to empire and eventually to war.

Japan was bent on expansion between the early 1870s and World War II, and by 1941, it controlled one of the world's most imposing empires (• Figure 7.4.8). Between 1875 and 1879, Japan absorbed the strategic outlying archipelagoes of the Kuril, Bonin, and Ryukyu Islands. Japan won its first acquisitions on the Asian mainland by defeating China in the Sino-Japanese War of 1894–1895, which broke out over disputes in Korea. In the humiliating 1895 Treaty of Shimonoseki, China ceded the island of Taiwan (Formosa) and Manchuria's Liaotung Peninsula to Japan and recognized the independence of Korea.

Feeling threatened by Japan's emerging strength so close to its vital port of Vladivostok, Russia demanded and won a Japanese withdrawal from the Liaotung Peninsula and leased the strategic peninsula from China. Japan went to war with Russia in 1904 after attacking the peninsula's main settlement of Port Arthur. Russia lost the ensuing Battle of Mukden, which involved more than a half-million soldiers, making it history's largest battle to that date. Russia's loss of the Russo-Japanese War of 1904 represented the first defeat of a European power by a non-European power. Victorious Japan regained the Liaotung Peninsula, secured southern Sakhalin Island, and established Korea as a Japanese protectorate; Korea was formally annexed to Japan in 1910.

At the end of World War I, the Caroline, Mariana, and Marshall Islands, in the western Pacific Ocean, were transferred from defeated Germany to Japan as mandated territory. Japanese encroachments on China's region of Manchuria followed in the 1930s. In 1937, Japan began an all-out attack that overran the most populous parts of China. Japanese troops committed many atrocities in the assault on China, even employing biological warfare by airdropping plague-infested fleas on the city of Ningbo. The most notorious was the **Rape of Nanking** (now known as Nanjing) in 1937, when the invaders took control of China's temporary capital, killing an estimated 300,000 Chinese civilians and soldiers and raping about 20,000 Chinese women. All told, 383 an estimated 2.7 million Chinese civilians were killed in Japan's effort to "pacify" China.

The encroachments continued. Japan seized French Indochina in 1940. After Japan's carrier-based air attack devastated the United States' Pacific Fleet at Pearl Harbor, Hawaii, on December 7, 1941 ("a day which will live in infamy," declared President Franklin Roosevelt), Japanese forces rapidly overran Southeast Asia as far west as Burma (Myanmar), as well as much of New Guinea and many smaller Pacific islands.

The motives for Japanese expansionism were mixed. They included a perception of national superiority and "manifest destiny," perceived security needs, a desire for recognition by the great powers, the wishes of military leaders to inflate their importance and gain control of the Japanese government, and the ambitions of industrialists to gain sources of raw materials (Manchurian and Korean coal, for example) and markets for Japanese industries.

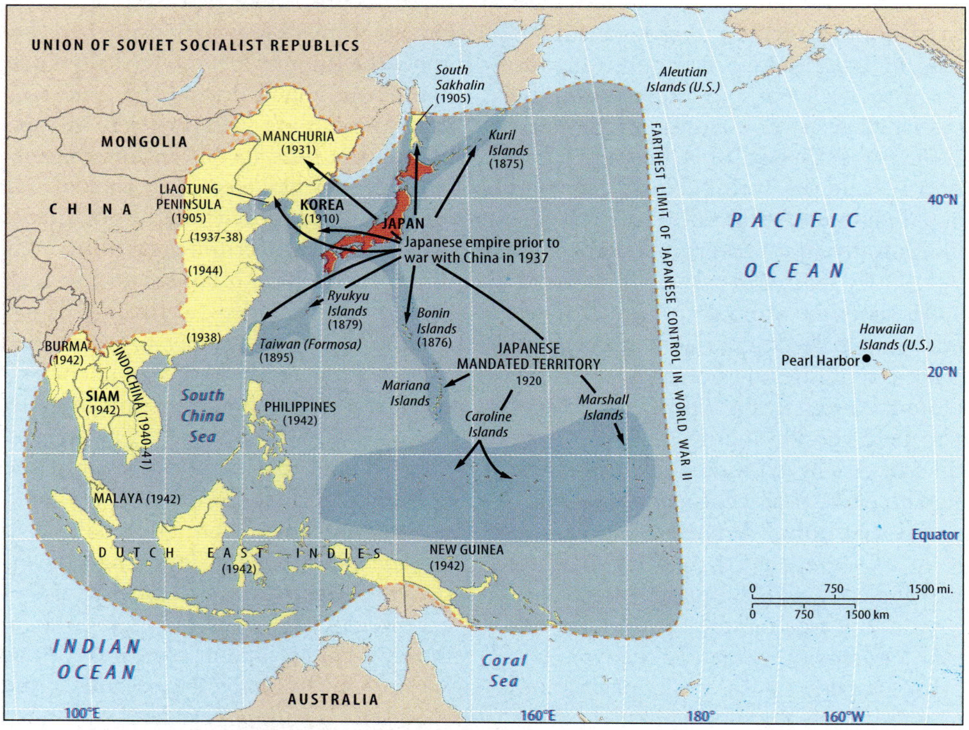

• **Figure 7.4.8** Overseas areas held by Japan prior to 1937 and the line of maximum Japanese advance in World War II.

The desire for materials and markets was based solidly on need. Expansion of industry and population on an inadequate base of domestic natural resources had made Japan dependent on sales of industrial products outside the homeland. Such sales provided, as they do now, the principal funds for buying necessary imported foods, fuels, and materials. As worldwide depression hurt the trade-dependent prewar Japanese economy, Japan chose to create a Japanese-controlled Asian realm that would insulate the Japanese economy from the volatile world economy and make the country a leading world power. In 1938, Japan therefore proclaimed its "New Order in East Asia," and in 1940 and 1941, this widened into the **Greater East Asia Co-Prosperity Sphere**, a euphemism for Japanese political and economic control over China and Southeast Asia.

Japan's militaristic and colonial enterprises proved disastrous at home and abroad. Japanese soldiers inflicted great suffering on civilians throughout the western Pacific. Even today, nations are demanding formal apologies from Japan. In August 1995, on the 50th anniversary of Japan's defeat in World War II, Japan's prime minister did issue a formal general apology for his nation's role in the war. Critics say Japan has not fully atoned for its war crimes and are trying, among other efforts, to block Japan's quest to gain a permanent seat on the United Nations Security Council. Annual visits by the Japanese prime minister to the Yasukuni War Memorial are accompanied by annual protests by the Chinese and Koreans, angry that 14 of the 2.5 million people enshrined there are convicted war criminals. There continue to be official protests by the governments of China, Korea, Taiwan, the Philippines, and other countries about Japan's reluctance to admit that it enslaved women as prostitutes for Japanese soldiers during World War II. An estimated 200,000 of these "comfort women" were put to work in brothels throughout the Japanese war empire.

August 1945 found Japan completely defeated. Its overseas territories, acquired during nearly 70 years of successful imperialism, were lost. Its great cities, including Tokyo—where 100,000 people died in napalm bombings—were in ruin. In addition to 1.8 million deaths among its armed services, Japan had suffered some 8 million civilian casualties from the American bombing of the home islands; most of its major cities were over half destroyed. Tokyo's population had fallen from nearly 7 million people to about 3 million, most of them living in shacks. In 1950, five years after the beginning of the U.S. military occupation of Japan, national production stood at only about one-third of its 1931 level, and annual per capita income was $32.

7.4.3 Japan's Postwar Miracle

Stripped of its colonies and its empire, Japan has astonishingly become an economic superpower since World War II. The nation's explosive economic growth after its defeat was one of the most remarkable developments of the late 20th century and is known widely as the "Japanese miracle."

Japan watchers cite different reasons for the country's
economic success. Proponents of dependency theory argue 45
that Japan, never having been colonized, escaped many of the debilitating relationships with Western powers that hampered many potentially wealthy countries. Some analysts believe that the country's postwar economic accomplishments grew from an intense spirit of achievement and enterprise among the Japanese. Notably, many Japanese attribute this industrious spirit to Japan's geography as a resource-poor island nation. To overcome the constraints nature has placed on them, the Japanese people feel they must work harder.

Still others attribute Japan's postwar achievements to its association with the United States following World War II. Some of the postwar U.S.-imposed reforms worked well economically, and relations between the two countries since the war have also generally stimulated Japan's industrial productivity. The U.S. military occupation of Japan from 1945 to 1952 brought about major changes in Japanese social and economic life. The divine status of the emperor was officially abolished. When Emperor Hirohito announced Japan's surrender on the radio, it was the first time the stunned Japanese public had ever heard the voice of this mythic figure. A new U.S.-written constitution (which has remained unchanged ever since) made Japan a constitutional monarchy with an elected parliamentary government. With the rejection of divine monarchy, the strongly nationalistic Shinto faith lost its status as Japan's official religion, although worship at Shinto shrines was allowed to continue; most Japanese observe a faith that is a syncretism
of Shinto and Buddhist beliefs and rituals (• Figure 7.4.9). 294
There was land reform in the countryside to do away with a feudal-style landlord system. Some major Japanese companies were broken up to reduce their monopolistic hold on the economy. Women were enfranchised. A democratic trade union movement was established.

Joe Hobbs

• **Figure 7.4.9** Prayers at the Meiji Jinju Shrine, Tokyo.This Shinto shine is dedicated to the deified spirits of the Emperor Meiji and his spouse, Empress Shoken, who died early in the 20th century.

Perhaps most significant for the economy, the United States forbade Japan to rearm, except for its small **Self-Defense Forces (SDFs).** When the occupation ended in 1952, leaving U.S. bases in Japan but returning control of Japanese affairs to the Japanese, Japan was placed under American military protection. Without large military expenditures, much of Japan's capital was freed to invest in economic development. Only recently has Japan sought a wider role for its military—for example, by posting some forces to serve in Iraq. As early as 2010, the Japanese may hold a referendum to change its pacifist constitution. The United States, which wrote the constitution, is in favor of this change because it would allow the two countries to form a stronger military counterweight against prospective foes in the region.

The U.S. military umbrella over Japan remains controversial today. Many Japanese regard it as an outmoded vestige of colonialism and the war. There have been several incidents of Japanese women being raped by U.S. soldiers stationed on the island of Okinawa, where three-quarters of the U.S. bases and more than half of the 50,000 U.S. troops in Japan are stationed. The resulting protests and diplomatic appeals led the United States to agree to return control of a major air base in Okinawa to Japan, to shift base locations on the island, and to relocate some troops from Okinawa to mainland Japan. In a subsequent referendum, the people of Okinawa voted overwhelmingly in favor of further U.S. military withdrawal from the island.

Japan's special relationship with the United States boosted the economy long after the American occupation ended. The postwar economic recovery accelerated when the United States called on Japanese production of steel, textiles, and clothing to support American forces in the Korean War of
406 1950–1953. American economic aid continued to flow to Japan. The United States permitted many Japanese products to have free access to the American market while allowing Japan to protect its economy from imports. Japanese firms also had relatively free access to American technologies, which they often improved on and marketed even before American firms did.

By the 1970s, Japan had become an industrial giant with a gross domestic product far exceeding that of any country except the United States and the Soviet Union. The United States and European nations found themselves at a disadvantage to Japanese competitors in many industries. They busily sought ways to match this competition or to protect themselves from it without doing too much damage to their exporting firms, consumers, and trade relations. By the 1980s, the United States and Japan were at odds over Japan's massive trade surplus with the United States.

Observers often point to several unique features of Japanese management and employment to explain the country's meteoric postwar gains. One has been recruitment through an extremely challenging (some say brutal) educational system that emphasizes technical training. While recent years have seen a move to a more lenient and more comprehensive education, a rigorous and stressful testing system still determines access to higher education. Japanese management strategies emphasize benevolence toward employees, encouragement of employee loyalty, and participation of workers in decision making. About 20 percent of Japanese workers enjoy guarantees of lifetime employment in their firms, and many large Japanese companies help provide housing and recreational facilities for their employees.

One essential factor in Japan's postwar economic growth was a high level of investment in new and efficient industrial plants. Investment capital was freed by government policies that cut expenditures on amenities and services such as roads, antipollution measures, parks, housing, and even higher education. Japanese industrial cities grew explosively and became highly polluted areas of dense and inadequate housing, with few public amenities and snarled transportation. The money "saved" by not being spent on amenities and services was made available for investment in industrial growth, but the costs to Japanese society were high.

Some analysts cite elements of Japan's political culture to explain the country's economic successes. One political party, the **Liberal Democratic Party (LDP),** has with few exceptions been elected to power since Japan regained its sovereignty in 1952. It is a conservative and strongly business-oriented and business-connected organization. The party has promoted Japanese exports with policies to keep the yen (the Japanese unit of currency)—and thus Japanese goods—inexpensive. The party has also cooperated closely with Japanese business in the development of new products and new industries. However, that very coziness between government and business—the "crony capitalism" seen in several other East and Southeast Asian countries—was partly 344
to blame for the economic malaise that spread over Japan in the 1990s, and support for the LDP began to erode.

Despite all of these factors in Japan's economic favor, the Japanese miracle did not last. The peak of Japan's postwar success came in the 1980s. A powerful economic boom led to speculative rises in stock prices and land prices. At the end of the 1980s, the total value of all land in Japan was four times greater than that in the United States. The Tokyo Stock Exchange was the world's largest, based on the market value of Japanese shares. Then the **bubble economy** burst, and real estate and stock prices fell by more than 50 percent. Japan had near-zero economic growth through most of the 1990s and into the early 2000s. Japanese industries whose growth had seemed unstoppable experienced an increasing loss of market share to U.S. and European producers. Part of the reason the economy could not regain its footing was official Japanese commitment to prop up faltering industries and farms with huge subsidies. Because of the disproportionate political influence of small towns and rural areas, the Japanese tried to modernize remote villages and islands with expensive and underutilized public works projects. Due to such projects, Japan is now the world's most indebted country. Overall, the system succeeds in subsidizing its past at the expense of its future.

7.4.4 Japanese Industry

The gradual evolution of industry over more than three centuries that characterized Europe and the United States has been compressed into little more than a century in Japan. Early iron and steel development was concentrated in northeastern Kyushu, on and near the country's main coal reserves. The products went mainly into a new railway system and into shipping. Japanese coal, supplemented increasingly by imports, powered the country's factories, railways, and ships. (•Figure 7.4.10). Hydroelectricity was developed on mountain streams and became important in the energy economy.

Japan's petrochemical and other oil-based industries reflect the country's great dependence on imports. Almost devoid of petroleum reserves, the country imports nearly all of its oil and gas. With about 75 percent of its oil coming from the Persian-Arabian Gulf region—and no sign that that level of dependence will decline—Japan downplays its alignment with the West in most Middle East disputes. Japan briefly committed ground troops after the 2003 Iraq War began and maintained a naval presence to support U.S.- and NATO-led operations in Iraq and Afghanistan for what it insisted were humanitarian reasons.

Despite the unique stigma Japan associates with the power of the atom, the country has come to rely on nuclear power, with 55 plants supplying about 30 percent of the country's electricity needs. Hydropower supplies 8 percent, but with fossil fuels almost absent, the critical lack of energy has compelled the nation to develop the world's most energy-efficient economy. The government imposes high taxes on fossil fuels and uses much of that revenue to invest in and subsidize energy alternatives such as solar power and voltaic cells. Japan has also persisted with the nuclear technology to which many of its people are opposed. Small fires, explosions, and other accidents in nuclear power and reprocessing plants in recent years—including a nuclear plant fire triggered by a 2007 earthquake—have prompted new domestic concerns about the safety of nuclear energy, forcing a slowdown in Japan's nuclear industry.

• **Figure 7.4.10** The Port of Tokyo. The island nation of Japan thrives on seaborne trade.

There are also international fears. Japan imports large quantities of plutonium, primarily from Europe, for use in its nuclear power industry, and the long-distance ocean shipment of this very hazardous material has contributed to Japan's poor reputation in environmental matters. International protests against the plutonium shipments have caused Japan to postpone the construction of a series of nuclear breeder reactors, which use and create recyclable plutonium.

The phenomenal advance of industry since the early 1950s has been marked by a series of booms and declines in various sectors. In the 1960s, electronics, cameras and optical equipment, petrochemicals, synthetic fibers, and automobiles (notably those produced by Mazda, Honda, Toyota, Mitsubishi, and Nissan) became boom industries. In the 1970s and 1980s, Japan stepped up the manufacture of computers and robots. By the early 1980s, Japan led the world in the robotization of industry. For the first time in the postwar era, Japanese industry in the 2000s—when Japan began offshoring a growing share of its manufacturing to China—was not characterized by world supremacy in any single manufacturing sector. Japanese industry must now look for new directions in which to establish leadership. As always, to overcome the severe limitations imposed by small, resource-poor islands, the Japanese must live by their wits.

7.4.5 The Industrious People behind Japanese Industry

Japan has one of the world's most homogeneous populations; it is 99.5 percent ethnic Japanese. The largest non-Japanese group is the 600,000 strong community of ethnic Koreans, the people and their descendants whom Japan imported during World War II to do mainly hard manual labor. These ethnic Koreans have long suffered from discrimination in Japan. So have the ethnic Japanese social outcasts known as the **Burakim**, descendants of people who practiced "impure" trades like butchering and street sweeping as early as the 8th century. They still tend to be marginalized in low-paying jobs and forced to live in ghettos. Other than Koreans, few non-Japanese have settled in the country. Only with the country's booming prosperity and growing labor shortages of the 1980s did Japan open its doors to a few unskilled and low-skilled immigrants from countries such as Pakistan, Bangladesh, Thailand, Peru, and Brazil.

This lack of racial and ethnic diversity has had mixed results for Japan. Many observers believe that it has helped the country achieve a sense of unity of purpose, allowing the Japanese to persist through periods of adversity, especially the postwar years of reconstruction. However, the Japanese have also earned a reputation for intolerance of ethnic minorities. This attitude may be costly for Japan's future economic development. One way to increase productivity in the face of Japan's declining population would be to import skilled workers from abroad, but the country remains averse to immigration. Inward-looking tendencies have also left Japan short of people capable of and interested in learning English, a language routinely sought by aspiring businesspeople in nearby South Korea and China.

Capitalist Japan has had a remarkably egalitarian society, with about 80 percent of the population solidly middle-class. The richest third of the population has a total income just three times as great as that of the poorest third (compared with a fivefold differential in the United States). The Japanese have historically embraced the concept of ***wa,*** or harmony, based in part on the principle of economic equality. However, the recessions of the 1990s and early 2000s, followed by an economic rebound, resulted in growing joblessness and homelessness in Japan and a widening gap between haves and have-nots (• Figure 7.4.11). The country's official unemployment rate stood at 3.8 percent in 2007, compared with 2.1 percent in 1990. There are growing ranks of "permanent temporary workers," mainly young people who live with their parents to keep their expenses low.

Joe Hobbs

• **Figure 7.4.11** In a society that has traditionally prided itself for its harmony and egalitarianism, some have fallen through the cracks in recent years.

The government has balked at many suggestions for economic reform because of the threat they pose to a socially harmonious Japan. There have been loud international appeals for Japan to reform its economic system by allowing inefficient businesses to collapse (rather than prop them up with expensive subsidies) and to resist the temptation to satisfy marginal rural populations with costly education and public works projects.

The legendary Japanese work ethic has had its advantages and drawbacks. It has helped the Japanese create a prosperous country, but it has also created a nation of workaholics beset with the same problems—stress, suicide, depression, and alcoholism—experienced by workers in the world's other MDCs. There is a growing incidence of what Japanese call ***karoshi,*** or death by overwork. The educational system encourages children to be highly successful but also to conform; critics say this stifles creativity and innovation. College graduates who are not hired during the annual recruiting season face difficult obstacles to entering the job market. And for those it affects, the practice of lifetime employment makes it difficult to change jobs. The country's welfare system, pensions, and social security are inadequate, compelling workers to work harder for savings. In their long hours at work, men are accustomed to spending little time with their spouses and children.

Women have not achieved parity with men in the workplace, and they complain increasingly of discrimination and sexism. In struggling to increase their footing in Japanese society, growing numbers of Japanese women are both working longer hours and marrying later, contributing to Japan's remarkably low birth rate (9 per 1,000 annually, one of the lowest in the world). Ironically, Japan's falling birth rates will only increase the burdens of its workers (• Figure 7.4.12). The population is both shrinking (it is expected to go from 127 million in 2007 to 100 million in 2050 and 67 million in 2100) and aging (already one in five people were over age 65 in 2007, and 1 million people a year join those ranks). The Japanese of working age will face an increasing load of taxes and family obligations to meet the needs of older citizens.

Crowding and overdevelopment, accompanied by a surprising lack of amenities, are inevitable facts of life in Japan. Officially designated parks comprise 14 percent of Japan's land area, but many of these are developed with roads, houses, golf courses, and resorts. There has been enormous growth in the popularity of winter sports destinations in northern Honshu and Hokkaido and of hot springs and mineral bath resorts throughout the country. Reflecting dissatisfaction with too much growth and development, the conservation ethic is strong and growing in Japan. However, it is difficult for the Japanese to truly "get away from it all" at home.

Daily life for the Japanese is devoid of many of the benefits associated with the MDCs. Apartments and homes are generally very small. Only about half of the country's homes are connected to modern sewage systems. Smog from automobile exhaust and industrial smokestacks is a persistent health hazard. The ever-present traffic jams are known locally as "traffic wars." Japan's much-vaunted rail system, including its high-speed *Shinkansen*, or bullet trains, that carry traffic between northern Honshu and Kyushu at speeds averaging 106 miles (170 km) per hour, is chronically overburdened with passengers (•Figure 7.4.13). Summarizing these living conditions, the Japanese politician Ichiro Ozawa described Japan as having "an ostensibly high-income society with a meager lifestyle." Japan has long emphasized production of manufactured exports to compete with the United States and other large economies rather than sought to raise the material standard of living of its own citizens.

Japan had a troubled period of almost continuous economic recession from the early 1990s to the early 2000s. By the middle of the latter decade, the economy's worst troubles seemed to be over, thanks largely to painful cuts in payrolls, in inefficient businesses, and in subsidies to rural communities. Japan appeared to turn recent adversity into opportunity, as it has in the past. Japan's highly educated and skilled workforce is a major resource. Although natural assets are lacking, the country's favorable geographic location should also help continue its recovery. As the western 46
Pacific region experiences the dynamic economic growth that is propelled by China (Japan's largest trading partner), Japan is well positioned to benefit from trade.

Japan will certainly continue to be an important member of the international community. Even during recession, it was the world's largest donor of foreign aid, eclipsed only in 2001 by the United States. Its long-term economic well-being is seen as crucial to the overall stability of the global economy. The Japanese consume more goods than the roughly 2 billion people of the rest of East Asia combined. Any slowdown in Japanese spending harms production in the factories of Indonesia, Malaysia, Thailand, and China that make goods for Japanese consumption and reexport; conversely, an economically stronger Japan raises all these economic boats. The United States supplies Japan with a wide range of goods, including semiconductors, software, heavy machinery, movies, apparel, and agricultural goods, and it buys

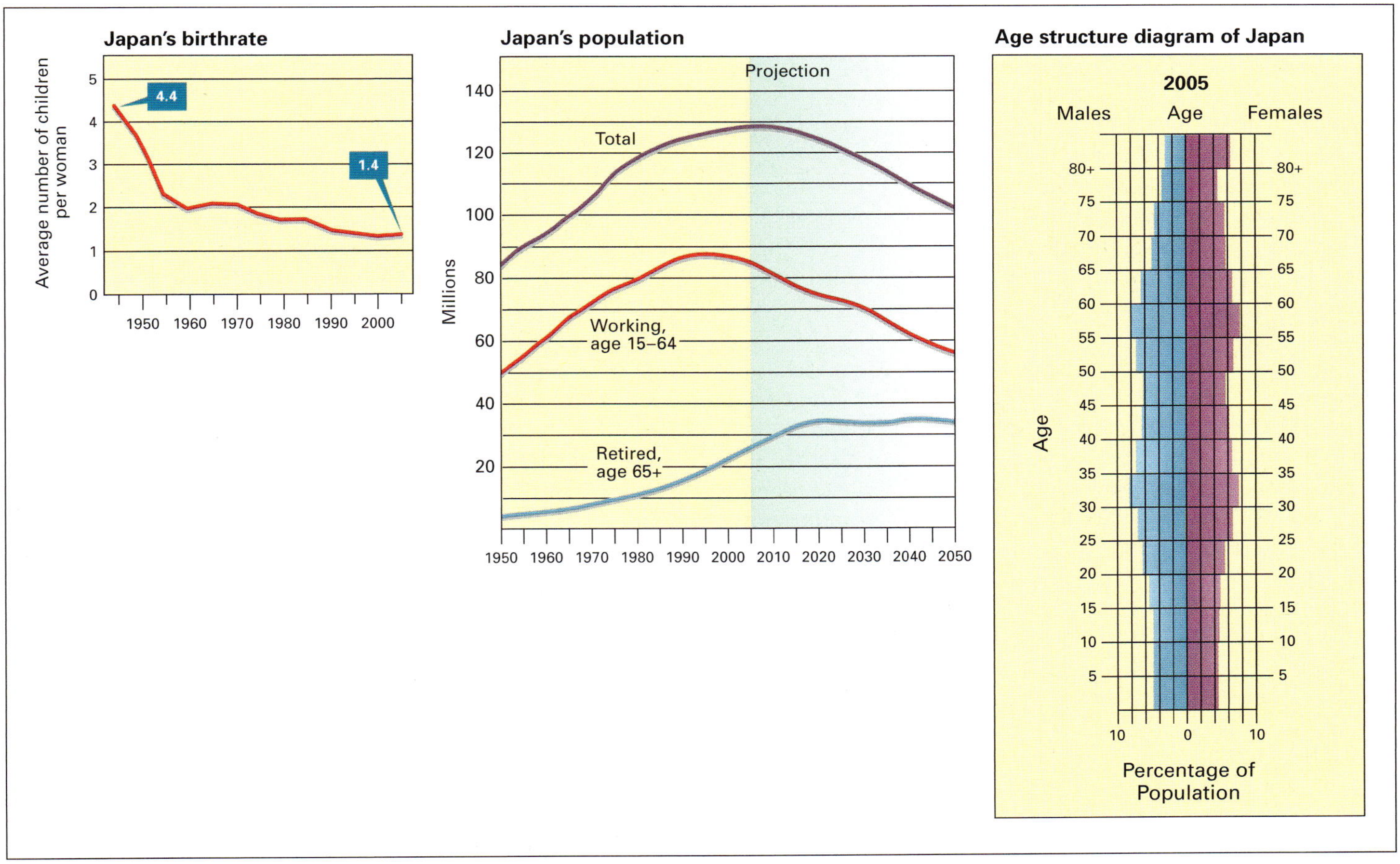

• **Figure 7.4.12** Many countries face problems of overpopulation, but Japan is concerned about not having enough people to keep its economic engine running in the future.

Joe Hobbs

• **Figure 7.4.13** The Tokyo subway.

even more goods from Japan, so these two economies—the world's largest and second largest—are also extremely interdependent on one another's health.

7.4.6 Unfortunately Located Korea

Just 110 miles (177 km) from westernmost Japan is the Korean Peninsula, roughly the size of Minnesota or Portugal (•Figure 7.4.14). Korea's political history during the 20th century, first as a colony of Japan and then as a divided land, has tended to obscure its distinctive culture and contributions to the world. Although the Koreans have been influenced by Chinese culture and to a smaller extent by Japanese culture, they are ethnically and linguistically a separate people that have made important contributions to world civilization.

The two Koreas have occupied an unfortunate location in historic geopolitical terms. North Korea adjoins China along a frontier that follows the Yalu and Tumen Rivers; it faces Japan across the Korea Strait; and in the extreme northeast, it borders Russia for a short distance. These small countries are thus located near larger and more powerful neighbors—China, Russia, and Japan—that have frequently been at odds with one another and with the Koreans. China and Russia traditionally feared Japanese rule in Korea because it might have served as a springboard for invasion of their home countries. Meanwhile, to Japan, Korea has been seen as a dagger aiming at its heart to be used by China or Russia. Japan therefore wanted to occupy or block Korea. For many centuries, the Korean Peninsula has served as a bridge between Japan and the Asian mainland. From an early time, both China and Japan have been interested in controlling this bridge, and Korea was often a subject or vassal state of one or the other.

• **Figure 7.4.14** Principal features of the Korean Peninsula.

Despite external ambitions of conquest and division, from the late 7th century to the mid-20th century, Korea was a unified state, sometimes invaded and forced to pay tribute but never destroyed as a political entity. The decline of Chinese power in the 19th century was accompanied by the rise of modern Japan, whose influence grew in Korea, and from 1905 until 1945, Korea was firmly under Japanese control. In 1910, it was formally annexed to the Japanese empire. A legacy of hostility and occupation continues to cast a shadow over relations between the Koreas and Japan, and disputes periodically emerge between them over issues such as control of islands and fishing rights (see Insights, page 407). 303

Japan lost Korea at the end of World War II, setting the stage for the division between North and South Korea. To understand Korea's split, it is important to consider the Korean War, sometimes in the United States called "America's Forgotten War." The conflict cost the lives of an estimated 2 million North Koreans, perhaps 300,000 South Koreans, and 33,629 Americans.

In the closing days of World War II, the Soviet Union entered the Pacific war as an ally of the United States against Japan. The two sides drew up plans to accept Japan's surrender on the Korean Peninsula, arbitrarily drawing a line

INSIGHTS

The Law of the Sea

Some 90 miles (145 km) between the shores of Japan and South Korea in the Sea of Japan (East Sea) lie 34 small, inhospitable islands known to outsiders as the Liancourt Rocks, Takeshima to the Japanese, and Dokdo (or Tokdo) to the Koreans (see Figure 7.4.14). South Korea has controlled the islands since 1956, but with only a handful of Korean civilians and some Korean coast guard personnel living on the islands, Japan has periodically asserted its right to them. Since the 1990s, these remote rocks have been the focus of a dispute between Japan and Korea, not because of any riches they contain but because of a 1970s United Nations treaty known as the **Convention on the Law of the Sea**. This treaty would permit their sovereign power to have greater access to surrounding marine resources.

The Law of the Sea began as an effort to apportion ocean resources as equitably as possible and to avoid precisely the kind of dispute that developed between Japan and South Korea. The treaty gives a coastal nation mineral rights to its own continental shelf, a territorial water limit of 12 miles (19.3 km) offshore, and the right to establish an **exclusive economic zone (EEZ)** of up to 200 miles (320 km) offshore (in which, for example, only fishing boats of that country may fish). The power that controls offshore islands such as Takeshima/Dokdo can extend its EEZ even further.

By early 1996, some 85 nations had ratified the treaty. South Korea ratified it late in 1995. Japan was preparing to ratify it in 1996 when news reached Tokyo that South Korea had plans to build a wharf on the islands. To avoid provoking Japan, North Korea, and China, South Korea avoided declaring an exclusive economic zone off its waters that would include the islands. Japan ratified the treaty. But fears that South Korea may build facilities and station more people on the islands as a step toward establishing an EEZ and thereby excluding Japanese fishermen from the area caused Japan to restate its claim to the islands in 1996. South Korean officials answered with military exercises near the islands, and South Korean civilians staged loud demonstrations outside Japan's embassy in Seoul. South Korea declared the islands a national monument in 2002. Long simmering, the issue boiled up again in 2006 when Japan announced it would send ships to survey the area and South Korea answered that it would form a naval blockade against the Japanese vessels. The problem may have to be resolved in the international legal arena.

Even more important than Japan's competition with South Korea over territorial waters and islands is its contest with China. Both lay claims to the islands in the East China Sea that the Japanese know as the Senkakus and the Chinese as the Diaoyu Islands (see Figure 7.4.1). Again, both are interested in staking out an EEZ in which oil reserves might be found. China does not accept the UN definition of the EEZ beginning from a country's coastline; it 302
insists on a wider zone that starts at the continental shelf.

The Senkaku or Diaoyu Islands are uninhabited, but their control is championed by nationalists on both sides and is therefore a potentially explosive issue. In 2004, Chinese activists landed on one of the islands to stake China's claim there, but Japanese authorities promptly arrested them, and Japan took formal possessions of the islands the following year. Since then, there have been several encounters between Japanese survey ships and Chinese coast guard vessels near the islands. Japan's formal claim to the islands dates to 1895, when Japan declared that the islands had previously been *terra nullius,* or no one's property. But China insists that there is Chinese documentation about control of the islands dating to the Ming Dynasty in the 15th century. The islands are close to Taiwan, which also lays claim to the islands, so any move by mainland China to secure access to them could also precipitate 308
conflict between the two Chinas.

at the 38th parallel. The Soviet Union would accept Japan's surrender north of that line and the United States south of that line. This was not meant to be a permanent boundary, but it became one. On either side of the line, the Soviet Union and the United States moved to set up governments that would be friendly to them. By 1948, the Soviet-style, Soviet-backed, Soviet-armed Democratic People's Republic of Korea was established in the north, while in the south, the Western-oriented Republic of South Korea was created under the auspices of the United Nations.

War came soon. On June 25, 1950, North Korean troops invaded the south and quickly took Seoul. In those Cold War days, the United States feared Soviet expansionism into and beyond South Korea. President Truman sent American forces in, and they soon fought under the United Nations flag with support from 16 other countries. Truman selected General Douglas MacArthur to serve as supreme commander of the effort. By September, MacArthur's forces had pushed the North Korean advance all the way back to the Yalu River on the Chinese border. The general proposed using nuclear weapons to press forward from there, but the president refused, and his dissent over the issue cost MacArthur his job.

China, seeing enemy forces on its doorstep, reacted by sending huge numbers of forces across the border, driving the allies south across the 38th parallel and taking Seoul on January 4, 1951. Seoul was recaptured on March 15, and the battlefront stabilized generally along the 38th parallel. An armistice was signed at the border site of Panmunjom on July 27, 1953, by the Chinese, the North Koreans, and the United Nations command. It was only a cease-fire, and to this day, there has not been a full treaty or reconciliation between the Koreas.

The border between the two Koreas, often cited as the tensest boundary on earth, still follows this armistice line.

Joe Hobbs

• **Figure 7.4.15** The DMZ between North and South Korea is an unlikely nature refuge, but wherever a half-century passes without significant human manipulation of the landscape, plants and animals thrive.

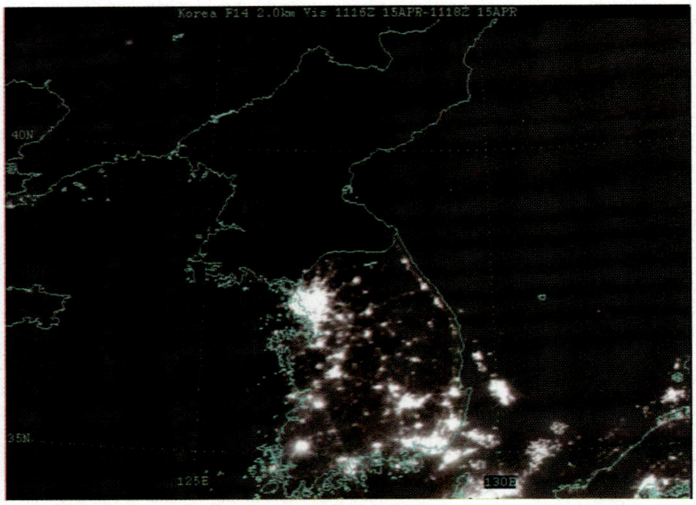

NASA

• **Figure 7.4.16** The two Koreas are like night and day. (a) A satellite image taken at night shows the profound economic distinctions, with prosperous South Korea awash with urban and industrial lights while North Korea is nearly dark.

The **demilitarized zone (DMZ)** between them, 150 miles (240 km) long and 2.5 miles (4 km) wide, is a virtual no-man's-land of mines, barbed wire, tank traps, and underground tunnels. The world's largest concentration of hostile troops faces off on either side of it. Remarkably, with the near absence of human activity, rare birds like the Manchurian crane and other endangered animal species have taken refuge in this narrow strip (•Figure 7.4.15). The DMZ thus has the potential to become an international "peace park" like that taking shape along the old Iron Curtain in
90 Europe.

7.4.7 Contrasts between the Two Koreas

The Korean armistice line divides one people, united by their ethnicity and language, into two very different countries (•Figure 7.4.16; also see Table 7.1, page 281). North Korea has an approximate area of 46,500 square miles (120,400 sq km) and is inhabited by about 23 million people, with a density of 501 per square mile (193/sq km). South Korea's corresponding figures are 48 million people in only 38,000 square miles (98,390 sq km), averaging 1,266 per square mile (489/sq km).

South Korea is a republic that has fluctuated between attempts at democracy and a repressive military dictatorship. It has a capitalist economy heavily dependent on relationships with the United States and Japan. North Korea is a rigid and very tightly controlled Communist state that promotes what it calls ***juche*** (pronounced "*djoo*'-cheh"), or self-reliance. North Korea's leadership has used this political philosophy's three principles of political independence, economic self-sustenance, and national self-defense to limit

Joe Hobbs

• **Figure 7.4.16** (b) Affluence, open trade, and freedom of expression are apparent in South Korea

Jeffrey Aaronson/Network Aspen

• **Figure 7.4.16** (c) while military security preoccupies austere Stalinist North Korea.

economic and other exchanges with the outside world. North Korea's principal Cold War ties shifted between 88 China and the Soviet Union With the collapse of the USSR, China became North Korea's main ally and benefactor, and the Russian Federation is emerging as one of North Korea's few trading partners.

From World War II until his death in 1994, the dictator Kim Il Sung, whom his citizens called "The Great Leader," governed North Korea. His son and successor, Kim Jong Il (dubbed "The Dear Leader"), has perpetuated the country's militaristic character. North Korea spends a staggering 23 percent of its gross domestic product on its military, compared with South Korea's expenditure of only 2.7 percent, China's 4.3 percent, and Japan's 0.8 percent. North Korea's lavish expenditures on defense have been one cause of its severe economic decline. North Korea's economy is only 7 percent the size of South Korea's. No GNI PPP data are available for North Korea, but South Korea's is $23,800, similar to that of Greece.

There are marked physical contrasts between the two countries. Although both are mountainous or hilly, North Korea is more rugged. North Korea has a humid continental long-summer climate, with hot summers and cold winters. South Korea is mostly a humid subtropical area with shorter and milder winters. Both are monsoonal, with precipitation concentrated in the summer.

Rice is the staple food in both countries. Rice-growing conditions are best in South Korea and diminish northward. Unlike the North, South Korea is able to double-crop irrigated rice fields by growing a dry-field winter crop such as barley after the rice has been harvested. The main supplementary crop in the North is corn, which must be grown in the summer and so cannot occupy the same fields as rice.

Beginning in 1995, successive waves of flood and drought brought famine to North Korea. The natural calamities only contributed to the already plummeting economic health of the country that came in the wake of the collapse of the Soviet Union, formerly North Korea's main benefactor and trading partner. North Korea's almost impenetrable veil of secrecy made it difficult to calculate the losses, but estimates of the number of people who died in the famine range from 900,000 to 2.4 million, or up to 10 percent of the country's prefamine population. As one crop after another failed, North Korea gradually and reluctantly sought food aid from abroad. It coaxed emergency supplies of wheat from the United States in part by threatening to withdraw from its 1994 agreement with the United States to halt its nuclear weapons program (see the discussion on pages 303–304 in Chapter 7). Although food aid from the United States, China, and South Korea helped alleviate the immediate crisis, the poor conditions of North Korea's health system, drinking water supplies, and electrical generation have perpetuated malnutrition and disease.

Another physical contrast between the two Koreas is that most of the nonagricultural natural resources are in North Korea. All of the peninsula's major resources—coal, iron ore, some less important metallic ores, hydropower potential, and forests—are more abundant in the North than in the South. North Korea was thus originally the more industrialized state, featuring a typical Communist emphasis on mining and heavy industry, together with hydroelectric production and timber products. Viewed strictly in terms of resource potential for development, North Korea should have become the more prosperous power. Its decision to close its doors to the outside world and insist on strict state control of industry and agriculture provides a fascinating 46 opportunity to appreciate how different political systems produce different economic results.

While the North focused on defense and strict socialism, the South took over industrial leadership in the 1970s and 1980s by developing a dynamic and diversified capitalist industrial economy.[1] South Korea's economy first began to boom between 1961 and 1980 under a state capitalism approach rather than free-market and free-trade capitalism. This was a time during which the government favored and strongly funded a few small companies that it nurtured into giant conglomerates called ***chaebols.*** Hyundai is a good example. Originally a small rice milk company, it diversified into trucks and buses. The government believed in its potential and supported Hyundai financially, supervising its growth into a world-class corporation renowned for the manufacture of automobiles and ships. Samsung, now one of the best electronics companies in the world, followed a similar path (•Figure 7.4.17). Altogether about 15 of these gigantic, interlocking, family-controlled conglomerates came to dominate the economy. Large investments from Japan and the United States, as well as access to markets in those countries, aided in the explosive development of *chaebol* industries. Also important was the availability of inexpensive and increasingly skilled Korean labor. Finally, the factor South Koreans most frequently point to in their success has been an emphasis on education. They sell valued possessions if need be to afford their children's educations at home or abroad, and their culture prizes teachers and professors. South Korea has one of the world's highest numbers of Ph.D.'s per capita.

Democratic and free-market challenges came to South Korea's economic, political, and social systems in the 1980s, with differences emerging between labor and management, military and civilian rule, and even men and women. In the late 1990s, an economic crisis occurred in part because of these challenges and also because South Korea's was hit by the shock waves of Asia's spreading economic crisis. The 344 pain was almost universal because so much of the country's wealth was in the hands of the *chaebols,* which employed the majority of South Korea's working population and owned most of the banks. During the country's explosive period of growth, the banks had lent money back to their parent companies for risky investments all over the world, particularly throughout Asia. As the Asian economic crisis spread, those projects were halted, and the borrowers defaulted on their loans. Foreign investors rushed to pull their capital out of South Korea, whose national currency and foreign reserves deflated.

Joe Hobbs

• **Figure 7.4.17** Directing a training session, a supervisor discusses Samsung's corporate work ethics and global ambitions.

Joe Hobbs

• **Figure 7.4.18** Seoul, capital of South Korea.

With financial help from the International Monetary Fund (IMF), South Korea responded quickly to the crisis, especially by getting rid of bad loans and by restructuring the companies that borrowed too much. In the early 2000s, the country was once again experiencing economic growth, bringing sighs of relief around the world. With the world's 11th-largest economy, South Korea is now among the top five countries worldwide in the production of automobiles, ships, steel, computers, and electronics. The development of its high-tech industries has concerned competitors in Japan and the United States; for example, Samsung has established an increasing market share in an industry dominated by Japanese firms. South Korea is the seventh-largest trading partner of the United States, which has long subscribed to a "domino theory" that if South Korea's economy fell, so would Japan's and perhaps even its own. In 2007, the United States and South Korea signed a free-trade pact that will even more strongly enmesh their economies.

South Korea's cities are visible evidence of the country's industrialization. The capital city of Seoul has a population of 22.6 million, ranking it among the five most populous metropolitan areas in the world (•Figure 7.4.18). Forty miles (64 km) east of Seoul, on an artificial island near the port of Inchon, is the free-trade zone of New Songdo, being built to capitalize on trade with China. Its designers call New Songdo a **ubiquitous city,** or **U-city,** where all major residential, medical, business, government, and other information systems are integrated. Other cities with more than 2 million population are the port of Pusan, which rivals Shanghai as the largest container port in northeastern Asia, and the inland industrial center of Taegu.

By contrast, the only large city in North Korea is the capital, Pyongyang, with a population of 3 million. In an effort to begin to close the gap, North Korea has built free-trade zones intended to attract foreign investment and industries to favorable locations near the borders with China, Russia, and South Korea. As unlikely as it may seem, the North Koreans have a vision for one of these, the city of Sinuiju (in the extreme northwest, just across the Yalu River from China), as a "new Hong Kong." Just north of the DMZ, North Korea has built a trade zone in the city of Kaesong. Many South Korean companies have taken advantage of North Korea's rock-bottom wages (5 percent of those of South Korea) and have set up factories there, producing kitchenware, semiconductor parts, and a range of other products.

In sum, South Korea is one of Asia's success stories, an Asian Tiger that made an extraordinary postwar recovery, especially by using brainpower to overcome its resource limitations, and has enjoyed periods of rocketing economic growth. South Korea is, however, still something of an "in-between" country, not entirely an MDC or an LDC, not all urban but not all rural either, and both industrial and agricultural. The government still lavishes huge subsidies to protect its rice growers, for example, while also promoting the virtues of free trade in industry. The transition from LDC to MDC has brought some growing pains, seen especially in the social changes of more women in the workforce, higher divorce rates, and falling birth rates. Like Japan, South Korea wants to maintain its cohesive society and does not want to invite foreigners to fill its impending
shortage of productive young people. The government once 402
encouraged family planning but is now asking its people to have more babies.

South Koreans tend to debate contemporary social and political issues more openly than many of their Asian neighbors, and raucous demonstrations for and against a wide range of causes are commonplace. South Korean pop culture and Christian mission movements are making inroads across China and Southeast Asia. The former, known as the **"Korean Wave"** and taking the form of music, television, films, and fashion, has generally been embraced but has produced a backlash in Japan, where some people steadfastly refuse to accept South Korea's rising clout.

Meanwhile, despite its experimental free-trade zones, North Korea is a poor country almost frozen in time, intent on pursuing the command economy model that has failed in every other nation that employed it. Despite its motto of self-reliance, it depends on huge imports of food and energy, especially from Russia. Its principal exports are apparently missiles and other military hardware; heroin, amphetamines, and other synthetic and semisynthetic drugs; and counterfeit money. Its people, whose sentiments and hopes are unknown because they are not allowed contact with the outside world, periodically hover close to famine. Only a select few are allowed access to the Internet, although the infrastructure is there: much of the country is wired with a fiber-optic intranet system. The North Korean leadership has a dilemma: If it gives up on *juche* and opens its doors, the economy might improve, but the people of this country often called the **Hermit Kingdom** will discover the bounties just across the border and might revolt.

7.4.8 Sunshine for Korea?

States of cold war and periodic border incidents have persisted between North and South Korea. Relations between the two countries have periodically thawed, only to ice over again. A warming trend came to be known as the **Sunshine Policy** in 2000, when the leaders of North and South Korea held a historic, first-ever summit meeting between the countries. Their substantive focus was on reconciliation rather than reunification. The South proclaimed that it wanted to help protect North Korea's national security and assist in its economic development. In exchange for this help, South Korea expected North Korea to cease threatening the South and to provide reassurances of abandoning its nuclear weapons and long-range missile programs. The two countries agreed to restore a long-defunct rail link built during Japan's occupation of Korea. South Korea is hoping its resurrection will provide a large economic boost: it could then ship its products overland to China and across the Trans-China Railway or Trans-Siberian Railway to Europe (Kore-
186 ans are thus wistfully calling the link the **Iron Silk Road**).

Koreans on both sides of the DMZ were delighted by the family reunions that occurred in the summit's wake. Seven million South Koreans who fled south before and during the Korean War have relatives in the North. Since 2000, small numbers of North Koreans have been able to make short visits with relatives in South Korea, and vice versa, enjoying emotional reunions that have given hope on both sides of better times to come. And as discussed earlier, South Korea has begun to capitalize on North Korean labor in the free-trade zone of Kaeson, an arrangement benefiting both countries.

The prospect for Korean reunification appears to be very remote, even as relations between the two countries improve. North Korea continues to argue that the two countries should unite peacefully and without foreign influence. Outside observers generally believe, however, that North Korea is not really interested in reunification because it would probably doom the country's regime. For its part, South Korea is fearful of the enormous economic cost it would have to pay for absorbing its much poorer neighbor, just as West Germany did in absorbing East Germany. Reunification would probably begin with a huge stream of 112
poor northerners moving south. The nearby great powers are also quietly pleased with Korea's lingering division. The two Koreas provide dual buffer zones between the historical adversaries China and Japan. And if the two Koreas united, finally taking advantage of North Korea's strong natural resource base, China and Japan could suddenly have a rival great power to contend with.

For now, there is little thought of Korean unification. Even with the progress of the Six Party Talks and North Korea's pledge to disable its nuclear weapons facilities (see page 304), there is much more concern with simply avoiding war. Any escalation of the tension between the United States and North Korea, or even an incident in the DMZ, could be a tripwire for a great conflict (•Figure 7.4.19). U.S. troops could not avoid being involved, and Chinese troops might be brought in. Some analysts fear that Korea may yet become a nuclear battlefield. Japan worries that it would also be drawn into any conflict between the Koreas. The economic and political stability of these three small countries on the western Pacific Rim is critical to the well-being of the global system.

Joe Hobbs

• **Figure 7.4.19** Two South Korean soldiers keep a wary eye on their North Korean counterpart, just steps away in the Joint Security Area of the Panmunjom "truce village" in the DMZ. The low concrete slab running left-to-right on the left of the photo marks the border between the two countries. From time to time, a North Korean defector runs across the border, sometimes pursued by North Korean soldiers. Firefights involving both U.S. and South Korean troops against North Korean soldiers have accompanied some of these incidents.

SUMMARY

- → Japan's economy is the world's second largest, after that of the United States.
- → There are four main islands in the Japanese archipelago: Hokkaido, Honshu, Shikoku, and Kyushu. The smaller Ryukyu Islands extend almost to Taiwan.
- → All of the main islands of Japan consist of volcanic mountains and are subject to frequent and sometimes severe earthquakes. Other natural hazards also plague Japan.
- → Japan went through periods of relative isolation. Then it opened up to the outside world and is today one of the most influential players in world affairs.
- → Japan experienced phenomenal economic growth after its crushing defeat in World War II, but it has recently suffered economic recession.
- → Due to its relative poverty in natural resources, Japan must import most of its raw materials, most energy supplies, and a large share of its food.
- → The people of Japan view the ocean as an important resource and a promising frontier. The Japanese have the world's highest per capita consumption of fish and other marine foods.
- → The quest for possession of offshore resources near islands is a source of tension between Japan and Korea and between Japan and China.
- → Most of Japan's economic activities and a majority of its people are packed into a corridor about 700 miles (c. 1,100 km) long. This megalopolis extends from Tokyo, Yokohama, and the surrounding Kanto Plain on the island of Honshu through northern Shikoku to northern Kyushu.
- → The hardworking Japanese are experiencing many symptoms of workaholism, and many complain about their low level of amenities relative to that of other prosperous countries.
- → The Korean Peninsula has an unfortunate location in geopolitical terms, sandwiched between the greater powers of Russia, China, and Japan. Korea has had a turbulent recent political history, first as a colony of Japan and then as a land divided between the Communist North and the more dynamic and democratic South.
- → Although the Koreans have been influenced by Chinese culture and to a lesser extent by Japanese culture, they are ethnically and linguistically a separate people.
- → There are marked physical and economic contrasts between North and South Korea, with North Korea being much more rugged and having more natural resources for industry. North Korea has a humid continental long-summer climate, with hot summers and cold winters, whereas South Korea is mostly a humid subtropical area, with much shorter and milder winters. Capitalist South Korea is far more prosperous than Communist North Korea.
- → North Korea has suffered famine recently but has been reluctant to open up to trade and assistance because North Koreans might revolt.
- → There are prospects for reconciliation and, more remotely, reunification but also for war between North and South Korea. Reunification is not favored by the South because it would be too costly. China and Japan would apparently be happier to have a weaker divided Korea than a stronger unified Korea.

KEY TERMS + CONCEPTS

active volcano (p. 392)
Ainu (p. 398)
bubble economy (p. 402)
Burakim (p. 403)
chaebol (p. 409)
Convention on the Law of the Sea (p. 407)
daimyo (p. 399)
demilitarized zone (DMZ) (p. 408)
double cropping (p. 394)
exclusive economic zone (EEZ) (p. 407)
Great East Asia Co-Prosperity Sphere (p. 401)
Hermit Kingdom (p. 411)
gunboat diplomacy (p. 399)
intertillage (p. 394)
Iron Silk Road (p. 411)
juche (p. 408)
kamikaze (p. 399)
karoshi (p. 404)
Liberal Democratic Party (LDP) (p. 402)
megalopolis (p. 395)
Meiji Restoration (p. 399)
Rape of Nanking (p. 400)
Ring of Fire (p. 396)
samurai (p. 399)
Self-Defense Forces (SDFs) (p. 402)
shogun (p. 399)
Sunshine Policy (p. 411)
Tokugawa Shogunate (p. 399)
ubiquitous city (U-city) (p. 410)
Utari (p. 398)
wa (p. 404)

REVIEW QUESTIONS

1. What are Japan's four main islands? What are the outlying islands? Which islands are contested with other countries?
2. What elements of Japan's physical geography are related to the islands' location on the Pacific Ring of Fire?
3. What are Japan's principal climate types? In which parts of the country do they occur?
4. What are the principal features of Japan's agriculture? What position does rice have in the agricultural and economic systems?

5. To what extent does Japan draw on the resources of the sea?
6. What area is defined as Japan's core? What urban, industrial, and population characteristics definite it as a core region?
7. What is Japan's ethnic makeup? What relations have the Japanese majority had with internal minorities and foreign groups at various times?
8. How has Japan been able to overcome its poverty in natural resource assets by commercial, military, or other means?
9. What are some of the features of Japanese society today, especially those related to the country's economy?
10. What precipitated the division of the Korean Peninsula into North and South Korea?
11. Why are relations tense between Japan and South Korea?
12. What are the major differences between North and South Korea in population numbers, natural resources, political systems, and economic development?
13. What are the prospects for reunification of the Koreas? How do some outside powers and the Koreans themselves view reunification?

NOTE

1. I thank the Korea Society for sponsoring my field study of Korea, on which much of this discussion is based. Professor Byong Man Ahn of the Hankuk University of Foreign Studies and Professor Taeho Bark of the Graduate School of International Studies at Seoul National University graciously provided much of the information on Korean economies and politics.

An image of Africa's physical beauty: The Uaso Nyiro River Samburu National Park, northern Kenya.

CHAPTER 9
A GEOGRAPHIC PROFILE OF SUB-SAHARAN AFRICA

Geographers typically recognize the continent of Africa as having two major divisions: North Africa, which is the predominantly Arab and Berber realm of the continent, and ethnically diverse sub-Saharan Africa. In this text, North Africa, including Sudan, is regarded as essentially Middle Eastern and is discussed in Chapter 6 and Module 6.1. This chapter sketches the broad outlines of the geography of sub-Saharan Africa, and Module 9.1 examines issues in the major African subregions. The basic reference map for the region's 47 countries is •Figure 9.1, and area and population data are in Table 9.1.

chapter outline

chapter objectives

This chapter should enable you to

→ Understand what caused sub-Saharan Africa to become and remain the world's poorest region

→ Know how the region came to have the highest HIV/AIDS infection rates in the world and how the epidemic could be reversed there

→ Appreciate the pressures on African wildlife and the unique approaches taken to protect the animals

→ Know what is uniquely African about African cultures

→ Recognize why, after more than a decade at the sidelines, sub-Saharan Africa is considered important again in geopolitical affairs

The region of sub-Saharan Africa is culturally complex, physically beautiful, and problem-ridden. This introduction to Africa surveys the region's diverse environments, peoples and ways of living, population distributions, European colonial legacies, and major current problems. Sub-Saharan Africa ranks at the very bottom of every statistical indicator of global quality of life. About half of this region's people subsist on less than $1 per day (the World Bank's benchmark for extreme poverty). This is the only world region that has grown poorer in the last 25 years.

It is also a region that has been plagued by strife, but there is good news here: while there were 16 conflicts in the region in 2002, there were just 5 in 2007. Africa's plight may seem ironic because by some measures, the region has the world's greatest variety of natural resources. But as this chapter will show, some of Africa's recent and current conflicts, in many cases fought by child soldiers, have not been about ideology or ethnicity but simply about control over resources such as diamonds, oil, and other minerals. A recent study of the world's civil wars since 1960 determined that there were three important risk factors for such conflicts: poverty, low economic growth, and high dependence on natural resources. Sub-Saharan Africa has been richly endowed with these ingredients of war.

For many people, sub-Saharan Africa is the most poorly understood or misunderstood of the world's regions. It seems to have so many countries, so much violence and disease, and so much poverty that it is simply too difficult to think about. Hollywood still perpetuates stereotyped images of an Africa rich in wildlife but with dangerous or backward peoples. This chapter will show that getting to know the real Africa is manageable, rewarding, and interesting. This region's diverse cultures and natural environments are sources of endless wonder, and conditions for many Africans are improving in a number of ways. There are many reasons for hope and much to celebrate in the study of this fascinating region. Just ask the people: a 2006 Gallup survey of 50,000 citizens around the world found that Africans are the most optimistic.[1]

9.1 Area and Population

Sub-Saharan Africa (including Madagascar and other nearby Indian Ocean islands) has the second-largest land area of all the major world regions described in this book. It covers 17.4 million square miles (21.8 million sq km) and so is more than twice the size of the United States (•Figure 9.2).

People overpopulation is apparent in some areas, and yet 61
much of the region is sparsely populated. With a population of 749 million as of 2007, the region's average population density is slightly more than that of the United States. Even with the loss of population due to AIDS, the rate of natural population increase in sub-Saharan Africa is 2.5 percent per year, or about four times that of the United States. As in most LDCs, African parents generally want large families for several reasons: to have extra hands to perform work; to be looked after when they are old or sick; and in the case of girls, to receive the "bride wealth" a groom pays in a marriage settlement. Large families also convey status. 51
However, there are signs of significant change in this pattern: birth rates have been dropping in every country in this region over the past two decades.

Most of this region's people live in a few small, densely populated areas (•Figure 9.3). The main areas are the coastal belt bordering the Gulf of Guinea in West Africa from the southern part of Africa's most populous country, Nigeria, westward to southern Ghana; the savanna lands of northern Nigeria; the highlands of Ethiopia; the highland region surrounding Lake Victoria in Kenya, Tanzania, Uganda, Rwanda, and Burundi; and the eastern coast and parts of the high interior plateau of South Africa.

Sub-Saharan Africa is the world's most rural region (69 percent), with rural populations of most countries between 65 and 85 percent. The most rural are two East African nations with very fertile soils: Burundi (8 percent urban) and Uganda (12 percent urban). The most urbanized are Djibouti (82 percent), where there is no arable land and a small number of people are clustered in a port city, and Gabon (73 percent), where people are flocking to partake in an oil boom that is benefiting urbanites. The region's major cities, discussed in Module 9.1, are magnets that attract many poor, rural people. But life in villages is the rule,

• **Figure 9.1** Sub-Saharan Africa.

where a typical rural home is a small hut made of sticks and mud, with a dirt floor, thatched roof, and no electricity or plumbing (•Figure 9.4).

Africa's Population Prospects

This is the world's youngest population, where 43 percent of the region's people are under 15 years old (compared with about a third of the populations of Latin America and Asia). This is another indicator of how rapidly the region's population should grow, barring the vagaries of disease or famine. But the Malthusian scenario does seem to loom over Africa. Analysts fear the consequences of what they call the **"1 percent gap"**: Since the 1960s, the population of sub-Saharan Africa has grown at a rate of about 3 percent annually, while food production in the region has grown at only about 2 percent annually. This is the only world region where per capita food production is declining. The

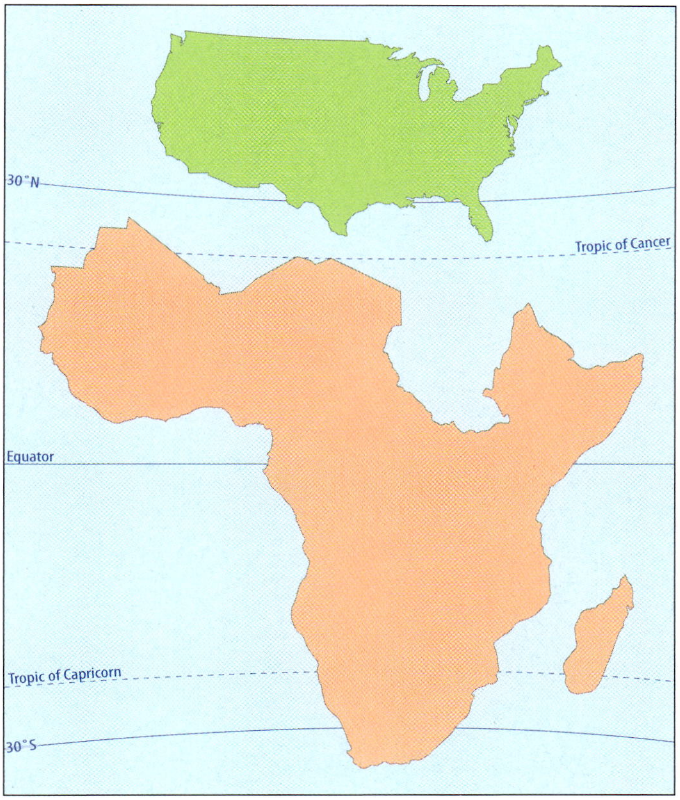

• **Figure 9.2** Sub-Saharan Africa compared in area and latitude with the conterminous United States.

wild card in Africa's population deck is the **human immunodeficiency virus (HIV)**, and the disease it causes—**AIDS (acquired immunodeficiency syndrome)**—has inevitably been identified by some as the Malthusian "check" to the region's population growth (see Medical Geography, pages 464–465).

9.2 Physical Geography and Human Adaptations

Sub-Saharan Africa is both rich in natural resources and beset with environmental challenges to economic development. It is home to some of the world's greatest concentrations of wildlife and to some of the most degraded habitats.

The Landscapes of Africa

Most of Africa is a vast plateau, actually a series of plateaus, with a typical elevation of more than 1,000 feet (c. 300 m) (•Figure 9.5). Near the Great Rift Valley in the Horn of Africa and in southern and eastern Africa, the general elevation rises 2,000 to 3,000 feet (c. 600 to 900 m), with many areas at 5,000 feet (1,520 m) and higher (see Regional Perspective, page 466). The highest peaks and largest lakes of the continent are located in this belt. The loftiest summits lie within 250 miles (c. 400 km) of Lake Victoria. They include Kilimanjaro (19,340 ft/5,895 m) and Kirinyaga (Mount Kenya, 17,058 ft/5,200 m; •Figure 9.6), which are volcanic cones, and the Ruwenzori range (up to 16,763 ft/5,109 m), a nonvolcanic massif produced by faulting. Lake Victoria, the largest lake in Africa, is surpassed in area among inland waters of the world only by the Caspian Sea and Lake Superior. It is relatively shallow, however, and the large numbers of people living on its shores are taxing its resources. Other very large lakes in East Africa include Lake Tanganyika and Lake Malawi.

The physical structure of Africa has influenced the character of African rivers. The main rivers, including the Nile, Niger, Congo, Zambezi, and Orange, rise in interior uplands and descend by stages to the sea. At some points, they descend abruptly, particularly at plateau escarpments, with rapids and waterfalls interrupting their courses. These often block navigation a short distance inland. Helping offset this problem, Africa's discontinuous inland waterways are interconnected by railroads and highways more than on any other continent. The Congo is used more for transportation than any other river in the region.

The many waterfalls and rapids do have a positive side: They represent a great potential source of hydroelectric energy. There are major power stations on the Zambezi River at the Cabora Bassa Dam in Mozambique and at the Kariba Dam (•Figure 9.7), which Zimbabwe and Zambia share; at the Inga Dam on the Congo River, just upstream from Matadi; at the Kainji Dam on the Niger River in Nigeria; and at the Akosombo Dam on the Volta River in Ghana. But only about 5 percent of Africa's hydropower potential has been realized (compared to about 60 percent in North America). Many of the best sites are remote from large markets for power. In some cases, geopolitical considerations pose obstacles to dam construction. Downstream Egypt, for example, has expressed concern and even hostile rhetoric about dams and water diversions of the Nile and its tributaries by upstream Sudan, Ethiopia, Uganda, Kenya, and Tanzania (see page 231 in Chapter 6).

Africa's Biomes and Climates

The equator bisects Africa, so about two-thirds of the region lies in the low latitudes, having tropical climates and biomes; Africa is the most tropical of the world's continents (•Figure 9.8). Areas of tropical rain forest climate center on the great rain forest of the Congo Basin in central and western Africa. The forest merges gradually into a tropical savanna climate on the north, south, and east. This is the climatic and biotic zone supporting the famous large mammals of Africa. The savanna areas in turn trend into steppe and desert on the north and southwest. A broad belt

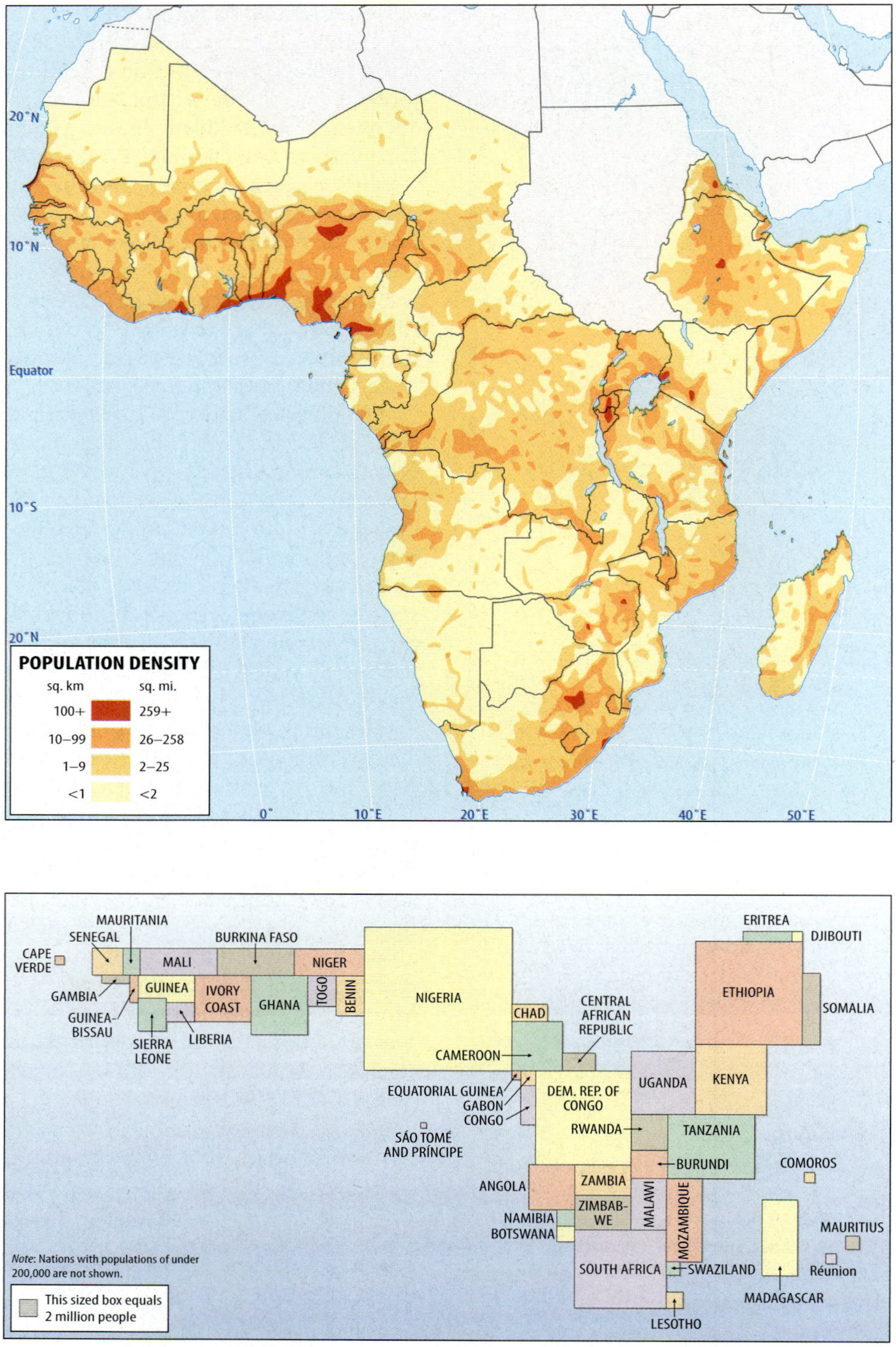

• **Figure 9.3** Population distribution (top) and population cartogram (bottom) of sub-Saharan Africa.

of drought-prone tropical steppe and savanna bordering the Sahara on the south is known as the Sahel. There is desert on the coasts of Eritrea, Djibouti, and Somalia in the Horn of Africa. In South Africa and Namibia, a coastal desert, the Namib, borders the Atlantic. The Kalahari Desert, which lies inland from the Namib, is better described as steppe or semidesert than as true desert. Along the northwestern and southwestern fringes of the continent are small but productive areas of Mediterranean climate, while eastern coastal sections and adjoining interior areas of South Africa have a humid subtropical climate. Bordering the subtropical climate region is an area of marine west coast climate.

Joe Hobbs

• **Figure 9.4** Homes in African villages are usually elevated above the ground to minimize the risk of flooding during the wet season. Their roofs are typically made of thatch or sheet metal. Most village roads are unpaved.

Total precipitation in the region is high but unevenly distributed; some areas are typically saturated, while others are perennially bone dry. Even in many of the rainier parts of the continent, there is a long dry season, and wide fluctuations occur from year to year in the total amount of precipitation. One of the major needs in Africa is better control over water. In the typical village household, women carry water from a stream or lake or a shallow (and often polluted) well. Use of more small dams would help provide water storage throughout the year.

Drought is a persistent problem in most of the countries. Although all droughts create problems, some last for years with devastating effects in this heavily agricultural region. Droughts have been particularly severe in recent decades in the Sahel and in the Horn of Africa. The number of food 488, 499

• **Figure 9.5** Major topographic features, rivers, lakes, and seas of sub-Saharan Africa.

Mark C. Ross/National Geographic Image Collection

Robert Mackinlay/Peter Arnold

• **Figure 9.6** Africa's highest mountains, both volcanic, are Tanzania's Kilimanjaro (above) and Kenya's Kirinyaga (right), also known as Mt. Kenya.

emergencies, most of them tied to drought, has tripled since the mid-1980s in sub-Saharan Africa. The consensus among scientists is growing that this region may pay the highest price for global climate change, even though it contributes the fewest greenhouses gases to the atmosphere. United Nations studies predict even more severe droughts and crop losses, along with the destruction of coastal infrastructure
34 that will accompany rising sea levels. Such dangerous environmental changes may be contributing to conflict in the region as well. Pastoralists and farmers already compete for scarce land and water resources in many places, and droughts intensify those struggles.

Living off the Land

The patterns of Africa's land use (•Figure 9.9) reveal that the most productive lands are on river plains, in volcanic regions (especially the East African and Ethiopian highlands), and in some grassland areas of tropical steppes (notably the High Veld in South Africa). Soils of the deserts and regions of Mediterranean climate are often poor. In the tropical rain forests and savannas, there are reddish, lateritic tropical soils that are infertile once the natural vegetation is re-
288 moved and can support only shifting cultivation.

To support growing populations, farmers across sub-Saharan Africa have shortened fallow periods and pressed their lands to yield more crops. The result has been an unprecedented degradation of the resource. In 2006, fully 75 percent of the region's farmland was severely low in the nutrients needed to grow crops, up from 40 percent a decade earlier. Studies suggest that at current rates, crop yields will fall as much as 30 percent by 2022. Most African farmers cannot afford fertilizers and are not familiar with the soil conservation techniques that would help reverse this ominous trend.

Africa's soils favor subsistence agriculture (people farming their own food but producing little surplus for sale) and pastoralism, and over half of the region's people practice these livelihoods. Women do a large share of the farm-

Joe Hobbs

• **Figure 9.7** The Kariba Dam straddles the Zambezi River between Zimbabwe and Zambia and provides hydropower to both countries.

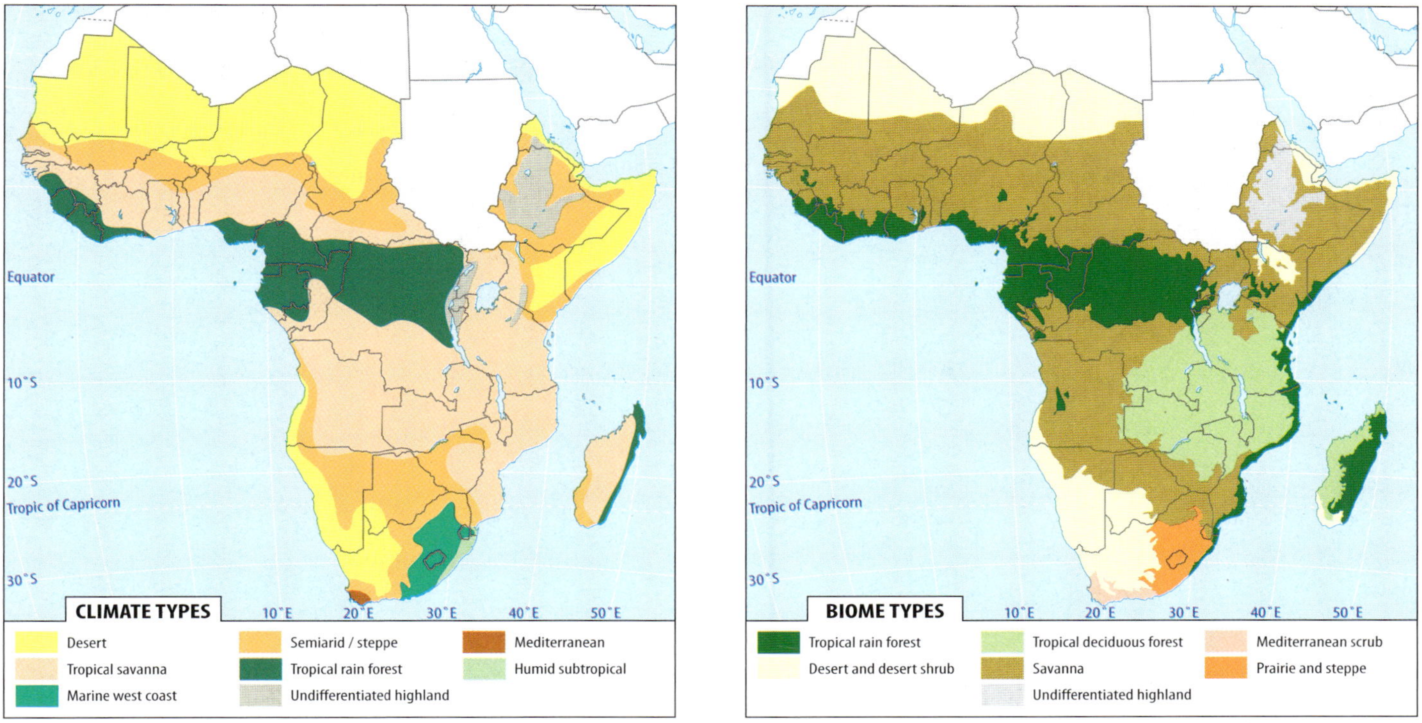

• **Figure 9.8** Climates (left) and biomes (right) of sub-Saharan Africa.

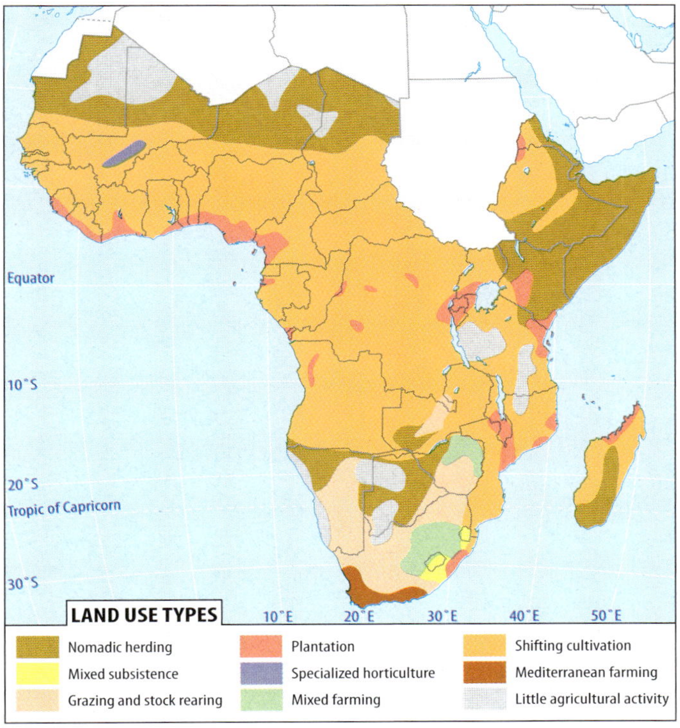

• **Figure 9.9** Land use in sub-Saharan Africa.

Joe Hobbs

• **Figure 9.10** Mother and child in Zimbabwe. Women plant and harvest most of Africa's food and care for most of its children.

work—they produce 80 to 90 percent of Africa's food—in addition to household chores and the bearing and nurturing of children (•Figure 9.10). Mechanization is rare, fertilizers are expensive, and so crop yields are low. In the steppe of the northern Sahel, both rainfall and cultivation are scarce. The more dependable rainfall of the southern Sahel creates a major area of rain-fed cropping, with unirrigated millet, sorghum, corn (maize), and peanuts the major subsistence crops. In the tropical savannas south of the equator, corn is a major subsistence crop in most areas, with manioc (cassava) and millet also widely grown. Corn, manioc, bananas, and yams are the major food crops of the rain forest areas.

TABLE 9.1 Sub-Saharan Africa: Basic Data

Political Unit	Area (thousands; sq mi)	Area (thousands; sq km)	Estimated Population (millions)	Estimated Population Density (sq mi)	Estimated Population Density (sq km)	Annual rate of Natural Increase (%)	Human Development Index	Urban Population (%)	Arable Land (%)	Per Capita GNI PPP ($US)
The Sahel	**2,047.6**	**5,303.3**	**69.6**	**31**	**12**	**3.1**	**0.372**	**27**	**3**	**1,370**
Burkina Faso	105.8	274.0	14.8	140	54	3.0	0.342	15	14	1,330
Cape Verde	1.6	4.1	0.5	313	121	2.5	0.722	53	9	5,980
Chad	495.8	1,284.1	10.8	22	8	3.1	0.368	24	3	1,230
Gambia	4.4	11.4	1.5	341	132	2.7	0.479	26	25	1,970
Mali	478.8	1,240.1	12.3	26	10	3.3	0.338	30	4	1,130
Mauritania	396.0	1,025.6	3.1	8	3	2.7	0.486	40	0	2,600
Niger	489.2	1,267.0	14.2	29	11	3.4	0.311	21	3	830
Senegal	76.0	196.8	12.4	163	63	3.0	0.460	43	12	1,830
West Africa	**812.4**	**2,104.1**	**224.1**	**255**	**98**	**2.5**	**0.444**	**37**	**20**	**1,340**
Benin	43.5	112.7	9.0	207	80	3.0	0.428	40	18	1,160
Ghana	92.1	238.5	23.0	250	96	2.3	0.532	44	16	2,640
Guinea	94.9	245.8	10.1	106	41	2.9	0.445	33	3	2,410
Guinea-Bissau	13.9	36.0	1.7	122	47	3.1	0.349	32	10	830
Ivory Coast	124.5	322.5	20.2	162	63	2.4	0.421	46	9	1,550
Liberia	37.2	96.3	3.8	102	39	3.1	N/A	45	4	N/A
Nigeria	356.7	923.9	144.4	405	156	2.5	0.448	36	31	1,050
Sierra Leone	27.7	71.7	5.3	191	74	2.8	0.335	37	7	850
Togo	21.9	56.7	6.6	301	116	2.8	0.495	33	46	1,490
West Central Africa	**1,576.0**	**4,081.8**	**90.8**	**53**	**20**	**2.8**	**0.423**	**35**	**4**	**1,200**
Cameroon	183.6	475.5	18.1	99	38	2.3	0.506	48	13	2,370
Central African Republic	240.5	622.9	4.3	18	7	1.9	0.353	39	3	1,280
Congo Republic	132.0	341.9	3.8	29	11	2.7	0.520	52	1	940
Congo, Democratic Republic of	905.4	2,345.0	62.6	69	27	3.0	0.391	30	3	720
Equatorial Guinea	10.8	28.0	0.5	46	18	2.3	0.653	45	4	10,150
Gabon	103.3	267.5	1.3	13	5	1.6	0.633	73	1	5,310
São Tomé and Príncipe	0.4	1.0	0.2	500	193	2.7	0.607	38	6	N/A
East Africa	**703.0**	**1,820.8**	**121.9**	**155**	**60**	**2.8**	**0.464**	**22**	**9**	**1,120**
Burundi	10.7	27.7	8.5	794	307	3.0	0.384	8	35	710
Kenya	224.1	580.4	36.9	165	64	2.8	0.491	36	8	1,300
Rwanda	10.2	26.4	9.3	912	352	2.7	0.450	17	40	1,270
Tanzania	364.9	945.1	38.7	106	41	2.6	0.430	22	4	740
Uganda	93.1	241.1	28.5	306	118	3.1	0.502	12	26	1,490

continued

TABLE 9.1 Sub-Saharan Africa: Basic Data *continued*

Political Unit	Area (thousands; sq mi)	Area (thousands; sq km)	Estimated Population (millions)	Estimated Population Density (sq mi)	Estimated Population Density (sq km)	Annual rate of Natural Increase (%)	Human Development Index	Urban Population (%)	Arable Land (%)	Per Capita GNI PPP ($US)
Horn of Africa	**727.0**	**1,882.9**	**91.9**	**152**	**59**	**2.6**	**0.340**	**17**	**6**	**1,200**
Djibouti	9.0	23.3	0.8	89	34	1.8	0.494	82	0	2,540
Eritrea	45.4	117.6	4.9	108	42	3.0	0.454	19	5	1,090
Ethiopia	426.4	1,104.4	77.1	181	70	2.5	0.371	15	10	1,190
Somalia	246.2	637.7	9.1	37	14	2.9	N/A	33	1	N/A
Southern Africa	**2,310.8**	**5,985.0**	**129.3**	**52**	**20**	**1.6**	**0.515**	**38**	**5**	**5,600**
Angola	481.4	1,246.8	16.3	34	13	2.7	0.439	33	2	2,360
Botswana	224.6	581.7	1.8	8	3	-0.1	0.570	54	1	12,240
Lesotho	11.7	30.3	1.8	154	59	0.3	0.494	17	11	4,340
Malawi	45.7	118.4	13.1	287	111	2.8	0.400	14	23	720
Mozambique	309.5	801.6	20.4	66	25	2.1	0.390	29	5	1,220
Namibia	318.3	824.4	2.1	7	3	1.5	0.626	33	1	8,110
South Africa	471.4	1,220.9	47.9	102	39	0.8	0.653	53	12	11,710
Swaziland	6.7	17.4	1.1	164	63	-0.1	0.500	25	10	5,170
Zambia	290.6	752.7	11.5	40	15	1.9	0.407	35	7	1,000
Zimbabwe	150.9	390.8	13.3	88	34	1.1	0.491	32	8	1,940
Indian Ocean Islands	**229.7**	**595.0**	**21.4**	**93**	**36**	**2.5**	**0.506**	**29**	**5**	**1,790**
British Indian Ocean Territory (U.K.)	0.02	0.05	0.003	150	58	N/A	N/A	100	0	N/A
Comoros	0.9	2.3	0.7	778	300	2.9	0.566	33	36	2,010
Madagascar	226.7	587.2	18.3	81	31	2.7	0.509	26	5	960
Mauritius	0.8	2.1	1.3	1,625	627	0.7	0.800	42	49	13,500
Mayotte (Fr.)	0.1	0.3	0.2	2,000	772	3.6	N/A	28	0	N/A
Réunion (Fr.)	1.0	2.6	0.8	800	309	1.3	N/A	89	13	N/A
Seychelles	0.2	0.5	0.1	500	193	1.0	0.842	50	2	16,590
Summary Total	**8,406.5**	**21,772.9**	**749.0**	**89**	**34**	**2.5**	**0.439**	**31**	**6**	**2,020**

Sources: World Population Data Sheet, Population Reference Bureau, 2007; Human Development Report, United Nations, 2006; World Factbook, CIA, 2007.

Many peoples, particularly in the vast tropical grasslands both north and south of the equator, are pastoral. Herding of sheep and hardy breeds of cattle is especially important in the Sahel. An increasing problem is that farmers often drive pastoralists from traditional grazing lands. Confined to smaller areas in which to browse and graze, the nomads' cattle, sheep, and goats often overgraze vegetation and compact the soil.

Although cattle raising is widespread throughout the savannas, cattle are largely excluded from extensive sections both north and south of the equator by the disease called **nagana,** which is carried by the tsetse flies that also transmit sleeping sickness to humans (see Insights, page 468). In tropical rain forests, tsetse flies are even more prevalent and few cattle are raised, but goats and poultry are common (as they are in tsetse-frequented savanna areas).

MEDICAL GEOGRAPHY

HIV and AIDS in Africa

AIDS is a global problem. It had killed over 25 million people worldwide by 2007, and there are an estimated 4 million new infections around the world each year. Sub-Saharan Africa is where the virus originated (apparently among chimpanzees; the first human case was reported in the Belgian Congo—today's Democratic Republic of Congo—in 1959) and is today the epicenter of this health crisis. The statistics are startling. In 2005 (the latest year for which data are available), 62 percent of the world's estimated 39 million persons infected with HIV lived in sub-Saharan Africa. In this region, 5.9 percent of adults were HIV-positive. In 2006, 2 million died in the region. Some public health authorities estimate that 25 percent of the region's more than 700 million people will have died of AIDS by 2010, leaving behind 20 million orphans and slashing the regional economy by 25 percent.

The epidemic is most severe in southern Africa. In some countries, including Botswana and Zimbabwe, as much as one-quarter of the adult population is HIV-positive. South Africa has more HIV cases (about 5 million) than any other country in the world. Eighteen percent of adult South Africans were HIV-positive in 2005. United Nations officials predict that if the epidemic continues apace, as many as half of South Africa's 15-year-olds will die of AIDS in the coming years. Similar impacts are predicted for South Africa's neighbors. Demographically, the effect would change the pyramid-shaped age-structure diagram typical of South Africa, Botswana, and Zimbabwe to a more tapered structure that demographers call "chimney-shaped" (•Figure 9.A). 54

The earth has not seen a **pandemic** (very widespread epidemic) like this since the bubonic plague devastated 14th-century Europe and smallpox struck the Aztecs of 16th-century Mexico. This scourge is causing sharp reductions in life expectancy in Africa and, unless contained, will dramatically alter projections of the region's population growth. Life expectancy in Botswana was 61 years at birth in 1993, but because of the virus, it was 34 in 2007 and is projected to be 29 in 2010. Life expectancies in Zimbabwe (37 years) and Namibia (52 years) are projected to be 33 and 42 years, respectively, in 2010; and in South Africa (51 years), it is projected to be 44. In Zimbabwe, the population growth rate in 1998 was 1.5 percent rather than the projected 2.4 percent because of AIDS-related deaths. By 2007, Zimbabwe's growth rate was down to 1.1 percent. If the disease stays on course, the populations of Zimbabwe and South Africa will soon begin declining; Botswana and Swaziland are already experiencing population declines.

HIV/AIDS is darkening the prospects for development in this least developed of the world's regions. Although it may seem logical that a lower population would mean greater prosperity, the incidence of HIV infection is particularly high among the region's most educated, skilled, and ambitious young urban professionals, including teachers, white-collar workers, and government employ-

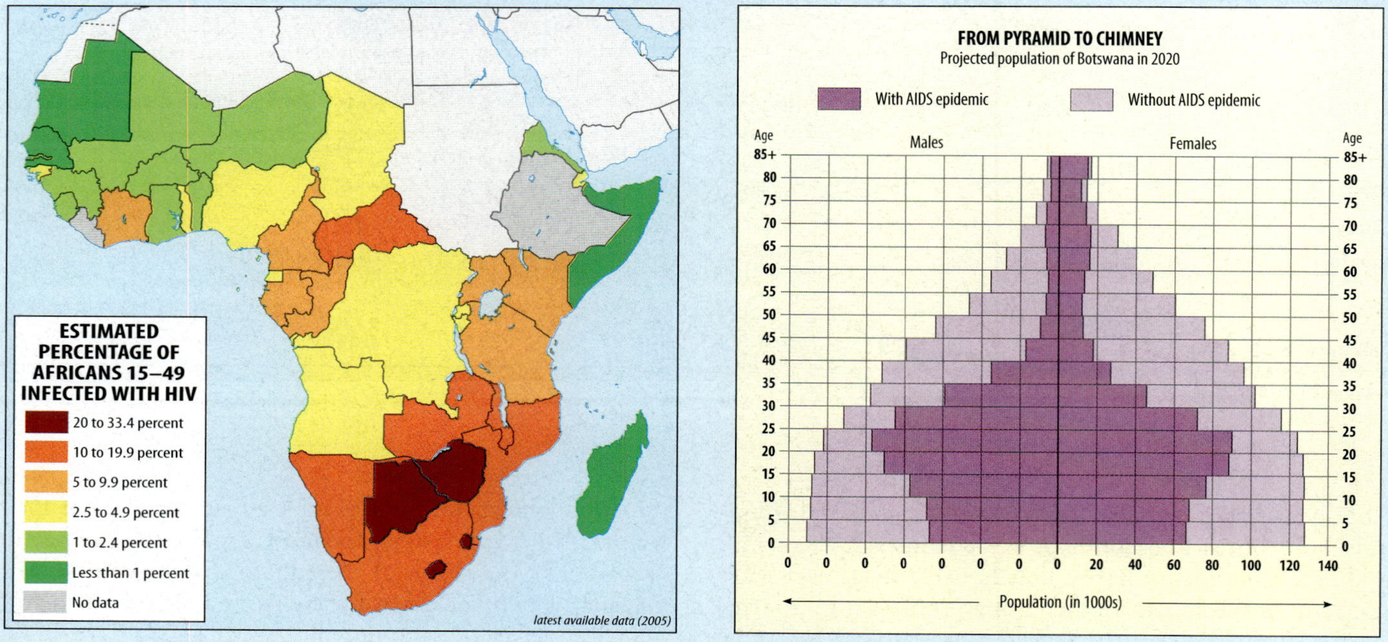

• **Figure 9.A** HIV/AIDS has had a devastating impact in sub-Saharan Africa and threatens to redraw the demographic profiles of some countries in unprecedented ways.

MEDICAL GEOGRAPHY

HIV and AIDS in Africa *continued*

ees. These are the ones, in addition to truckers and merchants, who travel the most, have higher incomes, and are therefore the most likely to have many sexual liaisons. The effects on human capital are incalculable but probably huge. They affect not only education but corporate profits, research, health care, tax revenues, and other indicators of progress.

AIDS's effects also contribute to underdevelopment. Sick people, notably women (who have the higher infection rates and who do most of Africa's farming) may be too weak to grow food. Sick workers are less productive. Family members must use precious funds to buy life-prolonging drugs rather than goods that would boost the economy. People anticipating short lives are less inclined to save and invest money and less inclined to send their children to school. They may join gangs or engage in other reckless behavior.

Why is the HIV infection rate so high in this region? A number of customs and cultural factors come into play, although the examples given here do not necessarily apply to every African culture and country. First, in some African societies, little stigma is attached to adultery or multiple sexual partners. Some countries are trying to crack down on the "sugar daddies" who seduce young women with money and promises, only to exploit and inadvertently infect them. Some male teachers, often city-educated and perceived as powerful, similarly transmit the disease to girls in the villages where they are sent to teach. Second, partners are often reluctant to use condoms. A tradition of subservience to male partners means that women often feel they have no choice when a man insists on unprotected sex. Both these factors help reveal why women are three times as likely as men to contract HIV in this region. Third, the stigma associated with having AIDS is pervasive. A person who is known to have the disease is often shunned, even by family members, and often suffers and dies in neglect and isolation. The family or community will insist that the person died of tuberculosis or some other ailment, so the real problem is never confronted and goes untreated. Fourth, despite the intensity of the problem, public awareness of the disease and ways to combat it still fall dramatically short. Where leadership on this issue is needed most, it is sometimes absent. South Africa's President Thabo Mbeki, for example, came to office professing that HIV does not cause AIDS, instead blaming poverty and malnutrition. His health minister discouraged the use of **antiretroviral (ARV) drugs** in favor of garlic to treat the infected. Finally, the health measures that have been taken to fight the epidemic have been inadequate. Simple tests to determine the presence of HIV are seldom available, so people who have the virus pass it on unknowingly (the United Nations estimates that 90 percent of HIV-positive Africans do not know they have the virus). And mainly because of their high cost, anti-AIDS drugs have been scarce.

What can be done about these difficult problems? Two countries that have had remarkable success in stemming the AIDS tide provide some answers. Uganda and Senegal have used relatively inexpensive public health tools, especially education, in the fight against the disease. To reduce the stigma associated with AIDS and to encourage counseling, testing, and condom use, the countries' political and religious leaders have been outspoken about AIDS. The results have been impressive. Uganda's adult infection rates dropped from 14 percent in the early 1990s to 8 percent by 2000 and 6.7 percent in 2005. Senegal's president spoke out forcefully about AIDS in the late 1980s, and infection rates have stayed below 2 percent ever since.

The wealthier countries outside Africa could invest more in combating the disease. The United Nations in 2001 established the Global Fund to Fight AIDS, Malaria and Tuberculosis, but struggles to raise enough resources to fight the epidemic. Canada and the European Union have been criticized for not donating more. The United States has its own effort to combat AIDS in 12 countries in Africa and 2 (Haiti and Guyana) in the Americas. The Bush administration pledged over $45 billion to the effort—an enormous amount, but one that is dwarfed by the scale of the problem: it is enough to treat only about one in four persons infected with HIV. Drug companies have a unique opportunity to address Africa's AIDS problems. Some of them manufacture life-prolonging ARV drugs containing protease inhibitors that suppress replication of HIV but do not kill the virus. These drugs reduce the likelihood of transmission of the virus to another person. The companies could help make such drugs more accessible; in 2005, only 18 percent of the 24 million HIV-positive Africans were receiving these medications. That would mean making the drugs inexpensive, which drug companies in the West have until recently been reluctant to do, fearing that there would be public pressure to reduce prices on all drugs.

However, other countries, notably Brazil, India, and Thailand, have taken advantage of a loophole in international patent rules to manufacture their own cheaper copies of the U.S.-made drugs. Called **compulsory licensing,** the loophole allows the countries to breach patents during national emergencies to manufacture generic versions of patented drugs. Citing AIDS as an emergency, these countries began producing ARV drugs at a small fraction of their cost in the West. Pressure grew on U.S. companies to lower their prices on anti-AIDS drugs sold in Africa and elsewhere in the developing world. Soon some U.S. drug firms relented. Merck, for example, offered to make one of its key drugs available in Africa at cost, making no profit. The U.S. administration promised to keep the door open for even cheaper anti-AIDS drugs by declaring it would not seek sanctions against poor countries stricken with AIDS, even if American patent laws were being broken. The U.S. Food and Drug Administration also approved some generic AIDS drugs to be produced inexpensively in the United States for use abroad. Such measures are helping to get ARVs to increasing numbers of the afflicted, but there are still big challenges to help the majority gain access to these drugs.

REGIONAL PERSPECTIVE

The Great Rift Valley

One of the most spectacular features of Africa's physical geography is the **Great Rift Valley,** a broad, steep-walled trough extending from the Zambezi Valley (on the border between Zimbabwe and Zambia) northward to the Red Sea and the valley of the Jordan River in southwestern Asia (see Figure 9.5 and also Figure 2.1, page 21). Its relationship to the tectonic movement of crustal plates is still poorly understood. However, most earth scientists believe it marks the boundary of two crustal plates that are rifting, or tearing apart, causing a central block between two parallel fault lines to be displaced downward, creating a linear valley. This movement will eventually cut much of southern and eastern Africa away from the rest of the continent and allow seawater to fill the valley.

The Great Rift Valley has several branches. Lakes, rivers, seas, and gulfs already occupy much of it. It contains most of the larger lakes of Africa, although Lake Victoria, situated in a depression between two of its principal arms, is an exception. Some, like Lake Tanganyika—at 4,823 feet (1,470 m), the world's second-deepest lake, after Russia's Lake Baikal—are extremely deep. Most have no surface outlet. Volcanic activity associated with the Great Rift Valley has created Kilimanjaro, Kirinyaga, and some of the other great African peaks, along with lava flows, hot springs, and other thermal features. Faulting along the Great Rift Valley in Ethiopia, Kenya, and Tanzania has also exposed remains of the earliest known ancestors of *Homo sapiens*.

Most Africans who live by tilling the soil also keep some animals, even if only goats and poultry. Among African peoples such as the Maasai of Kenya and Tanzania and the Tutsi (Watusi) of Rwanda and Burundi, livestock not only contribute to the daily diet but are also an indispensable part of customary social, cultural, and economic arrangements. Cattle are particularly important, with sheep and goats playing a smaller role. Traditional Maasai pastoralists are probably the most famous African example of close dependence on cattle. They milk and carefully bleed the animals for each day's food and tend them with great care. The Maasai give a name to each animal, and herds play a central role in the main Maasai social and economic events through the year.

Madagascar is a good example of an African country in which cattle represent status, wealth, and cultural identity. Malagasy livestock owners tend to want higher numbers of the animals, rather than better-quality stock, to enhance their standing (•Figure 9.11). A Malagasy family practicing the ritual commemoration of deceased ancestors sacrifices a large number of zebu cows for fellow villagers, and similar feasts accompany other important ritual dates. Due to such demands, the population of zebu cattle on the island is about 11 million. Their forage needs have grave consequences for Madagascar's rain forests and other wild habitats. People clear the forests and repeatedly set fire to the cleared lands to provide a flush of green pasture for their livestock, causing a rapid retreat of the island's natural
517 vegetation.

Joe Hobbs

• **Figure 9.11** Zebu cattle are extremely important in the cultures and household economies of rural Madagascar. There are about 11 million cattle in this country of 18 million people.

Africa's Wildlife

Africa has the planet's most spectacular and numerous populations of large mammals. The tropical grasslands and open forests of Africa are the habitats of large herbivorous animals, including the elephant, buffalo, zebra, giraffe, and many species of antelope, as well as carnivorous and scavenging animals, such as the lion, leopard, and hyena. The tropical rain forests have fewer of these "game" animals (as Africans call them); the most abundant species here are insects, birds, and monkeys, with the hippopotamus, the crocodile, and a great variety of fish in the streams and rivers draining the forests and wetter savannas.

• **Figure 9.12** Daggers are a nearly universal dress accessory for men in the Arabian Peninsula nation of Yemen. The most prized dagger handles are of rhino horn, a custom that has had a devastating impact on African rhinos thousands of miles away.

Although documentary films promote a perception outside Africa that the continent is a vast animal Eden, the reality is less positive. Human population growth, urbanization, and agricultural expansion are taking place in Africa, as elsewhere in the world, at the expense of wildlife. Hunting and competition with domesticated livestock also take their toll. Many species are safeguarded by law, especially in the protected areas that make up about 15 percent of the region. Such laws are difficult to enforce, however, and poaching has devastated some species. Still, Africa remains home to some of the world's most extraordinary and successfully managed national parks, including South Africa's Kruger National Park, Tanzania's Ngorongoro Crater National Park, and Kenya's Amboseli National Park (the location of Kilimanjaro). International tourism to these parks is a major source of revenue for some countries, particularly South Africa, Tanzania, Kenya, Namibia, and Botswana.

Elephants and rhinoceroses are Africa's largest herbivores and the most vulnerable to global market demand for wildlife products. In 1970, there were about 2.5 million African elephants living on the continent. Poaching and habitat destruction had reduced their numbers to an estimated 680,000 in 2007. Poaching for ivory was depleting the herds of African elephants at a rate of 10 percent annually when delegates of the 112 signatory nations of the Convention on International Trade in Endangered Species (CITES) met in 1989. The organization succeeded in passing a worldwide ban on the ivory trade, and since then, the precipitous decline has halted.

CITES member states won the ivory ban over the strong objections of several southern African states, led by Zimbabwe, that were experiencing what they saw as an elephant overpopulation problem. At the 1989 CITES meeting, these countries appealed for an exemption from the ivory ban so that they could earn foreign export revenue from a sustainable yield of their elephant populations. The majority of CITES members rejected this appeal, arguing that any loophole in a complete ban would subject elephants everywhere to illegal poaching.

CITES later reversed its policy, allowing Zimbabwe, Namibia, Botswana, and South Africa to sell ivory under rigorous monitoring as a reward for their positive wildlife policies. All but South Africa continue to cull (kill) their "excess" elephants, with the meat going to needy villagers and crocodile farms.

If elephant ivory is like gold, rhinoceros horns are like diamonds. Men in the Arabian Peninsula nation of Yemen prize daggers with rhino horn handles (• Figure 9.12). Although Western scientists deny the medicinal efficacy of powdered rhino horn, traditional medicine in East Asia makes wide use of it, including as an aphrodisiac. These demands, and the current black-market value of tens of thousands of dollars per horn, have led to a precipitous decline in population of black rhinoceroses in Africa. There were an estimated 65,000 black rhinos in Africa in 1982; in 2007, there were an estimated 3,700, but strident antipoaching programs are under way in the most critical countries.

9.3 Cultural and Historical Geographies

Many non-Africans are unaware of the achievements and contributions of the cultures of sub-Saharan Africa. The African continent was the original home of humankind.

INSIGHTS

Africa's Greatest Conservationist

Not much larger than the common housefly, the tsetse fly of sub-Saharan Africa packs a wallop. This insect carries two diseases, both known as **trypanosomiasis,** which are extremely debilitating to people and their domesticated animals. People contract **sleeping sickness** from the fly's bite, and cattle contract nagana. Since the 1950s, widespread efforts have been made to eradicate tsetse flies so that people can grow crops and herd animals in fly-infested wilderness areas (•Figure 9.B). Where the efforts have been successful, people have cleared, cultivated, and put livestock on the land. The results are mixed. While people have been able to feed growing populations in the process of opening up these lands, they have also eliminated important wild resources and in many cases caused erosion, desertification, and salinization of the land. Where the tsetse fly has been eliminated, so has the wilderness. The diminutive tsetse fly thus may be characterized as a **keystone species**—one that affects many other organisms in an ecosystem. The loss of a keystone species—in this case, a fly that keeps out humans and cattle—can have a series of destructive impacts throughout the ecosystem. For its role in maintaining wilderness in sub-Saharan Africa, some wildlife experts call the tsetse fly "Africa's greatest conservationist."

Joe Hobbs

• **Figure 9.B** Strenuous and successful efforts to eradicate tsetse flies have opened vast new areas of Africa to human use. This is a tsetse fly trap with two components: a liquid chemical called "simulated cow's breath" that attracts the fly and a pesticide-soaked tarp that kills it.

Recent DNA studies suggest that the first modern people (*Homo sapiens*) to inhabit Asia, Europe, and the Americas were descendants of a small group that left Africa via the Isthmus of Suez about 100,000 years ago. After about 5000 B.C.E., indigenous people were responsible for agricultural innovations in four culture hearths: the Ethiopian Plateau, the West African savanna, the West African forest, and the forest-savanna boundary of West Central Africa. Africans in these areas domesticated important crops such as millet, sorghum, yams, cowpeas, okra, watermelons, coffee, and cotton. From Africa, these diffused to populations in other world regions.

Civilizations and empires emerged in Ethiopia, West Africa, West Central Africa, and South Africa. In the first century C.E., a Christian empire based in the Ethiopian city of Axum controlled the ivory trade from Africa to Arabia. Ethiopian tradition holds that a shrine in Axum still con-
223 tains the biblical Ark of the Covenant and the tablets of the Ten Commandments, which disappeared from the Temple in Jerusalem in 586 B.C.E. Several Islamic empires, including the Ghana, Mali, and Hausa states, emerged in West Africa between the 9th and 19th centuries. All of these agriculturally based civilizations controlled major trade routes across the Sahara. They profited from the exchange of slaves, gold, and ostrich feathers for weapons, coins, and cloth from North Africa. Three kingdoms arose between the 14th and 18th centuries in what are now the southern Democratic Republic of Congo and northern Angola. These included the Kongo kingdom, which had productive agriculture and was the hub of an interregional trade network for food, metals, and salt. In what is now Zimbabwe, the Karanga kingdom of the 13th to 15th centuries built its capital city at the site known as Great Zimbabwe. Its skilled metalworkers mined and crafted gold, copper, and iron, and merchants traded these metals with faraway India and China.

The Languages of Africa

Sub-Saharan Africa exhibits great linguistic diversity (•Figure 9.13). The peoples of this region speak more than 1,000 languages, which generally belong to one of four broad language groups.

The **Niger-Congo language family** (sometimes considered a subfamily of the Niger-Khordofanian family) is the largest. It includes the many West African languages and the roughly 400 Bantu subfamily languages that fall into seven branches: **Benue-Congo, Kwa, Atlantic, Mandé, Gur, Adamawan,** and **Khordofanian.** Most of these are spoken south of the equator. The **Bantu** language of the Benue-Congo branch is the most widespread.

The **Afro-Asiatic language family** includes **Semitic** languages (such as the **Amharic** language of Ethiopia and **Arabic**) and tongues of the **Cushitic** (**Oromo** and **Somali** of the Horn of Africa, for example) and **Chadic** (especially the **Hausa** of

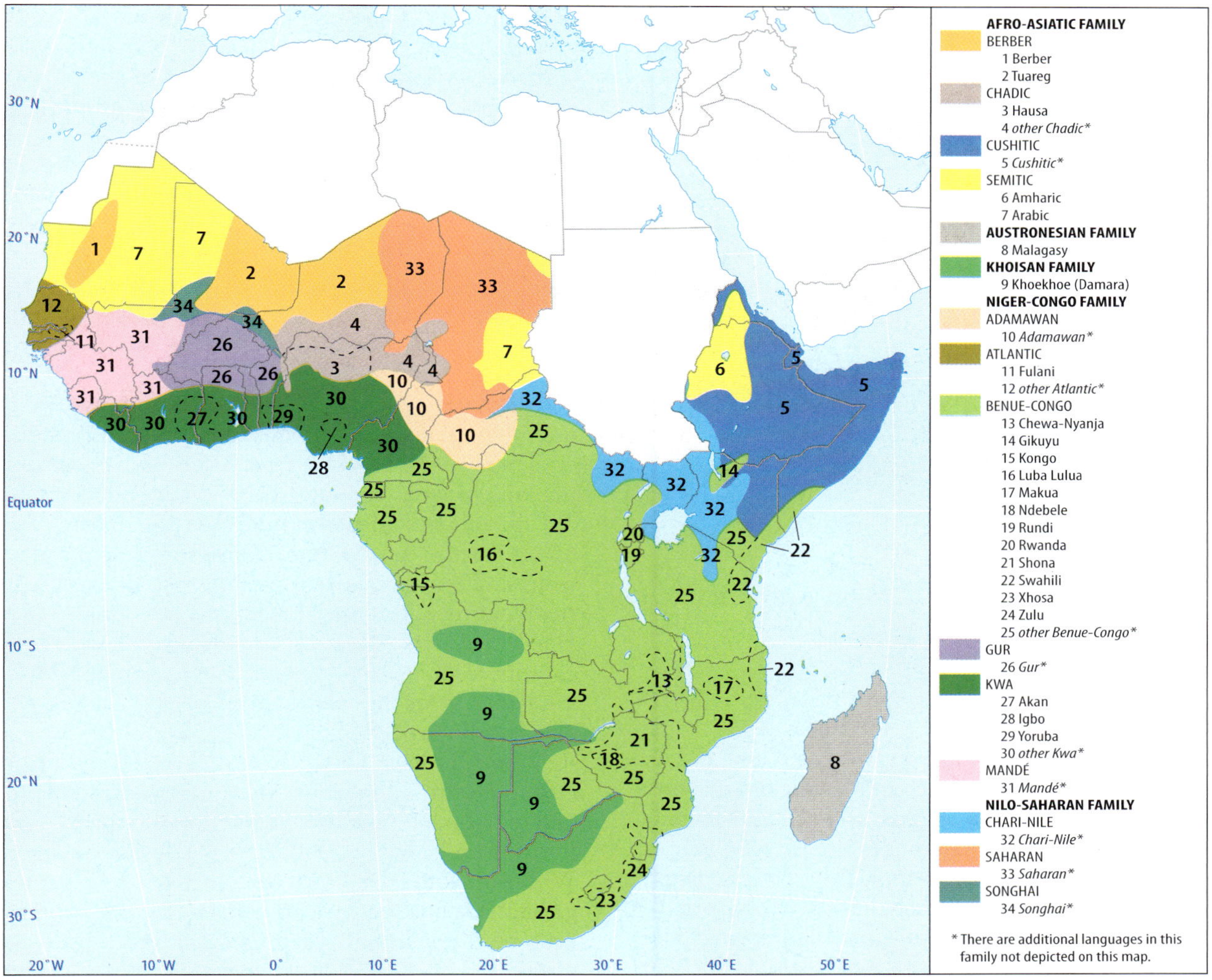

• **Figure 9.13** Languages of sub-Saharan Africa.

northern Nigeria) branches. People living in the area adjoining the Sahara, from West Africa to the Horn of Africa, speak these languages. Even some of the Niger-Congo languages originating south of the Sahara, such as the **Swahili (Kiswahili)** tongue spoken widely in East Africa, have borrowed much from Arabic and other languages with roots elsewhere. The prominence of Arabic words in Swahili reflects a long history of Arab seafaring along the Indian Ocean coast of Africa; in fact, Swahili means "coastal" in Arabic.

The **Nilo-Saharan language family** of the central Sahel region, the northern region of West Central Africa, and parts of East Africa includes about 100 languages in three branches: **Songhai** (a single language, spoken mainly in Mali) and the **Saharan** and **Chari-Nile** language groups.

The **Khoisan languages** of the San and related peoples in the western portion of southern Africa are nearly extinct. The largest, **Khoekhoe (Damara)**, is spoken by fewer than 200,000 people in Namibia.

Quite distinct from those four groups, the people of Madagascar speak a language without African roots. This is **Malagasy**, an **Austronesian** tongue that originated in Southeast Asia. 292

Africa's list of lingua francas—the languages most likely to be recognized on the continent and those that African can use to speak to the wider world—is provided by the six official languages of the African Union, the continent's supranational organization. They are English, French, Portuguese, Spanish, Swahili, and Arabic. Use of English is growing most rapidly, even in the formerly French-dominated "Francophone" countries. Countries that formerly were 493
British colonies have a distinct advantage in the world of outsourced services: South Africa, Kenya, and Ghana, for example, have already opened call centers that they hope

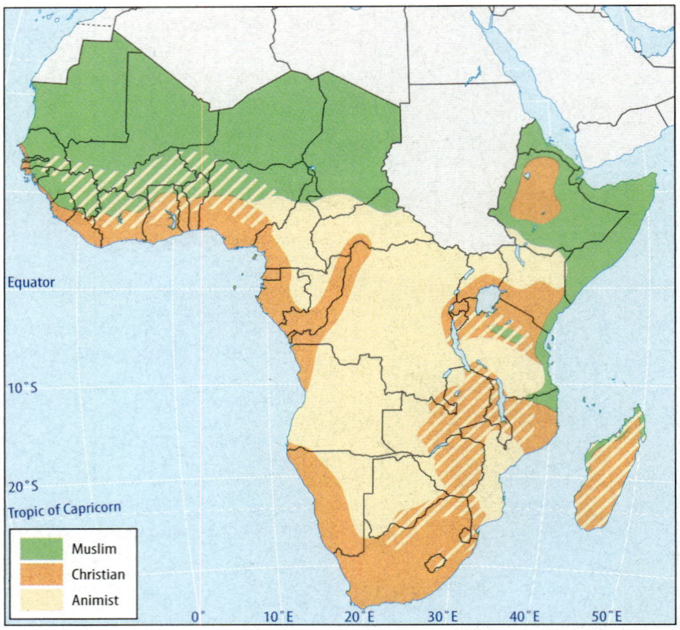

• **Figure 9.14** Religions of sub-Saharan Africa.

will rival those of India. Some Francophone countries also
300 have call centers catering to French industries and clients.

The dominance of foreign languages in digital technology could endanger African languages, many of which are already in peril because they are spoken by so few people. However, many linguists in sub-Saharan Africa see the computer as a tool for saving traditional languages. They argue that the computer will make it much easier for people to learn and preserve their traditional tongues.

Africa's Belief Systems

The religious landscape of Africa is complex and fluid. Spiritualism is extremely strong, but spiritual affiliations and practices are more interwoven and flexible than in most other world regions. It is not uncommon for family members to follow different faiths or for an individual to change his or her religious beliefs and practices in the course of a lifetime.

Broadly, however, some dominant patterns of religious
225 geography can be discerned (•Figure 9.14). Islam is the dominant religion in North Africa and the countries of the Sahel on the southern fringe of the Sahara. The **Ethiopian
504 Orthodox Church,** closely related to the Coptic Christian faith of Egypt, makes Ethiopia an exception to the otherwise Islamic Horn of Africa region. Islam is also the prevailing religion of the East African coast, where Arab traders introduced the faith. Muslims are a majority or strong minority in rural northern Nigeria and Tanzania, and there are minority Muslim populations in cities and towns across the continent. Christians are a majority in southern Nigeria, Uganda, Lesotho, and parts of South Africa. Both Christianity and Islam are strongest in the cities, whereas traditional religions prevail or overlap with these monotheistic faiths in rural areas.

Christians and Muslims have mounted efforts to dispose of "heretical" notions in African traditional belief systems, but these concepts are very resilient. Even where the large monotheistic faiths nominally prevail and their followers are devout, a strong substrate of indigenous beliefs has persisted and usually comfortably merged with the "official"
faith—a syncretism also present in other cultures of the 294,
world. 537

Indigenous African religions have many animistic elements, emphasizing that spiritual forces are manifested ev-
erywhere in the environment. Many spirits are tied to par- 294
ticular places in the landscape. Such beliefs contrast with the introduced Christian and Muslim faiths, which tend to see nature as separate from God and people as apart from and superior to nature.[2] Across the spectrum of African cultures, the use of mediums to contact spirits, either of deceased ancestors, creator beings, or earth genies that can intercede on one's behalf, is widespread. Reverence for ancestors is strong, reflected also in great respect for the elderly who are living.

Outsiders have tried, seldom successfully, to give accurate names to indigenous African belief systems, practices, and personnel: ancestor worship, animism, force vitale ("life force"), living dead, and witch doctor among them. While it would be gratifying to have precise names and generalizations for African beliefs, they are unique and can best be understood by careful reading, observation, and discussion on a case-by-case basis.

The Origins and Impacts of Slavery

Until about 1,000 years ago, the cultures of Africa south of the Saharan desert barrier remained largely unknown to the peoples north of the desert. Egyptians, Romans, and Arabs had contact with the northern fringes of this region, and some trade filtered across the Sahara, but to most outsiders, Africa was the "Dark Continent," a self-contained, tribalized land of mystery. Even at the opening of the 20th century, vast areas of interior tropical Africa were still little known to Westerners.

The tragic impetus for growing contact between Africa and the wider world was slavery (•Figure 9.15). Over a period of 12 centuries, as many as 25 million people from sub-Saharan Africa were forced to become slaves, exported as merchandise from their homelands. The trade began in the 7th century, with Arab merchants using trans-Saharan camel caravan routes to exchange guns, books, textiles, and beads from North Africa for slaves, gold, and ivory from sub-Saharan Africa. As many as two-thirds of the estimated 9.5 million slaves exported between the

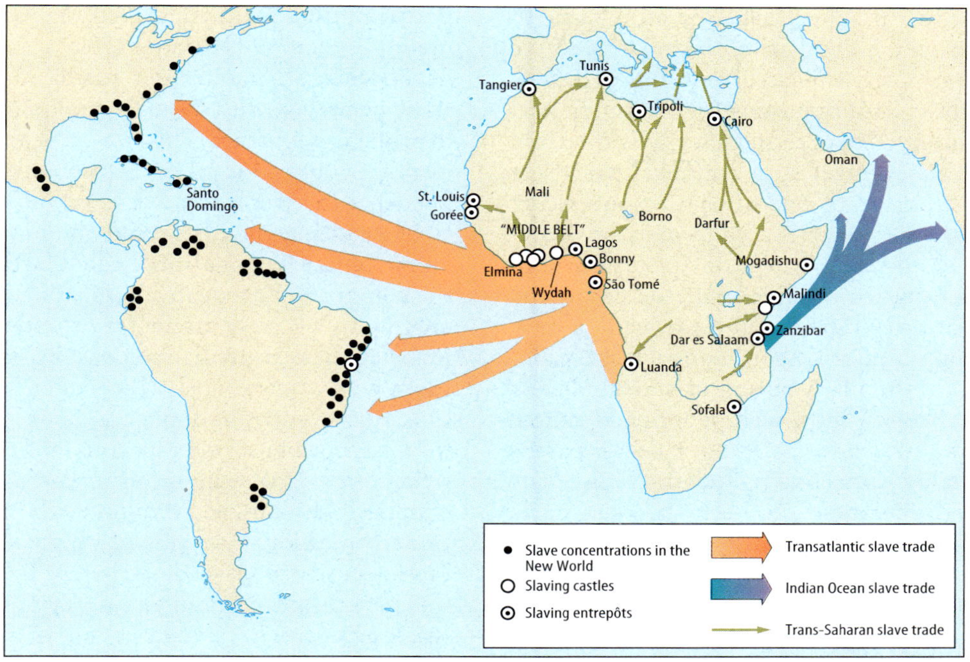

• **Figure 9.15** Slave export trade routes from sub-Saharan Africa.

years 650 and 1900 along this route were young women who became concubines and household servants in North Africa and Turkey. Male slaves usually became soldiers or court attendants (some of whom eventually assumed important political offices). From the 8th to 19th centuries, about 5 million more slaves were exported from East Africa to Arabia, Oman, Persia (modern Iran), India, and China. Again, most were women who became concubines and servants.

The notorious and lucrative traffic in slaves provided the initial motivation for European commerce along the African coasts, inaugurating the long era of European exploitation of Africa for profit and political advantage. The European-controlled slave trade was the largest by far, particularly with the development of plantations and mines in the New
528, 536, 537, 615 World. Between the 16th and 19th centuries, the capture, transport, and sale of slaves was the exclusive preoccupation of trade between the European world and West Africa. Portuguese and Spaniards began the trade in the 15th century, and within 100 years, English, Danish, Dutch, Swedish, and French slavers were participating. The peak of the transatlantic slave trade was between 1700 and 1870, when about 80 percent of an estimated total of 10 million slaves made the crossing. In escape attempts, in transit, and in the famines and epidemics that followed slave raids in Africa, probably more than 10 million others died.

Slaves were a prized commodity in the **triangular trade** linking West Africa with Europe and the Americas (see Figure 3.4, page 43). European ships carried guns, ammunition, rum, and manufactured goods to West Africa and exchanged them there for slaves. They then transported the slaves to the Americas, exchanging them for gold, silver, tobacco, sugar, cotton, rum, and tropical hardwoods to be carried back to Europe. As "raw material" and as the labor working the mines and plantations of Latin America, the
West Indies, and Anglo America, slaves generated much of 42
the wealth that made Europe prosperous and helped spark the Industrial Revolution.

While Europeans carried out the triangular trade, their physical presence was limited to coastal shipping points. Africans were the intermediaries who actually raided inland communities to capture the slaves and assemble them at the coast for transit shipment. West African kingdoms initially acquired their own slaves in the course of waging local wars. As the demand for slaves grew, these kingdoms increasingly went to war for the sole purpose of capturing people for the trade. As the exports grew, so did the practice of Africans keeping African slaves. Even after Britain (in 1807) and the other European countries abolished slavery (1870), it continued to flourish in Africa. By the end of the century, slaves made up half the populations of many African states.

Slavery has not yet died out in the region (and indeed, a modern-day form of slavery exists even in the United States and other MDCs; see pages 187 and 601). In Mauritania, some Moors still enslave blacks, although the national government has outlawed this practice several times. Enslavement of children persists in West Africa. The typical pattern is that impoverished parents in one of the region's poorer countries, such as Benin, are approached by an intermediary who promises to take a child from their care and see that the boy or girl is properly educated and employed. This involves a fee (as little as $14) that, unbeknown to the parents, is a sale into slavery. The intermediary sells the child (for $250 to $400, on average) to a trafficker who sees that the child is transported to one of the region's richer countries, such as Gabon. There the child ends up working without wages and under the threat of violence as a domestic servant, plantation worker (especially on cacao and cotton plantations), or prostitute. Benin is the leading slave supplier and trafficking center; Gabon, Ivory Coast, Cameroon, and Nigeria are the main buyers of slaves. Slavery also exists in Sudan. In the 1990s, well-intentioned church groups and other organizations in the West began paying hard cash to buy freedom for slaves in Sudan. While many real slaves were freed, Sudanese profiteers also moved in with "counterfeit slaves," ordinary people hired to act as slaves until their "rescue" had been paid for.

The Impact of Colonialism

In the 16th century, European colonialism began to overshadow and inhibit the growth of indigenous African civilizations. This is a diverse region, but one thing its peoples have in common is a sense that in the past they were humiliated and oppressed by outsiders.[3]

84 Portugal was the earliest colonial power to build an African empire. The epic voyage of Vasco da Gama to India in 1497–1499 via the Cape of Good Hope was the culmination of several decades of Portuguese exploration along Africa's western coasts. During the 16th century, Portugal controlled an extensive series of strong points and trading stations along both the Atlantic and Indian Ocean coasts of the continent. European penetration of the African interior began in 1850 with a series of journeys of exploration. Missionaries like David Livingstone, as well as traders, government officials, and adventurers and scientific explorers such as James Bruce, Richard Burton, and John Speke, undertook these expeditions. By 1881, when Africans still ruled about 90 percent of the region, foreign exploits had revealed the main outlines of inner African geography, and the European powers began to scramble for colonial territory in the interior. Much of the carving up of Africa took place at the Conference of Berlin in 1884 and 1885, when the French, British, Germans, Belgians, Portuguese, Italians, and Spanish established their respective spheres of influence in the region. By 1900, only Ethiopia and Liberia had not been colonized. For more than half a century, sub-Saharan Africa was a patchwork of European colonies (•Figure 9.16), and Europeans in these possessions were a privileged social and economic class.

At the outbreak of World War II in 1939, only three countries—South Africa, Egypt, and Liberia—were independent. The United Kingdom, France, Belgium, Italy, Portugal, and Spain controlled the rest. But after the war, mainly in the 1960s and 1970s, a sustained drive for independence was mounted. Ceasing to be a colonial region, Africa emerged with more than a quarter of the world's independent countries. This was a peaceful process in most instances, but bloodshed accompanied or followed independence in several countries.

Dependency theorists often point to Africa as a prime example of how colonialism created lasting disadvantages for the colonized. European colonization produced or perpetu- 45 ated many negative attributes of underdevelopment. These included the marginalization of subsistence farmers, notably those who colonial authorities, intent on cash crop produc- 46 tion, displaced from quality soils to inferior land. In addition, European use of indigenous labor to build railways and roads often took a high toll in human lives and disrupted countless families. Furthermore, the colonizers often corrupted traditional systems of political organization to suit their needs, sowing seeds of dissent and interethnic conflict.

The European colonial enterprise did have some positive impacts. The colonies, and the independent nations that succeeded them, were the beneficiaries of new cities and the transport links built with forced or cheap African labor; new medical and educational facilities (often developed through Christian missions); new crops and better agricultural techniques; employment and income provided by new mines and modern industries; new governmental institutions; and government-made maps useful for administration and planning. Such innovations were very helpful, but they were distributed unequally from one colony to another and were inadequate for the needs of modern societies when independence came.

One of the persistent problems in Africa's political geography is that many modern national boundaries do not correspond to indigenous political or ethnic boundaries. In most cases, this is another legacy of colonialism: British, French, and other occupying powers created arbitrary administrative units that were transformed into countries as the colonial powers withdrew. Nigeria is a good example of an ethnically complex, unnaturally assembled nation. On the political map, it appears as an integral unit, but its boundaries have no logical basis in physical or cultural geography. Some Nigerians still refer to the **"mistake of 1914,"** when British colonial cartographers created the country despite its ethnic rifts (see Figure 9.1B on page 491). The greatest divide is between Muslim north and Christian south, but there are at least 250 ethnic groups within the country. The British colonizers invested more in education and economic development in the south and built army ranks among northerners, whom they thought made better fighters; so the subsequent pattern is that the northern military leaders have tried to blunt the economic clout of southerners. Colonial administration and boundary drawing thus sowed seeds of modern conflict in Nigeria, a problem described in more detail in Module 9.1. 493

• Figure 9.16 Colonial rule in 1914. Germany lost its colonies to the British, French, and Belgians after World War I.

Although formal political colonialism has vanished, most countries still have important links with the colonial powers that formerly controlled them, and many foreign corporations that operated in colonial days still maintain a significant presence. France has a long history of postindependence intervention in the political and military affairs of its former African colonies. France is the only ex-colonial power to keep troops in Africa (with the highest numbers in Djibouti, Ivory Coast, the Central African Republic, and Chad). France took steps to ensure that most of its former colonies trade almost exclusively with France and in turn supported national currencies with the French treasury. But France found this paternalistic approach to be extremely expensive and has been reducing its military presence and other costly assistance to its African clients.

9.4 Economic Geography

Great poverty is characteristic of sub-Saharan Africa; 25 of the world's 30 poorest countries are there (see Table 9.1). All of the economies except South Africa's are underindustrialized. Africa's place in the commercial world is mainly that of a producer of primary products, especially cash crops and raw materials (particularly minerals), for sale outside the region. In most nations, one or two products supply more than 40 percent of all exports—for example, oil in Angola and coffee and tea in Kenya (•Figure 9.17). Such a country is vulnerable to international oversupply of an export on which it is vitally dependent. The value of imports far exceeds that of exports in sub-Saharan Africa, 45
with imports consisting mainly of manufactured goods, oil products, and food.

Social and structural problems contribute to the region's underdevelopment. Most African societies lack a substantial middle class and the prospect of upward economic mobility. Instead, most are hierarchical, and any significant income tends to flow into the hands of a small elite controlling the lion's share of the nation's wealth. There are not enough schools to promote the economic welfare that can accompany literacy, and attendance in many is poor. Even with recent growth in attendance, only 70 percent of all children attend primary school, the lowest percentage of any world region. Bureaucratic obstacles and corruption

• **Figure 9.17** Kenya's economy is highly dependent on the export of coffee. Cash crops and other raw materials are typical exports of Africa south of the Sahara.

can make starting up a new business a long, painful, and costly ordeal, discouraging investment both by Africans and foreigners.

Despite the overall grim picture of African economies, recent decades have seen some changes for the better, including improved health and literacy, growth in transportation infrastructure, and increased manufacturing. Such changes have affected some peoples and areas more than others, and their total impacts have only begun to lift the region from underdevelopment. Many problems remain unresolved, and much of the potential for development in agriculture, mining, and infrastructure for communication and transportation goes unexploited.

Agriculture

Per capita food output in most of sub-Saharan Africa has declined or remained flat since independence. The average African consumes 10 percent fewer calories than two decades ago, and malnutrition afflicts almost half the region's children. Rapid population growth and drought are partly responsible for this trend. Many regimes have also invested more in their militaries than in getting food to their citizens. Food shortages also relate to government preference for cash crops over subsistence food crops. Coffee, cotton, and cloves, for example, provide a means of gaining foreign exchange with which to buy foreign technology, industrial equipment, arms, and consumption items for the elite. The proportion of crops grown for export to overseas destinations and for sale in African urban centers has therefore risen significantly in recent times.

Most export crops are grown on small farms rather than on plantations and estates. Large plantations have never become as established as they are in Latin America and Southeast Asia. Political and economic pressures forced many of them out of business during the period of transition from European colonialism to independence. Governments of the newly independent countries nationalized and subdivided white-owned plantations, and many whites were obliged to sell or give their land to Africans. Now operated by Africans, many of these farms provide vital tax revenue and foreign exchange. One exception is South Africa, where the most important producers and exporters are the ethnic Europeans, who raise livestock, grains, fruit, and sugarcane. But throughout tropical Africa as a whole, the trend is toward export production from small, black-owned farms. The most valuable export crops are coffee, cacao, cotton, peanuts, and oil palm products. Secondary cash crops include sisal (grown for its fibers; see Figure 540
10.21), pyrethrum (used in insecticides), tea, tobacco, rubber, pineapples, bananas, cloves, vanilla, cane sugar, and cashew nuts.

Although cash crops are important sources of revenue in these poor countries, excessive dependence on them can

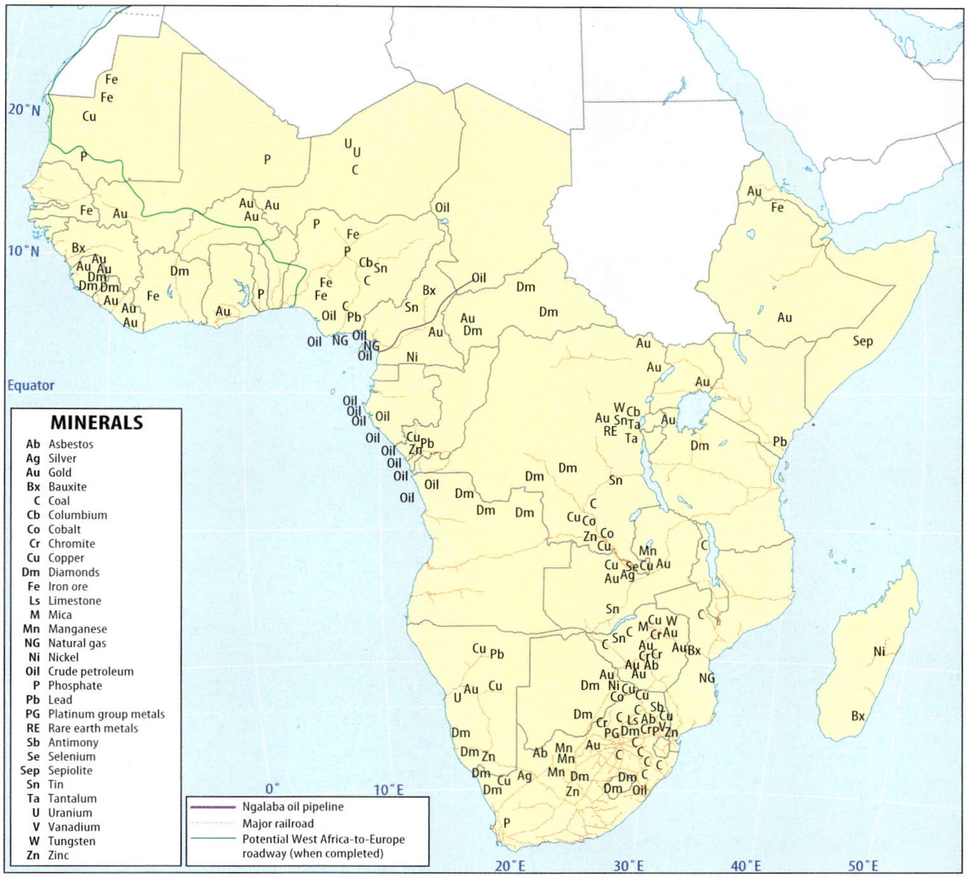

• **Figure 9.18** Minerals, oil pipelines, and transportation links in sub-Saharan Africa.

be harmful to a country's economy. The income they bring often goes almost exclusively to already prosperous farmers and corporations. The prices they fetch are vulnerable to sudden losses amid changing world market conditions. The plants are susceptible to drought and disease (see Biogeography, page 476). Finally, they are often grown instead of food crops, and cash crops do not feed hungry people.

Mineral Resources

Mineral exports have had a strong impact on the physical and social geographies of sub-Saharan Africa. The three primary mineral source areas are South Africa and Namibia; the Democratic Republic of Congo–Zambia–Zimbabwe region; and West Africa, especially the areas near the Atlantic Ocean (•Figure 9.18). Notable mineral exports from these regions include precious metals and precious stones (particularly diamonds; the region has 60 percent of the world's supply—see Problem Landscape, page 478), iron alloys, copper, phosphate, uranium, petroleum, and high-grade iron ore, all destined principally for Europe, the United States, and China.

Large multinational corporations, financed initially by investors in Europe or the United States and most recently by China, do most of the mining in Africa. Mining has attracted far more investment capital to Africa than any other economic activity. Money is invested directly in the mines, and many of the transportation lines, port facilities, power stations, housing and commercial areas, manufacturing plants, and other elements in the continent's infrastructure have been developed primarily to serve the needs of the mining industry.

Great numbers of workers in the mines are temporary migrants from rural areas, often hundreds of miles away. The recruitment of migrant workers has had important cultural and public health effects. Millions of Africans who work for mining companies have come into contact with Western ideas as well as those of other ethnic groups and have carried these back to their villages. Unfortunately, the phenomenon of migrant labor has also helped spread
HIV, the virus that causes AIDS. Most miners are single or 463
married men who spend extended periods away from their wives or girlfriends, and it is common that they contract the virus from prostitutes and then return home to their

BIOGEOGRAPHY

Crop and Livestock Introductions: "The Curse of Africa"

Many important food and export crops of Africa are not native to the continent. For example, corn (maize), manioc (cassava), peanuts, cacao, tobacco, and sweet potatoes were introduced from the New World. Cattle, sheep, chickens, and other domesticated animals now characteristic of sub-Saharan Africa also originated outside the region.

Some nonnative species, such as manioc and chickens, have adapted well to the environments and human needs in sub-Saharan Africa. Others are more problematic. Known as "mealie-meal," milled corn has become the major staple food for southern Africans. However, the people of southern Africa came to regard corn as more of a scourge than a blessing during the severe regional drought of 1992. Introduced from relatively well-watered Central America, corn has flourished in African regions as long as rain or irrigation water has been sufficient. But the crop failed completely during the 1992 drought. The situation was particularly critical in Zimbabwe, where the government had just sold its entire stock of corn reserves in an effort to repay international debts. Zimbabwe suddenly became a large importer of corn. Many in the country proclaimed corn to be the **"curse of Africa,"** lamenting that with its high water requirements, it should never have been allowed to become the principal crop of this drought-prone region. Agricultural specialists argued that native sorghum and cowpeas are much better adapted to African conditions and should be cultivated as the major foods of the future.

Corn was a major issue again in a severe drought in 2002. This time, the United States and other nations reacted early with major shipments of corn to especially hard-hit Zambia. Thousands of tons of American corn were refused by Zambian authorities or locked up in silos while Zambians went hungry. The reason: Zambia's concern that the U.S. corn was genetically modified (GM). Zambian authorities feared that if unmilled corn were to mix with Zambian seed stock, the anti-GM European Union might refuse future imports of corn from Zambia. They also worried about the potential health impacts of the milled GM corn. 89

Concern is also growing about whether the cow is an appropriate livestock species for Africa. Cattle frequently overgraze and erode their own rangelands. The traditional approach to improving cattle pastoralism in sub-Saharan Africa, and to boosting livestock exports, has focused on the need to develop new strains of grasses that can feed cattle and other non-indigenous strains of livestock because the native grasses of African savannas suit the wild herbivores but generally not the domesticated livestock.

A fresh approach turns this problem around by asking whether it might be more appropriate to commercially develop the indigenous animals already suited to the native fodder and soil conditions. Kenya, Zimbabwe, South Africa, and other countries are now involved in **game ranching** in which antelopes—such as 441 eland, impala, sable, waterbuck, and wildebeest—along with zebras, warthogs, ostriches, crocodiles, and even pythons are bred in semicaptive and captive conditions to be slaughtered for their meat, hides, and other useful products (•Figure 9.C). Safari hunting and tourism provide supplemental revenue in some of these operations.

So far, these enterprises have demonstrated that large native herbivores produce as much or more protein per area than domestic livestock do under the same conditions. In commercial terms, the systems that have incorporated both wild and domestic stock have proved more profitable than systems employing only domestic animals. The World Wildlife Fund and other proponents of this **multispecies ranch system** argue that it is an economically and ecologically sustainable option for land use in Africa. They emphasize that ranchers who keep both wild and domesticated animals will be better able financially to weather periods of drought and will require a much lower capital input than they would expend on cattle or other domesticated livestock alone.

© Ian Berry/Magnum

• **Figure 9.C** Game ranching in Africa. In addition to raising game as an alternative to domestic livestock, there are a number of ranches where tourists come to shoot animals for trophies with guns or bows and arrows.

Joe Hobbs

• **Figure 9.19** Many crucial African river crossings have no bridges, where people must rely on ferries instead. Here vehicles and people board a ferry to cross an estuary north of Mombasa on the Kenyan coast. Backup of traffic at these bottlenecks may take hours or even days to clear.

villages, where they spread it still further. The HIV infection rate among migrant miners working in South Africa is more than 40 percent.

Africa's Fragile Infrastructure

Poor transportation hinders development in sub-Saharan Africa. Few countries can afford to build extensive new road or railroad networks, and much of the colonial infrastructure has deteriorated (•Figure 9.19). The region critically needs a good international transportation network, together with a lowering of trade barriers, to enlarge market opportunities. Poor transportation is also often a contributing factor to famine. In Ethiopia, for example, the road network is so poor that it is extremely difficult to get food from the western part of the country, which often has crop surpluses, to the eastern part, which has chronic food shortages. On more than one occasion, it has proved cheaper to ship food from the United States to eastern Ethiopia than to truck it across the country! Similarly, a recent World Bank study showed that it cost $50 to ship a metric ton of corn from Iowa 8,500 miles (13,600 km) to the Kenyan port of Mombasa but $100 to move it from Mombasa 550 miles (880 km) inland to Kampala, Uganda. Transportation problems contribute to the high costs of agricultural inputs like fertilizers, which in this region cost two to three times what they do in Asia; indeed, fertilizer consumption in Asia is 10 times what it is in sub-Saharan Africa.

Poor communication, especially in digital technology applications, also hinders development. The expanding ability to receive radio and television broadcasts across the continent, even in remote villages, is providing Africans with an unprecedented wealth of information about the wider world. But more high-tech, interactive forms of communication are much in need. Telephone, fax, e-mail, and other technologies are critical to successful participation in the globalizing economy. Africa has not been able to take part because it has been critically short of most of these assets. For example, there are more telephone land lines in the city of Tokyo than in all of Africa.

Internet cafés are springing up, and the Internet may prove to be Africa's answer to many of its communications problems. There are hopes that voice-over-Internet technology will make telephone calling more affordable for hundreds of millions of Africans. In 2007, the entire continent of Africa had only 44 million Internet users (of 1.2 billion worldwide), but the number is growing quickly. Private companies and individuals cooperated to build a 20,000-mile (32,000-km) undersea fiber-optic cable system, completed in 2002, that provided a ring of connections around Africa.

Mobile phones may also help bridge the digital divide. So far, the main obstacle has been that many of the telecommunications companies are government-owned and therefore profitable for governments, which are reluctant to allow competition by private companies with cellular technologies. Some governments also like to monitor phone calls and e-mail, and more communication makes that a more challenging task. In some countries, the barriers are falling, however, and mobile phones are beginning to flourish in some of the world's most isolated places. Mobile phone network coverage was available to only 10 percent of sub-Saharan Africa's people in 1999, but that grew to 60 percent by 2007 and is expected to reach 85 percent in 2010, by which time there should be more than 100 million users. The simple but effective technology of text messaging is already improving

PROBLEM LANDSCAPE

Cleaning Up the Dirty Diamonds

As recently as 2004, as much as 15 percent of the world's annual production of rough diamonds was made up of **dirty diamonds** (also called **conflict diamonds** or **blood diamonds**), defined by the United Nations as "rough diamonds used by rebel movements or their allies to finance conflict aimed at undermining legitimate governments." Gems of such questionable origin financed at least three African wars. But diplomatic, nongovernmental, and business efforts have largely stemmed the tide of Africa's dirty diamonds.

In 2002, following four years of negotiations, 45 countries endorsed a UN-backed certification plan, called the **Kimberley Process,** designed to ensure that only legally mined rough diamonds, untainted by violence, reach the market. Rough diamonds must be sent in tamper-proof containers with a certificate guaranteeing their origin and contents. The importing country (the biggest of which are China and India, which have replaced Belgium and Israel as the world's leading diamond cutters and polishers) must certify that the shipments have arrived unopened and reject any shipments that do not meet the requirements. Only countries that subscribe to the Kimberley Process are allowed to trade in rough diamonds.

Corporate interests and consumer ethics have cleaned up the diamond trade. Diamond certification has been spearheaded by the world's leading diamond company (with about two-thirds of the market), the South African–based multinational named De Beers. De Beers was embarrassed by a report that it had bought $14 million worth of diamonds from Angolan rebels in a single year. Perceiving that a public relations debacle could lead to a business disaster, as happened with an organized boycott against fur products in the 1980s, De Beers seized the initiative. In the name of Africa's welfare, but certainly also a means of increasing demand and profit, De Beers introduced **branded diamonds,** certified as coming from nonconflict areas. Other diamond companies followed suit, especially in an effort to please Americans—who buy half the world's diamonds (• Figure 9.D).

Joe Hobbs

• **Figure 9.D** Many consumers are unaware that this symbol of love and beauty may have had a sinister origin in Africa.

farmers' access to markets through an integrated 12-country network in West Africa. Web-enabled cell phones could make it much easier for millions of Africans to jump to the Internet without having to use computers.

Africa's Place in World Markets and Economies

Africa's status as a supplier of raw materials to world markets has channeled significant new wealth into countries that have oil and other resources for which demand is high in China and other major consuming countries. In the decade 1996–2006, the commodities boom brought annual economic growth rates of about 5 percent to 16 sub-Saharan African countries. This windfall does not guarantee economic development, however, and there are several critical areas where reforms are needed.

One of the major obstacles to African economic development is that many countries outside the region have effectively closed their doors to African imports. These restrictions typically take the form of subsidies, high tariffs, or low quotas imposed on agricultural products (like cotton) or manufactured goods (like textiles). Potential importing countries, such as the United States, impose these re- 623
strictions to protect their own industries. For instance, the United States and Africa are the world's leading exporters of raw cotton. The U.S. government spends $2 billion each year to subsidize its 25,000 cotton farmers. This means higher U.S. production and exports and consequently lower cotton prices and incomes for African farmers.

An example of what reduced trade barriers might do for Africa is provided by the **African Growth and Opportunity Act (AGOA),** passed by the U.S. Congress in 2000 and due to expire in 2017. AGOA reduced or ended tariffs and quotas on more than 1,800 manufactured, mineral, and food items that could be imported from Africa. African exports to the United States have surged, especially in the clothing industry. Apparel made in Lesotho, Kenya, Mauritius, South Africa, Madagascar, and elsewhere in Africa is now sold in Target and Wal-Mart stores. The United States is considering extending AGOA's benefits with the group of countries comprising the **Southern African Customs Union (SACU):** South Africa, Lesotho, Swaziland, Botswana, and Namibia.

INSIGHTS

Microcredit

A promising alternative to the problem of national debt is **microcredit,** the lending of small sums to poor people to set up or expand small businesses. Typically, poor people have no collateral—assets to pledge against loans they seek—and therefore cannot borrow from commercial banks. This drives them into the arms of local loan sharks, who may charge huge interest rates of 10 to 20 percent per day. But in the interest of fostering development, microlending organizations have begun to encourage poor borrowers to form a small group that will cross-guarantee all members' loans. One member of the group may take out a loan of $25, for example, to help start up a small business such as a restaurant. Only when that individual has paid it back may the next person in the group receive a loan. Peer pressure minimizes the default rate, and the accumulated history of timely paybacks increases the entire group's creditworthiness.

Microlenders generally prefer to lend to women because women are more likely to use the money in ways that will feed, clothe, and otherwise benefit their children, whereas men may be more likely to spend the money on alcohol or other frivolous pursuits. One fear in Africa is that microloans will not be repaid if the borrower or someone in the borrower's family contracts HIV or some other debilitating illness. Therefore, the loans are often extended only after the borrower receives some amount of public health training, for example, in the use of condoms and in general food preparation hygiene.

Jobs and incomes for Africans have grown because of these new opportunities. Much of the profit, however, went to the Asian firms that set up clothing factories in Africa because of limitations on clothing exports from their home countries. Those export quotas were set by the World Trade Organization (WTO). After they expired in 2005, any country was able to export as much as it could from the home country. African countries now have to compete with Cambodia and other poor countries as manufacturers look
352, 428 for the cheapest labor they can find. When lower costs have been found elsewhere, the up-and-coming African factories have been threatened. The garment industry in Lesotho, for example, immediately plummeted in 2005, only to be resurrected by the AGOA and by cultivating a niche market in "sweatshop-free" garments.

For decades, almost all of the countries of sub-Saharan Africa were heavily in debt to foreign lenders. Young nations undertook costly development projects with borrowed money, particularly after the mid-1970s. Western financiers, planners, and contractors gave optimistic assessments of the benefits to be expected from such projects, and African leaders were ready to accept loans as a way to reap quick benefits from newly won independence. Countries soon found they had great difficulty in meeting even the interest payments on their debts, and their efforts to do so often resulted in further economic woes and destructive
46 environmental effects. The International Monetary Fund (IMF), World Bank, and other lenders generally resisted requests of African debtor nations for rescheduled interest payments and new loans.

As a condition of further support, foreign lenders often demanded that African debtors put their finances in better order by such measures as revaluing national currencies downward to make exports more attractive, increasing the prices paid to farmers for their products, and reducing corruption, especially among government officials and urban elites. Although many African debtor nations attempted to meet these demands, internal political factors slowed progress. Loss of economic privileges angered urban elites on whom governments depended for support. Relaxation of price controls on food to stimulate production was often met by rioting among city dwellers with tight budgets already allocated for other costs of living.

In the West, increased foreign aid in the form of gifts or grants to Africa has been controversial. Critics argue that foreign aid only increases dependence and deflects national attention away from problems underlying famine. International aid monies often free up funds from national treasuries to be spent on unfortunate uses. In the Horn of Africa, for example, when foreign money became available for famine relief, it allowed Ethiopia and Eritrea to spend more on weapons used in their war. Aid monies are often simply stolen, and monitoring by donors of how they are spent has been inadequate.

Since about 1990, donors have proceeded more cautiously, more often funneling money through private and religious groups working in health care, education, and small-business development using microcredit practices (see Insights, above). Foreign aid has been very effective in advancing primary and secondary education in the region. However, many African governments view universities as potential hotbeds of dissent and discourage investment in higher education. Africa's universities are crumbling, and
the classic brain drain of African talent to Western institu- 59
tions means there is that much less intellectual capital left at home to invest in the region's development.

Africa's untenable debt situation, and the slow flow of aid to the region, began to change around 2005. Two major factors were at work. First, in the West, awareness of the humanitarian costs of not directing more resources to Africa was growing. Media, political and corporate stars like the singer Bono (of U2), former U.S. President Bill Clinton, and Microsoft founder Bill Gates kept up a steady appeal for governments and citizens of the MDCs to be more generous to impoverished Africa. On the eve of the 2005 meeting of the G-8, the seven biggest Western economic powers plus Russia, a multicontinental live music event called "Live 8" rocked public opinion in favor of giving more. Under such scrutiny, the G-8 leaders decided to open their wallets, doubling their aid to the region by 2010. They agreed to forgive more than $40 billion of debt owed by 14 African countries recognized by the **Heavily Indebted Poor Countries (HIPC)** initiative. Those 14 qualified for debt relief because they met such standards as fighting corruption and having more transparent economic systems. Critics pointed out that this excluded critical, heavily indebted countries like Nigeria. But by 2007, it was clear that the West was falling behind on its promised course to double aid, and an Eastern giant stepped into the gap.

Thus the second and more powerful shift in Africa's debt, trade, and aid picture came after 2006 with the meteoric rise of China's engagement with the region. China's trade with sub-Saharan Africa increased 10-fold between 1999 and 2006, when China displaced Britain as the region's third most important trading partner (after the United States and France). In its 2007 dialogue with the **African Development Bank (AfDB),** China pledged $20 billion in infrastructure and trade financing in the region. Those funds, to be expended over a three-year period, represent a doubling of China's previous inputs and rival the value of those pledged by Africa's traditional financiers in the West. There is an important difference, however. As will be seen, the Western institutions tend to have strings attached to loans and aid, insisting that African countries adopt democratic and other reforms to be eligible for the monies. China, on the other hand, has a "no strings attached" policy. It is not insisting on political reforms, is not expressing concern about environmental or social responsibilities, and much to the chagrin of the West is even engaging with repressive governments that the West is trying to isolate. Western governments and financial institutions are increasingly alarmed that China is making rapid inroads into African markets, possibly at the expense of Western economies and geopolitical interests.

Even Africans look with some trepidation on China's advance. China is unashamed about its interests in sub-Saharan Africa: it needs and wants the region's raw materials—especially oil, iron ore, copper, and cotton—to feed its insatiable industrial appetite. To many Africans, China is building the same kind of exploitive mercantile relationship that characterized the colonial West: it takes Africa's raw materials, transforms them into value-added products, and sells them mainly to its own advantage.
South African President Thabo Mbeki warned in 2007 that
Africa risked becoming an economic colony of China. But 45
desperate for assistance, even the wary are inclined to accept China's terms.

A Legacy of Failed States

Africa scholars speak of the **failed-state syndrome,** a pernicious process of economic and political decay that is eating away at countries including the Democratic Republic of Congo, Ivory Coast, Somalia, and Zimbabwe. Some African countries are little more than **shell states.** A shell state appears to have all the institutions of a country: a constitution, a parliament, ministries, and so on. But most of the positions are occupied by friends and family of the country's ruler, who are happy to line their pockets rather than pursue development for their fellow citizens. Nine of the world's 15 most corrupt countries are in this region, according to Transparency International, a leading watchdog group on corruption. How bad can corruption be for a country's development? In Liberia, former President Charles Taylor passed a law allowing him to personally sell any of the country's "strategic commodities," including mineral resources, forest products, art, archaeological artifacts, fish, and agricultural products. The wealth taken by such countries' leaders generally does not find its way back into the economy but is either spent on grandiose personal accommodations or parked in foreign bank accounts.

One of the assumptions of international lenders and aid agencies is that democracy helps weed out such economic bloodletting. True democracy, they observe, is still in short supply in the region. Foreign lenders and donors want to see more, telling African leaders that if they do not institute democratic reforms, hold fair elections, or improve bad human rights records, there will no longer be development aid or lending. Often the leaders make just enough concessions to win the aid without instituting real reform—a phenomenon known as **donor democracy.** Some Africa observers say that positive development assistance must go where it is needed most, channeled directly into rebuilding destroyed institutions such as the civil service, courts, police, and the military.

Foreign powers might not be able to promote such reforms without becoming guilty of or being accused of neocolonialism, but new African institutions are emerging to confront the problems themselves. In 2002, the 53-member **African Union (AU)** was formed to replace the old and often inconsequential Organization of African Unity (OAU). One of the African Union's main articulated goals is to implement self-inspection or "peer review" to help promote democracy, good governance, human rights, gender equity, and development. It has formed an agency responsible for this monitoring, known as the **New Partnership for Africa's Development (NEPAD).** NEPAD's premise is that improvements in such human affairs will improve the climate for international investment and assistance to Africa.

The African Union has also created a 265-member **Pan-African Parliament** that will begin making regional laws by 2009. To help ensure the stability needed for economic and political progress, the African Union is also establishing a continental military force, the **African Standby Force,** that takes orders from the organization's Peace and Security Council. The force will, in theory, field peacekeepers or peacemakers to ensure that the map of Africa, so filled with deadly conflicts in the past, has far fewer trouble spots in the future. The African Union has plans to create an African Court on Human and People's Rights and, like the
90 European Union after which it is modeled, may also eventually seek a common currency.

9.5 Geopolitical Issues

Sub-Saharan Africa has waxed and waned as a theater of geopolitical interest since the end of World War II. Initially, the great powers were interested in the region. To boost their competing aims during the Cold War years, the Soviet Union and the United States played African countries against one another, arming them with weapons with
88 which to wage **proxy wars.** The superpowers also extended aid generously to many African nations.

Great power concerns about Africa changed with the end of the Cold War around 1990, and the nature of conflict in Africa changed. The great powers withdrew support, but their weapons remained to prolong smoldering conflicts. The United States helped build the arsenals of eight of the nine countries involved in the recent Democratic Republic of Congo conflict, for example. Cold War–era weapons like AK-47 assault rifles were "dumped"—sold at low prices—in Africa, where they were not considered obsolete and where they fanned the flames of conflict. Increasingly, fighting began to spread across international borders. Previously, the United States and the Soviet Union had maintained a kind of security balance that kept warfare from becoming internationalized, but that restraint no longer existed. The scenario of regional or even Africa-wide wars was realized with Africa's first "world war," centered on the Democratic
497 Republic of Congo. In the absence of regional or international powers to keep the peace, some countries (for example, Sierra Leone) simply imploded, fragmenting into fiefdoms run by factions that claimed to be revolutionaries but were essentially profiteers. Other national governments (as in Ethiopia) spent vast sums on expensive military aircraft while their people suffered from malnutrition.

The end of the Cold War constricted (and in the case of Russia, ruptured) aid pipelines. The United States cut its economic assistance to sub-Saharan Africa by 30 percent between 1985 and 1992 and reduced it another 10 percent between 1992 and 2001. Why the waning concern? U.S. President George W. Bush was quoted as saying, "While Africa may be important, it doesn't fit into the national strategic interests, as far as I can see them." [4]

Then came 9/11. Al-Qa'ida's attack on the United States and the U.S. counterstrike against al-Qa'ida in Afghanistan raised fears that the Islamist organization might find new recruits and training grounds in Africa. Al-Qa'ida had already bombed U.S. embassies in Kenya and Tanzania in 1998 and in 2002 struck again in the Kenyan city of Mombasa. The United States maintains a focus on **terrorism hot** 235
spots in the region, including Kenya, Somalia, Djibouti, Niger, Chad, and Mali, all of which are regarded as real or potential training grounds for al-Qa'ida–affiliated groups. With a relatively low profile, U.S. forces have made the tiny Horn of African country of Djibouti a major military center for potential action against nearby Yemen or other countries where Islamist militants may be active. So far, U.S. interest in the hot spots has translated into military expenditures and partnerships with African governments but not wider development assistance.

In addition to the terrorism hot spots, the United States has strong geopolitical interests in the region's oil-producing countries. These, too, have become more critical since 9/11 and the U.S. war on Iraq, which only heightened the sense that Middle Eastern oil supplies are vulnerable. The United States is seeking more stable oil supplies and has pinned many hopes on Africa. Petroleum-rich Nigeria and Angola together supply more than 15 percent of America's oil, a figure projected by the U.S. Central Intelligence Agency to rise to 25 percent by 2015. (Nigeria is also the United States' largest trading partner in Africa.)

Increasingly, the United States will be competing with China for these dwindling finite reserves, and given current trends, the geopolitical landscape of sub-Saharan Africa will see Asia on the rise and American clout in decline. Countries that are not pleased with Washington's wishes or demands will find partners elsewhere, particularly in Asia. Some military analysts describe the emerging U.S.-China rivalry in Africa as a modern Great Game, comparable to the 271
British-Russian contest over central Asia in the 19th century.

For many years following the deaths of U.S. service personnel in Somalia in 1993, there was great reluctance to 506
project U.S. military power in African trouble spots. In 2007, the Pentagon reshuffled its bureaucracy and created a command center dedicated exclusively to the region. There are plans to establish U.S. military bases in several key countries, beginning with Cameroon, Equatorial Guinea, and Gabon, and possibly to secure air base rights in Benin, Ivory Coast, and Nigeria. The U.S. Navy would like to establish bases in São Tomé and Príncipe and in Equatorial Guinea. Some Africa observers say the Iraq War will dampen U.S. ambitions in the region; not only are military resources stretched thin, but anti-American sentiment related to the war is entrenched, particularly in the Muslim countries.

The United States views South Africa as the region's economic powerhouse and as the key to regional stability. South Africa's military strength, which included nuclear weapons until they were dismantled in 1990, contributes to its global significance. It is strategically important for its frontages on both the Atlantic and Indian Oceans, its possession of Africa's finest transport network (vitally important to many African countries that trade with and through South Africa), and its diversified mineral wealth.

Finally, HIV/AIDS in the region is also a geopolitical concern. Despite Africa's apparent remoteness from the United States, the two are linked by hundreds of air traffic routes traveled by thousands of people each day. Hundreds of HIV-positive Africans, perhaps most of them unaware of their virus, pass through U.S. airport gateways daily. Some African strains (subtypes) of the virus are more easily transmitted by heterosexual contact than the strains prevalent in the United States, so the risk to the population at large is greater. There are other dimensions of the potential burden on the United States. As long as the epidemic rages in Africa, there may be large flights of "medical refugees" seeking asylum in the United States. There could also be AIDS-related political instability or civil wars that invite U.S. military intervention in the region. How the world should respond to poverty and recurrent crises is one of the key questions facing the huge, resource-rich, problem-ridden region of sub-Saharan Africa. The following module examines these challenges in greater detail.

SUMMARY

- Africa is the cradle of humankind where hominids originated and from where they diffused. The region has seen many indigenous civilizations and empires and is ethnically and linguistically diverse.
- The majority of the region's people are rural. There are several clusters of dense population.
- HIV/AIDS is taking a huge toll, dramatically lowering life expectancy and projections for population growth, especially in southern Africa.
- Sub-Saharan Africa has a relatively low population density overall, but a majority of this region's people live in a small number of densely populated areas.
- Most of Africa consists of a series of plateau surfaces dissected by prominent river systems.
- One of the most spectacular features of Africa's physical geography is the Great Rift Valley, a broad, steep-walled trough. The feature marks the boundary of two crustal plates that are rifting, or tearing apart.
- Although about two-thirds of the region lies within the low latitudes and has tropical climates and vegetation, sub-Saharan Africa contains a great diversity of climate patterns and biomes, some resulting from elevation rather than latitudinal position.
- Africa's diverse wildlife is often threatened by human population growth, urbanization, and agricultural expansion. Management of elephants and rhinoceroses involves international market demand and legal restrictions.
- In the four culture hearths of sub-Saharan Africa, early indigenous people were responsible for several agricultural innovations, including the domestication of millet, sorghum, yams, cowpeas, okra, watermelons, coffee, and cotton.
- The four major language families of the region are Niger-Congo, Afro-Asiatic, Nilo-Saharan, and Khoisan. Malagasy, spoken only on Madagascar, is an Austronesian language. Islam and Christianity are major faiths, but indigenous belief systems are often mixed with them or exist on their own in some locales.
- The tragic impetus for growing contact between Africa and the wider world was slavery. Pockets of slavery still exist in the region.
- Most of sub-Saharan Africa fell under European colonialism after the Conference of Berlin in 1884 and 1885.
- The majority of the people of sub-Saharan Africa are poor, live in rural areas, and practice subsistence agriculture. In most countries, per capita food output has declined or has not increased since independence. Export crops include coffee, cacao, cotton, peanuts, and oil palm products.
- Frequent droughts, lack of education, poor transportation, and serious public health issues have hindered development in sub-Saharan Africa. Most countries are underindustrialized and overly dependent on the export of a few primary products.
- The export of minerals has had a particularly strong impact on the physical and social geography of sub-Saharan Africa. The notorious trade in dirty diamonds has been largely cleansed.
- Exports of clothing from Africa to the United States boomed after the U.S. Congress lowered trade barriers on numerous products, but the suspension of WTO-imposed export quotas in 2005 has forced the region to compete with other low-cost producers.
- Most countries are heavily in debt to foreign lenders. Economic and humanitarian assistance to the region slowed considerably after the end of the Cold War, but it has picked up again since

9/11. China is the latest major economic power to engage in Africa and has a "no strings attached" policy when it comes to aid and trade with the region.

→ Although many countries have been under authoritarian governments since independence, there has been some progress toward democracy. Serious political instability is characteristic of many African countries. Important links with the colonial powers that formerly controlled them remain strong in many countries of the region.

→ The recently established African Union is a supranational organization dedicated to solving African problems without the intervention of outside powers.

→ So often judged to be marginal in world affairs, sub-Saharan Africa deserves and is receiving increased international attention because of its humanitarian problems, the global implications of its public health and environmental situations, problems in the management of its natural resource wealth, its oil reserves, and concerns about terrorism.

KEY TERMS + CONCEPTS

African Development Bank (AfDB) (p. 479)
African Growth and Opportunity Act (AGOA) (p. 477)
African Standby Force (p. 481)
African Union (AU) (p. 480)
Afro-Asiatic language family (p. 468)
- Semitic subfamily (p. 468)
  - Amharic (p. 468)
  - Arabic (p. 468)
- Chadic subfamily (p. 468)
  - Hausa (p. 468)
- Cushitic subfamily (p. 468)
  - Oromo (p. 468)
  - Somali (p. 468)

AIDS (acquired immunodeficiency syndrome) (p. 458)
antiretroviral (ARV) drugs (p. 463)
Austronesian language family (p. 468)
- Malagasy (p. 468)

blood diamonds (p. 478)
branded diamonds (p. 478)
compulsory licensing (p. 463)
conflict diamonds (p. 478)
"curse of Africa" (p. 476)
dirty diamonds (p. 478)
donor democracy (p. 480)
Ethiopian Orthodox Church (p. 469)
failed-state syndrome (p. 480)
game ranching (p. 476)
Great Rift Valley (p. 466)
Heavily Indebted Poor Countries (HIPC) (p. 479)
human immunodeficiency virus (HIV) (p. 458)
keystone species (p. 468)
Khoisan language family (p. 468)
- Khoekhoe (Damara) (p. 468)

Kimberley Process (p. 478)
microcredit (p. 480)
"mistake of 1914" (p. 473)
multispecies ranch system (p. 476)
nagana (p. 461)
New Partnership for Africa's Development (NEPAD) (p. 480)
Niger-Congo language family (p. 468)
- Adamawan subfamily (p. 468)
- Atlantic subfamily (p. 468)
- Benue-Congo subfamily (p. 468)
  - Bantu (p. 468)
  - Swahili (Kiswahili) (p. 468)
- Gur subfamily (p. 468)
- Khordofanian subfamily (p. 468)
- Kwa subfamily (p. 468)
- Mandé subfamily (p. 468)

Nilo-Saharan language family (p. 468)
- Chari-Nile subfamily (p. 468)
- Saharan subfamily (p. 468)
- Songhai (p. 468)

"1 percent gap" (p. 456)
Pan-African Parliament (p. 481)
pandemic (p. 462)
proxy war (p. 481)
shell state (p. 480)
sleeping sickness (p. 468)
Southern African Customs Union (SACU) (p. 478)
terrorism hot spot (p. 481)
triangular trade (p. 470)
trypanosomiasis (p. 468)

REVIEW QUESTIONS

1. Where are the five principal areas of population concentration in sub-Saharan Africa?
2. What factors account for the high HIV infection rates in sub-Saharan Africa? What can be done to fight the epidemic? What is HIV/AIDS doing to life expectancy and the age structure profiles in countries where it is epidemic?
3. What are the principal climatic zones and biomes of sub-Saharan Africa?
4. Where were the major slave trade routes in and from sub-Saharan Africa?
5. Where were the major colonial possessions of various European powers in sub-Saharan Africa?
6. What is the significance of cattle in many cultures of sub-Saharan Africa?
7. What are the region's major export crops? Why does dependence on them make a country vulnerable?
8. What is a keystone species? Why is the tsetse fly considered a keystone species in sub-Saharan Africa?
9. What is the "1 percent gap"? What are its implications?

10. Why have clothing exports from Africa to the United States boomed in recent years?
11. What do the terms *shell state, failed-state syndrome, donor democracy,* and *donor fatigue* mean, and what do they reveal about Africa's political geography?
12. What impacts did the end of the Cold War have on many African countries?
13. How is China's ascendancy affecting Africa's economies and geopolitical status?

NOTES

1. Lydia Polgreen, "Misery Loves Optimism in Africa." *New York Times,* March 5, 2006, p. WK-1.
2. Robert Stock, *Africa South of the Sahara,* 2nd ed. (New York: Guilford Press, 2004), p. 41.
3. Ibid.
4. Quoted in Ian Fisher, "Africans Ask if Washington's Sun Will Shine on Them." *New York Times,* February 8, 2001, p. A3.

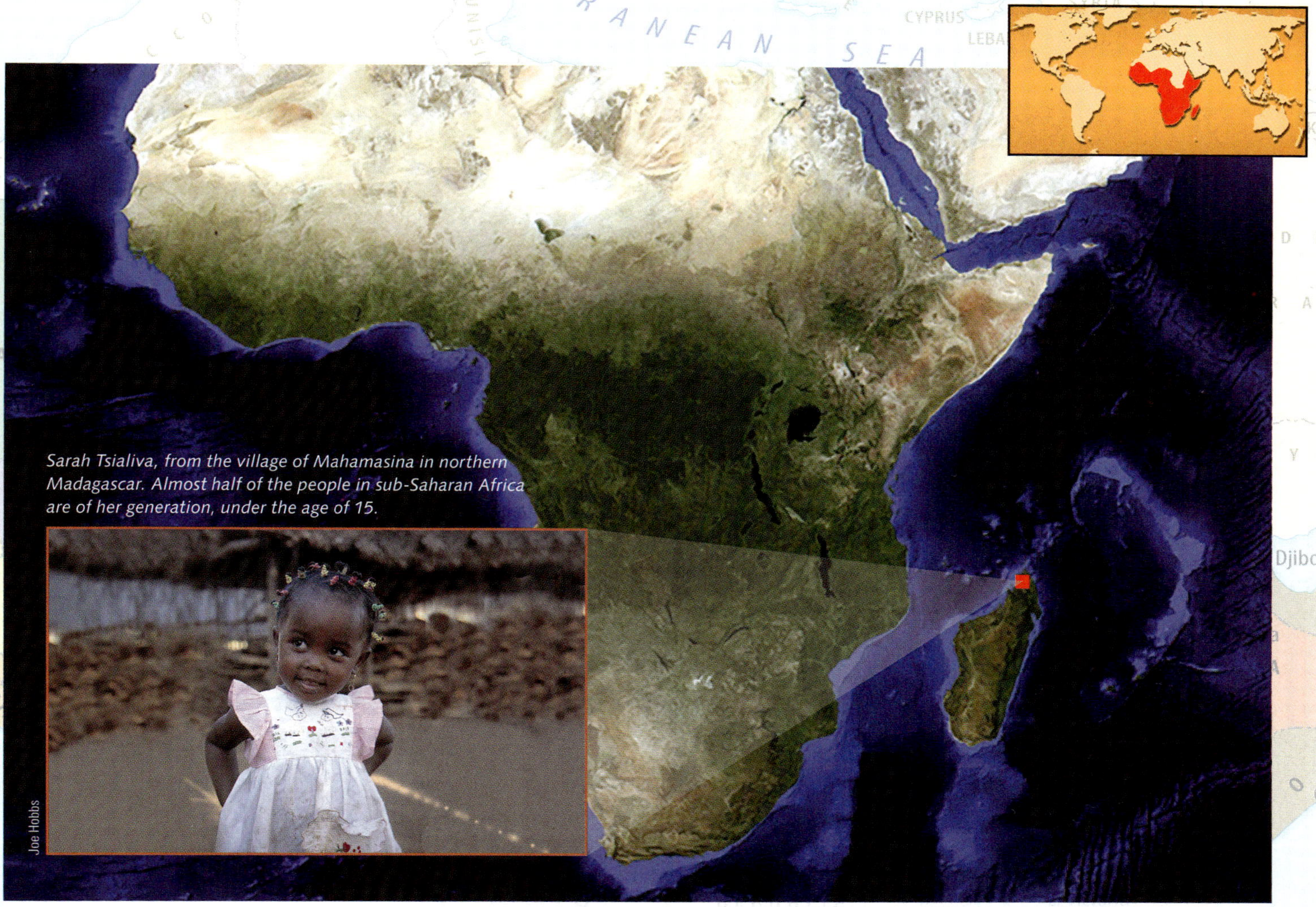

Sarah Tsialiva, from the village of Mahamasina in northern Madagascar. Almost half of the people in sub-Saharan Africa are of her generation, under the age of 15.

Joe Hobbs

MODULE 9.1 THE ASSETS AND AFFLICTIONS OF THE SUB-SAHARAN COUNTRIES

In this chapter, sketches of the major subregions and their countries illustrate some of the characteristics and unique challenges of sub-Saharan Africa. The major subregions are the Sahel, West Africa, West Central Africa, East Africa, the Horn of Africa, southern Africa, and the Indian Ocean islands (●Figure 9.1.1). The relevant area and population data are in Table 9.1 on pages 457–458.

module outline

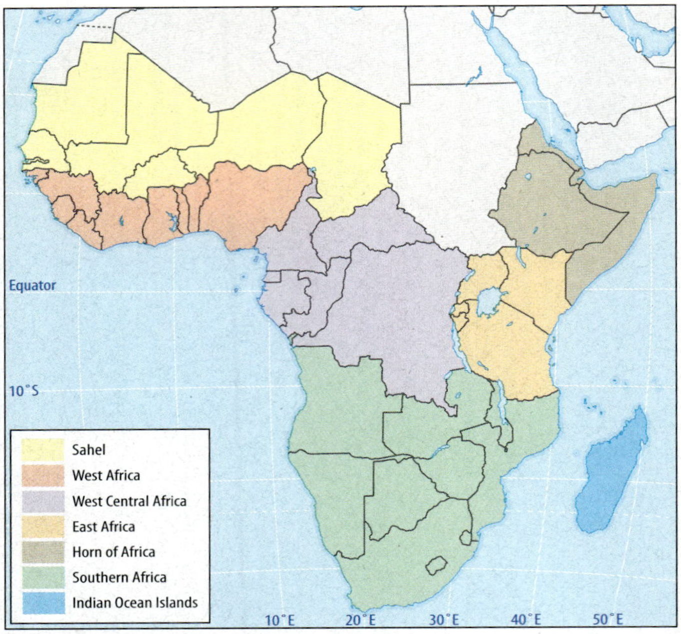

• **Figure 9.1.1** Regional groupings of the countries of sub-Saharan Africa.

module objectives

This module should enable you to

- → Recognize how European colonial favoritism of some ethnic groups over others sowed seeds of modern strife and warfare
- → Appreciate what corrupt leadership has done to impoverish people in countries with enormous oil and other natural resource wealth
- → Recognize the political, economic, and demographic clout of a select number of countries—notably South Africa and Nigeria—in African affairs
- → Understand how drought regularly triggers environmental degradation and famine and becomes a political as well as humanitarian issue in national and international affairs
- → Consider regional and international efforts to prevent recurrence of the genocide and "ethnic cleansing" that marred Africa in the 1990s
- → Evaluate the efforts to redistribute farmland from minority white to majority black control

9.1.1 The Sahel: On the Shore of a Great Desert

The Sahel region extends eastward from the Cape Verde Islands to the Atlantic shore nations of Mauritania, Senegal, and the Gambia and inland to Mali, Burkina Faso ("Land of the Upright Men," formerly Upper Volta), Niger, and Chad (•Figure 9.1.2). In colonial times, the Gambia was a British dependency; Mauritania, Senegal, Mali, Burkina Faso, Niger, and Chad were French; and Cape Verde was Portuguese. Cape Verde's dry volcanic islands lie well out in the Atlantic off Senegal, but in population, culture, economy, and historical relationships, the islands are so akin to the adjacent mainland that they fit the Sahelian context.

Environments, Resources, and Settlements

Climatically and floristically, the Sahel region ranges from desert (parts of the Sahara) in the north through belts of tropical steppe and dry savanna in the south (see Figure 9.8). One country is exceptional: nearly surrounded by Senegal, the Gambia consists of a narrow strip of savanna grassland and woodland only 20 miles (32 km) wide, extending inland 300 miles (c. 500 km) along the banks of the its namesake river. The name Sahel in Arabic means "coast" or "shore," referring to the region as a front on the great desert "sea" of the Sahara. Since the late 1960s, the area has been subjected to severe droughts, which, in combination with increased human pressure on resources, have prompted a process of desertification (see Regional Perspective, pages 488–489).

This area has seen many dramatic changes in climate and vegetation. There is abundant evidence, particularly in the form of prehistoric rock drawings, that much of the now extremely arid northern Sahel and Sahara region was a grassy, well-watered savanna some 6,000 to 10,000 years ago. Many of the large mammals now associated with East Africa frequented the region. Now, even though under much less favorable conditions, the raising of sheep, goats, camels, and cattle, combined where possible with subsistence farming, is the major livelihood for peoples of the Sahel. There are new hopes in the region that farmers will be able to expand cultivation of a drought-tolerant plant called jatropha in order to feed the world's growing demand for biofuels. Jatropha beans, from which the fuel is made, can be 589, 619
produced with far less water, fertilizers, and pesticides than the corn, soybeans, and sugarcane conventionally used to make plant-based fuels. European aid agencies and investors are already working with Sahelian farmers to boost their jatropha output.

One of the most prominent features of the Sahel's landscape is Lake Chad, located north of landlocked Chad's capital, N'Djamena (population 1.4 million). This large,

• **Figure 9.1.2** Principal features of the Sahel.

shallow lake, only 7 to 23 feet (2 to 7 m) deep, is situated in a vast open plain. It has no outlets and yet is not very saline because groundwater carries most of the salts away. Its waters are important economically for fish and irrigated agriculture. Characteristic of shallow lakes in arid lands, its surface area historically increased dramatically in rainy periods and shrank in times of drought. However, the expansion of irrigation along the northward-flowing rivers that feed the lake has caused a huge and enduring shrinkage. Lake Chad's surface area in 2007 was 500 square miles (1,300 sq km), compared to 9,700 square miles (25,125 sq km) in 1963; that is a 95 percent loss. Four countries share the lake and its drainage area, so decisions to reverse the lake's shrinkage are difficult to reach, and the likelihood of future conflict brought about by water shortages is high.
202 The situation is reminiscent of that of the Aral Sea in central Asia, which is slowly disappearing.

Mining, manufacturing, and urbanization are modest in this subsistence-oriented region. Mauritania exports iron ore, and Senegal exports phosphates. Mali's cotton production has faded, but the country is now Africa's third-largest producer of gold and is hopeful of producing oil soon. France controls virtually all of Niger's uranium, which is used for France's nuclear power program. Salt is mined on Niger's sterile landscape, but its importance today pales in comparison with centuries ago, when a bar of Niger salt could be traded for a bar of gold. Niger ranks last on the UN's Human Development Index.

The most remarkable asset in the region's economy is the oil of southern Chad (see Figure 9.18, page 475). These reserves, with an estimated life span of 25 years, began to be exported in 2003 through a controversial World Bank–financed pipeline to Cameroon's Atlantic coast. Chad's government promised the World Bank that almost all the oil revenues would be used to assist the country's poor or be set aside in a "future generations fund." Critics warned the World Bank that Chad's government was among the most corrupt on the planet and should not be trusted. In a classic example of the **"resource curse,"** Chad's government proved them correct by nullifying the laws it had passed to satisfy the World Bank's demands, redirecting the oil wealth away from the poor. The World Bank responded by freezing all future loans.

The Sahel region was historically a crossroads for caravan traffic (especially northbound slaves, food, and gold and southbound salt, weapons, and cloth) between West Africa and North Africa. Ancient Mali was an especially important trading intermediary. Its legendary port of Timbuktu on the Niger River was an intellectual as well as a commercial center. In 1989, it was recognized by the United Nations Educational, Scientific and Cultural Organization (UNESCO) as a **world patrimony site.** Today, much of the traffic through this region is a northward migration of poor people from Nigeria, Niger, Guinea, Ghana, Cameroon, and the Democratic Republic of Congo seeking work in oil-rich Libya. Many subsequently use Libya as a jumping-off point for settlement in Europe. Some Europeans are fearful of a potential new stream of immigrants who might use a highway that, when complete, will for the first time allow vehicles to travel from Nigeria, Mali, and Mauritania all the way to the Strait of Gibraltar (see Figure 9.18).

The largest metropolises are Senegal's main port and capital of Dakar (population 2.4 million) and Bamako (population 1.7 million), the capital and main city of Mali. Dakar was once the capital of the immense group of eight colonial territories known as French West Africa. It is the commercial outlet for much of the peanut-producing area, known as the peanut belt, in the Sahelian and West African savanna and is a considerable industrial center by African standards. Dakar is also a tourist and resort destination with fine beaches and an exotic blend of French and Senegalese cultures.

REGIONAL PERSPECTIVE

Drought and Desertification in the Sahel

The Sahel region, like much of sub-Saharan Africa, experiences periodic drought. This is a naturally occurring climatic event in which rain fails to fall over an area for an extended period, often years. When rain does return, the arid and semiarid ecosystems of the Sahel come to life with a profusion of flowering plants, insects, and herbivores like gazelles, whose populations climb when foods are abundant.

These ecosystems have high **resilience,** meaning they are able to recover from the stress of drought and have mechanisms to cope with a natural cycle that includes periods of dryness and rain.

Desertification is the destruction of that resilience and the bio-
368 logical potential of arid and semiarid ecosystems. It is an unnatural, human-induced condition that has afflicted the Sahel periodically and severely since the late 1960s, and thus it is different from the natural phenomenon of drought (•Figure 9.1.A). However, the recent history of the Sahel suggests that drought can be the catalyst that initiates the process of desertification. During decades of good rains prior to the late 1960s, the Fulani (Fulbe), Tuareg, and other Sahelian pastoralists allowed their herds of cattle and goats to grow more numerous. The animals represent wealth, and people naturally wanted their numbers to increase beyond the immediate subsistence needs of their families. However, due to declining death rates, the numbers and sizes of families keeping livestock were also very large. Thus the unprecedented numbers of livestock that built up during the rainy years made the Sahelian ecosystem vulnerable to the impact of drought on an unprecedented scale.

Steve McCurry/Magnum

• **Figure 9.1.A** Desertification in progress is shown graphically in this photo from Mali in the Sahel. As growing human populations intensify the impact of naturally occurring drought, the Sahara crept southward in the 1970s and early 1980s.

Ethnic Tensions

The Sahel is the border region between mainly Arab and Berber North Africa and mainly black sub-Saharan Africa. Relations between the ethnic groups are good in most countries, but there are problems in some. In Mauritania, for example, the government—led by the country's majority Arabic-speaking **Moors**—expelled black Mauritanians across the border into Senegal in 1995 and has not allowed them to come back. Analysts continue to fear that the Mauritanian government aims eventually to "cleanse" the country of its 40 percent black population. Laws passed to outlaw the enslavement of blacks have yet to be enforced.

452, Chad has north-south tensions that are reminiscent of
493 those of Sudan and Nigeria. The northerners, who now dominate Chad's political life, are mostly Arabic-speaking Muslims whose historical mainstay has been cattle pastoralism. The southerners are non-Arab Christians (mainly of the **Sara** ethnic group) whose livelihood is farming. The French favored the Christian southerners and empowered them more than the northerners, and now that the tables are turned, the northern elite is discriminating against southerners. Historically, northerners have enslaved southerners, and there are reports that they continue to do so. There are frequent land disputes as northern pastoralists overrun southern farms. The government's recent backpedaling on its promise to share oil wealth with the poor also fuels fears that as in Sudan and Nigeria, the dominant northern power

REGIONAL PERSPECTIVE

Drought and Desertification in the Sahel *continued*

Drought struck the region in 1968, persisted through 1973, and returned over many years through 1985. With the drought, annual plants failed to grow, so the large herds of cattle and goats turned to acacia trees and other perennial sources of fodder. They ate all of the palatable vegetation. This destruction was a critical problem because plants play an important role in maintaining soil integrity by helping intercept moisture and funnel it downward through the root system. Plant litter and decomposer organisms working around the plant contribute to soil fertility and help stabilize the soil. With hungry livestock in the region eating the plants, the landscape changed. No longer anchored and replenished, good soils eroded. "Junk plants" like Sodom apple (*Calotropis*) replaced palatable plants, and an almost impermeable surface formed on the land. This degraded ecosystem lost its resilience so that even when rains returned, vegetation did not recover.

Meteorologists studying the Sahel drought discovered another important link between precipitation and the removal of vegetation. They noticed that when people and livestock reduced the vegetative cover, they increased the **albedo,** the amount of the sun's energy reflected by the ground. Fewer plants meant more solar energy was deflected back into the atmosphere, lowering humidity and reducing local precipitation. This connection is known as the **Charney effect,** named for Jules Charney, the scientist who documented it in the early 1970s. It was a tragic sequence: People responding to drought actually perpetuated further drought conditions.

The 1968–1973 drought devastated the great herds of the Sahel. An estimated 3.5 million cattle died. Two million pastoral nomads lost at least 50 percent of their herds, and many lost as much as 90 percent. The Niger and Senegal Rivers dried up completely. The cost in human lives was estimated at 100,000 to 250,000. Environmental refugees poured into towns unprepared to deal with such a large and sudden influx. Many pastoral nomads gave up herding and settled down to become wage earners and, where possible, farmers.

This crisis resulted in an international effort to combat desertification using recommendations developed by the United Nations in 1977. The UN report, however, soon became a case study in how *not* to deal with an environmental crisis in the developing world. Most of the recommendations were universal, high-technology solutions that proved impossible to implement in the villages and degraded pastures of the Sahel. For example, many international agencies attempted to relieve the suffering of Sahelian pastoralists using techniques such as digging deep wells to water livestock. But these agencies failed to anticipate that these wells would act like magnets for great numbers of people and animals; although there was plenty to drink, many animals starved to death when they decimated what vegetation remained around the new water supplies.

Since the disappointment of the 1977 UN plan, and with the lessons relief organizations learned as a result, there has been greater focus on sustainable development, with its local rather than universal solutions. In Niger, for example, hit by a devastating 63
drought as recently as 2005, farmers have ceased the traditional practice of clearing trees before planting crops and are now protecting trees and growing crops under and between them. They are also harvesting and selling many useful tree products. A key factor for the farmers' new concern for trees is that Niger's government released its long-held claim that all trees belonged to the state; now farmers recognize the trees as private property. New attitudes and practices like these, and a sustained period of good rains (which unfortunately have prompted periodic locust swarms), have combined to produce a remarkable regreening of much of the Sahel since 1985. Satellite imagery shows significant regrowth of vegetation in southern Mauritania, northern Burkina Faso, northwestern Niger, central Chad, and southern Sudan. Annual millet and sorghum harvests have risen in some cases as much as 70 percent above their pre-1985 levels. This change in the Sahel's fortunes is an important lesson that desertification is not inevitable or irreversible.

will skim off the benefits from the newly developing oil industry and leave the oil-bearing region in poverty.

Chad is deeply engaged in the struggles of neighboring
255 Sudan. Chad's president is from the same Zaghawa ethnic group as some of the rebels fighting Sudanese government forces and militia in Sudan's Darfur province. He has allowed refugees from Darfur to settle in camps inside Chad. Sudanese military and paramilitary forces have made periodic raids on those camps and are also accused of supporting an internal rebellion against the Chadian government. In explaining why he is diverting oil wealth away from Chad's poor, the president said the funds are needed to fight this rebellion.

9.1.2 West Africa: Populous and Struggling to Leave Strife Behind

West Africa extends from Guinea-Bissau eastward to Nigeria (•Figure 9.1.3). Its nine political units make up about 800,000 square miles (c. 2 million sq km), or nearly one-fourth of the area of the United States. These nine countries are Guinea-Bissau, Guinea, Sierra Leone, Liberia, Ivory Coast, Ghana, Togo, Benin, and Nigeria. The spatial, demographic, political, and economic giant of the region is Nigeria.

● **Figure 9.1.3** Principal features of West Africa.

Environments and Resources

Climatic and biotic contrasts within this region are extreme, ranging from tropical steppe, dry savanna, and wetter savanna to areas of tropical rain forest along the southern and southeastern coasts (see Figure 9.8). An estimated 90 percent of the region's tropical forest has been lost to commercial and subsistence deforestation. Most of the countries' more populous areas are in the zone of tropical savanna climate.

As is typical of much of Africa, the West African countries depend heavily on subsistence farming and livestock grazing, along with selected agricultural or mineral exports. In the wetter south, subsistence agriculture relies mainly on root crops such as manioc and yams and on maize, the oil palm, and in some areas, irrigated rice. In the drier, seasonally rainy grasslands of the north (in northern Nigeria and northern Ghana, for example), nomadic and seminomadic herding of cattle and goats coexists with subsistence farming of millet and sorghum.

The major export specialties of West African agriculture are cacao (from which cocoa and chocolate are made), coffee, and oil palm products in the wetter, forested south and peanuts and cotton in the drier north. Ivory Coast leads the world in cacao exports, producing almost half the global total in 2007. Nigeria was once the world leader in exports of palm oil and peanuts, but these exports almost vanished as Nigerian agriculture was neglected during the oil boom of recent decades.

The other main export industry of West Africa is mining (see Figure 9.18). The most important resource is petroleum, with OPEC member Nigeria leading the region in production and export. In 2007, when Nigeria stood as the world's sixth-largest oil exporter (and the source of 9 percent of the United States' imports), petroleum accounted for xx percent of the country's foreign export earnings. Oil drilling takes place mainly in the mangrove-forested Niger River Delta and offshore in the adjacent Gulf of Guinea (see Problem Landscape, page 491).

Other prominent mineral products of West Africa include iron ore, bauxite, diamonds, phosphate, and uranium. Iron ore exports are very important to Liberia. The main bauxite resources lie near the southwestern bulge of the West African coast; Guinea is the main producer and exporter. Togo exports phosphate. Gold, bauxite, and manganese are important to Ghana, as are bauxite, titanium, and diamonds to Sierra Leone.

Population and Settlements

West Africa has an impressive total population of 224 million, or about three-fourths that of the United States. The most striking single aspect of West Africa's population distribution is that about two-thirds of the region's people reside in one country, Nigeria (whose 144 million also represent 19 percent of the population of the entire region of sub-Saharan Africa; see Figure 9.3). This unusual concen- 460
tration of people, which existed even before European contact, is related in part to the region's agricultural productivity and resource wealth.

Only about 37 percent of West Africa's people are city dwellers, but the urban population is increasing rapidly. The world's large cities are associated mostly with trade, centralized administration, and large-scale manufacturing, but West Africa's big cities are different. They include national capitals and seaports, but as the region has only minor manufacturing, they have developed mainly in areas of productive commercial agriculture. In nearly every country
in the region, the capital is a primate city, far larger than 111
the nation's second-largest city.

Nigeria has the most impressive urban development: Of the 11 largest metropolitan cities in West Africa, three are

PROBLEM LANDSCAPE

The Poor, Oil-Rich Delta of Nigeria

Oil is the economic lifeblood of populous Nigeria. Most of the production is concentrated in the Niger River Delta, home to about 7 million mostly Christian people (Figure 9.1.B). They belong to several different ethnic groups who have one thing in common: they say they have derived few benefits and have suffered greatly from oil development in their homeland. They complain that for more than 30 years, oil spills have tainted their croplands and water, destroying their crops and fisheries, while the flaring off of natural gas has polluted their air and caused acid rain. They note that despite the enormous revenue generated from oil drilled on their land, little money has returned to the area; most goes to foreign oil companies like Shell and Chevron and to the Muslim-dominated government in the north. Meanwhile, most of them live in palm-roofed mud huts; half of the delta region lacks adequate roads, water supplies, and electricity. Schools have few books, and clinics have few medical supplies.

Since 1990, several ethnically based organizations have been agitating for change. **Ogoni** activists founded the **Movement for the Survival of the Ogoni People (MOSOP),** issuing a bill of rights in which they declared the right to a safe environment and more federal support of their people. One of the movement's leaders, a popular author, playwright, and television producer named Ken Saro-Wiwa, also called for self-determination for the Ogoni. Despite international condemnation, Nigeria's government executed him after a questionable murder trial.

Daniel Lainé/Corbis

Figure 9.1.B Nigeria's ethnic landscape.

Despite the return to civilian rule in Nigeria that came after Saro-Wiwa's death, dissent in the delta continued to grow. Ogoni defiance and contempt of government and the oil companies spread to the region's other major ethnic groups: the **Ibo (Igbo), Yoruba,** and **Ijaw.** The Ijaw (who make up about 5 percent of Nigeria's population) have been especially strident in demanding that some of the oil money be used to fund roads, electricity, running water, and medical clinics. Ijaw activists have spoken of secession and self-determination if their demands are not met. They have periodically registered dissatisfaction by seizing onshore stations and offshore rigs belonging to foreign oil companies, temporarily disrupting Nigeria's oil exports on both occasions. In 2006, a mainly Ijaw group calling itself the **Movement for the Emancipation of the Niger Delta (MEND)** escalated these tactics into frequent kidnappings of foreign oil workers. Opportunistic profiteers without well-established political views also joined in the action, and kidnapping became big business. An oil worker is typically freed unharmed upon payment of a ransom of several hundred thousand dollars.

This militancy in the delta has sent shockwaves through the world economy. Nigeria had touted itself as the safe alternative to Middle Eastern oil but has had to cut production by as much as 25 percent when tensions have been particularly high. The resulting shortages on world markets have helped send oil and gasoline prices to record levels.

Questions of equity in a potentially rich country thus pose a threat to national, regional, and international stability. The peoples of the delta are waiting for implementations of former President Obasanjo's promise to allot the oil-producing states 13 percent of onshore oil revenues (compared with 5 percent previously), along with some portion of offshore receipts, and to shut down the region's foul gas flares. The government insists the money is coming in, but locals say it is not reaching them, and watchdog groups say corrupt officials are siphoning it off.

One of Nigeria's ironies is that in this oil-rich country, there are continuous shortages of gasoline and electricity. Neglect, mismanagement, corruption, and theft are all responsible. One of the cruelest and most repetitive Nigerian news stories is the immolation of scores and even hundreds of poor delta villagers who, in an activity known as "scooping" or "bunkering," illegally puncture gasoline or oil pipelines in an effort to sell the hydrocarbons, only to inadvertently ignite the fires that consume them.

clustered in the Yoruba-dominated southwestern part of Nigeria, in and near the densely populated belt of commercial cacao production. The largest by far is Lagos (population 11.6 million; Figure 9.1.4). It had long been Nigeria's political capital, but in the 1990s, most governmental functions were moved to a new capital at Abuja in the interior. The growth of Lagos surged in the early 20th century when the British colonial administration chose the city as the ocean terminus for a trans-Nigerian railway linking the Gulf of Guinea with the north. During the oil boom of recent

Daniel Lainé/Corbis

• **Figure 9.1.4** Lagos is Nigeria's vibrant and chaotic capital.

decades, migrants from all parts of Nigeria (and many from outside the country) flooded into the city. With its severe traffic congestion, rampant crime, inflated prices, and inadequate public facilities, Lagos gives the impression of urbanism out of control. It is also indicative of a disorderly national economy.

Close to the border with Niger is Kano, the most important commercial and administrative center of Nigeria's north and a major West African focus of Islamic culture. Immediately west of Nigeria on the Guinea Coast lies Cotonou, the main city and port of Benin. Farther west, in southern Ghana—another agricultural area with a high degree of commercialization—is Accra (population 3.5 million), Ghana's national capital. Until recently, it was the main seaport, despite the lack of deep-water harbor facilities. A modern deep-water port was built east of Accra at Tema, and that city is now Ghana's main port and industrial center. Well inland from Accra is the country's second-largest city, Kumasi, in the heart of Ghana's cacao belt, which was once the world's greatest cacao-producing area.

Today, the world's leading cacao producer and exporter is Ivory Coast (the country's official name is its French name, Côte d'Ivoire), with another of West Africa's largest cities, Abidjan (population 5.3 million; •Figure 9.1.5). Although the inland city of Yamoussoukro is the official capital, Abidjan (dubbed the "Paris of Africa" for its architectural and cultural graces) is the de facto capital of a country that in the 1990s—prior to its descent into chaos—vowed to achieve economic prosperity comparable to that

Charles O. Cecil/Visuals Unlimited

• **Figure 9.1.5** Only about one-third of the people in sub-Saharan Africa live in cities, but with considerable rural-to-urban migration under way, that number is growing. This is very modern Abidjan, the leading city of Ivory Coast, with Saint Paul's Cathedral in the foreground.

of the Asian Tiger countries. Also on the Guinea Coast are 298
three more seaport capitals: Monrovia in Liberia; Freetown in Sierra Leone, on West Africa's finest natural harbor; and Conakry in Guinea.

Many of these port cities have poor connections with their countries' hinterlands. Rapids and lower water during the dry season limit the utility of rivers for transport. Some rivers do carry traffic, notably the Niger and its major tributary, the Benue. Railways are more important than rivers in the region's transport structure, but they are few and widely spaced. Instead of forming a network, the rail pattern is a series of individual fingers that extend inland from different ports without connecting with other rail lines (see Figure 9.18).

Ethnic and Political Complexity and Conflicts

West Africa was an advanced part of sub-Saharan Africa when the European Age of Discovery began. A series of strong pre-European kingdoms and empires developed there, both in the forest belt of the south and in the grasslands of the north. In colonial times, it became a French and British realm, except for a small Portuguese dependency and independent Liberia. Of the present countries, Sierra Leone, Ghana (formerly Gold Coast), and Nigeria were British colonies; Guinea, Ivory Coast, Togo, and Benin (formerly Dahomey) were French; and Guinea-Bissau was Portuguese. Liberia is unique, having become Africa's first republic in 1822, when 5,000 freed American slaves sailed there and settled with support from the U.S. Treasury. The country's capital, Monrovia, was named for U.S. President James Monroe. Its flag is a variant of the U.S. stars and stripes, and its currency is the Liberian dollar. Many Liberians speak fondly of the United States as their "big brother."

In West Africa today, the English, French, and Portuguese languages continue in widespread use, are taught in the schools, and are official languages in the respective countries. European languages are a useful means of communication in this region, where hundreds of indigenous
471 languages and dialects, often unrelated, are spoken. Economic, political, and cultural relationships between West African countries and the respective European powers that formerly controlled them continue to be close. Britain maintains ties with **Anglophone** (English-speaking) **Africa** through meetings of the Commonwealth of Nations, to which all its former colonies in sub-Saharan Africa belong. France supplies economic and military aid to its former colonies in this and other parts of **Francophone Africa.** France perpetuates relations with its former dependencies by inviting their leaders to summit conferences to discuss matters of common interest. France also has some colonial-style economic relations with some of its former colonies.

As is often the case in Africa, the governments of the region must deal with serious problems related to ethnicity. European governments contributed initially to some present troubles by drawing arbitrary boundary lines around colonial units without proper concern for ethnicity. As a result, the typical West African country today, like most countries elsewhere in sub-Saharan Africa, contains a collection of ethnic groups that have little sense of identification with the national unit. Nigeria, as described on page 473 in Chapter 9, is a good example.

The largest of Nigeria's 250 or more ethnic groups are the mostly Muslim Hausa and Fulani (together, 29 percent of the population) mainly in the north, the mostly Christian Yoruba (21 percent) in the southwest, the Ibo (Igbo) (18 percent) mainly in the southeast, and the Ijaw (10 percent) of the southern delta region. Tensions and violence have erupted between some of these groups, especially between the mainly Muslim Hausa and mainly Christian Yoruba, and each has formed militias. Muslim northerners have traditionally dominated Nigerian politics, and the long rule of Christian Yoruba President Obasanjo was a concession by the northerners to prevent the country from fragmenting. For the 2007 elections, the northern Muslim power base insisted that one of theirs should be president, and Obasanjo handpicked a successor from their ranks. The subsequent election of his man, Umaru Yar'Adua, was widely viewed inside and outside Nigeria as fraudulent. Polls show that Nigerians have gone from overwhelmingly optimistic to extremely pessimistic about their future, and the rigged election raises serious concerns about the country's future stability.

Mainly as a means of blunting former President Obasanjo's perceived Christian leanings, most of the Muslim northern states in 1999 adopted sharia (Islamic) law. The Christian minorities of these regions fear persecution and intolerance (see Figure 9.1.B, page 491). For their part, Christians have attacked and killed Muslims elsewhere in the country, and violence between the groups has taken thousands of lives. Some of the unrest here is related to livelihood as well as religion: longtime residents are mainly Christian farmers, and they resent a recent influx of Muslim cattle herders. Cattle theft and crop burning are typically part of their skirmishes. The prospect of violent devolution hangs over Nigeria, reviving memories of deaths of a million people in 1967–1970 when the country's ethnic Ibo popula- 83
tion struggled to establish the separate nation of Biafra.

A diamond purge and civil war plagued Sierra Leone between 1991 and 2002, taking more than 50,000 lives. Almost two-thirds of the country, including all of the diamond-producing area, came under the control of the rebel **Revolutionary United Front (RUF).** The RUF projected its power 478
with a brutal combination of mutilation, rape, and murder. The country's prospects improved when its first democratically elected president, Ahmed Tejan Kabbah, took office in 1996. However, a military junta assisted by the RUF seized power a year later, forcing Kabbah into exile. Asserting its self-appointed identity as West Africa's regional superpower, and perhaps with an eye on Sierra Leone's mineral wealth, Nigeria responded in 1998 with a military attack on Sierra Leone that in turn forced the junta to flee and restored President Kabbah to office. RUF resistance continued. The 1999 **Lomé Agreement** gave the RUF some seats in the national cabinet and established RUF leader Foday

Sankoh as the country's vice president. But the RUF broke the peace agreement and took hundreds of United Nations peacekeepers hostage, reigniting the war. As the fighting continued, British troops intervened, capturing Sankoh (who died while awaiting a war crimes trial) and helping Kabbah reassert government authority over this devastated country. In 2002, Kabbah was reelected to office, and the war was declared over; all UN peacekeepers had left the country by 2005.

In neighboring Liberia, a regional kingpin, Charles Taylor, long provided arms to Sierra Leone's RUF in exchange for diamonds. In 1989, the U.S.-educated Taylor ignited a civil war that persisted for seven years. In 1997, Liberians elected him president because, analysts say, they feared that Taylor would restart the war if not elected. After that, he led the country on a path to ruin. Public services such as electricity were all but nonexistent. Schools and hospitals were downsized or closed. Taylor's family enriched itself by selling the county's assets, including its tropical hardwoods, and in exchange for Guinean diamonds, Taylor supplied
480 guns to rebels in neighboring Guinea. By 2003, anti-Taylor rebels calling themselves the **Liberians United for Reconciliation and Democracy (LURD)** had mounted an effective insurgency, and Liberia was once again plunged into war. This time, the United States opposed its onetime ally Taylor, who fled the country, and briefly fielded a peacekeeping force to Liberia. An interim administration, in which the government and the rebels shared power, was established to control the country. An election in 2005 brought to power the world's first elected black female head of state, President Ellen Johnson-Sirleaf. Her principal task is to rebuild the economy of a country with 25 percent unemployment. There is much discussion of reviving the timber industry, but under strict supervision to prevent the rapid depletion of the rain forest that covers nearly one third of Liberia.

There was violence in Ivory Coast too. The conflict there was particularly unfortunate because it took place in a country long heralded as a beacon of hope and relative prosperity, with the region's second-strongest economy (behind Nigeria's). Ironically, Ivory Coast's economic promise helped sow the seeds of conflict and economic decline. The country grew prosperous from cacao and coffee exports in the 1990s. There were new roads, electricity and phone services, and schools. The middle class grew. Ivory Coast's fortunes made it a magnet for settlement by people from nearby, poorer nations. The government actively encouraged immigration because it needed extra hands and skills to assist in the country's development. Large numbers of mainly Muslim immigrants poured in from Burkina Faso and Mali; most of them settled in the predominantly Muslim northern part of Ivory Coast. By the end of the decade, these newcomers made up an astonishing one-third to one-half of the country's population of 16 million.

Cacao prices took a nosedive after 2000, and so did Ivory Coast's economic good times. The once-welcomed Muslim immigrants in the north began to be resented and even despised by the mainly Christian southerners. The government, essentially a mirror of that southern population, began an active program of discrimination against the north. Southerners were portrayed as the only "true" Ivoirians—a concept they call ***Ivoirité***—and laws were passed to exclude northerners from participation in many aspects of the country's economic and political life.

By 2002, three rebel groups in the north were sufficiently organized to pose a challenge to the government. They were poised to take control of the rest of the county and might well have succeeded. But France, the country's former colonial power, intervened, sending 2,500 Legionnaires to block the rebel advance. The legendary fighting force quickly separated the warring sides and was joined by peacekeepers from the **Economic Community of West African States (ECOWAS).** The French government managed to persuade leaders from both sides to attend peace talks in Paris as a means of averting all-out war and perhaps the country's partition.

In a peace treaty signed in Paris in 2003, the southern-based government under President Laurent Gbagbo made large concessions to the rebels, especially in agreeing to share power with them. The new "national reconciliation" government was supposed to allow the rebels to achieve their main aim in a nonmilitary fashion: an end to the official discrimination against the mainly Muslim northerners. In a subsequent pact signed in 2007, the main rebel leader became the country's prime minister. A degree of normalcy has returned to Ivory Coast; road networks have reopened, and cotton and sugar produced in the north is again being trucked to Abidjan. The European Union is promising generous economic aid if the peace can hold.

9.1.3 West Central Africa: Colonial "Heart of Darkness"

The subregion of West Central Africa is flanked by Cameroon and the Central African Republic to the north and the Democratic Republic of Congo to the south (•Figure 9.1.6). It contains seven countries: Cameroon, Gabon, Central African Republic, Congo Republic (often known by the shorthand Congo-Brazzaville), Democratic Republic of Congo (known by the shorthand Congo-Kinshasa or DRC), São Tomé and Príncipe, and Equatorial Guinea.

Environments and Resources

Parts of West Central Africa are mountainous, with the most prominent mountain ranges lying in the eastern part of the Democratic Republic of Congo near the Great Rift Valley and in the west of Cameroon along and near the border with Nigeria. Volcanism has been important in mountain building in both instances (see Natural Hazards, page 496).

• **Figure 9.1.6** Principal features of West Central Africa.

As in West Africa, the coast is fairly straight and has few natural harbors. Most of the region is composed of rolling or hilly plateaus. An exception to the generally uneven terrain is a broad area of flat land in the inner Congo Basin, once the bed of an immense lake. The greater part of West Central Africa lies within the drainage basin of the Congo River.

The heart of the region has tropical rain forest climate and vegetation grading into tropical savanna to the north and south. The biological resources are immense. The tropical forests that cover much of the two Congos, Gabon, Cameroon, Equatorial Guinea, and a small corner of the Central African Republic make up one-fifth of the world's remaining tropical rain forests (•Figure 9.1.7).

One of the region's most problematic environmental and economic issues is the fate of these forests. They are being cut, mainly for commercial logging, at a rate of about 3,000 square miles (8,000 sq km) per year. At that pace, they will be gone by 2020. In an effort to slow the destruction, the World Wildlife Fund has reached an agreement with the World Bank whereby the bank will not extend loans in this region unless the borrowing governments can demonstrate they are using, or planning to use, their tropical forest resources in a sustainable manner.

These forested countries do not like being, as they see it, held hostage by Western environmentalists, but they are
586 making concessions. A good example is Gabon, which has an extensive area of almost untouched tropical rain forest. As its oil reserves dwindle, Gabon sees its future in selling tropical hardwoods. But in a recent pact, it has promised that part of its pristine rain forest, in the Lopé Reserve of central Gabon, will not be harvested by the Malaysian and French logging companies that have extensive concessions in the country.

Subsistence agriculture supports people in most areas of West Central Africa. There is small-scale production of export crops, mainly in rain forest areas, and some cotton is exported from savanna areas. Southern Cameroon is an important cash crop region, with coffee and cacao the main export crops, supplemented by bananas, oil palm products, natural rubber, and tea. The farms occupy an area of volcanic soil at the foot of Mount Cameroon, a volcano standing 13,451 feet (4,100 m) tall that is still intermittently active. The mountain has extraordinarily heavy rainfall, averaging around 400 inches (1,000 cm) a year.

The island of Bioko has long been notable for its cacao exports and more recently for its oil boom. Bioko (earlier known as Fernando Pó) and the adjacent mainland territory of Río Muni, both former Spanish colonies, received independence in 1968 as the Republic of Equatorial Guinea. Cacao production in Bioko developed on European-owned plantations, but the newly independent country's leaders expelled the plantation owners.

Exports of minerals, principally oil, aid the financing of some countries, notably Gabon, Congo-Brazzaville, Cameroon, and most recently, Equatorial Guinea, where oil revenues since production began in 1996 have spurred enormous economic growth rates. Oil exploration began in the waters off São Tomé and Príncipe in 2004, and the 150,000 people of this cocoa-producing country are hoping for a bonanza.

Cameroon won a protracted oil-related dispute with Nigeria over possession of the Bakassi Peninsula, which juts into the Gulf of Guinea near the border between the two countries. The two nations nearly went to war in 1981 over control of Bakassi; the International Court of Justice ruled that it was part of Cameroon in 2002. Control of the peninsula gives Cameroon ownership of promising offshore oil reserves of the Gulf of Guinea. Cameroon is also receiving royalties and lost-land compensation for the shipment of Chadian petroleum in a pipeline to the Atlantic through Cameroon's territory. Copper from the mineral-rich Ka- 477
tanga province in the southeast part of the Democratic Republic of Congo was long the country's main export, but civil war disrupted the industry. There are also rich cobalt deposits there, and diamonds are in the adjoining Kasai region (see Figure 9.18). Another important resource in the DRC is coltan, a rare mineral used in the manufacture of cell phones and fighter jets.

Population, Settlements, and Infrastructure

In West Central Africa as a whole, population densities are low and populated areas are few and far between. Rough terrain, thick forests, wetlands, and areas with tsetse flies 468

NATURAL HAZARDS

Deadly Lakes

The Cameroonian volcanoes region is also its "deadly lakes" region. In two volcanic-cone lakes, Nyos and Monoun, carbon dioxide gas generated by volcanic activity deep in the earth builds up in the watery depths below a layer of gas-free freshwater (• Figure 9.1.C). The boundary between these layers is called a **chemocline.** Sometimes an event such as a windstorm or landslide disturbs the top layer deeply enough to puncture the chemocline and release the carbon dioxide from its solution. It erupts explosively, rocketing gas and water upward and outward. On August 21, 1986, such an explosion in Lake Nyos killed 1,700 people who were suffocated by a roving cloud of carbon dioxide.

In recent years, U.S. and Cameroonian scientists and engineers have cooperated to install pipes that dissipate the carbon dioxide from these deadly lakes before it builds up to potentially catastrophic levels. The authorities are pleased with the progress but uncertain that enough venting pipes have been installed. Similar steps have yet to be taken in East Africa's Lake Kivu, where the same threat exists.

Volcanism also poses more conventional hazards in this region. In 2002, the eruption of Mount Nyiragongo, near the northern shore of Lake Kivu, sent rivers of molten lava into the Congolese city of Goma, incinerating the city and causing almost its entire population of 400,000 to flee as refugees.

© Peter Turnley/CORBIS

• **Figure 9.1.C** Cameroon's Lake Nyos.

Gallo Images/Corbis

• **Figure 9.1.7** Tropical rain forest covers a large portion of West Central Africa. This bridge provides rare access to the forest canopy in Gabon.

tend to isolate clusters of people from each other. The savanna lands and highlands are more densely populated than the rain forests.

Most of the countries have an overall population density well below that of sub-Saharan Africa as a whole. The Democratic Republic of Congo is sub-Saharan Africa's largest country; it is about the size of western Europe, or Alaska, Texas, and Colorado combined, but its estimated population in 2007 was only 62 million, or well under one-half that of the continent's most populous country, Nigeria. Within the subregion, however, it is by far the most populous country. Nearly half of the Democratic Republic of

Congo is an area of tropical rain forest in which the population is extremely sparse.

Cities are few in these countries. Most of the larger ones are political capitals or seaports, including Cameroon's capital, Yaoundé (population 1.9 million), and main seaport, Douala (1.8 million), and the Congo Republic's capital city of Brazzaville (population 1.3 million). Brazzaville, located directly across the Congo River from the much larger capital city of Kinshasa in the Democratic Republic of Congo, was once the administrative center of French Equatorial Africa. Kinshasa (population 8.4 million), once known as Leopoldville, is the DRC's capital, largest city, and principal manufacturing center. Another important city is Lubumbashi (formerly Elisabethville), in the copper-mining region of the southeast.

Transportation infrastructure is generally poor due to political mismanagement, warfare, neglect, and poverty. Aside from a handful of rail fingers reaching inland from seaports, West Central Africa is largely devoid of railways, and roads only poorly serve most areas. Isolation and poor transport are major factors limiting economic development. There is not yet a paved road between Brazzaville and Pointe Noire, the Congo Republic's main seaport, but the 320-mile (515-km) Congo-Océan Railway does connect the cities. There are only 300 miles (c. 500 km) of paved roads in the massive Democratic Republic of Congo. The country's major transportation artery is a combination of railroad and river connecting Kinshasa with Lubumbashi.

Colonialism and Modern Struggles

In colonial days, France held three of the present units—Central African Republic, Congo-Brazzaville (its colonial name was Middle Congo), and Gabon—in the federation of colonies called French Equatorial Africa. Cameroon consisted of two trust territories administered by France and Britain. Belgium held what is now the Democratic Republic of Congo, Spain held Equatorial Guinea, and Portugal held the islands of São Tomé and Príncipe. During the last quarter of the 19th century, the Congo Basin was virtually a personal possession of Belgium's King Léopold II, whose agents ransacked it ruthlessly for wild rubber, ivory, and other tropical products gathered by Africans. This lawless era inspired Joseph Conrad's famous novel *Heart of Darkness* (1902), in which the trader Kurtz, at the point of death, evokes the ravaged Congo region with the cry "The horror! The horror!" In 1908, the Belgian government formally annexed the greater part of the Congo Basin, creating the Belgian Congo colony.

Pressure for independence built rapidly in 1959, and Belgium yielded to the demands of Congolese leaders. In 1960, three months before adjacent French Equatorial Africa became the independent People's Republic of the Congo, Belgium's Congo colony became the independent Republic of the Congo. In 1971, it took the name Zaire, meaning "river." Following the overthrow of Zaire's government in 1997, the country was renamed Democratic Republic of Congo.

Hundreds of ethnic groups speaking many languages and dialects live in West Central Africa. The majority speak Bantu languages of the Niger-Congo family, while Nilo- 471
Saharan language speakers predominate in the drier areas of grassland north of the rain forest. Small bands of Pygmies (Twa) live in the rain forest of the Congo Basin. The population of the Democratic Republic of Congo is especially complex, with some 250 ethnolinguistic groups speaking more than 75 languages.

In recent decades, the region's troubles have focused on its major power, the DRC. During his long tenure in office (1965–1997), Zaire's autocratic ruler, Mobutu Sese Seko, had neglected the country's development but acquired a massive fortune. Political opposition and public discontent with unemployment and inflation led to unrest in the early 1990s. All-out war began after 1997, when Zairean army troops attempted to expel a force of indigenous Tutsis living in the country's far eastern region (for discussion of the region's Tutsis and Hutus, see pages 501–502). The assault backfired when these Tutsis, supported by neighboring Tutsi-dominated governments in Rwanda and Burundi and by other sympathetic countries in the region, took up yet more arms and initiated fierce attacks on government forces.

The Tutsi counterattack quickly widened into an anti-government revolt comprised of many ethnic factions and led by a non-Tutsi guerrilla named Laurent Kabila. Kabila's forces pushed westward virtually unopposed through the huge country and in May 1997 occupied Kinshasa with little loss of life. President Mobutu fled on the eve of Kabila's triumph. Kabila then declared himself president of the country he now called the Democratic Republic of Congo and promised that political reforms would live up to the country's new name. Foreign companies rushed to sign mining contracts with the fledgling government, and throughout the country, there was hope that decades of neglect and poverty could be reversed in this resource-rich giant of West Central Africa.

But it was not to be. President Kabila's new regime proved to be a bitter disappointment to most of the DRC's citizens and neighbors and to the international community. Like his predecessor Mobutu, he funneled much of the country's wealth and power into the hands of family and friends. He also obstructed international human rights investigations into the deaths and disappearances of thousands of Hutu refugees during the heady days of rebellion against Mobutu. And he failed to bring stability to the far east of his huge country, where an environment of chaos had given birth to his own bid for power.

In 1998, Hutu forces based in the eastern DRC launched successive devastating attacks on Tutsi interests in neighboring Rwanda. Although the Tutsis of Rwanda had helped bring Kabila to power, they now turned against him, forming the backbone of a rebel alliance bent on bringing down Kabila's young regime. The unrest led to something rather rare in the region, a broader fight dubbed **"Africa's first world war"** that involved numerous countries, some of which were not even the DRC's neighbors. Each player became involved to protect its own national, separatist, or economic interests, largely at the expense of anything that would resemble national cohesion and stability in the Democratic Republic of Congo.

On one side was the so-called Rebel Coalition, which sent forces to oppose the government in Kinshasa: Rwanda, Uganda, and Angola's rebel UNITA forces. On the other side, the Congo Coalition sent forces to support the government in Kinshasa: Zimbabwe, the Angolan government, Zambia, Namibia, Sudan, and Chad. Although the fighting was sometime ferocious, it was not warfare that caused the loss of more than 3 million lives—the most in any war since World War II—in this struggle between 1998 and 2001. Most victims were civilians who died from starvation and disease or in widespread massacres directed against ethnic groups, usually Hutus killing Tutsis but sometimes the reverse.

The country appeared to be on the verge of dissolution when Laurent Kabila was assassinated by a bodyguard in 2001. He was succeeded by his son Joseph, a Congolese leader seemingly, finally, committed to halting the country's decline. A formal peace treaty was reached in April 2003 when Joseph Kabila agreed to have several formal rebel groups and political parties join him in a government of national unity, meant to pave the way for national elections in 2006. Kabila kept his post as president in what was widely seen as a fair election.

Conflict in the Democratic Republic of Congo was enormously profitable for the warring factions, foreign governments, and some foreign companies. United Nations investigators singled out Uganda and Rwanda for looting the eastern part of the country of its gems, minerals, timber, agricultural produce, and wildlife, including elephant ivory from some of the country's national parks. African, European, and U.S. companies were able to exploit the country's mineral wealth without paying corporate taxes. Despite fighting by Rwandan and Ugandan rebels operating inside the country and by forces loyal to the presidential candidate who lost the 2006 election, hopes rose that the long nightmare of Congo's plunder would end. A United Nations peacekeeping mission known by the French acronym MUNOC stayed on the ground to help ensure the country's stability, but rebellion simmered in the east.

9.1.4 East Africa: Mauled but Healing

Five countries make up East Africa: Kenya and Tanzania, which front the Indian Ocean, and landlocked Uganda, Rwanda, and Burundi (•Figure 9.1.8). Their total area is 703,000 square miles (1.8 million sq km), roughly the size of the U.S. state of Texas, with a total population in 2007 of 122 million (more than one-third the population of the United States).

• **Figure 9.1.8** Principal features of East Africa.

Environments and Resources

East Africa's terrain is mainly plateaus and mountains, with the plateaus generally at elevations of 3,000 to 6,000 feet (c. 900 to 1,800 m). All five countries include portions of the Great Rift Valley. International frontiers follow the floor of the Western Rift Valley for long distances and divide Lakes Tanganyika, Malawi, and Albert and smaller lakes in different countries. The more discontinuous Eastern Rift Valley crosses the heart of Tanzania and Kenya. Lake Victoria, lying in a shallow downwarp between the two major rifts, is divided between Uganda, Kenya, and Tanzania.

The two highest mountains in Africa, Kilimanjaro (19,340 ft/895 m) and Kirinyaga (Mount Kenya, 17,058 ft/5,200 m; see Figure 9.6), are extinct volcanoes. Both are majestic peaks crowned by snow and ice and visible for great distances across the surrounding plains; recent scientific observations, however, suggest that Kilimanjaro's legendary snows may be melted completely by 2015. Most 465
of East Africa's productive agricultural districts are in the fertile soils of these volcanic areas (•Figure 9.1.9).

Volcanism was also important in building the fertile highlands of Rwanda and Burundi. The Virunga Mountains along the Rwanda-Uganda border include many active volcanoes. One of the region's most active is Mount Elgon (14,176 ft/4,321 m), on the Uganda-Kenya border. On the Uganda-DRC border is Margherita Peak (16,762 ft/5,109 m), part of a nonvolcanic range known as the Ruwenzori. This range is often identified as the "Mountains of the Moon" depicted by early Greek geographers on their maps of interior Africa, but it should be noted that these geographers had little real knowledge of the region. Tanzania's nonvolcanic Eastern Arc mountains, although cleared of more than 70 percent of their forests, remains one of the world's greatest concentrations of endemic plant and animal species and is a biodiversity hot spot (see page 31).

The climate and biome types of most of East Africa are
467 classified broadly as tropical savanna (see Figure 9.8). In Uganda, Kenya, and Tanzania, park savanna—composed of grasses and scattered flat-topped acacia trees—stretches over broad areas and is the habitat for some of Africa's largest populations of large mammals. Great herds of herbivores and their predators and scavengers migrate seasonally across these countries to follow the changing availability of food and water. The countries manage many parks and wildlife reserves, and in Kenya and Tanzania, these are responsible for a huge share of the nations' hard-currency revenues. Among the most famous are the Serengeti Plains and Ngorongoro Crater of Tanzania and the Maasai Mara and Amboseli National Parks of Kenya (•Figure 9.1.10).

Joe Hobbs

• **Figure 9.1.9** East Africa is blessed with extensive areas of volcanic soil where commercial and subsistence crops thrive. This is a wheat field in the central highlands of Kenya.

Over much of East Africa, moisture is too scarce for nonirrigated agriculture. There are great variations in the quantity, effectiveness, and dependability of precipitation and in the length and time of occurrence of the dry season. Long and sometimes catastrophic droughts occur. From 1998 to 2001 and again in 2006, Kenya suffered historic droughts. Millions of people were in need of food aid as crops shriveled and livestock died. As the country's hydropower reserves were depleted, Kenyan cities suffered long periods without electricity. Tanzanian, Kenyan, and Ugandan efforts to stave off future emergencies in part by damming and redistributing waters of the Nile Basin for irrigation, drinking water, and hydropower have aroused the ire of downstream Egypt (see discussion and map on page 233).

The main subsistence crops are maize (corn), millet, sorghum, sweet potatoes, plantains, beans, and manioc (cassava). Elevation has such a strong effect on temperatures that in some areas, midlatitude crops like wheat, apples, and strawberries do well. Kenyan tourist brochures boast that Nairobi (elevation 5,500 ft/1,676 m) has a springlike climate year round. Some cultivators depend exclusively on crops for subsistence, and others also keep livestock. Cattle are the most important livestock animals.

The most valuable export crop for Kenya, the region's leading exporter, is tea, followed closely by cut flowers and fresh vegetables air-freighted to Europe. Coffee exports are also very important to Kenya. A niche market in high-quality gourmet coffee, produced in ideal conditions, has

Joe Hobbs

• **Figure 9.1.10** Savanna grasslands in eastern and southern Africa host the world's greatest concentrations of large mammals. This is Kenya's Maasai Mara National Park in part of the greater Serengeti ecosystem that also extends into northern Tanzania.

helped Rwanda's economy experience robust economic growth since 1995. Many specialty coffee stores abroad sell the coffees with pitches of benefits for people and wildlife in Rwanda. Rwandans themselves do not drink coffee; they see it as a bitter reminder of the colonial Belgians who forced them to grow it. However, they are thankful that the Europeans did not uproot some of the world's best arabica coffee trees, which were planted in that time. Kenya and Tanzania produce sisal for its valuable fiber. The islands of Zanzibar and Pemba support the millions of clove trees that provide much of world's supply for spices and for Indonesia's aromatic kretek cigarettes.

Tanzania, Kenya, and Uganda in 2001 revived the **East African Community,** vowing to reduce trade barriers among its member states, establish a stock exchange, adopt a common currency for a unified market (by 2009), and even elect a common president (by 2010). Many tariffs between the countries have been eliminated, and products, services, and people can move more freely between them. First established in 1967, the community fell apart 10 years later because of regional strife. The bloc showed confidence that Rwanda's and Burundi's years of conflict were over by admitting these countries in 2007.

The East African nations generally lack mineral reserves and production. A project to mine titanium, very controversial because it will require the relocation of thousands of people, is under way near the Kenyan coast. Manufacturing is limited mainly to agricultural processing, some textile milling, and the making of simple consumer items. There is almost no petroleum. There is a large cotton textile mill at Jinja, Uganda, that uses hydroelectricity from a station at the nearby Owen Stanley Dam on the Nile. The Tana River project in Kenya produces hydropower and supplies irrigation waters.

Populations, Settlements, and Infrastructure

The highest population densities are in a belt along the northern, southern, and eastern shores of Lake Victoria; in south central Kenya around and north of Nairobi; and in Rwanda and Burundi, which have among the world's highest population densities, aside from those of city-states
460 and some islands (see Figure 9.3a). As recently as the early 1990s, population growth rates in this region were so high that East Africa was a common academic example of the
60 Malthusian scenario. But after sharp reductions as a result of falling birth rates and increasing AIDS-related deaths, the population growth rates of the five East African countries moderated, averaging 2.4 percent as of 2007. There were some exceptions. Rwandan women bear an average of six children, and the government was delicately trying to bring the birth rate down in a country that lost nearly a million people in the 1994 genocide (to be discussed shortly). The region's populations are overwhelmingly rural. Only about 8 and 12 percent of the people of Burundi and Uganda, respectively, live in cities. Kenya and Tanzania are the most urbanized at 36 and 22 percent, respectively.

The most agriculturally productive parts of East Africa are bound together by railways that form a connected system leading inland from the seaports of Mombasa in Kenya and Dar es Salaam and Tanga in Tanzania. Mombasa is the most important seaport in East Africa. From Mombasa, the main line of the Kenya-Uganda Railway leads inland to Kenya's capital, Nairobi (population 3.9 million), the largest city and most important industrial center in East Africa. It is also the busiest crossroads of international air traffic and the main outfitting and departure point for safaris into East Africa's world-renowned national parks. It is a gritty, crime-ridden city that many locals call "Nairobbery." Nairobi was founded early in the 20th century as a construction camp on the railway, which continues westward to Kampala (population 1.7 million), the capital and largest city of Uganda. Dar es Salaam ("House of Peace" in Arabic; population 2.9 million) is the main city, main port, and main industrial center of Tanzania, with rail connections to Lake Tanganyika, Lake Victoria, and Zambia. Dar es Salaam had been the capital of Tanzania since colonial times, but the government is in the process of moving to the small inland city of Dodoma. Zanzibar wants to become a free economic zone, modeled after Singapore and Hong Kong, that would boast the region's most modern city. Rwanda's capital is in the highlands at Kigali (population 800,000), and Burundi's capital is on the northernmost point of Lake Tanganyika at Bujumbura (population 400,000). All of the East African capitals are primate cities.

Political and Social Issues

The five East African countries face many social and political dilemmas. None has had much internal peace, order, or prosperity, and none has had easy relations with its neighbors. Ethnic rivalries and conflicts have beset all of them, but prospects are also now better for most of them than at any time in recent decades.

Aside from small minorities of Asians, Europeans, and Arabs, the population of East Africa is composed of a large number of African ethnic groups; for example, Tanzania alone has about 120. Among the large groups are the Kikuyu and Luo of Kenya, the Baganda of Uganda, the Sukuma of Tanzania, the pastoral Maasai of Kenya and Tanzania, and the Tutsi and Hutu of Rwanda and Burundi. Most East African peoples speak Bantu tongues of the Niger-Congo language family, but some in northern Uganda, southern and western Kenya, and northern Tanzania speak Nilo-Saharan languages. Swahili, a Bantu language draw- 471
ing heavily on Arabic for vocabulary, is a widespread lingua franca (and is the national language of Tanzania), as is English. Christian, Muslim, and animist faiths thrive in all the countries. The Muslim religion reflects a long history of Arab commercial enterprise in the region, including slave trading.

All of the countries are former European colonies. Rwanda, Burundi, and mainland Tanzania (known as Tanganyika) were part of German East Africa until the end of World War I. At the end of the war, Tanganyika joined Kenya and Uganda as British dependencies, and Rwanda and Burundi, at the time the single Belgian colony known as Ruanda-Urundi, became independent countries in 1962. The islands of Zanzibar and Pemba, which lie north of the city of Dar es Salaam and some 25 to 40 miles (40 to 64 km) off the coast, were once a British-protected Arab sultanate. The city of Zanzibar was the major political and slave-trading center and entrepôt of an Arab empire in East Africa. While Arabs formerly controlled political and economic life, the majority of the population was, and is, African. Africans have been in control since 1964, when a revolution overthrew the ruling sultan. That same year, the political union of Tanganyika and Zanzibar was formed, taking the name United Republic of Tanzania. However, Zanzibar and Pemba have a parliament and president separate from Tanzania's.

Prior to independence, East Africa was home to about 100,000 Europeans, 60,000 Arabs, and 300,000 Asians. Arabs were primarily a mercantile class. Europeans were a professional, managerial, and administrative class. In the so-called White Highlands of southwestern Kenya, about 4,000 European families owned and operated estates that were worked by African laborers to produce export crops and cattle. With independence in 1963, Kenya made a point of inviting the British and other Europeans to maintain their presence in the commercial crop sector and to train the black majority in the managerial and technical skills that had served white interests well. Many Europeans left, and Africans acquired their farms, but those who stayed formed the core of the still-influential white Kenyans of today. Some still have vast tracts of land coveted by the Maasai and black Kenyan communities. With Zimbabwe's economically disastrous policies of confiscation of white lands in mind, however, Kenya's government refuses to help push
515 white owners off their lands. It is urging that land leases to whites—some written to last as long as 1,000 years—be revised dramatically downward to less than 100.

Collectively known as the **Wahindi** in Swahili, the Asians included Indians, Pakistanis, and Goans (from the former Portuguese colony of Goa on the west coast of India). They dominated retail trade, owned factories, and produced export crops. Some were wealthy, and practically all were more prosperous than most Africans. Since independence, the Asian minority has lost much of its commercially predominant position. The Ugandan leader Idi Amin abruptly expelled most of the 80,000 ethnic Indians and Pakistanis from Uganda in 1972. Because they were so vital to the country's economy, it nearly collapsed. Many thousands have returned in recent years, have successfully reclaimed lost land and other assets, and once again own and operate successful service and manufacturing businesses. The more recent economic decline of Kenya led to a slow but steady exodus of Asians, mainly to Australia, Britain, and the United States.

Episodes of bloodshed have punctuated East Africa's history. From 1905 to 1907, when Tanganyika was under German control, German authorities put down a great revolt against the colonial power, costing 75,000 African lives. Immediately prior to its independence, Kenya experienced the bloody guerrilla **Mau Mau Rebellion**, led by a secret society within the large Kikuyu ethnic group. Antagonism directed against British colonialism in Kenya cost the lives of about 11,000 Africans and 95 white settlers over the period of rebellion from 1951 to 1956.

From the 1960s through the 1980s, Kenya was regarded as a relative economic success story in sub-Saharan Africa. However, corruption, ethnic strife, crime, terrorism (including the 1998 bombing of the U.S. embassy in Nairobi, apparently by associates of al-Qa'ida, and another al-Qa'ida-linked bombing in Mombasa in 2002), drought, and floods then combined to reduce Kenya's prospects for economic development. After a sharp drop, tourism rebounded after 2005, with over 1.5 million people visiting 235
the country in 2007, bringing in some much-needed cash and helping the Kenyan economy grow by 6 percent before turmoil resumed.

Through much of the latter half of former President Daniel arap Moi's 24 years in office, which ended in 2002, Kenya suffered from corruption and human rights abuses. Britain, the United States, and other international lenders cut off economic aid because Moi refused to implement democratic reforms. Mwai Kibaki, the candidate opposing Moi's political party, won the 2002 presidential election with a vow of "zero tolerance" of corruption. Hopes evaporated, however, when his administration proved to be as corrupt as his predecessor's. Wary international aid agencies focussed on strictly monitored projects in Kibaki's Kenya. Kenya plunged into anarchy following a 2008 election, widely denounced as fraudulent, in which Kibaki was declared victorious over opposition leader Raila Odinga.

In postcolonial Uganda, a savage orgy of murder, rape, torture, and looting took place under the notorious regimes of Idi Amin and Milton Obote. The ravaged country had a respite after 1986 when Yoweri Museveni, the leader of a southern guerrilla force that had developed in opposition to Obote, took over the government. Despite some simmering conflicts among Uganda's 40 ethnic groups and despite the country's controversial interventions in regional conflicts, Uganda under Museveni has generally been regarded as a relatively progressive model for development in sub-Saharan Africa. International scrutiny focuses on Uganda's government to ensure that it will not promote ethnic cleansing of the Acholi people. Some Acholi supported the rebel **Lord's Resistance Army (LRA)**, which fought government interests in the country until a 2006 cease-fire.

Rwanda and Burundi have had tragic disputes between their majority and minority populations. About 85 percent of the population in Burundi and 90 percent in Rwanda are composed of the **Hutu (Bahutu),** and most of the remainder of the two populations is **Tutsi (Watusi).** These were not originally separate tribal or ethnic groups; the peoples speak the same language, share a common culture, and often intermarry. The distinction between Hutu and Tutsi was instead based on socioeconomic classification. Historically, the Tutsi were a ruling class who dominated the Hutu majority. Tutsi power and influence were measured especially by the vast numbers of cattle they owned. Tutsi who lost cattle and became poor came to be identified as Hutu, and Hutu who acquired cattle and wealth often became Tutsi.

Colonial rule in East Africa attempted to polarize these groups, increasing antagonism between them. German and Belgian administrators differentiated between them exclusively on the basis of cattle ownership: Anyone with fewer than 10 animals was Hutu, and anyone with more was Tutsi. Europeans mythologized the wealthier Tutsis as "black Caucasian" conquerors from Ethiopia who were a naturally superior, aristocratic race whose role was to rule the peasant Hutus. The colonists replaced all Hutu chiefs with Tutsi chiefs. These leaders carried out colonial policies that often imposed forced labor and heavy taxation on the Hutus. Education and other privileges were reserved almost exclusively for the Tutsis.

Ferocious violence between these two peoples marred Rwanda and Burundi intermittently after the states became independent in the early 1960s. After independence, the majority Hutus came to dominate the governments of both nations. In 1994, the death of Rwanda's Hutu president in a plane crash (in which Burundi's president also died) sparked civil war in Rwanda between the Hutu-dominated government and Tutsi rebels of the **Rwandan Patriotic Front (RPF),** based in Uganda. Hutu government paramilitary troops, still vengeful about decades of domination by Tutsis, systematically massacred Tutsi civilians with automatic weapons, grenades, machetes, and nail-studded clubs. Women, children, orphans, and hospital patients were not spared. An estimated 800,000 Tutsis and "moderate" Hutus (those who did not support the genocide) died. International media tracked these horrors, but outside nations did nothing to stop the fighting.

Well-organized and motivated Tutsi rebels seized control of most of the country and took power in the capital, Kigali, in 1994. That precipitated a flood of 2 million Hutu refugees mainly into neighboring Zaire, now the Democratic Republic of Congo. Although the Tutsi-dominated government promised to work with the Hutus to build a new multiethnic democracy, Hutu insurgents, based mainly in eastern Zaire, continued to carry out attacks against Tutsi targets in Rwanda. That violence and Rwandan interests in the Democratic Republic of Congo's civil war kept Rwandan forces active outside its borders until 2002.

Rwanda now has a national unity policy aimed at reconciling Hutus and Tutsis; both occupy important government posts, and although the Tutsis still hold the political upper hand, they have encouraged the resettlement of Hutu refugees who fled the country during its troubles (•Figure

SEAN SPRAGUE/Peter Arnold

• **Figure 9.1.11** Burundian Hutu refugees at Likole camp, Ngar.

9.1.11). In 2004, marking the 10-year anniversary of the genocide with a new policy meant to prevent its recurrence, Rwanda's government outlawed ethnicity. Now any Rwandan who speaks or writes of Hutus and Tutsis may be fined or imprisoned for practicing "divisionism." The country's past nevertheless continues to shape its future; for example, there is an ongoing defection of the faithful from Rwanda's majority Catholic church to other religions, especially Islam but also smaller Christian denominations. Most of the country's worst massacres took place in Catholic churches, with the apparent knowledge and even participation of church leaders.

In Burundi, the Tutsi minority traditionally dominated politics, the army, and business. In 1993, there was a brief shift of power to the Hutu majority when Burundians elected the country's first Hutu president. Tutsis murdered him in October 1993, and a large-scale Hutu retaliatory slaughter of Tutsis ensued. An estimated 300,000 died in a decade of violence. Burundi's Tutsis and Hutus now have a power-sharing agreement in which the presidency rotates between them. The peace has been maintained by the first-ever deployment of the African Union's African Standby Force. 481

Collective shame and embarrassment in the international community about the failure to stop the killing in Rwanda and Burundi led in subsequent years to numerous new antigenocide measures. These include a United Nations "early warning" system to prevent genocide and a new forum for crimes against humanity in the **International Criminal Court,** established in 2002 in The Hague, Netherlands.

9.1.5 The Horn of Africa: Refuge for Judaism, Christianity, Islamist Militancy

In the extreme northeastern section of sub-Saharan Africa, a great volcanic plateau rises steeply from the desert. This highland and adjacent areas occupy the greater part of the

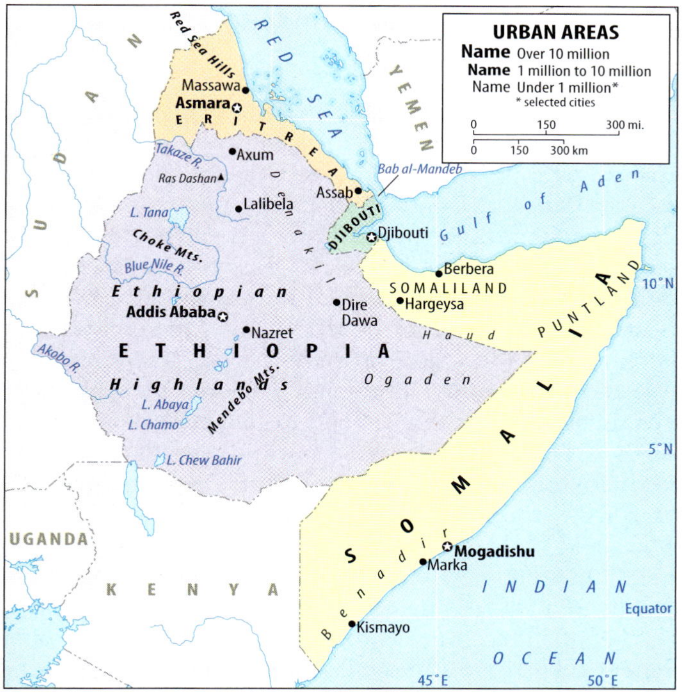

• **Figure 9.1.12** Principal features of the Horn of Africa.

Horn of Africa, named for its projection from the continent into the Indian Ocean. This subregion includes the countries of Ethiopia, Eritrea, Somalia, and Djibouti (•Figure 9.1.12).

Environments and Resources

Much of the Horn of Africa lies at elevations above 10,000 feet (c. 3,000 m), and one peak in Ethiopia reaches 15,158 feet (4,620 m). The highlands where the Blue Nile, Atbara, and other Nile tributaries rise receive their rainfall during the summer. Temperatures change from tropical to temperate as elevation increases. Bananas, coffee, dates, oranges, figs, temperate fruits, and cereals can be produced without irrigation. Highland pastures support sheep, cattle, and other livestock. The region's people are concentrated in the relatively well-endowed highlands; most of Ethiopia's 77 million people live there (•Figure 9.1.13).

East of the mountain mass, in Eritrea, Djibouti, and Somalia, lower plateaus and coastal plains descend to the Red Sea, the Gulf of Aden, and the Indian Ocean. Extreme heat and aridity prevail at these lower levels, and nomadic and seminomadic peoples make a living by herding camels, goats, and sheep. Dwellers of scattered oases carry on a precarious agriculture. The arid lowland sections have many characteristics more typical of the Middle East than other areas of sub-Saharan Africa. The populations of these countries are small compared to Ethiopia: Somalia with 9 million, Eritrea with 5 million, and Djibouti with less than a million.

Ethiopia has substantial resources, perhaps including significant mineral wealth. However, its resource base re-

• **Figure 9.1.13** The rugged Ethiopian highlands have many good soils, but frequent drought and poor transportation infrastructure set back Ethiopia's agricultural productivity.

mains largely undeveloped, and the country is poor. Coffee and animal hides are the main exports. With foreign assistance, a few textile and food-processing factories have been built, and with Chinese help, the massive Tekaze Dam is
being built on a tributary of the Blue Nile to harness part of 233
the country's large hydroelectric potential.

A poor transportation system—whose best feature is the 3,000 miles (c. 4,800 km) of good roads built by the Italians during their occupation in the 1930s—is a major factor keeping the country isolated and underdeveloped. The French-built railroad from Djibouti to Ethiopia's capital, Addis Ababa (population 2.8 million), lacks feeder lines and has never been as successful as hoped. Many roads into the Ethiopian heartland are rough tracks, although better connections have gradually been developing. Air services help overcome the deficiencies in ground transportation. Many of Ethiopia's people live in villages on uplands cut off from each other and from the outside world by precipitous chasms. These formidable valleys were cut by streams carrying huge amounts of fertile volcanic silt to the Nile. Before the Aswan High Dam was built, they provided ma-
terial for the nourishing floods in Egypt. 252

Eritrea has gold, copper, and marble reserves but lacks oil, and little of its mineral potential has been developed. Its Red Sea coastline has superb coral reefs, and the country is hoping to attract international tourists to dive and snorkel among them. There is some cotton and oilseed agriculture, and fish are a mainstay in the diet.

Djibouti supports pastoralists but has almost no arable land. The country has little to offer economically except its strategic position near the mouth of the Red Sea, and this has been a windfall for Djibouti since the terrorist attacks on
the United States in 2001. Its fine port, also named Djibouti, 234
is a major staging ground in the U.S. military forces in the region. France has long maintained a significant military presence and has pumped much-needed economic aid into the country.

Somalia is extremely arid, and only 2 percent of the country is cultivated. Nomadic and seminomadic pastoralism is the main economic activity. Until an outbreak of livestock disease in the 1990s, Somalia's main exports were live animals (goats, sheep, camels, cattle), shipped mainly to Saudi Arabia, Yemen, and the United Arab Emirates. Fish products and bananas are still shipped, but despite a ban in effect since 2002, charcoal is the main export to the Arabian Peninsula. The resulting environmental impacts on Somalia are severe because trees never were abundant, and there are fights between the tree cutters and the pastoralists, whose camels feed on tree fodder. An estimated 11 percent of Somalia is forested.

Droughts have stricken this region, particularly Ethiopia, almost incessantly since the late 1960s. Ethiopian villagers are especially vulnerable to drought because of their heavy dependence on rain-fed agriculture. The volcanic soils are fertile, but the streams are entrenched so far below the upland fields that irrigation is virtually impossible. In 1984 and 1985, drought in Ethiopia reached the point of widespread catastrophe. An estimated 1 million people perished. Only a massive and sustained international relief effort, some of it promoted by British and American rock musicians, helped reduce the suffering. Successive droughts have revisited the country, and chronic hunger reportedly affects between 4 and 5 million Ethiopians each year. Ethiopia's multimillion-dollar effort to resettle 2.2 million farmers from the apparently overcrowded highlands to the less populous lowlands has run into major problems. The lowlands were less crowded to begin with because of malaria and sleeping sickness, which have taken a big toll on settlers.

Local Cultures, European Imperialism, and Regional Struggles

Ethiopia's people are ethnically and culturally diverse. About 45 percent, including the politically dominant **Amhara** peoples, practice **Ethiopian Orthodox Christianity,** an ancient branch of Coptic Christianity that came to Ethiopia in the 4th century from Egypt. The entire area has had important cultural and historical links with Egypt, the Fertile Crescent, and Arabia. The Ethiopian monarchy based its origins and legitimacy on the union of the biblical King Solomon and the Queen of Sheba, who, tradition holds, gave birth to the first Ethiopian emperor, Menelik. Until a Marxist coup brought an end to the emperorship in the 1970s, Ethiopia's rulers were always Christian. Ethiopia has many outstanding Christian artistic and architectural treasures, including the 11 churches of Lalibela, carved from solid rock in the 12th and 13th centuries (•Figure 9.1.14). Most of the rest of Ethiopia's people are either Muslims (who make up about 40 percent of the population) or members of Protestant, Evangelical, and Roman Catholic churches. There are still small numbers of **Falashas,** or **Ethiopian Jews,** in Ethiopia, a remnant of a very ancient and isolated Jewish population. The majority, about 100,000, have fled to Israel since the mid-1980s. Because this mountainous country has long served an isolated refuge for such unique groups, it has been nicknamed the **"Galápagos Islands of Religion."** 578

Eritrea's population, like Ethiopia's, is about half Muslim and half Christian. Djibouti is about 95 percent Muslim and 5 percent Christian. In Somalia, about 99.8 percent of

Joe Hobbs

• **Figure 9.1.14** The churches of Lalibela, carved from volcanic rock in the 12th and 13th centuries, are among Ethiopia's many Christian cultural treasures. Note the man at the left center for scale.

the people are Sunni Muslims and 85 percent are ethnic Somalis. They share the same language and the same nomadic culture. This is one of the most homogeneous populations in the world, a fact that would suggest peace and stability, but rivalries between clans tore the country apart in the 1990s, as will be related shortly.

Colonialism had a big impact on the region's peoples. European powers seized coastal strips of the Horn of Africa in the late 19th century (see Figure 9.16). Britain was first with British Somaliland in 1882, and then France annexed French Somaliland in 1884. Finally, Italy took control of Italian Somaliland and Eritrea in 1889. These areas along the Suez–Red Sea route have never had much economic importance, but they have had great strategic significance.

Italy tried to extend its domain from Eritrea over more attractive and potentially valuable Ethiopia in 1896, but Ethiopians annihilated the Italian forces at Aduwa. Forty years later, in 1936, the Italians succeeded in taking Ethiopia. In World War II, however, British Commonwealth forces defeated Italian troops throughout the region. Ethiopia regained independence, and Eritrea was federated with it as an autonomous unit in 1952. Ethiopia incorporated Eritrea as a province in 1962.

After World War II, Italian Somaliland was returned to Italian control to be administered as the trust territory of Somalia under the United Nations, pending independence in 1960 as a republic. The present country of Somalia officially includes both the former Italian territory and the former British Somaliland, which chose to unite with Somalia. The people of much smaller French Somaliland remained separate and eventually became independent as the Republic of Djibouti, with its capital at the seaport of the same name.

In 1974, Ethiopia's aging Emperor Haile Selassie (whom
537 **Rastafarians,** who live mainly in Jamaica, regard as God) was deposed by a civilian and military revolt against the country's feudal order. A Marxist dictatorship, strongly oriented to the former Soviet Union, emerged. American influence, which had been strong under Haile Selassie, was eliminated. Munitions and advisers from the Soviet bloc poured in to equip and train one of Africa's largest armies, and the Soviets gained access to base facilities in Ethiopian (now Eritrean) ports. The United States, which had previously supported Ethiopia in its struggles against neighboring Somalia, now switched sides to back Somalia. Such foreign interventions and proxy conflicts contributed to the strife that has hindered development in the region.

Ethiopia had many internal troubles following independence. In what was then the province of Eritrea, both Muslims and non-Amhara Christians resented their political subjugation to Ethiopia's Christian Amhara majority. Terrorism and guerrilla actions escalated into civil war in the early 1960s. The Ethiopian army tried but failed to stamp out the revolts of two Eritrean "liberation fronts." At the same time, there was also a rebellion in Tigre province adjoining Eritrea. In 1991, these combined rebel forces captured Addis Ababa and took over the Ethiopian government. The rebels in Eritrea established an autonomous regime, and Eritrea became independent in 1993. Eritrea's main city and manufacturing center, Asmara (population 1 million), is located in highlands at an elevation of more than 7,000 feet (2,130 m) and has a rail link with the Red Sea port of Massawa.

Many geographers subsequently pointed to Eritrea as a model for sustainable development in sub-Saharan Africa. After independence, the country reduced its foreign debt, cut its population growth rate, and took other promising steps toward eliminating the burdens of underdevelopment. However, in 1998, Eritrea dimmed its bright prospects by launching a war against Ethiopia. The conflict emerged from economic rivalries and unresolved territorial issues between the nations. With Eritrea's independence, Ethiopia had agreed to become landlocked, as long as it could freely use Eritrea's Red Sea ports of Massawa and Assab. The concession was in large part Ethiopia's reward to Eritreans for helping with the overthrow of Ethiopia's Marxist regime. The countries' leaders in effect agreed on an economic union in which Eritrea would be the industrial power and Ethiopia the agricultural one. They promised to allow the free movement of people and goods between the two states and to share a single currency, the Ethiopian birr.

Trouble began when Eritrea introduced its own currency and announced steep duties and fees on goods shipped to and from Ethiopia through Eritrean ports. Ethiopia then turned to Djibouti and Kenya to find new outlets to the sea. Deprived of an anticipated major source of revenue, Eritrea attacked Ethiopia. The war finally ended in 2000 when Ethiopia gained the upper hand by securing a border region disputed between the two countries. A demilitarized zone monitored by United Nations troops was established along the border. An estimated 100,000 Eritreans and Ethiopians had perished in the conflict. Eritrea's economy was hit particularly hard. With the Ethiopian market gone, Eritrea's exports declined 80 percent from their 1997 high. The port of Assab shut down for some time. As the years passed, the two countries failed to agree on the demarcation of their common border. Anti-Ethiopian forces became active inside Eritrea, and war threatened to engulf these neighbors again.

Somalia has been in ruin since a rebellion overthrew the government in 1991 and ignited a civil war between regional Somali clans. The country has six major clans and numerous subclans. These are family-based units based on a patrilineal (male-line) descent system, essentially an extended 219
family that defines one's identity, loyalties, and obligations. There is little sense of nationhood in Somalia because a person's primary affiliations are with clan rather than any larger political entity. Most people are educated, socialized, and married within their clan. There is a history of antagonism between the clans, worsened by easy access to light arms and more deadly weapons.

In the 1991 conflict, contending clan militias inflicted great damage on the capital, Mogadishu (population 2.8 million), from which large numbers of people fled to the countryside as refugees. Massive famine threatened the devastated country in mid-1992. Law and order disintegrated, and humanitarian assistance became difficult. United States military forces intervened in **Operation Restore Hope,** a mission first designed to help protect relief workers and aid shipments. However, policymakers of the Clinton administration decided to use military force to neutralize the power of one of the country's leading clan leaders, or "warlords," Mohamed Farrah Aidid. In a gruesome sequence of events remembered today as the **Blackhawk Down incident,** 18 U.S. servicemen stranded in Aidid's territory in Mogadishu were killed in a single day in 1993.

Osama bin Laden's al-Qa'ida organization had supplied the weapons used to assault the American troops. American public outcry at the tragedy brought an end to this U.S. military mission in 1994. Blackhawk Down had two lasting repercussions. It made the United States more reluctant to
237 assist in humanitarian crises. It also emboldened al-Qa'ida, which became convinced that the United States could not stomach the loss of American lives. Osama bin Laden frequently cited the American withdrawal from Somalia as evidence of his group's successful *jihad* against the United States.

For its part, the United States kept a vigilant watch on Somalia as a potential base for al-Qa'ida and on its waters as a dangerous, pirate-ridden threat to international trade. In 2006, a loose coalition of Islamist groups, including a small number of al-Qa'ida personnel, succeeded in gaining control of Mogadishu and some other cities in Somalia. With their characteristic motto of "Islam is the solution," they argued that clan differences should be subservient to a broader unity of faith. This philosophy attracted a growing number of war-weary Somalis. The United States was alarmed that Somalia could fall to radical Islam. Neighboring Ethiopia worried that Somali Islamists, with support from Ethiopia's enemy Eritrea, would encourage Muslim rebellion within its borders. In late 2006, from bases in Ethiopia, Djibouti, and Kenya, the United States mounted air assaults on the Somalia Islamists, while Ethiopian troops moved in on the ground. As the Islamist forces were killed or melted back into the countryside, and hundreds of thousands of civilians fled by sea to Yemen, Ethiopian troops took control of Mogadishu. Somali resentment of the occupation by the county's historical Ethiopian enemy led to a rapid withdrawal of Ethiopian troops and the deployment of an African Union peacekeeping force made up mainly of Ugandans. A provisional government was established in Mogadishu with the aim of restoring order to the country. But with the country's clan-based factionalism again at the fore, the prospect of peace seemed very distant.

Somalia since 1991 has fractured into three de facto states (see Figure 9.1.12). The southernmost, centered on Mogadishu (still much in ruin and without services), is Somalia, the only one with international recognition as a sovereign country. To the north is Puntland, which includes the tip of the Horn of Africa and considers itself autonomous (with a president of its own) but not independent of Somalia. To the northwest, bordering Djibouti and Ethiopia, is Somaliland (the former British Somaliland). It proclaimed itself independent of Somalia in 1991 and has a president and broadly representative government based in the city of Hargeysa. Fearing fragmentation of their own countries, even African nations do not recognize its independence. Its port city of Berbera has served as Ethiopia's main trade outlet to the sea since Ethiopia's conflict with Eritrea began. Somaliland's economic mainstay was exports of cattle to the Muslim nations of the Arabian Peninsula until an outbreak of Rift Valley fever led to a complete cessation of the trade in 2001. Now Somaliland is pinning its hopes on exports of gemstones, frankincense, vegetable dyes made from henna, and products made from qasil plants, such as hair conditioners and body cleansers.

9.1.6 Southern Africa: Resource-Rich, Finally Free

In the southern part of sub-Saharan Africa, five countries—Angola, Mozambique, Zimbabwe, Zambia, and Malawi—share the basin of the Zambezi River and have long had important relations with each other. The first two were former Portuguese colonies, and the latter three were formerly British. Still farther south are South Africa and four countries whose geographic problems are closely linked to South Africa: the three former British dependencies of Botswana, Swaziland, and Lesotho, and the former German colony of Namibia.

These 10 states are grouped here as the countries of southern Africa (Figure 9.1.15). Other than their geographic location, little economic or political integration binds these diverse states together. This may change as a result of goals established by the **Southern African Development Community (SADC),** a regional bloc composed of these 10 countries plus Tanzania, Democratic Republic of Congo, and Mauritius. The SADC's member nations have established an economic free-trade zone in which goods can cross borders without tariffs and taxes. The dominant power in that group, in the region, and on the continent is South Africa, and so it is treated in greatest detail here (with a dedicated map, Figure 9.1.16).

Environment and Resources: Water, Natural Regions, and Agriculture

Water is all-important in this drought-prone region of a dry continent. South Africa is fairly well endowed with water, as two river systems—the Orange (including its tributary, the Vaal) and the Limpopo—drain much of the country's high (3,000 to 6,000 ft/c. 900 to 1,800 m) interior plateaus. Many dams have been constructed on these and

• **Figure 9.1.15** Principal features of southern Africa.

• **Figure 9.1.16** Principal features of South Africa.

other South African rivers to provide irrigation water and hydroelectricity.

The Zambezi is another of the region's great rivers. Rising in Zambia, it flows eastward to empty into the Indian Ocean on the Mozambican coast. On the middle course of the Zambezi River, separating Zambia from Zimbabwe, is the spectacular Victoria Falls, regarded traditionally as one of the seven natural wonders of the world. It is a significant international tourist destination today. Downstream from the falls, hydropower from the Kariba Dam (completed in 1962; see Figure 9.7, page 466) flows to both countries.

The driest countries are Namibia and Botswana. They can support pastoralism and, in some areas, limited subsistence agriculture but not commercial agriculture. The Namib Desert extends the full length of Namibia in the coastal areas (•Figure 9.1.17). Inland is a broad belt of semiarid plateau country, merging at the east with the Kalahari semidesert of Botswana. The dry zone extends southward into South Africa's Cape provinces and northward into western Angola. Chronic shortages of freshwater in Namibia have led the government to consider desalinating Atlantic Ocean

Gallo Images/Corbis

• **Figure 9.1.17** A surprising abundance and diversity of wild animals and plants are adapted to the harsh conditions of the Namib Desert in Namibia. This is an oryx antelope known as the gemsbok.

waters (this option was rejected as too expensive); building the Epupa Dam on the Cunene River, which divides Namibia from Angola (this project was shelved after being opposed by environmentalists and by human rights groups, who argued that the homeland of the Himba ethnic group would be inundated); and diverting water by pipeline southward from the Namibian portion of the Okavango River to Windhoek (widely regarded as Africa's cleanest and safest capital; population 300,000). Arguing that the World Heritage Site of its Okavango Delta would be damaged or destroyed, Botswana is strongly opposed to this last Namibian plan. With a wealth of wildlife attractions in the Okavango Delta and adjacent national parks, Botswana is a growing destination for international nature-loving tourists. Namibia has its own wetlands resource of international importance: large wildlife populations responding to dramatic fluctuations in wet and dry conditions make Etosha National Park an attraction for the international tourists, who represent an important source of hard currency for Namibia.

Not being landlocked, Namibia enjoys an additional resource that Botswana lacks. The arid western coast of southern Africa, swept by the cold, northward-moving Benguela Current, supports Africa's most important fisheries. **Upwelling** cool water brings up nutrients from the ocean floor and provides a fertile habitat for plankton, the basic
532 food of fish. Namibia has established an exclusive economic zone (EEZ; see page 407) in the Atlantic stretching 200 miles (320 km) offshore, in which fishing and fish processing based at Walvis Bay have become important industries.

More than 90 percent of the region's annual rainfall comes in the cooler six months from November to April. The long dry period creates ideal conditions for the tropical savanna prevailing over most of southern Africa, including much of Zimbabwe, Zambia, Malawi, Mozambique, eastern Angola, and northern South Africa (see Figure 9.8, page 461). Malawi had extensive tropical forests, but at least half have fallen to the axes of woodcutters and charcoal burners in recent years. This extremely poor country, where only 2 percent of the population has electricity, has the unfortunate distinction of having the region's highest deforestation rate.

There is tallgrass savanna on the High Veld of Zimbabwe. The High Veld lies generally at elevations of 4,000 feet (c. 1,200 m) and higher and forms a band about 50 miles (80 km) broad by 400 miles (c. 650 km) long, trending southwest-to-northeast across the center of Zimbabwe. The High Veld was historically the main area of European settlement, and most of Zimbabwe's postcolonial population of 50,000 whites lived on it. Until the year 2000, these included about 4,500 white farmers whose large holdings, worked by black laborers, produced most of the country's agricultural surplus to feed urban populations and to export. Tobacco from these farms was the country's most important export, followed by cotton, cane sugar, corn (maize), and roses. As discussed on page 515, that export sector collapsed with the subsequent eviction of the white farmers.

South Africa's physical geography has played a major role in its unusual course of development. Lying almost entirely in the middle latitudes, it is a cooler land than most of sub-Saharan Africa, with annual temperatures generally between 60°F and 70°F (16°C–21°C). In combination with the country's natural resource wealth, this climate made the area very attractive for European settlement; it was one of the "neo-Europes" of the colonial era and was historically 45
Africa's main center of European settlement and advanced economic development.

Europeans favored the extreme east of the country's interior plateau, a region also known as the High Veld. The original vegetation here was a grassy steppe, similar to the midwestern prairies of the United States. This is South Africa's leading area of crop and livestock production. The principal farm animals are beef cattle, dairy cattle, and

Stone/Getty Images

• **Figure 9.1.18** The Mediterranean climate of Western Cape province is ideal for viticulture. This is a vineyard near Stellenbosch.

Greatstock Photographic Library/Alamy

• **Figure 9.1.19** Gold miners at work near Johannesburg, South Africa.

sheep, and the main crops are maize and wheat. Maize is the staple diet for South Africa's blacks. Surpluses are exported and used as livestock feed. In KwaZulu-Natal province, there is hilly terrain between the Indian Ocean and the Drakensberg, a mountainous escarpment at the edge of the interior plateau. The leading commercial crop of that area is sugarcane.

South Africa's Western Cape province has a Mediterranean climate. It is the only African area south of the Sahara with this climate type (• Figure 9.1.18). Average temperatures are similar to those of the same climate area in southern California. The early settlers established vineyards in this region, and today, the area's grapes and wines sell both in South Africa and abroad, where they are acquiring a growing share of the world market. The province also produces oranges and other citrus fruits, deciduous fruits, and pineapples. Sheep ranching is a major occupation in the semiarid interior.

Mining and Manufacturing

Minerals, including petroleum, are by far the most valu-
477 able material assets of this region. South Africa has the greatest mineral endowment, and the country's relative prosperity is owed in large part to that wealth. Immense British and other foreign investments pour into South Africa's mining industry. South Africa's diverse and abundant mineral wealth has fueled its continental leadership in manufacturing. The country has about a third of Africa's manufacturing and a third of its electric generation. Its diversified industries include iron and steel, machinery, metal manufacturing, weapons, automobiles, tractors, textiles, chemicals, and processed foods. Many thousands of jobs in Western countries are directly linked to mining and manufacturing in South Africa. The mines employ mainly black African workers recruited domestically and from other African countries, particularly Lesotho, Swaziland, Mozambique, Botswana, and Malawi. There is a large illegal immigration to South Africa, too; as many as 10 million undocumented migrant workers may be in this relatively prosperous country.

South Africa is the world's largest producer of gold (• Figure 9.1.19), and gold has often been its most important export (platinum and coal exports have recently brought in more revenues). The gold-mining industry is concentrated in the High Veld, especially the Witwatersrand (formerly Rand), an area in the province of Gauteng. Johannesburg, South Africa's largest city (population 7.3 million), the adjacent impoverished and overcrowded black township of Soweto (population 1.8 million), and nearby Pretoria, the national capital (population 2.1 million), are all located in Gauteng. Limpopo and Mpumalanga provinces and adjacent areas make up one of the world's most vital mineral-exporting regions; exports include copper, asbestos, chromium, platinum, vanadium, antimony, and phosphate. Recovery of uranium from mine tailings and as a by-product of current gold mining has made South Africa the largest uranium producer on the continent. And since 1867, the country has been an important source of diamonds from alluvial deposits along the Vaal and Orange Rivers and from kimberlite "pipes" (circular rock formations of volcanic origin). Kimberley in Northern Cape province is the administrative center for the diamond industry.

South Africa's mining industries require large amounts of electricity. Most of this is supplied by generating plants powered by bituminous coal, which is present in vast quantities near the surface on the interior plateau. These coal deposits, the largest in Africa, are critical to South Africa, as the country has only small petroleum and natural gas reserves and modest hydroelectric potential. Coal is a major export, and it has facilitated development of Africa's leading iron and steel industry, concentrated at Pretoria and around Johannesburg. Iron ore comes by rail from large deposits in the northeast and central parts of the country.

Some of South Africa's neighbors also have consider-
able mineral wealth (see Figure 9.18). Partly for this reason, 477
South Africa is much engaged in their affairs, especially in neighboring Swaziland and in Lesotho, which South Africa

entirely surrounds. Botswana and Lesotho gained independence from Britain in 1966, and Swaziland, the last absolute monarchy in southern Africa, became independent in 1968. South Africa has occasionally exerted power on behalf of its interests in these weaker, mineral-rich countries, particularly Lesotho. In 1998, for example, the prime minister of Lesotho perceived that a coup might be imminent in his nation and called on South Africa for assistance. South African troops immediately invaded Lesotho, but rather than quelling the unrest, this intervention led to widespread condemnation of South Africa for meddling in the internal affairs of a sovereign nation. Mining of coal, gold, and tin is an important source of revenue in Swaziland, as is tourism—which features gambling casinos—and sugar production. Lesotho's economy depends largely on exports of shoes and textiles, as well as remittances from miners working in South Africa.

Diamonds make up about four-fifths of the total value of exports from Botswana, which after South Africa has the region's strongest economy. South African, British, and U.S. mining companies export Namibian diamonds, uranium, copper, and various other metals. These exports contribute to the high (by African standards) GNI PPP levels of Botswana and Namibia ($12,240 and $8,110, respectively).

Zimbabwe has an extraordinary geological feature called the Great Dike, an intrusion of younger, mineral-rich material into older rock. Ranging in width from 2 to 7 miles (3 to 11 km), the Great Dike crosses the entire nation and is readily visible from space as a giant stripe on the landscape. Along and near the Great Dike are the principal areas where commercial farms thrived and mines producing gold, nickel, asbestos, copper, chromium, titanium, and other minerals provided valuable exports until Zimbabwe plunged into political and economic crisis (described on page 515). Along the Great Dike, the historical axis of economic development in Zimbabwe, lies a railway connecting the largest city and capital, Harare (formerly Salisbury; population 2.9 million), in the northeast with the second-largest city, Bulawayo, in the southwest. Iron ore mined at Kwekwe, about midway between Harare and Bulawayo, was used there to produce steel. The Wankie coalfield in western Zimbabwe supplied coal to smelt the ore. Until 1998, Zimbabwe's manufacturing industries were more important, both in value of output and in diversity of production, than those of any other African nation south of the Sahara except Nigeria and South Africa. Manufacturing since then has been reduced to a trickle because of the country's crisis, which has made it very difficult for factories to purchase fuel and spare parts.

Zambia's economy depends heavily on copper exported from several large mines developed with British and American capital. An urban area has developed at each mine, and the mining area, known as the Copperbelt, consists of a series of separate population nodes strung close to Zambia's frontier with the adjacent copper-mining Shaba region of southeastern Democratic Republic of Congo. The Copperbelt, which contains the majority of Zambia's manufacturing plants and copper mines, is a primarily urban region. Boosted by a global commodities boom and Chinese and other foreign investment, copper (now produced exclusively by private companies) made up 70 percent of Zambia's exports in 2007. The government wants to diversify the country's economy so that it does not remain overly reliant on this single product. Other mineral production includes cobalt, a by-product of copper mining; zinc and lead mined along the railway between the Copperbelt and Zambia's capital and largest city, Lusaka (population 2.3 million); and coal production from extreme southern Zambia.

The region's leader in petroleum production is Angola, which is increasingly important on the global energy stage and is OPEC's newest member. One of its vital oil-producing 228
areas is the coastal exclave of Cabinda, separated from the rest of the country by the mouth of the Congo River (which is controlled by the Democratic Republic of Congo). Other fields are scattered along the Angolan coast between the Congo mouth and the city of Lobito. Major new reserves have been discovered recently offshore Angola; in fact, more new petroleum reserves have been detected there in the past decade than in any other place in the world. In addition to oil and its products, Angola exports small amounts of diamonds, coffee, and other items. Diamonds are concentrated in the northeast, where UNITA rebels mined them for years to finance their antigovernment operations (see below).

Ethnicity, Colonialism, and Strife

South Africa: A Special Case

Visitors from western Europe and North America will find in South Africa most of the institutions and facilities to which they are accustomed. However, the Europeanized cultural landscape does not reflect the majority culture of this unusual country. Whites represent only about 10 percent of the total population, and blacks make up 79 percent. There are two other large racial groups: the so-called **coloreds,** of mixed origin (9 percent), and the Asians (2 percent), most of whom are Indians. Many peoples make up the black African majority. The **Zulu** and **Xhosa** (pronounced "*khoh*-suh") are both concentrated in hilly sections of eastern South Africa near the Indian Ocean and are the most populous of the nine officially recognized tribal groups.

A huge economic gulf separates South Africa's impoverished black Africans from about 5 million whites of European descent who dominate the economy and enjoy a lifestyle comparable to that of people in western Europe and North America. South Africa's overall per capita GNI PPP of $11,710 is comparable to that of Russia. For the unskilled and semiskilled labor needed to operate their mines, factories, farms, and services, the whites of South Africa have always relied mainly on low-paid black workers in most of the country and colored workers in Western Cape province. The white-dominated economy could not operate without them, and the Africans are in turn extremely dependent on white payrolls. Although the races are interlocked economically, relations between them have historically been very poor and have been structured to benefit the whites. A country whose people were legally segregated by race, South Africa was one of the world's most controversial nations until its new beginning in 1994. Following is a summary of how the country's racial structures—a critical issue in its modern geography—evolved.

South Africa's white population is split between the British South Africans and the **Afrikaners,** or **Boers** (Dutch for "farmers"; pronounced "*boorz*'"). The Afrikaners speak Afrikaans, a derivative of Dutch, as their preferred language. They outnumber the British by approximately three to two. The Afrikaners are the descendants of Dutch, French Huguenot, and German settlers who began coming to South Africa more than three centuries ago. Their impact on indigenous Africans was catastrophic. The white settlers killed, drove out, and through the unintentional introduction of smallpox decimated most of the original **Khoi** and **San** inhabitants (derisively called **"Hottentots"** and **"Bushmen,"** respectively, by the settlers). They put the survivors to work as servants and slaves. When these proved to be insufficient in number, the colonists brought slaves from West Africa, Madagascar, East Africa, the Malay Peninsula, India, and Ceylon (Sri Lanka).

Most of the workers in the cane fields of the Natal colony were indentured laborers brought from India beginning in the 1860s. The majority chose to remain in South Africa when their terms of indenture ended, and some Indians came as free immigrants. About 85 percent of South Africa's Indian population is still in KwaZulu-Natal province, mainly in and near Durban (population 3.3 million), the province's largest city and main industrial center and port. Most Indians today are employed in commercial, industrial, and service occupations.

Like KwaZulu-Natal, the Cape provinces have a distinctive racial group: South Africa's coloreds (widely known as the "Cape coloreds"). Nearly nine-tenths live in the three provinces, primarily in and near Cape Town (population 4.7 million), the country's second-largest city and one of its four main ports. This group originated in the early days of white settlement as a product of relationships between Europeans (Dutch East India Company employees, settlers, and sailors) and non-Europeans, including slaves and Khoi, Malagasy, West African, and various South Asian peoples. The resulting mixed-race people vary in appearance from those with pronounced African features to others who are physically indistinguishable from Europeans. About nine-tenths speak Afrikaans as a customary language, and most of the others speak English.

Culturally, the coloreds are much closer to Europeans than to Africans, and many with light skins have "passed" into the European community. Coloreds have always had a higher social standing and greater political and economic rights than black South Africans, although they have ranked below Europeans in these respects. Most coloreds work as domestic servants, factory workers, farm laborers, and fishers and perform other types of unskilled and semiskilled labor. There is a small but growing professional and white-collar class of coloreds. White owners of grape vineyards are donating portions of their fields to an increasing number of the coloreds who once worked as indentured servants on the white lands, leading to a sharp increase in the region's output and exports of wines. Such progress is a sign of how truly different the new South Africa is from the old.

Racial issues in South Africa have a long history of producing strife, even among the ruling white powers. Britain occupied South Africa—which it called the Cape Colony—in 1806, during the Napoleonic Wars. Friction developed between many Boers and the British authorities, who imposed tighter administrative and legal controls than the Boers were accustomed to. The British abolished slavery throughout their empire in 1833, contributing to the anti-British sentiments of the slaveholding Boers. Boer discontent resulted in the **Great Trek,** a series of northward migrations by which groups of Boers, primarily from the eastern part of the Cape Colony, sought to find new interior grazing lands and establish new political units beyond British reach. After some earlier exploratory expeditions, the main trek by horse and ox-wagon began in 1836. It resulted in the founding of the Orange Free State, the Transvaal, and Natal as Boer republics. Britain annexed Natal in 1845 but recognized Boer sovereignty in the Transvaal and the Orange Free State in 1852 and 1854, respectively.

While Boer disaffection with Britain was building, British settlers were coming to South Africa in increasing numbers. The area's natural resource wealth helped shape subsequent events. The British colonies and Boer republics might have developed peaceably side by side if diamonds had not been discovered in the Orange Free State in 1867 and gold in the Transvaal in 1886. The discoveries set off a rush to these republics of prospectors and other fortune hunters and entrepreneurs from outside the region, particularly from Britain. Ill feeling led to the Anglo-Boer War in 1899. British troops defeated the Boer forces decisively, ending the war in 1902 but leaving behind a reservoir of animosity that exists to this day.

The Union of South Africa was organized in 1910 from four British-controlled units as a self-governing constitutional monarchy under the British crown. Dutch became an official language on par with English (this was later altered to specify Afrikaans rather than Dutch). Pretoria in the Transvaal became the administrative capital as a concession to Boer sentiment (however, the national parliament now meets at Cape Town and the supreme court sits at Bloemfontein in the Free State). In 1961, following a close majority vote by the ruling white population, South Africa became a republic outside the British Commonwealth. After 1910, and particularly after the Nationalist Party took power in 1948, Afrikaners dominated the political life of South Africa.

Racial segregation characterized South African life from 1652 onward, but the Afrikaners after 1948 systematized it with laws supporting the official policy of "separate development of the races," subsequently called "multinational development" and commonly known as **apartheid** (pronounced "uh-*pahr*-tayt"). The new laws imposed racially based restrictions and prohibitions on the entire population, but they weighed most heavily on black Africans and denied them political power. The laws fragmented and displaced families and ethnic groups, rearranging the country's cultural landscape, and culminated in the establishment of makeshift tribal states, or **homelands** (also known as **native reserves, Bantustans,** and **national states**), which were nominally intended to achieve "independence" as black nations. Whites ejected several million Africans not wanted in "white" South Africa from their homes and transferred them to the homelands, which were located on the most undesirable and least productive lands. Thus great numbers of blacks were removed from cities and white farms and dumped in poverty-stricken, crowded, resource-poor areas, often far from where they had ever lived before.

The apartheid system drew furious criticism from many other nations. Most of the world community ostracized South Africa, and many countries, including the United States, imposed economic sanctions against it. Black unrest directed against apartheid and the general underdevelopment of the African majority became so widespread and violent between 1984 and 1986 that the government declared a state of emergency. But even after the state of emergency was lifted in 1990, violence continued. Much fighting took place between two large, tribally based rival factions: the Xhosa-dominated **African National Congress (ANC),** led by Nelson Mandela, and the Zulu-dominated **Inkatha Freedom Party (IFP),** led by Chief Mangosuthu Buthelezi.

With the country on the brink of anarchy, a complete official turnabout on the issue of apartheid resolved South Africa's ongoing racial crisis. It began in 1989 after Frederik W. de Klerk, an insider in the Afrikaner political establishment, became president. Urged on by white South African business leaders, antiapartheid activists, the powerful force of South African and international political opinion, the impact of international economic boycotts against South Africa, and his own stated convictions concerning the injustices and unworkability of apartheid, de Klerk launched a broad program to repeal the apartheid laws and put South Africa on the road to revolutionary governmental changes under a new constitution. Years of negotiations led to an all-race election in 1994, in which voters turned out in massive numbers to elect a new national parliament. The parliament then chose Nelson Mandela as president. (The white government had freed Mandela only four years earlier after a 27-year period of political imprisonment.) This historic election ended white European political control in the last bastion of European colonialism on the African continent. The apartheid laws became null and void, and the homelands were abolished.

Mandela, "South Africa's George Washington," retired in 1999 and was succeeded by a second black president, Thabo Mbeki. The country's political landscape is remarkably stable, but challenges remain. The races are continuing on their long path of reconciliation. Some of the healing process has been formalized in the national **Truth and Reconciliation Commission,** which allows those who were party to racial violence during the apartheid era to confess their misdeeds and, in a sense, be absolved of them.

There is still a huge economic gulf between the haves and have-nots, but there are now more blacks—over 10 million—in the ranks of the middle and upper classes. Growing black wealth has boosted racial integration in neighborhoods. The black underclass continues to grow, however, and an overall unemployment rate of 25 percent fuels an ongoing epidemic of petty and violent crime, especially in the cities. Crime, the world's worst HIV/AIDS epidemic, a low level of education among workers, and relatively high labor costs have dampened foreign investment and economic growth in South Africa. No longer restrained by apartheid-era economic sanctions, South Africa has at the 462
same time become a major investor in other African economies, buying banks, railways, cell phone networks, power plants, and breweries across the region.

There is a real need for land reform in South Africa because the white minority owns about 70 percent of the productive land. A plan is in place to help reduce the imbalance by redistributing 30 percent of the country's farmland from white to black hands by 2014 on a willing-seller, willing-buyer basis. Progress is slow, however, and some landless blacks vow a takeover of white farms as occurred in Zimbabwe (see page 515). A program is also under way to compensate the estimated 3.5 million blacks forcibly displaced by the government to the homelands between 1960 and 1982. More broadly, South Africa has an aggressive affirmative action program to help redress the economic imbalance between the races. The **Employment Equity Act** does not impose quotas but requires employers to move toward "demographic proportionality" based on the national proportions of race and gender. In another effort to secure more income for the black majority, the South African government assumed control of all the country's mineral resources in 2002. Mining companies now can exploit these resources only under state license and, presumably, with more supervision to ensure that profits are not drained excessively abroad or to whites.

Troubles for Portugal's Ex-Colonies and Independent Namibia

Although Portugal was the first colonial power to establish interests in sub-Saharan Africa, stronger European powers came to prevail in the region. However, Portugal managed to retain footholds along both the Atlantic and Indian Ocean coasts. By 1964, Portugal's principal African possessions, Portuguese West Africa (now Angola) and Portuguese East Africa (now Mozambique), were the most populous

AP Photo/Bruce Stanley

• **Figure 9.1.20** Andre Lina, a training manager for the oil company TotalFinaElf, points out a Spain-bound tanker taking a cargo of crude oil from the production platform at Girasso.

European colonies still remaining in the world. Portugal fielded a large military force to combat guerrilla opposition in both areas. In 1974, Portugal's resistance broke, and an army coup in Portugal installed a regime in Lisbon that granted the two colonies independence.

When the Portuguese withdrew from Angola in 1975, a Marxist-Leninist liberation movement supported by the Soviet Union and Cuba took over the central government in the seaport capital and largest city, Luanda (population 4 million). But this movement's authority was not recognized by other liberation movements, one of which, the **National Union for the Total Independence of Angola (UNITA)**—rooted in the country's largest ethnic group, the Ovimbundu—came to control large parts of the country. South Africa and the United States supplied aid to this anti-Marxist movement. Large quantities of Soviet arms, thousands of Cuban troops, and many Soviet and East German military advisers supported the government in Luanda.

88 This Cold War proxy conflict raged for years. Finally, in 1988, representatives of Angola, Cuba, and South Africa signed a U.S.-mediated agreement under which Cuba and South Africa would withdraw militarily from Angola, and South Africa would grant independence to Namibia. Fighting continued, but in a 2002 peace agreement, UNITA agreed to demobilize and disarm, bringing an end to 27 years of civil war. Efforts are ongoing to clear the country
274 of millions of land mines.

Like Nigeria, Angola is a resource-rich country that so far has little to show for its oil wealth (•Figure 9.1.20). Oil revenues have made the country's wealthy elite even wealthier but have done little for most Angolans, 65 percent of whom live in poverty. Multinational oil companies, including Chinese and American companies, are at work developing this resource; Angola is China's leading oil supplier. Crude oil and refined products made up 90 percent of Angola's exports in 2007, and in that year, Angolan oil held the fastest-growing share of all U.S. oil imports as well. Desperate for low-interest loans and more investment, Angola has promised the international community to triple its expenditures on health, education, and other social services. With its "no strings attached" approach, China is already heavily invested there, helping with the construc- 480
tion of both a new railway and a new capital city south of Luanda.

Namibia became independent from South Africa only recently. South African forces overran this former German colony during World War I, and after the war, the League of Nations mandated it to South Africa under the name of South West Africa. Only the war in Angola released South Africa's grip. Backing UNITA, South African military forces had actively intervened in the Angolan struggle in the mid-1970s. Eventually, South African forces withdrew but continued to attack Angolan bases of the **South West Africa People's Organization (SWAPO)**, a guerrilla group fighting for Namibia's independence. In 1988, South Africa agreed to grant independence to Namibia, but only if the would-be Namibian government under SWAPO leadership could convince its main source of support, the government of Angola, to expel Cuban troops from Angola. The South African government granted independence to Namibia in 1990.

SWAPO is now the major political party, but its Marxist leanings have not prevented the country from promoting capitalist investment. White Namibians look anxiously at the confiscation of white farmlands in nearby Zimbabwe;

just 6 percent of Namibia's whites own nearly half of Namibia's arable land. Like South Africa, Namibia has a willing-seller, willing-buyer policy designed to encourage the sale of white farms to blacks, but not much land has changed hands to date. Black farmers warn that Zimbabwe-style land invasions could ensue (see page 515).

The anomalous sliver of territory in Namibia's extreme northeast known as the Caprivi Strip, which extends 280 miles (450 km) eastward to the waters of the Zambezi River, was important to the German colonists who ruled Namibia early in the 20th century. They insisted on this access to the Zambezi in the mistaken belief that it would allow navigation all the way to the Indian Ocean (alas, Victoria Falls stands in the way). Independent Namibia was permitted to retain the strip, despite protests from neighboring Botswana. The Caprivi is home to the **Barotse (Lozi)** ethnic group, whose tribal lands also extend into Botswana, Angola, and Zambia. Barotse efforts both in the Caprivi Strip and in Zambia to establish an autonomous or independent homeland are strongly opposed by both Namibia and Zambia. With its scattered livestock forage, the Kalahari section of Namibia is peopled mainly by Bantu-speaking herders and by remnants of the Khoi and San populations. Groups of San are among the world's last remaining hunters and gatherers, but the majority have given up the nomadic way of life in recent decades and have settled as farmers and wage laborers.

Portugal's other ex-colony, Mozambique, is one of the world's poorest countries, but it has rather bright prospects. It suffered from years of guerrilla warfare, which first involved African insurgencies against the Portuguese and then warfare by African dissidents against the Marxist government that took over from the Portuguese in Maputo, the country's capital (population 1.9 million). The highly destructive civil war between the Maputo government and the South African– and Zimbabwean-backed rebel organization called **RENAMO (Mozambique National Resistance)** dragged on, at the cost of 1 million lives, until 1992, when a peace accord was signed and the United Nations sent a military peacekeeping force to Mozambique. Elections took place peacefully in 1994, and a large share of the 5 million people displaced by the war began returning to their villages. Only periodic disputes between the government and RENAMO have occurred since their reconciliation.

Mozambique is being rebuilt, although recurrent devastating floods bring setbacks. Transportation facilities in the Beira Corridor to Zimbabwe are being upgraded, and a regional economic link with South Africa has been established. Known as the **Maputo Development Corridor,** this toll road joins Johannesburg's industrial and mining center with the deep-water port at Maputo. Low-tax zones were created to attract new businesses to the corridor. The country's estimated 1 million land mines are being cleared, and war refugees are returning to farm this fertile land. Remarkably, Mozambique has become self-sufficient in food production. Major exports are prawns, cashew nuts, cotton, tea, sugar, and timber. Multinational efforts are under way to process titanium for export and to develop the country's pristine beaches into world-class tourist destinations. Impressed with Mozambique's progress, foreign lenders have agreed to cancel almost all of the country's debt. Mozambique has acted aggressively on this windfall, making dramatic investments in health, education, and other social services.

Zimbabwe's Struggles

If Mozambique symbolizes the new, hopeful Africa, Zimbabwe represents a slide backward into the abyss. It is a cruel reversal of fortune for a country that was one of the region's beacons of postcolonial progress.

Prior to the post–World War II movement for African independence, the countries of Zimbabwe, Zambia, and Malawi (then known as Southern Rhodesia, Northern Rhodesia, and Nyasaland, respectively) were British colonies. In 1953, the three became linked politically in the Federation of Rhodesia and Nyasaland, also known as the Central African Federation. It had its own parliament and prime minister but did not achieve full independence. The racial policies and attitudes of Southern Rhodesia's white-controlled government caused serious strains within the federation, which led to its dissolution in 1963. Northern Rhodesia then achieved independence as the Republic of Zambia, and Nyasaland became the independent Republic of Malawi.

The government of Southern Rhodesia, the main area of European settlement in the federation, was unable to come to an agreement with Britain concerning British demand for political participation of the colony's African majority. In 1965, Southern Rhodesia issued a "unilateral declaration of independence," and in 1970, it took a further step in separation from Britain by recasting itself as Rhodesia, a republic that would no longer recognize the symbolic sovereignty of the British crown.

Britain and the Commonwealth responded with a trade embargo. Negotiators for Britain and the Rhodesian government unsuccessfully attempted to arrive at a formula to allow political rights and participation for country's 95 percent black population. While economic sanctions were ineffective, guerrilla fighting against the white government increased in the mid-1970s. Independence as a parliamentary state within the Commonwealth of Nations finally came in 1980, when Rhodesia became Zimbabwe.

From the start, Zimbabwe has faced several ethnic dilemmas. First, the course of relations between the country's two largest ethnic groups—the majority **Shona** (82 percent of the total population) and the **Ndebele** (14 percent)—remains uncertain. Many Ndebele people want autonomy for the Ndebele stronghold in the southwest known as Matabeleland. In addition, although most whites left Zimbabwe when it became independent, about 100,000 remained. The black government allowed whites to participate in Zimbabwe's political life, setting an early example to South Africa of an effective transition to majority-black rule. But the apparent harmony has since shattered, particularly over the issue of land ownership. By the mid-

1990s, millions of rural blacks were crowded onto agriculturally inferior **communal lands,** while whites owned about 70 percent of Zimbabwe's arable land. A poorly handled effort to address this imbalance has threatened to tear the country apart.

The problem of disproportionate white land ownership became explosive in 1997, when Zimbabwe's president, Robert Mugabe (of the Shona ethnic group), used the issue to boost his party's prospects in impending parliamentary elections. He announced that the government would seize half of Zimbabwe's white-owned farms and, without compensating the farmers, simply distribute the land to black peasants living in the overcrowded communal areas. Mugabe argued that the former colonial power, Britain, illegally occupied black lands in the first place and should now pay compensation to whites losing those lands. Insisting that the seizures violated its agreements with the government and angry about Zimbabwe's support of the Kinshasa government in the Democratic Republic of Congo's civil war, the International Monetary Fund (IMF) and most other foreign lenders cut off aid to Zimbabwe in 1999.

An economic crisis, so far unabated, followed; the value of the Zimbabwean dollar crashed, inflation soared (at the blistering rate of 7,600 percent in 2007), gold mines shut down, and tourism all but disappeared. Unemployment reached 80 percent, and a tide of illegal Zimbabwean immigrants flowed into South Africa and other countries. Mugabe won a fraudulent election, postponed another, and cracked down hard on political opponents.

Meanwhile blacks, mostly either Mugabe loyalists or war veterans having little or no experience with farming, moved onto the white-owned farms. Virtually all of the country's 4,500 white farmers had been uprooted by 2007, and the 350,000 black workers on the former white farms had lost their jobs. Zimbabwe's supreme court ruled that the farm seizures were illegal, but Mugabe encouraged the squatters to stay. Few had the requisite farming abilities or materials, and the country's agricultural output plummeted. Some white farmers fled to Zambia, where the government welcomed the new agricultural investors with open arms. Zambia hopes that their expertise with tobacco production in particular will boost exports and revenues. Zimbabwe's loss in tourism numbers has also been Zambia's gain, as visitors can view Victoria Falls from either country, and few now think of going to Zimbabwe.

9.1.7 The Indian Ocean Islands: Former Edens

The islands and island groups off the Indian Ocean coast of Africa are unique in their cultures and natural histories. Madagascar, Comoros, Réunion, Mauritius, and the Seychelles have had African, Asian, Arab, European, and even Polynesian ethnic and cultural influences (•Figures 9.1.21 and 9.1.22). As island ecosystems, they are home to many endemic plant and animal species.

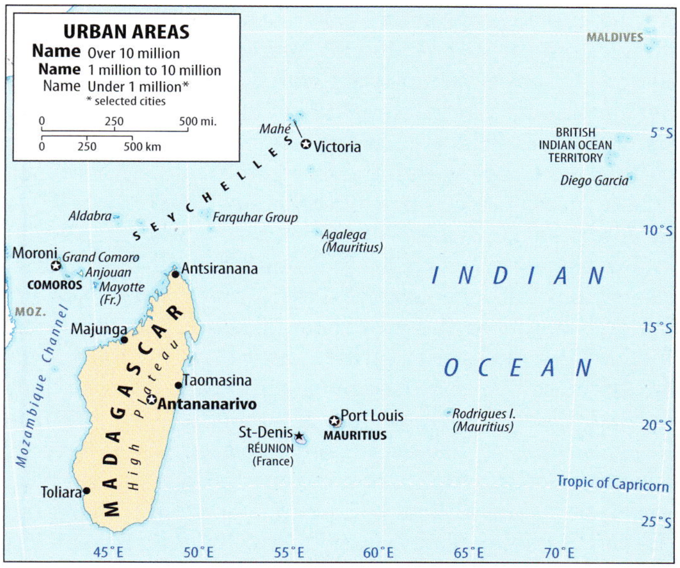

• **Figure 9.1.21** Principal features of the Indian Ocean islands.

• **Figure 9.1.22** The people of Madagascar have a unique culture, with roots in Southeast Asia, the Pacific, and Africa.

Madagascar and the tiny Comoro Islands are former colonies of France, and Réunion remains an overseas possession of France. Mauritius became independent of Britain in 1968 as a constitutional monarchy under the British crown. The Seychelles, which gained independence from Britain in 1976, is a republic within the Commonwealth of Nations.

Madagascar (known in French as La Grande Île) is the fourth-largest island in the world, nearly 1,000 miles (1,600 km) long and about 350 miles (c. 560 km) wide. It lies off the southeast coast of Africa and has geologic formations similar to those of the African mainland. Its distinctive flora and fauna include most of the world's lemur and chameleon species (see Biogeography, page 517). Some of Madagascar's early human inhabitants migrated from the southwestern Pacific region, bringing with them the cultivation of irrigated rice and other culture traits. There was also a later influx of Africans from the mainland. The official language of the country's 18 million people is Malagasy, an Austronesian tongue that has acquired many
292 French, Arabic, and Swahili words.

The east coast of Madagascar rises steeply from the Indian Ocean to heights over 6,000 feet (1,830 m). Because the island lies in the path of trade winds blowing across the Indian Ocean, the east side receives the heaviest rain and has a natural vegetation of tropical rain forest. The remainder of the island has vegetation consisting primarily of savanna grasses with scattered woody growth. An eastern coastal strip is backed by hills and then by escarpments of the central highlands. The most densely populated part of the island is the central High Plateau, where the economy is a combination of growing rice in valleys and raising cattle on higher lands. Antananarivo (formerly Tananarive; population 1.7 million), the capital and largest city, is located in the central highlands. It is connected by rail with Toamasina (formerly Tamatave), the main seaport, located on the east coast.

Madagascar is not as economically developed as other large tropical islands such as Java, Taiwan, and Sri Lanka. Deep economic troubles and political dissent have marked its recent history under a socialist government. Unwise decision making made Madagascar fertile ground in the 1990s for international drug money launderers, loan sharks, and con artists who purloined much of this poor nation's financial assets. Paddy rice is the principal food, and coffee, vanilla (Madagascar is the source of 70 percent of the world supply), cloves, and sugar represent most of the country's limited exports. There is a growing clothing export industry, thanks in large part to the American AGOA initiative (see pages 477–478). Given Madagascar's extraordinary wildlife and scenic beauty, the ecotourism sector could grow considerably, but so far, the government has taken little interest in developing an infrastructure to encourage and sustain tourism.

The Comoro Islands, located about midway between northern Madagascar and northern Mozambique, are of volcanic origin. They have a complex mixture of African, Arab, Malaysian, and European influences. Aside from the island of Mayotte, which has a Christian majority and which, by its own choice, remains a dependency of France, the Comoro group forms the Union of Comoros (usually called just the Comoros), in which Islam is the official religion and both Arabic and French are official languages. The capital is Moroni (population 40,000) on the island of Grand Comoro. France has a defense treaty with the Comoros, which it invoked in 1995 to put down a coup led by a French mercenary. In 1997, two more Comoros islands (Mohéli and Anjouan) declared secession from the national government in Moroni and pleaded to become French dependencies in the hope that they could enjoy the somewhat higher standard of living prevailing on nearby Mayotte. The country implemented a new constitution in 2002, giving each island its own president and self-governance on most issues. The federal presidency is rotated among the three island presidents.

Fishing, subsistence agriculture, and exports of vanilla provide livelihoods to the 700,000 inhabitants of these islands. The coelacanth (pronounced "*see*-luh-kanth"), a primitive fish thought to have become extinct soon after the end of the Mesozoic era (the age of the dinosaurs), has been observed alive in Comoro Islands waters since 1938 (and more recently in Indonesian waters).

Réunion (capital, Saint-Denis; population 130,000), east of Madagascar in the Mascarene island group, is the tropical island home to 800,000 people mainly of French and African origin. The island was uninhabited until the French arrived in the 17th century. Elevations reach 10,000 feet (3,050 m). The volcanic soils are fertile, and tropical crops of great variety are grown. Cane sugar from plantations supplies 75 percent of all exports by value, with rum and molasses (both derived from cane) supplying additional export revenues. Réunion also has a significant international tourist industry.

Sugar is the mainstay of the larger island of nearby Mauritius (capital, Port Louis; population 150,000), also in the Mascarenes. Grown mainly on plantations, sugarcane occupies most of the cultivated fields on this volcanic landscape. Mauritius has a larger proportion of lowland area than Réunion. The island has developed a clothing industry that supplies more export value than sugar. International tourism also produces valuable revenues (about $250 million annually) for its 1.3 million people. The people of Mauritius are mainly descendants of Indians brought in by the British to work the cane fields, along with descendants of 18th-century French planters as well as some blacks, Chinese, and racial mixtures. Mauritius is also home to about 5,000 Ilois, or "islanders," relocated from their homes in the Chagos Islands 1,250 miles (2,000 km) to the northwest. The Chagos make up a British protectorate known as the British Indian Ocean Territory (BIOT). The British administered the islands from Mauritius until 1965, when at the height of the Cold War, it leased the largest island, Diego Garcia, to the U.S. military and required all the Ilois to move out. Since then, U.S. forces have used Diego Garcia

BIOGEOGRAPHY

Madagascar and the Theory of Island Biogeography

The entomologist Edward Wilson and the biologist Paul Ehrlich have forecast that if tropical rain forests continue to be cut down at the present rate, a quarter of all of the plant and animal species on earth will become extinct by 2040.* They base their estimate on a model that correlates habitat area with the number of species living in the habitat. This **theory of island biogeography** emerged from observations of island ecosystems in the West Indies. It states that the number of species found on an individual island correlates with the island's area, with a 10-fold increase in area normally resulting in a doubling of the number of species. If island A, for example, is 10 square miles in area and has 50 species, 100-square-mile island B may be expected to support 100 species.

What makes the theory useful in projecting species losses is the inverse of this equation: a 10-fold reduction in area will result in a halving of the number of species; therefore, 1-square-mile island C can be expected to hold only 25 species. In applying the model, ecologists treat habitat areas as if they were islands. Thus if people cut down 90 percent of the tropical rain forest of the Amazon Basin, for example, the theory of island biogeography suggests that they would eliminate half of the species of that ecosystem. Scientists caution that the theory is only a tool meant to help in making rough estimates; the actual number of species lost with habitat removal may be higher or lower.

As a rough guideline, the theory of island biogeography is useful in projecting and attempting to slow the rate of extinction in
31 the world's biodiversity hot spots, such as Madagascar. More than 90 percent of Madagascar's plant and animal species are endemic, occurring nowhere else on earth. Extinction of species was well under way soon after people arrived on the island; the giant flightless elephant bird (*Aepyornis*) was among the early casualties. But human activities, particularly the clearing of forests to grow rice and provide pasture for zebu cattle, are eliminating habitat areas on the island at a faster rate than ever before. Meanwhile, scientists are anxious to learn whether some of Madagascar's remaining plants might be useful in fighting diseases such as AIDS and cancer. Already Madagascar's rosy periwinkle has yielded compounds effective against Hodgkin's disease and lymphocytic leukemia. Other species could become extinct before their useful properties ever become known.

How urgent is the task to study and attempt to protect plant and animal species in Madagascar? Scientists turn to the theory of island biogeography for an answer. Although people have lived on Madagascar for less than 2,000 years, they have succeeded in removing 90 percent of the island's forest, setting the stage for some of the most ruinous erosion seen anywhere on earth (• Figure 9.1.D). The theory of island biogeography suggests that in the process, they have caused the extinction of roughly half of the island's species. With Madagascar's human population on track to double in 23 years, and with pressure on the island's remaining wild habitats expected to increase accordingly, the task of conservation is extremely urgent.

*Edward O. Wilson, "Threats to Biodiversity." *Scientific American, 261*(3), 1989, pp. 60–66.

Joe Hobbs

Joe Hobbs

• **Figure 9.1.D** The subsistence needs of a growing human population have had ruinous effects on Madagascar's landscapes and wildlife. Before people came some 2,000 years ago, most of the island was forested, as in the Montagne d'Ambre National Park (top). Today, less than 10 percent of the island is forest, and a characteristic landscape feature of its High Plateau is the erosional feature known locally as *lavaka* (bottom).

on many military forays, particularly against Iraq during the Gulf War and the later Iraq War and against Afghanistan in the wake of the 9/11 attacks. Now many of the Ilois want to return to their home islands, a move opposed by the United States.

The unusual core of the Seychelles island group is composed of granites, which typically are found only on continental mainlands. Many foods have to be imported to the islands, which are generally steep and lack cultivable land. The country's largest exports consist of petroleum products, although the islands have no crude petroleum and no refining industry; the products are simply imported and then reexported. Fish products are the only other important exports. The Aldabra atoll, far to the south of the main group of islands, is home to a population of giant tortoises and other endemic animals (•Figure 9.1.23).

A tourist boom began in 1971 when an airport was built in the Seychelles (capital, Victoria; population 20,000). The exceptional physical beauty of the islands and the renowned friendliness of its 100,000 people of French, Mauritian, and African descent have made the Seychelles a prime destination among tourists wealthy enough to afford travel to this remote area. Outsiders are drawn to the Seychelles, as they have long been to much of sub-Saharan Africa, as an imagined Eden. The realities of land and life in the region actually exceed the imagination and are indeed worthy of the world's attention.

• **Figure 9.1.23** A giant Aldabran tortoise on the beach of Aldabra atoll in the Seychelles. As in the Galápagos Islands, unique life forms have evolved through long periods of isolation from other animal populations.

SUMMARY

- Sub-Saharan Africa can be divided into seven subregions: the Sahel, West Africa, West Central Africa, East Africa, the Horn of Africa, southern Africa, and the Indian Ocean islands.
- The Sahel region extends eastward from the Cape Verde Islands to the Atlantic shore nations of Mauritania, Senegal, and the Gambia and inland to Mali, Burkina Faso, Niger, and Chad. The name Sahel in Arabic means "coast" or "shore," referring to the region as a front on the great desert "sea" of the Sahara.
- Between the late 1960s and 1985, the Sahel was subjected to severe droughts, which, in combination with increased human pressure on resources, prompted a process of desertification. Human and natural factors have helped restore the region recently.
- Oil production in Chad is the most significant recent development in the Sahel countries' economies.
- West Africa is the region extending from Guinea-Bissau eastward to Nigeria. Most West African countries maintain strong economic, political, and cultural relationships with the European powers that formerly controlled them.
- The West African countries are heavily dependent on subsistence crop farming, livestock grazing, and a few agricultural and export specialties.
- The most significant mining development in West Africa has been Nigeria's emergence as a producer and major exporter of oil. There are serious conflicts among ethnic, religious, and political groups in Nigeria, some resulting from the maldistribution of income from the country's oil wealth.
- West Africa's largest cities are also national capitals and seaports and have developed primarily in its areas of major commercial agriculture.
- Civil wars have recently taken place in Liberia, Sierra Leone, and Ivory Coast, but all have ended or subsided.
- The subregion of West Central Africa is flanked by Cameroon and the Central African Republic to the north and the Democratic Republic of Congo to the south.
- Subsistence agriculture supports people in most areas of West Central Africa. Exports of cash crops and minerals aid the financing of some countries.
- Originally a colony of Belgium, the Democratic Republic of Congo has faced a number of challenges since its independence in 1960. Due to political mismanagement and war, the resource-rich country is in a state of extreme poverty. It was the epicenter of "Africa's first world war," which concluded with a hopeful peace.

- East Africa includes the countries of Kenya, Tanzania, Uganda, Rwanda, and Burundi.
- Subsistence farming is a mainstay of East Africa. Export crops, such as coffee, tea, and cut flowers, provide important revenues. East African nations generally lack mineral reserves.
- Ethnic rivalries and conflicts have beset all of the East African countries. Warfare between the Hutus and Tutsis of Rwanda and Burundi in the 1990s was the most serious.
- Kenya suffers from corruption that has prevented the long-suspended resumption of significant foreign aid.
- Comprised of a great volcanic plateau and adjacent areas, the Horn of Africa includes the countries of Ethiopia, Eritrea, Somalia, and Djibouti.
- After nearly 30 years of terrorism, guerrilla actions, and civil war, the former Ethiopian province of Eritrea became independent in 1993.
- In addition to the devastation wrought by war with Eritrea, droughts have stricken Ethiopia periodically since the late 1960s.
- The United States and Ethiopia recently intervened militarily to suppress Eritrean-backed Islamist forces in Somalia, a country divided along clan lines.
- Southern Africa includes the countries of Angola, Mozambique, Zimbabwe, Zambia, Malawi, South Africa, Botswana, Swaziland, Lesotho, and Namibia.
- The dominant power in southern Africa is South Africa. Its economic power, based on its industries and mineral wealth, is comparable to that of the rest of Africa combined.
- Racial segregation characterized South African life from 1652 onward, but it was systematized after 1948 under a body of laws known as apartheid.
- In 1994, Nelson Mandela was elected president of South Africa after an all-race election was held in the country. After years of conflict, this historic election ended apartheid and white European political control. Reconciliation between the races is continuing. The ranks of middle- and upper-class blacks are growing, but the gap between the haves and have-nots in South Africa is huge.
- After a long civil war, peace has returned to Angola. This country is becoming an important supplier of oil to the West, but oil exports have benefited few Angolans so far.
- Zimbabwe has experienced an economic crisis of its own making, triggered when President Mugabe redistributed commercial farms owned by the white minority to blacks who were not farmers.
- The Indian Ocean islands of Madagascar, the Comoros, Réunion, Mauritius, and the Seychelles exhibit African, Asian, Arab, European, and even Polynesian ethnic and cultural influences. Madagascar has a wealth of endemic species under tremendous pressure by humans, who have cleared 90 percent of the original vegetation cover in the past 2,000 years.

KEY TERMS + CONCEPTS

African National Congress (ANC) (p. 513)
"Africa's first world war" (p. 497)
Afrikaners (p. 512)
albedo (p. 489)
Amhara (p. 504)
Anglophone Africa (p. 493)
apartheid (p. 513)
Bantustans (p. 513)
Barotse (p. 511)
Blackhawk Down incident (p. 506)
Boers (p. 512)
"Bushmen" (p. 512)
Charney effect (p. 489)
chemocline (p. 496)
coloreds (p. 512)
communal lands (p. 515)
desertification (p. 488)
East African Community (p. 500)
Economic Community of West African States (ECOWAS) (p. 494)
Employment Equity Act (p. 514)
Ethiopian Orthodox Christianity (p. 504)
Falashas (Ethiopian Jews) (p. 504)
Francophone Africa (p. 493)
"Galápagos Islands of Religion" (p. 504)
Great Trek (p. 512)
homelands (p. 513)
"Hottentots" (p. 512)
Hutu (Bahutu) (p. 501)
Ibo (Igbo) (p. 491)
Ijaw (p. 491)
Inkatha Freedom Party (IFP) (p. 513)
International Criminal Court (p. 502)
Ivoirité (p. 494)
Khoi (p. 512)
Liberians United for Reconciliation and Democracy (LURD) (p. 494)
Lomé Agreement (p. 493)
Lord's Resistance Army (LRA) (p. 501)
Lozi (p. 511)
Maputo Development Corridor (p. 511)
Mau Mau Rebellion (p. 501)
Moors (p. 488)
Movement for the Emancipation of the Niger Delta (MEND) (p. 491)
Movement for the Survival of the Ogoni People (MOSOP) (p. 491)
National Union for the Total Independence of Angola (UNITA) (p. 510)
national states (p. 513)
native reserves (p. 513)
Ndebele (p. 515)
Ogoni (p. 491)
Operation Restore Hope (p. 506)
Rastafarians (p. 505)
RENAMO (Mozambique National Resistance) (p. 511)
resilience (p. 488)
"resource curse" (p. 487)
Revolutionary United Front (RUF) (p. 493)
Rwandan Patriotic Front (RPF) (p. 502)
San (p. 512)
Sara (p. 488)
Shona (p. 515)

Southern African Development Community (SADC) (p. 506)
South West Africa People's Organization (SWAPO) (p. 511)
theory of island biogeography (p. 517)
Truth and Reconciliation Commission (p. 513)
Tutsi (Watusi) (p. 501)
upwelling (p. 508)
Wahindi (p. 501)
world patrimony site (p. 487)
Xhosa (p. 512)
Yoruba (p. 491)
Zulu (p. 512)

REVIEW QUESTIONS

1. What are the seven subregions of sub-Saharan Africa? What are some of the most significant countries in each of these subregions?
2. In what ways is the Sahel both a physical and cultural boundary zone in Africa?
3. What country has the greatest demographic, political, and economic clout in West Africa? What unique difficulties does this country face?
4. What are the notable natural resource assets of the Democratic Republic of Congo? Why isn't this country richer than it is? What kinds of relationships has it had with its neighbors?
5. How did Idi Amin bring economic ruin to Uganda? What did Daniel arap Moi do to reverse the bright prospects of Kenya? Has his successor lived up to his promises?
6. What was the original distinction between the Hutus and the Tutsis? What issues brought them into conflict in the 1990s?
7. Why is Ethiopia known as the "Galápagos Islands of Religion"?
8. How has Somalia's instability involved foreign powers?
9. What are the main ethnic groups in South Africa? What were their traditional social and economic roles? How have these changed since the end of apartheid?
10. What were the main causes of the Anglo-Boer War of 1899?
11. What tensions emerged between blacks and whites over the issue of land tenure in Zimbabwe? What effects has government policy had on land reform?
12. Who owns and who uses the island of Diego Garcia? What is its importance in military affairs? What are the principal characteristics and livelihoods of the other Indian Ocean islands?

glossary

Absolute (mathematical) location Determined by the intersection of lines, such as latitude and longitude, providing an exact point expressed in degrees, minutes, and seconds.

Acid rain Precipitation that mixes with airborne industrial pollutants, causing the moisture to become highly acidic and therefore harmful to flora and bodies of water on which it falls. Sulfuric acid is the most common component of this acid precipitation.

Adaptive reuse Finding new uses for older buildings and stores, often accompanied by a shift from decline to steady renewal in an urban neighborhood.

Age of Discovery (Age of Exploration) The three to four centuries of European exploration, colonization, and global resource exploitation and trading led largely by European mercantile powers. It began with Columbus at the end of the 15th century and continued into the 19th century.

Age-structure profile (population pyramid) The graphic representation of a country's population by gender and 5-year age increments.

Agricultural (Neolithic or **New Stone Age** or **Food-Producing Revolution)** The domestication of plants and animals that began about 10,000 years ago.

Albedo The amount of the sun's energy reflected by the ground. Less vegetation cover correlates with high albedo, and vice versa.

Alfisol soils These productive soils found in steppe regions are among the world's more fertile. Also known as *chestnut soils*.

Altitudinal zonation In a highland area, the presence of distinctive climatic and associated biotic and economic zones at successively higher elevations. Ethiopia and Bolivia are examples of these kinds of zones. See also *Tierra caliente, Tierra fría, Tierra helada, Tierra templada*.

Anticyclone An atmospheric high-pressure cell. In the cell, the air is descending and becomes warmer. As it warms, its capacity to hold water vapor increases, and the result is minimal precipitation.

Anti-Semitism Anti-Jewish sentiments and activities.

Apartheid The Republic of South Africa's former official policy of "separate development of the races," designed to ensure the racial integrity and political supremacy of the white minority.

Arable Suitable for cultivation.

Archipelago A chain or group of islands.

Arid China In a climatic division of China approximately along the 20° isohyet, Arid China lies to the west of the line. It characterizes more than 50 percent of China's territory but accommodates less than 10 percent of the population.

Arms race Usually associated with the competition between the United States and the Soviet Union, it refers to rival and potential enemy powers increasing their military arsenals—each in an open-ended effort to stay ahead of the other.

Atmospheric pollution The modification of the blanket of gases that surround the earth largely through the airborne products of industrial production and the human consumption of fossil fuels.

Atoll Low islands made of coral and usually having an irregular ring shape around a lagoon.

Autonomy Self-rule, generally with reference to Palestinians' rights to run their own civil (and some security) affairs in portions of the West Bank and Gaza Strip allocated to them in the 1993–2000 peace agreements.

Balkanization The fragmentation of a political area into many smaller independent units, as in former Yugoslavia in the Balkan Peninsula.

Barrios Densely settled neighborhoods in city space that is characteristically inhabited by migrants of Latino or Hispanic origin.

Barter The exchange of goods or services in the absence of cash.

Basin irrigation The ancient Egyptian system of cultivation using fields saturated by seasonal impoundment of Nile floodwaters.

Bazaar (suq) The central market of the traditional Middle Eastern city, characterized by twisting, close-set lanes and merchant stalls.

Belief systems The set of customs that an individual or culture group has relating to religion, social contracts, and other aspects of cultural organization.

Benelux The name used to collectively refer to Belgium, the Netherlands, and Luxembourg.

Biodiversity hot spots A ranked list of places scientists believe deserve immediate attention for flora and fauna study and conservation.

Biological diversity (biodiversity) The number of plant and animal species and the variety of genetic materials these organisms contain.

Biomass The collective dried weight of organisms in an ecosystem.

Biome A terrestrial ecosystem type categorized by a dominant type of natural vegetation.

Birth rate The annual number of live births per 1,000 people in a population.

Black-earth belt An important area of crop and livestock production spanning parts of Russia, Ukraine, Moldova, and Kazakhstan. The main soils of this belt are mollisols.

Boat people Refugees who flee by sea because of conflict or political circumstances, including refugees from conflict in Indochina, Cuba, and Haiti.

Bourgeoisie In Marxist doctrine, the capitalist class.

Brain drain The exodus of educated or skilled persons from a poor to a rich country or from a poor to a rich region within a country.

Break-of-bulk point A classic geographic term describing a point in transit when bulk goods must be removed from one mode of transport and installed on another. A trainload of grain carried to a port for transshipment on cargo boats or barges is a common example.

Broadleaf deciduous forest Forests typical of middle-latitude areas with humid subtropical and humid continental climates. As cool fall temperatures set in, broadleaf trees shed their leaves and cease to grow, thus reducing water loss. They produce new foliage and grow vigorously during the hot, wet summer.

Buffer state A generally smaller political unit adjacent to a large, or between several large, political units. Such a role often enables the smaller state to maintain its independence because of its mutual use to the larger, proximate nations. Uruguay, between Brazil and Argentina, is an example.

Capital goods Goods used to produce other goods.

Carrying capacity The size of a population of any organism that an ecosystem can support.

Carter Doctrine President Jimmy Carter's declaration, following the Soviet Union's invasion of Afghanistan in 1979, that the United States would use any means necessary to defend its vital interests in the Persian Gulf Region. The "vital interests" were interpreted to mean oil, and "any means necessary" interpreted to mean that the United States would go to war with the Soviet Union if oil supplies were threatened.

Cartogram A special map in which an area's shape and size are defined by explicit characteristics of population, economy, or distribution of any stated product.

Cartography The craft of designing and making maps, the basic language of geography. In recent years, this traditional manual art has been changed profoundly through the use of computers and Geographic Information Systems (GIS) and through major improvements in machine capacity to produce detailed, colored, map products.

Cash (commercial) crops Crops produced generally for export.

Caste The hierarchy in the Hindu religion that determines a person's social rank. It is established by birth and cannot be changed.

Charney Effect Observed by an atmospheric scientist named Charney, this states that the less plant cover there is on the ground, the higher the albedo—solar energy deflected back into the atmosphere—and therefore the lower the humidity and precipitation.

Chemocline The boundary between lower, carbon dioxide-laden waters and higher, gas-free, fresh water in some African lakes. The puncture of this boundary can cause eruptions that are fatal to humans and other life around the lakes.

Chernobyl The site in the Ukraine where, in April 1986, the worst nuclear power plant accident in history occurred. It is thought that approximately 5,000 people died, and a zone with a 20-mile (32-km) radius is still virtually uninhabitable; 116,000 people were moved from the area, and cleanup continues to this day.

Chernozem A Russian term meaning "black earth." It is a grassland soil that is exceptionally thick, productive, and durable.

Chestnut soils Productive soils typical of the Russian steppe and North American Great Plains.

Chokepoint A strategic narrow passageway on land or sea that may be closed off by force or threat of force.

Choropleth maps Maps that are drawn to show the differing distribution of goods or geographic characteristics (including population) across a broad area. Such maps are good for generalizations but often mask significant local variations in the presence of the item being mapped.

Chunnel (Eurotunnel) The 16-mile (26-km) tunnel that links Britain with the European continent. It was completed in 1994 at a cost of more than $15 billion, making it the single most costly project in landscape transformation ever undertaken.

Civilization The complex culture of urban life.

Climate The average weather conditions, including temperature, precipitation, and winds, of an area over an extended period of time.

Climatology The scientific study of patterns and dynamics of climate.

Coal The residue from organic material compressed for a long period under overlying layers of the earth. Coals vary in hardness and heating value. Anthracite coal burns with a hot flame and almost no smoke. Bituminous coal is used in the largest quantities; some can be used to make coke by baking out the volatile elements. Coke is largely carbon and burns with intense heat when used in blast furnaces to smelt iron ore. Peat, which is coal in the earliest stages of formation, can be burned if dried. Concern over atmospheric pollution by the sulfur in coal smoke has recently increased the demand for low-sulfur coals.

Cold War The tense but generally peaceful political and military competition between the United States and the Soviet Union, and their respective allies, from the end of World War II until the collapse of the Soviet Union in 1991.

Collective farm A large-scale farm in the former Soviet Union that usually incorporated several villages. Workers received shares of the income after the obligations of the collective had been met.

Collectivization The process of forming collective farms in Communist countries.

Colonization The European pattern of establishing dependencies abroad to enhance economic development in the home country.

Colored A South African term for persons of mixed racial ancestry.

Command economy A centrally planned economy typical of the Soviet Union and its Communist allies, in which the government rather than free enterprise determines the production, distribution, and sale of economic goods and services.

Common Market An earlier name given to the (current) 25 countries that make up the European Union. In 1957, an initial six countries combined to form the European Economic Community (EEC), and this supranational community has grown to have considerable economic and political importance in Europe. See *European Union*.

Computer cartography Mapmaking using sophisticated software and computer hardware. It is a new, actively growing career field in geography.

Computer-controlled robots Microtechnology has allowed the creation of highly precise, robotic units that can be used in delicate manufacturing. Although expensive to create, they ultimately reduce per unit manufacturing costs because of predictable higher quality output and easier maintenance than human work organizations.

Concentric zone model A generalized model of a city. A city becomes articulated into contrasting zones arranged as concentric rings around its central business district.

Coniferous vegetation Needleleaf evergreen trees; most bear seed cones.

Consumer goods Goods that individuals acquire for short-term use.

Consumer organisms Animals that cannot produce their own food within a food chain.

Consumption overpopulation The concept that a few persons, each using a large quantity of natural resources from ecosystems across the world, add up to too many people for the environment to support.

Containerization The prepackaging of items into larger standardized containers for more efficient transport.

Continental islands Once attached to nearby continents, these are the islands north and northeast of Australia, including New Guinea.

Convectional precipitation The heavy precipitation that occurs when air is heated by intense surface radiation, then rises and cools rapidly.

Convention on the Law of the Sea A 1970s United Nations treaty permitting a sovereign power to have greater access to surrounding marine resources.

Coordinate systems A means of determining exact or absolute location. Latitude and longitude are most often used.

Cottage industry Handwork that is done in the evenings or in slack seasons in the rural world. Products created during such activity can include woven and sewn goods or piecework for local textile or manufacturing industries. This work brings cash into the rural economy, especially to those not completely involved in full-time farming activity.

Council for Mutual Economic Assistance (COMECON) An economic organization, now disbanded, consisting of the Soviet Union, Poland, East Germany, Czechoslovakia, Hungary, Romania, Bulgaria, Cuba, Mongolia, and Vietnam.

Crop calendar The dates by which farmers prepare, plant, and harvest their fields. Farmers who are involved in new agricultural patterns necessitated by greater use of chemical fertilizers and new plant strains are sometimes unable to adjust to the demands of a much more exact crop calendar than has been traditional.

Crop irrigation Bringing water to the land by artificial methods.

Crusades A series of European Christian military campaigns between the 11th and 14th centuries aimed at recapturing Jerusalem and the rest of the Holy Land from the Muslims.

Cultural diffusion One of the most important dynamics in geography, cultural diffusion is the engine of change as crops, languages, culture patterns, and ideas are transferred from one place to other places, often in the course of human migration.

Cultural geography The study of the ways in which humankind has adopted, adapted to, and modified the face of the earth, with particular attention given to cultural patterns and their associated landscapes. It also includes a culture's influence on environmental perception and assessment.

Cultural landscape The landscape modified by human transformation, thereby reflecting the cultural patterns of the resident culture.

Cultural mores The belief systems and customs of a culture group.

Culture The values, beliefs, aspirations, modes of behavior, social institutions, knowledge, and skills that are transmitted and learned within a group of people.

Culture hearth An area where innovations develop, with subsequent diffusion to other areas.

Culture System A system in which Dutch colonizers required farmers in Java to contribute land and labor for the production of export crops under Dutch supervision.

Cyclone A low-pressure cell that composes an extensive segment of the atmosphere into which different air masses are drawn.

Cyclonic (frontal) precipitation The precipitation generated in traveling low-pressure cells that bring different air masses into contact.

Death rate The annual number of deaths per 1,000 people in a population.

Debt-for-nature swap An arrangement in which a certain portion of international debt is forgiven in return for the borrower's pledge to invest that amount in nature conservation.

Deciduous trees Broadleaf trees that lose their leaves and cease to grow during the dry or the cold season and resume their foliage and grow vigorously during the hot, wet season.

Deflation A fall in prices, such as that caused by currency devaluations.

Deforestation The removal of trees by people or their livestock.

Delta A landform resulting from the deposition of great quantities of sediment when a stream empties into a larger body of water.

Demographic shift Major population redistributions as people move from the countryside to the city or from one region to another.

Demographic transition A model describing population change within a country. The country initially has a high birth rate, a high death rate, and a low rate of natural increase, then moves through a middle stage of high birth rate, low death rate, and high rate of natural increase, and ultimately reaches a third stage of low birth rate, low or medium death rate, and low or negative rate of population increase.

Dependency theory A theory arguing that the world's more developed countries continue to prosper by dominating their former colonies, the now independent less developed countries.

Desert An area too dry to support a continuous cover of trees or grass. A desert generally receives less than 10 inches (25 cm) of precipitation per year.

Desert shrub vegetation Scant, bushy plant life occurring in deserts of the middle and low latitudes where there is not enough rain for trees or grasslands. The plants are generally xerophytic.

Desertification The expansion of a desert brought about by changing environmental conditions or unwise human use.

Development A process of improvement in the material conditions of people often linked to the diffusion of knowledge and technology.

Devolution The process by which a sovereign country releases or loses more political and economic control to its constituent elements, such as states and provinces.

Diaspora The scattering of the Jews outside Palestine beginning in the Roman Era.

Digital divide The divide between the handful of countries that are the technology innovators and users and the majority of nations that have little ability to create, purchase, or use new technologies.

Dissection The carving of a landscape into erosional forms by running water.

Distributary A stream that results when a river subdivides into branches in a delta.

Domestication The controlled breeding and cultivation of plants and animals.

Donor democracy Typical of countries in Africa south of the Sahara, this is a situation in which a government makes just enough concessions on voting rights or human rights to win foreign loans and aid, without instituting any serious democratic reform.

Donor fatigue Public or official weariness of extending aid to needy people.

Double cropping Growing two crops a year on the same field.

Drought avoidance Adaptations of desert plants and animals to evade dry conditions by migrating (animals) or being active only when wet conditions occur (plants and animals).

Drought endurance Adaptations of desert plants and animals to tolerate dry conditions through water storage and heat loss mechanisms.

Dry farming Planting and harvesting according to the seasonal rainfall cycle.

Ecologically dominant species A species that competes more successfully than others for nutrition and other essentials of life.

Ecology The study of the interrelationships of organisms to one another and to the environment.

Economic shock therapy Russia's economic transformation in the early 1990s from a command economy to a free-market economy. Overseen by Boris Yeltsin, this transformation was difficult for a country accustomed to government direction in all economic matters, thus the "shock."

Ecosphere (biosphere) The vast ecosystem composed of all of Earth's ecosystems.

Ecosystem A system composed of interactions between living organisms and nonliving components of the environment.

Ecotourism Travel by people who want to both see and save the natural habitats remaining on earth.

Ejido An agricultural unit in Mexico characterized by communally farmed land or common grazing land. It is of particular importance to indigenous villages.

Endemic species A species of plant or animal found exclusively in one area.

Energy crisis The petroleum shortages and price surges sparked by the 1973 oil embargo.

Environmental assessment The process of determining the condition and value of a particular environmental setting. Used both in aspects of landscape change and in evaluating environmental perception.

Environmental determinism The concept that the physical environment has played a sovereign role in the cultural development of a people or landscape. Also known as *environmentalism.*

Environmental perception The concept that how people view the world and its landscapes and resources influences their uses of the earth and therefore the condition of the earth.

Environmental possibilism A concept that rejects environmental determinism, arguing that although environmental conditions do influence human and cultural development, people choose from various possibilities in how to live within a given environment.

Equinox On or about September 23, and again on or about March 20, Earth reaches the equinox position. Its axis does not point toward or away from the sun, so days and nights are of equal length at all latitudes on Earth.

Escarpment A steep edge marking an abrupt transition from a plateau to an area of lower elevation.

Estuary A deepened ("drowned") river mouth into which the sea has flooded.

Ethnic cleansing The relocation or killing of members of one ethnic group by another to achieve some demographic, political, or military objective.

Ethnocentrism Regarding one's own group as superior and as setting proper standards for other groups.

Euro Part of the authority of the European Union has been the institution of a new currency that has become "coin of the realm" since early 2002. Not all EU nations accepted the euro.

European Community The name that was replaced in 1993 by the term European Union. The European Community continues to serve as a governing and administrative body for the EU.

European Economic Community An economic organization designed to secure the benefits of large-scale production by pooling resources and markets. The name has been changed to Economic Union. See also *Common Market.*

European Free Trade Association (EFTA) An organization that maintains free trade among its members but allows each member to set its own tariffs in trading with the outside world. Members are Iceland, Norway, Sweden, Switzerland, Austria, and Finland.

European Union The current organization begun in the 1950s as the Common Market. It now is made up of 25 nations (see Chapter 3).

Evaporation The loss of moisture from the earth's land surfaces and its water bodies to the air through the ongoing influence of solar radiation and transpiration by plants.

Exotic species A nonnative species introduced into a new area.

Extensive land use A livelihood, such as hunting and gathering, that requires the use of large land areas.

External costs (externalities) Consequences of goods and services that are not priced into the initial cost of those goods and services.

Fall line A zone of transition in the eastern United States where rivers flow from the harder rocks of the Piedmont to the softer rocks of the Atlantic and Gulf Coastal Plain. Falls and/or rapids are characteristic features.

Fault A break in a rock mass along which movement has occurred. A break due to rock masses being pulled apart is a tensional or "normal" fault, whereas a break due to rocks being pushed together until one mass rides over the other is a compressional fault. The processes of faulting create these breaks.

Feral animals Domesticated animals that have abandoned their dependence on people to resume life in the wild.

Fertile Crescent The arc-shaped area stretching from southern Iraq through northern Iraq, southern Turkey, Syria, Lebanon, Israel, and western Jordan, where plants and animals were domesticated beginning about 10,000 years ago.

Final status issues Issues deferred to the end of the Oslo Peace Process between Israel and the PLO. Finally dealt with at Camp David in 2000, they included the status of Jerusalem, the fate of Palestinian refugees, Palestinian statehood, and borders between Israel and a new Palestinian state. The negotiations broke down over these final status issues.

Fjord A long, narrow extension of the sea into the land usually edged by steep valley walls that have been deepened by glaciation.

Folding The process creating folds (landforms resulting from an intense bending of rock layers). Upfolds are known as anticlines, and downfolds are called synclines.

Food chain The sequence through which energy, in the form of food, passes through an ecosystem.

Formal region See *Region.*

Fossil waters Virtually nonrenewable freshwater supplies, the product of ancient rainfall stored in deep natural underground aquifers, especially in the Middle East and North Africa.

Four Modernizations The effort of the Chinese after the death of Mao Zedong in 1976 to focus on economic development in agriculture, industry, science and technology, and defense.

Front A contact zone between unlike air masses. A front is named according to the air mass that is advancing (cold front or warm front).

Frontal precipitation See *Cyclonic precipitation.*

Fuelwood crisis Deforestation in the less developed countries caused by subsistence needs.

Functional region See *Region.*

Gaia hypothesis A hypothesis stating that the ecosphere is capable of restoring its equilibrium following any disturbance that is not too drastic.

Gentrification The social and physical process of change in an urban neighborhood by the return of young, often professional, populations to the urban core. These people are often attracted by the substantial nature of the original building stock of the place and the proximity to the city center, which generally continues to have major professional opportunities. While this process brings an urban landscape back into a primary role as a tax base, it does dispossess a considerable number of minority peoples.

Geographic analysis By giving attention to the spatial aspects of a distribution, geographic analysis helps to explain distribution, density, and flow of a given phenomenon.

Geographic Information Systems (GIS) The growing field of computer-assisted geographic analysis and graphic representation of spatial data. It is based on superimposing various data layers that may include everything from soils to hydrology to transportation networks to elevation. Computer software and hardware are steadily improving, enabling GIS to produce ever more detailed and exact output.

Geography The study of the spatial order and associations of things. Also defined as the study of places, the study of relationships between people and environment, and the study of spatial organization.

Geologic hot spot A small area of Earth's mantle where molten magma is relatively close to the crust. Hot spots are associated with island chains and thermal features.

Geomorphology The scientific analysis of the landforms of the earth; sometimes called *physiography*.

Geopolitics The study of geographic factors in political matters, including borders, political unity, and warfare.

Glacial deposition In the process of continental and valley glaciation, the deposition of moraines that become lateral or terminal in the act of glacial retreat. This same process also leads to glacial scouring as moving ice picks up loose rock and reshapes the landscape as the glacier moves forward or retreats.

Glacial scouring See *Glacial deposition*.

Globalization The spread of free trade, free markets, investments, and ideas across borders and the political and cultural adjustments that accompany this diffusion.

GNI PPP Per capita gross national income purchasing power parity (per capita GNI PPP) combines gross national income (GNI) and purchasing power parity (PPP) as a method of comparing the real value of output between different countries' economies.

Graben (rift valley) A landform created when a segment of Earth's crust is displaced downward between parallel tensional faults or when segments of the crust that border it ride upward along parallel compressional faults.

Great Rift Valley The result of tectonic processes, this is a broad, steep-walled trough extending from the Zambezi Valley in southern Africa northward to the Red Sea and the valley of the Jordan River in southwestern Asia.

Great Trek In what is now South Africa, a series of northward migrations in the 1830s by which groups of Boers, primarily from the eastern part of the Cape Colony, sought to find new interior grazing lands and establish new political units beyond British reach.

Green Revolution The introduction and transfer of high-yielding seeds, mechanization, irrigation, and massive application of chemical fertilizers to areas where traditional agriculture has been practiced.

Greenhouse effect The observation that increased concentrations of carbon dioxide and other gases in Earth's atmosphere cause a warmer atmosphere.

Gross domestic product (GDP) The value of goods and services produced in a country in a given year. It does not include net income earned outside the country. The value is normally given in current prices for the stated year.

Gross national product (GNP) The value of goods and services produced internally in a given country during a stated year plus the value resulting from transactions abroad. The value is normally expressed in current prices of the stated year. Such data must be used with caution in regard to developing countries because of the broad variance in patterns of data collection and the fact that many people consume a large share of what they produce.

Growth pole A new city, a resource, or some development in a heretofore undersettled area that begins to attract population growth and economic development.

Guano Seabird excrement that is also found in phosphate deposits.

Guest workers Migrants (mainly young and male) from less developed countries who are employed, sometimes illegally, in more developed countries.

Gulf Stream The strong ocean current originating in the tropical Atlantic Ocean that skirts the eastern shore of the United States, curves eastward, and reaches Europe as a part of the broader current called the North Atlantic Drift.

Gulf War The 1990–1991 confrontation between a large military coalition, spearheaded by the United States, and the Iraqi forces that had invaded Kuwait. The allies successfully drove Iraqi forces from Kuwait. Earlier, the Gulf War had referred to the 1980–1988 war between Iraq and Iran.

Hacienda A Spanish term for large rural estates owned by the aristocracy in Latin America.

Haj The pilgrimage to Mecca, the principal holy city in the Islamic religion. Every Muslim is required to make this pilgrimage at least once in a lifetime, if possible.

Han Chinese The original Chinese peoples who settled in North China on the margins of the Yellow River (Huang He) and who were central to the development of Chinese culture. Han Chinese make up more than 90 percent of China's population.

Hierarchical rule A system by which leadership for a culture group is defined by traditional and sometimes frustrating patterns of age–sex distinction. It is evident in most cultures and slowly modified by an increasing role of democratic elections and the establishment of a broader opportunity base for individual advancement.

High islands Generally the result of volcanic eruptions, these are the higher and more agriculturally productive and densely populated islands of the Pacific World.

High pressure cell (anticyclone) An air mass descending and warming because of increased pressure and weight of the air above. High pressure typically means low relative humidity and minimal precipitation.

Hinterland The region of a country that lies away from the capital and largest cities and is most often rural or even unsettled; it is often seen in the minds of economic planners as an area with development potential.

Historical geography A concern with the historical patterns of human settlement, migration, town building, and the human use of the earth. Often, this subdiscipline best blends geography and history as a perspective on human activity.

Holocaust Nazi Germany's attempted extermination of Jews, Roma (Gypsies), homosexuals, and other minorities during World War II.

Homelands Ten former territorial units in South Africa reserved for native Africans (blacks). They had elected African governments, and some were designated as "independent" republics, although they were not recognized outside South Africa. Formerly known as *Bantustans*. They were abolished in 1994. Also known as *native reserves* and *national states*.

Horizontal migration The movements of pastoral nomads over relatively flat areas to reach areas of pasture and water.

Hot money Short-term and often volatile flows of investment that can cause serious damage to the "emerging market" economies of less developed countries.

Human Development Index (HDI) A United Nations–devised ranked index of countries' development that evaluates quality of life issues (such as gender equality, literacy, and human rights) in addition to economic performance.

Humboldt (or Peru) Current The cold ocean current that flows northward along the west coast of South and North America. It plays a significant role in regional fish resources.

Humid China The eastern portion of China that is relatively well watered and where the great majority of the Chinese population has settled. An arc from Kunming in southwest China to Beijing in North China describes the approximate western margin of this zone.

Humid continental A climate with cold winters, warm to hot summers, and sufficient rainfall for agriculture, with the greater part of the precipitation in the summer half-year.

Humid pampa A level to gently rolling area of grassland centered in Argentina with a humid subtropical climate.

Humid subtropical A climate that generally occupies the southeastern margins of continents with hot summers, mild to cool winters, and ample precipitation for agriculture.

Humus Decomposed organic soil material. Grasslands characteristically provide more humus than forests do.

Hunting and gathering A mode of livelihood, based on collecting wild plants and hunting wild animals, generally practiced by preagricultural peoples. Also known as *foraging*.

Hydraulic control The ability to control water in irrigation systems, rivers, urban settlements, and for the generation of electricity. It has been central to the development of major civilizations and to culture groups throughout human history.

Ice cap A climate and biome type characterized by permanent ice cover on the ground, no vegetation (except where limited melting occurs), and a severely long, cold winter. Summers are short and cool.

Ideology A system of political and/or economic beliefs such as communism, capitalism, autocracy, or democracy.

Igneous (volcanic) rock Rock formed by the cooling and solidification of molten materials. Granite and basalt are common types. Such rocks tend to form uplands in areas where sedimentary rocks have weathered into lowlands. Certain types break down into extremely fertile soils. These rocks are often associated with metal-bearing ores.

Industrial Revolution A period beginning in mid-18th century Britain that saw rapid advances in technology and the use of inanimate power.

Inflation A rise in prices.

Information technology (IT) The Internet, wireless telephones, fiberoptics, and other technologies characteristic of more developed countries. Generally seen as beneficial for a country's economic prospects, IT is also spreading in the less developed countries.

Intensive land use A livelihood requiring use of small land areas, such as farming.

Intertillage Growing two or more crops simultaneously in alternate rows. Also known as *interplanting*.

Irredentism Movement by an ethnic group in one country to revive or reinforce kindred ethnicity in another country—often in an effort to promote succession there.

Irrigation The artificial placement of water to produce crops, generally in arid locations.

Island chain A series of islands formed by ocean crust sliding over a stationary hot spot in the earth's mantle.

Karst Landscape features associated with the dissolution of limestone or dolomite including caves, sinkholes, and towers.

Keystone species A species that affects many other organisms in an ecosystem.

Kyoto Protocol A treaty on climate change signed by 160 countries in Kyoto, Japan, in 1997, and put into force in 2005. It requires MDCs to reduce their greenhouse gas emissions by more than 5 percent below their 1990 levels by the year 2012.

Lacustrine plain The floor of a former lake where glacial meltwater accumulated and sediments washed in and settled. An extremely flat surface is characteristic.

Landscape A portion of the earth's land surface. Geographers are interested in the transformation of natural landscapes into cultural landscapes.

Landscape transformation The human process of making over the earth's surface. From initial human shelters to contemporary massive urban centers, humans have been driven to change their environmental settings.

Large-scale map A map constructed to show considerable detail in a small area.

Laterite A material found in tropical regions with highly leached soils; composed mostly of iron and aluminum oxides which harden when exposed and make cultivation difficult.

Latifundia (sing. latifundio) Large agricultural Latin American estates with strong commercial orientations.

Latitude A measurement that denotes position with respect to the equator and the poles. Latitude is measured in degrees, minutes, and seconds, which are described as parallels. Places near the equator are said to be in low latitudes; places near the poles are in high latitudes. The Tropic of Cancer and the Tropic of Capricorn, at 23.5°N and 23.5°S, respectively, and the Arctic Circle and Antarctic Circle, at 66.5°N and 66.5°S, respectively, form the most commonly recognized boundaries of the low and high latitudes. Places occupying an intermediate position with respect to the poles and the equator are said to be in the middle latitudes. There are no universally accepted definitions for the boundaries of the high, middle, and low latitudes.

Law of Return An Israeli law permitting all Jews living in Israel to have Israeli citizenship.

Law of the Sea A United Nations treaty or convention permitting coastal nations to have greater access to marine resources.

Less developed countries (LDCs) The world's poorer countries.

Lifeboat ethics Ecologist Garrett Hardin's argument that, for ecological reasons, rich countries should not assist poor countries.

Location Central to all geographic analysis is the concept of location. Where something is relates to all manner of influences, from climate to migration routes. It is a crucial component in trying to understand patterns of historic and economic development.

Loess A fine-grained material that has been picked up, transported, and deposited in its present location by wind; it forms an unusually productive soil.

Long March The 1935–1936 flight of approximately 100,000 Communist troops under Mao Zedong from south central China to the mountains of North China. Mao was fleeing attack by General Chiang Kai-shek, and the success of the Long March led to the establishment of Mao's troops as a force with the capacity to overcome the much larger and better armed troops of Chiang.

Longitude A measurement that denotes a position east or west of the prime meridian (Greenwich, England). Longitude is measured in degrees, minutes, and seconds, and meridians of longitude extend from pole to pole and intersect parallels of latitude.

Low islands Made of coral, these are the generally flat, drier, less agriculturally productive, and less densely populated islands of the Pacific World.

Malthusian scenario The model forecasting that human population growth will outpace growth in food and other resources, with a resulting population die-off.

Map projection A way to minimize distortion in one or more properties of a map (direction, distance, shape, or area).

Map scale The actual distance on the earth that is represented by a given linear unit on a map.

Maquiladoras Operations dedicated to the assembly of manufactured goods, generally in Mexico and Central America, from components initially produced in the United States or other places. With the enactment of NAFTA in 1994, there was a massive expansion of maquiladora operations.

Marginalization A process by which poor subsistence farmers are pushed onto fragile, inferior, or marginal lands that cannot support crops for long and that are degraded by cultivation.

Marine west coast A climate occupying the western sides of continents in the higher middle latitudes; it is greatly moderated by the effects of ocean currents that are warm in winter and cool in summer relative to the land.

Marshall Plan The plan designed largely by the United States after the conclusion of World War II by which U.S. aid was focused on the rebuilding of the very Germany that had been its enemy in the war just concluded. Secretary of State George Marshall (who had been Chief of Staff of U.S. Army from 1939–1945) was central to the plan's design and implementation.

Material culture The items that can be seen and associated with cultural development, such as house architecture, musical instruments, and tools. In the study of a cultural landscape, items of material culture add considerable personality to cultural identity.

Medical geography The study of patterns of disease diffusion, environmental impact on public health, and the interplay of geographic factors, migration, and population. With the increasing ease of international movement, medical geography is becoming more important as the potential for disease diffusion increases.

Medina An urban pattern typical of the Middle Eastern city before the 20th century.

Mediterranean (dry-summer subtropical) A climate that occupies an intermediate location between a marine west coast climate on the poleward side and a steppe or desert climate on the equatorward side. During the high-sun period, it is rainless; in the low-sun period, it receives precipitation of cyclonic or orographic origin.

Mediterranean scrub forest The xeroyphytic vegetation typical of hot, dry summer, Mediterranean climate regions. Local names for this vegetation type include *maquis* and *chaparral.*

Meiji Restoration The Japanese revolution of 1868 that restored the legitimate sovereignty of the emperor. *Meiji* means "Enlightened Rule." This event led to a transformation of Japan's society and economy.

Melanesia (black islands) A group of relatively large islands in the Pacific Ocean bordering Australia.

Mental maps Every individual's mind has a series of locations, access routes, physical and cultural characteristics of places, and often a general sense of the good or bad of locales. The term *mental map* is used to define such geographies.

Mercantile colonialism The historical pattern by which Europeans extracted primary products from colonies abroad, particularly in the tropics.

Meridian See *Longitude.*

Mestizo In Latin America, a person of mixed European and Native American ancestry.

Metamorphic rock Rock formed from igneous or sedimentary rock through changes occurring in the rock structure as a result of heat, pressure, or the chemical action of infiltrating water. Marble, formed of pure limestone, is a common example.

Metropolitan area A city together with suburbs, satellites, and adjacent territory with which the city is functionally interlocked.

Microcredit The lending of small sums to poor people to set up or expand small businesses.

Micronesia (tiny islands) Thousands of small and scattered islands in the central and western Pacific Ocean, mainly north of the equator.

Microstate A political entity that is tiny in area and population and is independent or semi-independent.

Middle Eastern ecological trilogy The model of mostly symbiotic relations among villagers, pastoral nomads, and urbanites in the Middle East.

Migration A temporary, periodic, or permanent move to a new location.

Milpero A Latin American farmer who engages in swidden or shifting cultivation in forest lands. The plot is often called the *milpa.*

Minifundia (sing. minifundio) Small Latin American agricultural landholdings, usually with a strong subsistence component.

Mixed forest A transitional area where both needleleaf and broadleaf trees are present and compete with each other.

Mobility The pattern of human movement, typically changing because of continual improvement in means of transportation.

Mollisols Thick, productive, and durable soils, such as the *chernozem,* whose fertility comes from abundant humus in the top layer.

Monoculture The single-species cultivation of food or tree crops, usually very economical and productive but threatening to natural diversity and change.

Monsoon A current of air blowing fairly steadily from a given direction for several weeks or months at a time. Characteristics of a monsoonal climate are a seasonal reversal of wind direction, a strong summer maximum of rainfall, and a long dry season lasting for most or all of the winter months.

Montreal Protocol A 1989 international treaty to ban chlorofluorocarbons.

Moor A rainy, deforested upland, covered with grass or heather and often underlain by water-soaked peat. Also, Muslim inhabitants of Spain.

Moraine The unsorted material deposited by a glacier during its retreat. Terminal moraines are ridges formed by long-continued deposition at the front of a stationary ice sheet.

More developed countries (MDCs) The world's wealthier countries.

Mulatto A person of mixed European and black ancestry.

Multinational companies Companies that operate, at least in part, outside their home countries.

Multiple cropping Farming patterns in which several crops are raised on the same plot of land in the course of a year or even a season. Common examples are winter wheat and summer corn or soybeans in North America or several crops of rice in the same plot in Monsoon Asia.

Nation Commonly, a nation is a term describing the citizens of a state—or that state—but it also refers to an ethnic group existing with or without a separate political entity. See *Nation-state.*

Nation-state A political situation in which high cultural and ethnic homogeneity characterizes the political unit in which people live.

Nationalism The drive to expand the identity and strength of a political unit that serves as home to a population interested in greater cohesion and often expanded political power.

Natural levee A strip of land immediately alongside a stream that has been built up by deposition of sediments during flooding.

Natural replacement rate The highest rate at which a renewable resource can be used without decreasing its potential for renewal. Also known as *sustainable yield.*

Natural resource A product of the natural environment that can be used to benefit people. Resources are human appraisals.

Near Abroad Russia's name for the now-independent former republics, other than Russia, of the USSR.

Neocolonialism The perpetuation of a colonial economic pattern in which developing countries export raw materials to, and buy finished goods from, developed countries. This relationship is more profitable for the developed countries.

Neo-Malthusians Supporters of forecasts that resources will not be able to keep pace with the needs of growing human populations.

Newly industrializing countries (NICs) The more prosperous of the world's less developed countries.

Nonblack soil zone Areas of poor soil in cool, humid portions of the Slavic Coreland, suitable for cultivation of rye.

North Atlantic Drift A warm current, originating in tropical parts of the Atlantic Ocean, that drifts north and east, moderating temperatures of Western Europe. Also known as *Gulf Stream.*

North Atlantic Treaty Organization (NATO) A military alliance formed in 1949 that included the United States, Canada, many European nations, and Turkey.

North European Plain The level to rolling lowlands that extend from the low countries on the west through Germany to Poland. These are areas rich in agricultural development, dense human settlements, and a number of major industrial centers. The plain is broken by a number of rivers flowing from the Alps and other mountain systems in central Europe into the North and Baltic Seas.

Occupied Territories The territories captured by Israel in the 1967 War: the Gaza Strip and West Bank, captured respectively from Egypt and Jordan, and the Golan Heights, captured from Syria.

Oil embargo The 1973 embargo on oil exports imposed by Arab members of the Organization of Petroleum Exporting Countries (OPEC) against the United States and the Netherlands.

Open borders Political boundaries that have minimal political or structural impediment to easy crossing between two countries.

Organization for Economic Cooperation and Development (OECD) See *Organization for European Economic Cooperation (OEEC).*

Organization for European Economic Cooperation (OEEC) An organization created in 1948 to organize and facilitate the European response to the 1947 Marshall Plan. In the early 1960s, the OEEC became the Organization for Economic Cooperation and Development (OECD), and this served as a multinational base for continued planning in economic and social development. The Common Market came to overshadow this organization, and finally, the European Union, in 1993, became the most powerful and influential multinational organization in Europe.

Orographic precipitation The precipitation that results when moving air strikes a topographic barrier, such as a mountain, and is forced upward.

Ozone hole Areas of depletion of ozone in Earth's stratosphere, caused by chlorofluorocarbons.

Palestine Liberation Organization The Palestinian military and civilian organization created in the 1960s to resist Israel and recognized in the 1990s as the sole legitimate organization representing official Palestinian interests internationally.

Parallel A latitude line running parallel to the equator.

Patrilineal descent system A kinship naming system based on descent through the male line.

People overpopulation The concept that many persons, each using a small quantity of natural resources to sustain life, add up to too many people for the environment to support.

Per capita For every person, or per person.

Per Capita Gross Domestic Product Purchasing Power Parity (GDP PPP) In this figure, annual per capita gross domestic product (GDP)—the total output of goods and services a country produces for home use in a year—is divided by the country's population. For comparative purposes, that figure is adjusted for purchasing power parity (PPP), which involves the use of standardized international dollar price weights that are applied to the quantities of final goods and services produced in a given economy. The resulting measure, per capita GDP PPP, provides the best available starting point for comparisons of economic strength and well-being between countries.

Perennial irrigation The year-round irrigated cultivation of crops, as in the Nile Valley following construction of barrages and dams.

Permafrost Permanently frozen subsoil.

Photosynthesis The process by which green plants use the energy of the sun to combine carbon dioxide with water to give off oxygen and produce their own food supply.

Physical geography The subdiscipline of geography most concerned with the climate, landforms, soils, and physiography of the earth's surface.

Piedmont A belt of country at an intermediate elevation along the base of a mountain range.

Pillars of Islam The five fundamental tenets of the faith of Islam.

Pivotal countries Those countries whose collapse would cause international refugee migration, war, pollution, disease epidemics, or other international security problems.

Place identity In geographic analysis of a given locale, the nature of place identity becomes a means of understanding people's response to that particular place. Determination of the environmental and cultural characteristics that are most frequently associated with a certain place helps to establish that "sense of place" for a given location.

Plain A flat to moderately sloping area, generally of slight elevation.

Plantation A large commercial farming enterprise, generally emphasizing one or two crops and utilizing hired labor. Plantations are commonly found in tropical regions.

Plate tectonics The dominant force in the creation of the continents, mountain systems, and ocean deeps. The steady, but slow, movement of these massive plates of the earth's mantle and crust has created the positions of the continents that we have and major patterns of volcanic and seismic activity. Areas where plates are being pulled under other plates are called subduction zones.

Plateau An elevated plain, usually lying 2,000 feet (610 m) above sea level. Some are known as tablelands. A dissected plateau is a hilly or mountainous area resulting from the erosion of an upraised surface.

Pleistocene overkill A hypothesis stating that hunters and gatherers of the Pleistocene Era hunted many species to extinction.

Podzol Soil with a grayish, bleached appearance when plowed, lacking in well-decomposed organic matter, poorly structured, and very low in natural fertility. Podzols are the dominant soils of the taiga.

Polder An area reclaimed from the sea and enclosed within dikes in the Netherlands and other countries. Polder soils tend to be very fertile.

Polynesia (many islands) A Pacific island region that roughly resembles a triangle with its corners at New Zealand, the Hawaiian Islands, and Easter Island.

Population change rate The birth rate minus the death rate in a population.

Population density The average number of people living in a square mile or square kilometer. It is a very handy statistic for generalized comparisons but often fails to provide a detailed sense of the real distribution of people.

Population explosion The surge in Earth's human population that has occurred since the beginning of the Industrial Revolution.

Postindustrial Description of an economy or society characterized by the transformation from manufacturing to information management, financial services, and the service sector.

Prairie An area of tall grass in the middle latitudes composed of rich soils that have been cleared for agriculture. The original lack of trees may have been due to repeated burnings or periodic drought conditions.

Primary consumers (herbivores) Consumers of green plants.

Primate city A city that dominated a country's urban scene, and usually defined as being larger than the country's second and third largest cities combined. Primate cities are generally found in developing countries, although some developed countries, such as France, have them.

Prime meridian See *Latitude*.

Producers (self-feeders) In a food chain, organisms that produce their own food (mostly, green plants).

Projection The distortion caused by the transfer of three-dimensional space on Earth's surface to the two dimensions of a flat map.

Push and pull forces of migration Emigration may be caused by so-called push factors, as when hunger or lack of land "pushes" peasants out of rural areas into cities, or by pull factors, as when an educated villager responds to a job opportunity in the city. People responding to push factors are often referred to as nonselective migrants, whereas those reacting to pull factors are called selective migrants. Both push and pull forces are behind the rural to urban migration that is characteristic of most countries.

Pyramid of biomass A diagrammatic representation of decreasing biomass in a food chain.

Pyramid of energy flow A diagrammatic representation of the loss of high quality, concentrated energy as it passes through the food chain.

Qanat A tunnel used to carry irrigation and drinking water from an underground source by gravity flow. Also known as *foggara* or *karez*.

Rain shadow A condition creating dryness in an area located on the lee side of a topographic barrier such as a mountain range.

Region A "human construct" that is often of considerable size, that has substantial internal unity or homogeneity, and that differs in significant respects from adjoining areas. Regions can be classed as formal (homogeneous), functional, or vernacular. The formal region, also known as a uniform region, has a unitary quality that derives from a homogeneous characteristic. The United States is an example of a formal region. The functional region, also called the nodal region, is a coherent structure of areal units organized into a functioning system by lines of movement or influence that converge on a central node or trunk. A major example would be the trading territory served by a large city and bound together by the flow of people, goods, and information over an organized network of transportation and communication lines. Vernacular or perceptual regions are areas that possess regional identity, such as the "Sun Belt," but share less objective criteria in the use of this regional name. General regions, such as the major world regions in this text, are recognized on the basis of overall distinctiveness.

Remote sensing Through the use of aerial and satellite imagery, geographers and other scientists have been able to get vast amounts of data describing places all over the face of the earth. Remote sensing is the science of acquiring and analyzing data without being in contact with the subject. It is used in the study of patterns of land use, seasonal change, agricultural activity, and even human movement along transport lines. This process relates closely to GIS. See also *Geographic Information Systems (GIS)*.

Renewable resource A resource, such as timber, that is grown or renewed so that a continual supply is available. A finite resource is one that, once consumed, cannot be easily used again. Petroleum products are a good example of such a resource, and because it takes too much time to go through the process of creation, they are not seen as renewable.

Resources Resources become valuable through human appraisal, and their utilization reflects levels of technology, location, and economic ambition. There are few resources that have been universally esteemed (water, land, defensible locales) throughout history, so patterns of resource utilization serve as indexes of other levels of cultural development. See also *Natural resource*.

Rift valley. See *Graben*.

Right of return, Palestinian The Palestinian Arab principle that refugees (and their descendants) displaced from Israel and the Occupied Territories in the 1948 and 1967 wars be allowed to return to the region.

Ring of Fire The long horseshoe-like chain of volcanoes that goes from the southern Andes Mountains in South America up the west coast of North America and arches over into the northeast Asian island chains of Japan, the Philippines, and Southeast Asia. This zone of seismic instability and erratic vulcanism is caused by the tension built up in plate tectonics. See also *Plate tectonics*.

Riparian A state containing or bordering a river.

Russification The effort, particularly under the Soviets, to implant Russian culture in non-Russian regions of the former Soviet Union and its Eastern European neighbors.

Sacred space (sacred place) Any locale that people hold in reverence, such as places of worship, cemeteries, and battlefields.

Salinization The deposition of salts on, and subsequent fertility loss in, soils experiencing a combination of overwatering and high evaporation.

Sand sea A virtual "ocean" of sand characteristic of parts of the Middle East, where people, plants, and animals are all but nonexistent.

Savanna A low-latitude grassland in an area with marked wet and dry seasons.

Scale The size ratio represented by a map; for example, a map with a scale of 1:12,500 is portrayed as 1/12,500 of the actual size.

Scorched earth The wartime practice of destroying one's own assets to prevent them from falling into enemy hands.

Scrub and thorn forest The low, sparse vegetation in tropical areas where rainfall is insufficient to support tropical deciduous forest.

Seamount An underwater volcanic mountain.

Second law of thermodynamics A natural law stating that high-quality, concentrated energy is increasingly degraded as it passes through the food chain.

Secondary consumers (carnivores) Consumers of primary consumers.

Sedentarization Voluntary or coerced settling down, particularly by pastoral nomads in the Middle East.

Sedimentary rock Rock formed from sediments deposited by running water, wind, or wave action either in bodies of water or on land and which have been consolidated into rock. The main classes are sandstone, shale (formed principally of clay), and limestone, including the pure limestone called chalk.

Segmentary kinship system The organization of kinship groups in concentric units of membership, as of lineage, clan, and tribe among Middle Eastern pastoral nomads.

Service sector The labor sector made up of employees in retail trade and personal services; it is the sector most likely to increase in employment significance in postindustrial society.

Settler colonization The historical pattern by which Europeans sought to create new or "neo-Europes" abroad.

Shatter belt A large, strategically located region composed of conflicting states caught between the conflicting interests of great powers.

Shi'a Islam The branch of Islam regarding male descendants of Ali, the cousin and son-in-law of the Prophet Muhammad, as the only rightful successors to the Prophet Muhammad.

Shifting cultivation A cycle of land use between crop and fallow years that is needed to work around the infertility of tropical soils. After clearing tropical forest, a farmer may get only a few years of crops before there is no fertility in the soil and must move on to clear more land. In the meantime, forest reclaims the previously farmed plots. Where new lands are not available, fallow periods on old fields are reduced or eliminated, resulting in soil deterioration. Also known as *swidden* or *slash-and-burn cultivation*.

Site and situation Site is the specific geographic location of a given place, while situation is the accessibility of that site and the nature of the economic and population characteristics of that locale.

Slavic Coreland (Fertile Triangle or **Agricultural Triangle)** The large area of the western former Soviet region containing most of the region's cities, industries, and cultivated lands.

Small-scale map A map constructed to give a highly generalized view of a large area.

Soil The earth mantle made of decomposed rock and decayed organic material.

Sovereign state A political unit that has achieved political independence and maintains itself as a separate unit.

Spatial Geographers recognize spatial distributions and patterns in Earth's physical and human characteristics. The term *spatial* comes from the noun "space," and it relates to the distribution of various phenomena on Earth's surface. Geographers portray spatial data cartographically—that is, with maps.

Special Economic Zones (SEZs) In China, the urban areas designated after the 1970s to attract investment and boost production through tax breaks and other incentives.

Spodosols Acidic soils that have a grayish, bleached appearance when plowed, lack well-decomposed organic matter, and are low in natural fertility. Also known as *podzols*.

St. Lawrence Seaway Project This seaway has a total length of 2,342 miles (3,796 km) between the Atlantic Ocean and its western terminus in Lake Superior. It allows major oceanic vessels to reach the Great Lakes, thus enabling ports as far inland as the west side of Lake Superior to serve as international ports. The seaway was completed in 1959, although Canada and the United States began to anticipate such a waterway project in the last years of the 19th century.

State A political unit over which an established government maintains sovereign control.

State farm (sovkhoz) A type of collectivized state-owned agricultural unit in the former Soviet Union; workers receive cash wages in the same manner as industrial workers.

Steppe (temperate grassland) A biome composed mainly of short grasses. It occurs in areas of steppe climate, which is a transitional zone between very arid deserts and humid areas.

Subarctic A high-latitude climate characterized by short, mild summers and long, severe winters.

Subduction zone See *Plate tectonics*.

Subsidence The settling of land by the compression of lower layers of rock and soil, usually accelerated by the removal of water from below the surface. Of particular importance in urban centers, with Mexico City having one of the most persistent problems.

Summer solstice On or about June 22, the first day of summer in the Northern Hemisphere, the northern tip of Earth's axis is inclined toward the sun at an angle of 23.5° from a line perpendicular to the plane of the ecliptic. This is the summer solstice in the Northern Hemisphere. On or about December 22, the first day of winter in the Northern Hemisphere, the southern tip of Earth's axis is inclined toward the sun at an angle of 23.5° from a line perpendicular to the plane of the ecliptic. This is the summer solstice in the Southern Hemisphere.

Sun Belt A U.S. area of indefinite extent, encompassing most of the South, plus the West at least as far north as Denver, Salt Lake City, and San Francisco. This region has shown faster growth in population and jobs than the nation as a whole during the past several decades.

Sunni Islam The branch of Islam regarding successorship to the Prophet Muhammad as a matter of consensus among religious elders.

Sustainable development (ecodevelopment) Concepts and efforts to improve the quality of human life while living within the carrying capacity of supporting ecosystems.

Sustainable yield (natural replacement rate) The highest rate at which a renewable resource can be used without decreasing its potential for renewal.

Swidden See *Shifting cultivation*.

Symbolization The representation of distinct aspects of information shown on a map, such as stars for capital cities.

Taiga A northern coniferous (needleleaf) forest.

Technocentrists (cornucopians) Supporters of forecasts that resources will keep pace with or exceed the needs of growing human populations.

Tectonic processes Processes that derive their energy from within the earth's crust and serve to create landforms by elevating, disrupting, and roughening the earth's surface.

Tertiary consumers Consumers of secondary consumers.

Theory of island biogeography A theoretical calculation of the relationship between habitat loss and natural species loss, in which a 90 percent loss of natural forest cover results in a loss of half of the resident species.

Tierra caliente A Latin American climatic zone reaching from sea level upward to approximately 3,000 feet (914 m). Crops such as rice, sugarcane, and cacao grow in this hot, wet environment.

Tierra fría A Latin American climatic zone found between 6,000 feet (1,800 m) and 12,000 feet (3,600 m) above sea level. Frost occurs and the upper limit of agriculture and tree growth is reached.

Tierra helada A Latin American climatic zone above 10,000 feet (3,048 m) that has little vegetation and frequent snow cover.

Tierra templada A Latin American climatic zone extending from approximately 3,000 to 6,000 feet (914–1,829 m) above sea level. It is a prominent zone of European-induced settlement and commercial agriculture such as coffee growing.

Tokugawa Shogunate The period 1600 to 1868 in Japan, characterized by a feudal hierarchy and Japan's isolation from the outside world.

Tradable permits A proposed mechanism for reducing total global greenhouse gases in which each country would be assigned the right to emit a certain quantity of carbon dioxide, according to its population size, setting the total at an acceptable global standard. The assigned quotas could be traded between underproducers and overproducers of carbon dioxide.

Trade winds Streams of air that originate in semipermanent high-pressure cells on the margins of the tropics and are attracted equatorward by a semipermanent low-pressure cell.

Transhumance The movement of pastoral nomads and their herd animals between low-elevation winter pastures and high-elevation summer pastures. Also known as *vertical migration*.

Triangular trade The 16th- to 19th-century trading links between West Africa, Europe, and the Americas, involving guns, alcohol, and manufactured goods from Europe to West Africa exchanged for slaves. Slaves brought to the Americans were exchanged for the gold, silver, tobacco, sugar, and rum carried back to Europe.

Trophic (feeding) levels Stages in the food chain.

Tropical deciduous forest The vegetation typical of some tropical areas with a dry season. Here, broadleaf trees lose their leaves and are dormant during the dry season and then add foliage and resume their growth during the wet season.

Tropical rain forest A low-latitude broadleaf evergreen forest found where heat and moisture are continuously, or almost continuously, available.

Tropical savanna climate A relatively moist low-latitude climate that has a pronounced dry season.

Tundra A region with a long, cold winter, when moisture is unobtainable because it is frozen, and a very short, cool summer. Vegetation includes mosses, lichens, shrubs, dwarf trees, and some grass.

Underground economy The "black market" typical of the former Soviet region and many LDCs.

Undifferentiated highland A climate that varies with latitude, altitude, and exposure to the sun and moisture-bearing winds.

Vernacular region See *Region*.

Vertical migration Movements of pastoral nomads between low winter and high summer pastures.

Virgin and idle lands (new lands) Steppe areas of Kazakhstan and Siberia brought into grain production in the 1950s.

Warsaw Pact A military alliance, now dissolved, consisting of the Soviet Union and the European countries of Poland, East Germany, Czechoslovakia, Hungary, Romania, and Bulgaria.

Waterhead In attempting to control water for the generation of electricity or a gravity flow irrigation project, it is essential to pond up water in relative quantity so that there is an adequate force—waterhead—to achieve the desired goal. Dam structures are the most common engineering responses to this requirement.

Weather The atmospheric conditions prevailing at one time and place.

Weathering The natural process that disintegrates rocks by mechanical or physical means, making soil formation possible.

West The shorthand term for the countries of the world that have been most influenced by Western civilization and that link most closely with the United States and Western Europe.

Westerly winds An airstream located in the middle latitudes that blows from west to east. Also known as *westerlies*.

Western civilization The sum of values, practices, and achievements that had roots in ancient Mesopotamia, as well as Palestine, Greece, and Rome, and that subsequently flowered in western and southern Europe.

Westernization The process whereby non-Western societies ac-quire Western traits, which are adopted with varying degrees of completeness.

Winter solstice On or about December 22, the first day of winter in the Northern Hemisphere, the southern tip of Earth's axis is inclined toward the sun at an angle of 23.5° from a line perpendicular to the plane of the ecliptic. This is the winter solstice in the Northern Hemisphere. On or about June 22, the first day of summer in the Northern Hemisphere, the northern tip of Earth's axis is inclined toward the sun at an angle of 23.5° from a line perpendicular to the plane of the ecliptic. This is the winter solstice in the Southern Hemisphere.

Xerophytic Literally, "dry plant," referring to desert shrubs having small leaves, thick bark, large root systems, and other adaptations to absorb and retain moisture.

Zero population growth (ZPG) The condition of equal birth rates and death rates in a population.

Zionist movement The political effort, beginning in the late 19th century, to establish a Jewish homeland in Palestine.

Zone of transition An area where the characteristics of one region change gradually to those of another.

pronounciation guide

The authors acknowledge with thanks the assistance of David Allen, Eastern Michigan University, in preparing an earlier pronunciation guide that the authors expanded and modified for this book. In general, names are those appearing in the text; pronunciations of other names on maps can be found in *Goode's World Atlas* (Rand McNally) or in standard dictionaries and encyclopedias. Vowel sounds are coded as: ă (uh), ā (ay), ĕ (eh), ē (ee), ī (eye), ĭ (ih), o͞o (oo).

Afghanistan (af-*gann'*-ih-stann')
- Helmand R. (*hell'*-mund)
- Hindu Kush Mts. (*hinn'*-doo *koosh'*)
- Kabul (*kah'*-b'l)
- Khyber Pass (*keye'*-ber)

Albania (al-*bay'*-nee-a)
- Tirana (tih-*rahn'*-a)

Algeria (al-*jeer'*-ee-a)
- Ahaggar Mts. (uh-*hahg'*-er)
- Sahara Desert (suh-*hahr'*-a)
- Tanezrouft (*tahn'*-ez-rooft')

American Samoa
- Pago Pago (*pahng'*-oh *pahng'*-oh)
- Tutuila I. (too-too-*ee'*-lah)

Andorra (an-*dohr'*-a)

Angola (ang-*goh'*-luh)
- Benguela R.R. (ben-*gehl'*-a)
- Cabinda (exclave) (kah-*binn'*-duh)
- Luanda (loo-*an'*-duh)
- Lobito (luh-*bee'*-toh)

Anguilla (an-*gwill'*-a)

Antigua and Barbuda (an-*teeg'*-wah, bahr-*bood'*-a)

Argentina (ahr'-jen-*teen'*-a)
- Buenos Aires (*bwane'*-uhs *eye'*-ress)
- Córdoba (*kawrd'*-oh-bah)
- Patagonia (region) (patt'-hu-*goh'*-nee-a)
- Rio de la Plata (*ree'*-oh duh lah *plah'*-tah)
- Rosario (roh-*zahr'*-ee-oh)
- Tierra del Fuego (region) (tee-ehr'-uh dell *fway'*-goh)
- Tucumán (too'-koo-*mahn'*)

Armenia (ahr'-*meen'*-y-uh)
- Yerevan (yair'-uh-*vahn'*)

Aruba (ah-*roob'*-a)

Australia (aw-*strayl'*-yuh)
- Adelaide (*add'*-el-ade)
- Brisbane (*briz'*-bun)
- Canberra (*can'*-bur-a)
- Melbourne (*mell'*-bern)
- Sydney (*sid'*-nih)
- Tasmania (state) (tazz-*may'*-nih-uh)

Austria (*aw'*-stree-a)
- Graz (*grahts'*)
- Linz (*lints'*)
- Salzburg (*sawlz'*-burg)
- Vienna (vee-*enn'*-a)

Azerbaijan (ah'-zer-*beye'*-jahn)
- Baku (bah-*koo'*)

Bahamas (buh-*hahm'*-uz)
- Nassau (*nass'*-aw)

Bahrain (bah-*rayn'*)

Bangladesh (bahng'-lah-*desh'*)
- Dacca (Dhaka) (*dahk'*-a)

Barbados (bar-*bay'*-dohss)

Belarus (*bell'*-uh-roos)
- Minsk (*mensk'*)

Belgium
- Antwerp (*ant'*-werp)
- Ardennes Upland (ahr-*den'*)
- Brussels (*brus'*-elz)
- Charleroi (*shahr*-luh-*rwah'*)
- Ghent (*gent'*)
- Liège (lee-*ezh'*)
- Meuse R. (*merz'*)
- Sambre R. (*sohm*-bruh)
- Scheldt R. (*skelt'*)

Belize (buh-*leez'*)

Benin (beh-*neen'*)
- Cotonou (koh-toh-*noo'*)

Bhutan (boo-*tahn'*)
- Thimphu (*thim'*-poo')

Bolivia (boh-*liv'*-ee-uh)
- Altiplano (region) (al-tih-*plahn'*-oh)
- La Paz (lah *pahz'*)
- Santa Cruz (sahn-ta *krooz'*)
- Sucre (*soo'*-kray)
- Lake Titicaca (tee'-tee-*kah'*-kah)

Bosnia and Herzegovina (*boz'*-nee-uh herts-uh-goh-*veen'*-uh)
- Sarajevo (sahr-uh-*yay'*-voh)

Botswana (baht-*swahn'*-a)
- Gaborone (gahb'-uh-*roh'*-nee)

Brazil (bra-*zill*)
- Amazonia (region) (am-uh-*zohn'*-ih-a)
- Belo Horizonte (*bay'*-loh haw-ruh-*zonn'*-tee)
- Brasília (bruh-*zeel'*-yuh)
- Caatinga (region) (kay-*teeng'*uh)
- Manaus (mah-*noose'*)
- Minas Gerais (*meen'*-us zhuh-*rice*)
- Natal (nah-*tal'*)
- Paraíba Valley (pah-rah-*ee'*-bah)
- Paraná R. (pah-rah-*nah'*)
- Pôrto Alegre (por-too' al-*egg'*-ruh)
- Recife (ruh-*see'*-fee)
- Rio de Janeiro (*ree'*-oh dih juh-*nair'*-oh)
- Salvador (sal'-vah-*dor'*)
- São Paulo (sau *pau'*-loh)

Brunei (*broo'*-neye)

Bulgaria
- Sofia (*soh'*-fee-a)

Burkina Faso (bur-*kee'*-na *fah'*-soh)
- Ouagadougou (wah'-guh-*doo'*-goo)

Burma (*bur'*-ma)

Burundi (buh-*roon'*-dee) ("oo" as in "wood")
- Bujumbura (boo-jem-*boor'*-a) (second "oo" as in "wood")

Cambodia
- Phnom Penh (*nom' pen'*)

Cameroon (kam'-uh-*roon'*)
- Douala (doo-*ahl'*-a)
- Yaoundé (yah-onn-*day'*)

Canada
- Montreal (mon'-tree-*awl'*)
- Nova Scotia (*noh'*-vuh *scoh'*-shuh)
- Ottawa (*aht'*-ah-wah)
- Quebec (kwih-*beck'*; Fr., kay-*beck'*)
- Saguenay R. (*sag'*-uh-nay)

Cape Verde (*verd'*)

Central African Republic
- Bangui (*bahng'*-ee)

Chile (*chill'*-ee)
- Atacama Desert (ah-tah-*kay'*-mah)
- Chuquicamata (choo-kee-*kah'*-mah-tah)
- Concepción (con-sep'-*syohn'*)
- Santiago (sahn'-tih-*ah'*-goh)
- Valparaíso (vahl'-pah-rah-*ee'*-soh)

China
- Altai Mts. (al-*teye*)
- Amur R. (ah-*moor'*) ("oo" as in "wood")
- Anshan (ahn-*shahn'*)
- Baotou (*bau'*-toh')
- Beijing (bay-*zhing'*)
- Canton (kan'-*tahn'*)
- Chang Jiang (chung jee-*ahng'*)
- Chengdu (chung-*doo'*)
- Chongqing (chong-*ching'*)
- Dzungarian Basin (zun-*gair'*-ree-uhn)
- Guangdong (gwahng-*dung'*)
- Guangzhou (gwahng-*joh'*)
- Hainan I. (*heye'*-*nahn'*)
- Harbin (*hahr'*-ben)
- Huang He (*hwahng' huh'*)
- Lanzhou (lahn-*joh'*)
- Lhasa (*lah'*-suh)
- Li R. (*lee'*)
- Liao (lih-ow')

Lüda (*loo'*-dah)
Nanjing (nahn-*zhing'*)
Peking (pea-*king'*)
Shandong (shahn-*dung'*)
Shanghai (shang-*heye'*)
Shantou (shahn-*toh'*)
Shenyang (shun-*yahng'*)
Shenzhen (shun-*zún*)
Sichuan (zehch-*wahn'*)
Taklamakan Desert (*tah'*-kla-mah-*kahn'*)
Tarim Basin (*tah'-reem'*)
Tiananmen (*tyahn'*-an-men)
Tianjin (tyahn-*jeen'*)
Tien Shan (*tih'*-en *shahn'*)
Tientsin (*tinn'-tsinn'*)
Tsinling Shan (*chinn'*-leeng *shahn'*)
Ürümqi (oo-*room'*-chee)
Wuhan (*woo'-hahn'*)
Xian (shee-*ahn'*)
Xinjiang (shin-jee-*ahng'*)
Xizang (shee-*dzahng'*)
Yangtze R. (*yang'*-see)
Yunnan (yoo-*nahn'*) ("oo" as in "wood")
Colombia (koh-*lomm'*-bih-uh)
Barranquilla (bah-rahn-*keel'*-yah)
Bogotá (boh'-ga-*tah'*)
Cali (*kah'*-lee)
Cartagena (kahrt'-a-*hay'*-na)
Cauca R. (*kow'*-kah)
Magdalena R. (mahg'-thah-*lay'*-nah)
Medellín (med-deh-*yeen'*)
Comoros (*kahm'*-o-rohs')
Congo
Brazzaville (*brazz'*-uh-vill'; Fr.: brah-zah-*veel'*)
Pointe Noire (pwahnt *nwahr'*)
Costa Rica (*kohs'*-tuh *ree'*-kuh)
San José (sahn hoh-*zay'*)
Croatia (kroh-*ay'*-shuh)
Zagreb (*zah'*-grebb)
Cyprus (*seye'*-pruhs)
Nicosia (nik'-o-*see'*-uh)
Czech Republic (check)
Bratislava (bratt'-ih-*slahv'*-a, braht-)
Brno (*ber'*-noh)
Moravia (muh-*rayve'*-ih-uh)
Ostrava (*aw'*-strah-vah)
Prague (*prahg'*)

Denmark
Copenhagen (*koh'*-pen-hahg'-gen)
Djibouti (jih-*boot'*-ee)
Dominica (dahm'-i-*nee'*-ka)
Dominican Republic (duh-*min'*-i-kan)
Santo Domingo (sant'-oh d-*min'*-goh)

Ecuador (*ec'*-wa-dawr)
Guayaquil (gweye'-uh-*keel'*)
Quito (*kee'*-toh)
Egypt
Aswan (*ahs'*-wahn)
Cairo (*keye'*-roh)
Port Said (sah-*eed'*)
El Salvador (el *sal'*-va-dor)
Equatorial Guinea (*gin'*-ee)
Estonia (ess-*toh'*-nee-a)
Tallin (*tahl'*-in)
Ethiopia (ee'-thee-*oh'*-pea-a)
Addis Ababa (*ad'*-iss *ab'*-a-ba)
Asmara (az-*mahr'*-a)

Fiji (*fee'*-jee)
Suva (*soo'*-va)
Finland
Helsinki (*hel'*-sing-kee)
Karelian Isthmus (kuh-*ree'*-lih-uhn)
France
Alsace (region) (al-*sass'*)
Ardennes (region) (ahr-*den'*)
Bordeaux (bawr-*doh'*)
Boulogne (boo-*lohn'*-yuh)
Calais (kah-*lay'*)
Carcassonne (kahr-kuh-*suhn'*)
Champagne (sham-*pahn*-yuh)
Garonne R. (guh-*rahn'*)
Jura Mts. (*joo'*-ruh)
Le Havre (luh *ah'*-vruh)
Lille (*leel'*)
Loire R. (*lwahr'*)
Lorraine (loh-*rane'*)
Lyons (Fr., Lyon), both (*lee*-aw)
Marseilles (mahr-*say'*)
Massif Central (upland) (mah-*seef'* saw-*trahl*)
Meuse R. (*merz'*)
Moselle R. (moh-*zell'*)
Nancy (naw-*see'*)
Nantes (*nawnt'*)
Nice (*neece'*)
Oise R. (*wahz'*)
Pyrenees (peer'-uh-*neez'*)
Riviera (riv'-ih-*ehr'*-a)
Rouen (roo-*aw'*)
Saône R. (*sohn'*)
Seine R. (*senn'*)
Strasbourg (strahz-*boor'*)
Toulon (too-*law'*)
Toulouse (too-*looz'*)
Vosges Mts. (*vohzh'*)
French Guiana (gee'-*ahn*-a)
Cayenne (*keye'*-enn)

Gabon (gah-*baw'*)
Gambia (*gamm'*-bih-uh)
Banjul (*bahn*-jool)
Georgia
Abkhazia (ahb-kahz'-zee-uh)
Adjaria (ah-*jahr'*-ree-uh)
Caucasus Mts. (*kaw'*-kuh-suhs)
South Ossetia (oh-see'-shee-uh)
Tbilisi (tuh-*blee'*-see)
Germany
Aachen (*ah'*-ken)
Berlin (ber-*lin'*)
Bremen (*bray'*-men)
Chemnitz (*kemm'*-nits)
Cologne (kuh-*lohn'*)
Dresden (*drez'*-den)
Düsseldorf (*doo'*-sel-dorf)
Elbe R. (*ell'*-buh)
Erzgebirge (*ehrts'*-guh-beer-guh)
Frankfurt (*frahngk'*-foort')
Karlsruhe (*kahrls'*-roo-uh)
Leipzig (*leyep'*-sig)
Main R. (*mane'*; Ger., *mine'*)
Mannheim-Ludwigshafen (*man'*-hime *loot'*-vikhs-hah-fen)
Munich (*myoo'*-nik)
Neisse R. (*nice'*-uh)
Oder R. (*oh'*-der)
Ruhr (region) (*roor'*)
Stuttgart (*stuht'*-gahrt)
Weser R. (*vay'*-zer)
Wiesbaden-Mainz (*vees'*-bahd-en *meyents*)
Wuppertal (*voop'*-er-tahl) ("oo" as in "wood")
Ghana (*gahn'*-a)
Accra (a-*krah'*)
Kumasi (koo-*mahs'*-ee)
Tema (*tay'*-muh)
Volta R. (*vohl'*-tuh)
Greece
Aegean Sea (uh-*jee'*-un)
Piraeus (peye-*ree'*-us)
Thessaloniki (thess'-uh-loh-*nee'*-kih)
Grenada (gruh-*nayd'*-a)
Guadeloupe (gwah'-duh-*loop'*)
Guatemala (gwah'-tuh-*mah'*-luh)
Guinea (*gin'*-ee)
Conakry (*kahn'*-a-kree)
Guinea-Bissau (biw-*ow'*)
Guyana (geye-*an'*-a; geye-*ahn*-a')

Haiti (*hayt'*-ee)
Port-au-Prince (pohrt'-oh-*prants'*)
Honduras (hahn-*dur'*-as)
Tegucigalpa (tuh-goo'-sih-*gahl'*-pah)
Hungary (*hun'*-guh-rih)
Budapest (boo-duh-*pesht'*)

Iceland
Akureyri (ah'-koor-*ray'*-ree)
Reykjavik (*ray'*-kyuh-vik')
India
Agra (*ah'*-gruh)
Ahmedabad (ah'-mud-uh-*bahd'*)
Assam (state) (uh-*sahm'*)
Bengal (region) (benn-*gahl'*)
Bihar (state) (bih-*hahr'*)
Brahmaputra R. (brah'-muh-*poo'*-truh)
Deccan (region) (*deck'*-un)
Delhi (*deh'*-lih)
Ganges R. (*gan'*-jeez)
Eastern, Western Ghats (mountains) (*gahts'*)
Himalaya Mts. (himm-*ah'*-luh-yuh, -ah-*lay*-yuh)
Jaipur (*jeye'*-poor)
Jamshedpur (juhm-*shayd'*-poor)
Kashmir (region) (*cash'*-meer)
Kerala (state) (*kay'*-ruh-luh)
Madras (muh-*drass'*, -*drahs'*)
Punjab (region) (*pun'*-jahb, -jab; pun-*jahb'*, -*jab'*)
Rajasthan (state) (*rah'*-juh-stahn)
Uttar Pradesh (state) (*oo'*-tahr pruh-*desh'*)

Indonesia
- Bali (*bah'*-lih)
- Bandung (*bahn'*-doong)
- Celebes (Sulawesi) I. (*sell'*-uh-beez; sool-uh-*way'*-see)
- Irian Jaya (ihr'-ee-ahn *jeye'*-uh)
- Jakarta (yah-*kahr'*-tah)
- Java (*jah'*-vuh)
- Kalimantan (kal-uh-*mann'*-tann)
- Medan (muh-*dahn'*)
- Palembang (pah-lemm-*bahng'*)
- Sumatra (suh-*mah'*-truh)
- Surabaya (soo'-ruh-*bah'*-yah)
- Ujung Pandang (oo'-jung pahn-*dahng'*)

Iran (ih-*rahn'*)
- Abadan (ah-buh-*dahn'*)
- Elburz Mts. (ell-*boorz'*)
- Isfahan (izz-fah-*hahn'*)
- Meshed (muh-*shed'*)
- Tabriz (tah-*breez'*)
- Tehran (teh-huh-*rahn'*)
- Zagros Mts. (*zah'*-gruhs)

Iraq (i-*rahk'*)
- Baghdad (*bag'*-dad)
- Basra (*bahs'*-ra)
- Euphrates R. (yoo-frate'-eez)
- Kirkuk (Kihr-*kook'*) ("oo" as in "wood")
- Mosul (moh-*sool'*)
- Shatt al Arab (R.) (*shaht'* ahl ah-*rahb'*)
- Tigris R. (*teye'*-griss)

Israel
- Gaza Strip (*gahz'*-uh)
- Eilat (*ay'*-laht)
- Haifa (*heye'*-fa)
- Tel Aviv (tel' a-*veev'*)

Italy
- Adriatic Sea (ay-drih-*at'*-ik)
- Apennines (Mts.) (*app'*-uh-nines)
- Bologna (boh-*lohn'*-ya)
- Genoa (*jen'*-o-a)
- Milan (mih-*lahn'*)
- Naples (*nay'*-pels)
- Palermo (pah-*ler'*-moh)
- Turin (*toor'*-in)

Ivory Coast (Côte d'Ivoire) (koht-d'-vwahr)
- Abidjan (abb'-ih-*jahn'*)

Jamaica (ja-*may'*-ka)

Japan
- Fukuoka (foo'-koo-*oh'*-kah)
- Hiroshima (heer'-a-*shee'*-mah)
- Hokkaido (island) (hah-*keye'*-doh)
- Honshu (island) (*honn'*-shoo)
- Kitakyushu (kee-*tah'*-*kyoo'*-shoo)
- Kobe (*koh'*-bay)
- Kyoto (kee-*oht'*-oh)
- Kyushu (island) (kee-*yoo'*-shoo)
- Nagoya (nay-*goy'*-ya)
- Osaka (oh-*sahk'*-a)
- Sapporo (sahp-*pohr'*-oh')
- Shikoku (island) (shih-*koh'*-koo)
- Tokyo (*toh'*-kee-oh)
- Yokohama (yoh'-ko-*hahm'*-a)

Jordan (*jor'*-dan)
- Amman (a-*mahn'*)

Kazakhstan (kuh-*zahk'*-stahn)
- Almaty (al-*mah'*-tee)
- Aral Sea (uh-*rahl'*)
- Karaganda (kahr'-rah-*gahn'*-dah)
- Lake Balkhash (bul-*kahsh'*)
- Syr Darya (river) (*seer'* dahr-*yah'*)

Kenya (*ken'*-ya)
- Mombasa (mahm-*bahs*-a)
- Nairobi (neye-*roh'*-bee)

Kiribati (keer'-ih-*bah'*-tih)

Korea, North
- Pyongyang (pea-ong-*yahng'*)
- Yalu River (*yah'*-loo)

Korea, South
- Inchon (*inn'*-chonn)
- Pusan (*poo'*-sahn)
- Seoul (*sohl'*)
- Taegu (teye-*goo'*)

Kuwait (koo-*wayt'*)

Kyrgyzstan (keer'-geez-*stahn'*)

Laos (*lah'*-ohs)
- Vientiane (vyen-*tyawn'*)

Latvia
- Riga (*reeg'*-a)

Lebanon (*leb'*-a-non)
- Beirut (bay-*root'*)

Lesotho (luh-*soh'*-toh)

Libya (*lib'*-yah)
- Benghazi (benn-*gah'*-zee)
- Tripoli (*tripp'*-uh-lee)

Liechtenstein (*lick'*-tun-stine')

Lithuania (lith'-oo-*ane'*-ee-uh)
- Vilnius (*vill'*-nee-us)

Luxembourg (*luk'*-sem-burg)

Madagascar
- Antananarivo (an-ta-nan'-a-*ree'*-voh)

Malawi (muh-*lah'*-wee)
- Lilongwe (lih-*lawng'*-way)

Malaysia
- Kuala Lumpur (*kwah'*-luh *loom'*-poor) ("oo" as in "wood")
- Sabah (*sah'*-bah)
- Sarawak (suh-*rah'*-wahk)
- Strait of Malacca (muh-*lahk'*-uh)

Maldives (*mawl'*-deevz)

Mali (*mah'*-lee)
- Bamako (*bam'*-ah-koh')

Malta (*mawl'*-tuh)

Martinique (mahr'-tih-*neek'*)

Mauritania (mahr-ih-*tay'*-nee-a)

Mauritius (muh-*rish'*-us)

Mexico
- Acapulco (a-ca-*pool*-ko)
- Ciudad Juárez (see-yoo-*dahd' wahr'*-ez)
- Guadalajara (gwad'-a-la-*har'*-a)
- Monterrey (mahnt-uh-*ray'*)
- Puebla (poo-*eb'*-la)
- Tijuana (tee-*hwah'*-nah)

Moldova (mohl-*doh'*-vah)
- Kishinev (kish'-in-*yeff'*)

Monaco (*mon'*-a-koh')

Mongolia (mon-*gohl'*-yuh)
- Ulan Bator (oo-*lahn' bah'*tor)

Montserrat (mont'-suh-*rat'*)

Morocco (moh-*rah'*-koh)
- Casablanca (kas-a-*blang'*-ka)
- Rabat (rah-*baht'*)

Mozambique (moh'-zamm-*beek'*)
- Beira (*bay'*-rah)
- Cabora Bassa Dam (kah-*bore'*-ah *bah'*-sah)
- Maputo (mah-*poot'*-oh')

Namibia (nah-*mib'*-ee-a)
- Windhoek (*Vind'*-hook)

Nauru (nah-*oo*-roo)

Nepal (nuh-*pawl'*)
- Katmandu (kat'-man-*doo'*)

Netherlands
- Ijsselmeer (*eye'*-sul-mahr')
- The Hague (*hayg'*)
- Utrecht (*yoo'*-trekt')
- Zuider Zee (*zeye'*-der zay')

Netherlands Antilles
- Curaçao (*koo'*-rahs-ow')

New Zealand
- Auckland (*aw'*-klund)

Nicaragua (nik'-a-*rahg'*-wah)

Niger (*neye'*-jer; Fr., nee'-*zhere'*)
- Niamey (nee-*ah'*-may)

Nigeria (neye-*jeer*-ee-a)
- Abuja (a-*boo'*-ja)
- Ibadan (ee-*bahd'*-n)
- Kano (*kahn'*-oh)
- Lagos (*lahg'*-us)

Norway
- Bergen (*berg'*-n)
- Oslo (*ahz'*-loh)
- Stavanger (stah-*vahng'*-er)
- Trondheim (*trahn'*-haym')

Oman (oh-*mahn'*)
- Muscat (*mus'*-kat')

Pakistan (pahk-i-*stahn'*)
- Hyderabad (*heyed'*-er-a-bahd')
- Islamabad (is-*lahm*-a-bahd')
- Karachi (kah-*rahch'*-ee')
- Karakoram (Mts.) (kahr'-a-*kohr*-um)
- Lahore (lah-*hohr'*)
- Peshawar (puh-*shah'*-wahr)
- Punjab (see India)

Papua New Guinea (*pap'*-yoo-uh new *gin'*-ee)
- Port Moresby (*morz'*-bee)

Paraguay (*pair'*-uh-gweye')
- Asuncíon (a-soon'-see-*ohn'*)

Peru (puh-*roo'*)
- Callao (kah-*yow'*)
- Cerro de Pasco (*sehr'*-roh day *pahs'*-koh)
- Chimbote (chimm-*boh'*-te)
- Cuzco (*koos'*-koh)
- Iquitos (ee-*kee'*-tohs)
- Lima (*lee'*-ma)
- Machu Picchu (ma-chu *pee*-chu)

Philippines (*fil*-i-peenz')
- Luzon (island) (loo'-*zahn'*)
- Mindanao (island) (minn'-dah-*nah'*-oh)
- Mt. Pinatubo (pea'-nah-*too'*-buh)
- Visayan Islands (vih-*sah'*-y'n)

Poland
- Gdansk (ga-*dahntsk'*)

Katowice (kaht-uh-*veet'*-sih)
Krakow (*krahk'*-ow)
Lodz (*looj'*)
Posnan (*pohz'*-nan-ya)
Silesia (region) (sih-*lee'*-shuh)
Szczecin (*shchet'*-seen')
Wroclaw (*vrawt'*-slahf)
Portugal (*pohr'*-chi-gal)
Lisbon (*liz'*-bon)
Oporto (oh-*pohrt'*-oh)

Qatar (*kaht'*-ar)

Romania (roh-*main'*-yuh)
Bucharest (boo'-kuh-*rest'*)
Carpathian Mts. (kahr-*pay'*-thih-un)
Russia
Altai Mts. (*al'*-teye)
Amur R. (ah-*moor'*) ("oo" as in "wood")
Angara R. (ahn'-guh-*rah'*)
Arkhangelsk (ahr-*kann'*-jelsk)
Astrakhan (as-trah-*khahn'*)
Baikal, Lake (beye-*kahl'*)
Bashkiria (bahsh-*keer'*-ee-uh)
Bratsk (*brahtsk'*)
Buryatia (boor-*yaht'*-ee-uh)
Caucasus (kaw'-kah-suss)
Chechen-Ingushia (*chehch'*-yenn enn-*gooshia'*)
Chelyabinsk (*chell'*-yah-binnsk)
Cherepovets (*chehr'*-yuh-puh-vyetz')
Chuvashia (choo-*vahsh'*-i-a)
Dagestan Rep. (dag'-uh-*stahn'*)
Irkutsk (ihr-*kootsk'*)
Ivanovo (ih-*vahn'*-uh-voh)
Izhevsk (ih-*zhehvsk'*)
Kama R. (*kah'*-mah)
Karelia (kah-*reel'*-yuh)
Kazan (kuh-*zahn'*)
Khabarovsk (kah-*bah'*-rawfsk)
Kola Pen. (*koh'*-lah)
Kolyma R. (kuh-*lee'*-muh)
Krasnoyarsk (kras'-nuh-*yahrsk'*)
Kuril Is. (*koo'*-rill)
Kuznetsk Basin (kooz'-*netsk'*)
Ladoga, Lake (*lad'*-uh-guh)
Lena R. (*lee*-nah, *lay*-nah)
Lipetsk (lyee'-*petsk'*)
Magnitogorsk (mag-*nee'*-toh-gorsk')
Moscow (*mahs'*-koh or *mahs'*-kow)
Murmansk (moor'-*mahntsk'*)
Nakhodka (nuh-*kawt*-kah)
Nizhniy Novgorod (*nizh'*-nee nahv-guh-*rahd'*)
Nizhniy Tagil (tah-*geel*)
Novokuznetsk (naw'-voh-*kooznetsk'*)
Novaya Zemlya (*naw'*-vuh-yuh zem-*lyah'*)
Novosibirsk (naw-vuh-suh-*beersk'*)
Oka R. (oh-*kah'*)
Okhotsk, Sea of (oh-*kahtsk'*)
Omsk (*ahmsk'*)
Pechora R. (peh-*chore'*-a)
Rostov (rahs-*tawf'*)
Sakhalin (sahk'-uh-*leen'*)
Samara (suh-*mahr'*-a)
Saratov (suh-*raht'*-uff)
Sayan Mts. (sah-*yahn'*)
Tatarstan (Tatar Rep.) (tah-*tahr'*-stahn')
Togliatti (tawl-*yah'*-tee)
Tuva Rep. (*too'*-va)
Ufa (oo-*fah'*)
Ussuri R. (oo-*soor'*-i)
Vladivostok (vlah'-dih-vahs-*tawk'*)
Volga R. (*vahl*-guh, *vohl'*-guh)
Volgograd (vahl-guh-*grahd'*, vohl'-)
Voronezh (vah-*raw'*-nyesh)
Yakutsk (yah-*kootsk'*)
Yekaterinburg (yeh-*kahta'*-rinn-berg)
Rwanda (roo-*ahn'*-da)

St. Kitts and Nevis (*nee'*-vis)
St. Lucia (*loo'*-sha)
St. Vincent-Grenadines (gren'-a-*deenz'*)
Samoa (sa-*moh'*-a)
Apia (ah-*pea'*-uh)
São Tomé and Príncipe (sau too-*meh'*, pren-*see'*-puh)
Saudi Arabia (*saud'*-ee)
Jiddah (*jid'*-a)
Riyadh (ree-*adh'*)
Senegal (sen'-i-*gahl'*)
Dakar (da'-*kahr'*)
Seychelles (say'-*shelz'*)
Sierra Leone (see-*ehr'*-a lee-*ohn'*)
Slovenia (sloh-*veen'*-i-a)
Ljubljana (*lyoo'*-b'l-yah-nah)
Somalia (so-mahl'-ee-a)
Mogadishu (mahg'-uh-*dish'*-oo)
South Africa
Bloemfontein (*bloom'*-fonn-*tayn'*)
Bophuthatswana (homeland) (boh-poot'-aht-*swahna'*)
Drakensberg (escarpment) (*drahk*-n's-burg')
Durban (*durr'*-bun)
Johannesburg (joh-*hann'*-iss-burg)
Kwazulu (homeland) (kwah-zoo'-loo)
Natal (province) (nuh-*tahl'*)
Pretoria (prih-*tohr'*-ee-uh)
Transkei (homeland) (trans-*keye'*)
Transvaal (province) (trans-*vahl'*)
Spain
Barcelona (bahr'-suh-*lohn*-uh)
Bilbao (bill-*bah'*-oh)
Catalonia (region) (katt'-uh-*lohn'*-ee-a)
Madrid (muh-*dridd'*)
Meseta (plateau) (muh-*say'*-tuh)
Seville (suh-*vill'*)
Valencia (vull-*enn'*-shee-uh)
Sri Lanka (sree *lahn'*-*kah*)
Sudan (*soo'*-dann)
Gezira (guh-*zeer'*-uh)
Juba (*joo'*-bah)
Khartoum (kahr'-*toom'*)
Omdurman (ahm-*durr'*-man)
Wadi Halfa (*wah-di' hawl'*-fah)
Suriname (*soor'*-i-nahm')
Swaziland (*swah'*-zih-land')
Sweden
Göteborg (yort'-e-*bor'*)
Skåne (*skohn'*)
Småland (*smoh'*-lund')
Switzerland
Basel (*bah'*-z'l)
Zürich (*zoor'*-ick)
Syria (*seer'*-i-uh)
Aleppo (a-*lep*-oh')
Damascus (da-*mas'*-kus)

Tajikistan (tah-*jick'*-ih-stahn')
Tanzania (tan'-za-*nee'*-a)
Dar es Salaam (dahr' es sa-*lahm'*)
Thailand (*teye'*-land')
Bangkok (*bang'*-kok')
Togo (*toh'*-goh)
Lomé (loh-*may'*)
Tonga (*tawng'*-a)
Trinidad and Tobago (to-*bay'*-goh)
Tunisia (too-*nee'*-zhee-a)
Tunis (*too'*-nis)
Turkey
Anatolian Peninsula (ann'-ah-*tohl'*-yun)
Ankara (*ang'*-kah-rah)
Bosporus (strait) (*bahs'*-puhr-us)
Dardanelles (strait) (dahr'-duh-*nelz'*)
Istanbul (iss'-tann-*bool'*)
Izmir (izz'-*meer'*)
Sea of Marmara (*mahr'*-muh-ruh)
Turkmenistan (turk'-menn-ih-*stahn'*)
Ashgabat (*ash'*-guh-baht)

Ukraine (yoo'-*krane'*)
Crimea (cry'-*mee'*-uh)
Dnepropetrovsk (d'nyepp-pruh-pay-*trawfsk'*)
Dnieper R. (duh-*nyepp'*-er)
Donetsk (duhn-*yehtsk'*)
Kharkov (*kahr'*-kawf)
Kiev (*kee'*-yeff)
Krivoy Rog (*kree'*-voi *rohg'*)
Sevastopol (sih-*vass'*-tuh-*pohl'*)
Yalta (*yahl'*-tuh)
Zaporozhye (zah-puh-*rawzh'*-yuh)
United Arab Emirates
Abu Dhabi (*ah'*-boo dah'-bee)
Dubai (doo-*beye'*)
United Kingdom
Edinburgh (*ed'*-in-buh-ruh)
Glasgow (*glass'*-goh)
Thames R. (*timms'*)
United States
Albuquerque (*al'*-buh-ker'-kih)
Aleutian Is. (uh-*loo'*-sh'n)
Allegheny R. (*al'*-ih-gay'-nih)
Appalachian Highlands (ap'-uh-*lay'*-chih-un)
Coeur d'Alene (*kurr'* duh-*layn'*)
Des Moines (duh *moin'*)
Grand Coulee Dam (*koo'*-lee)
Juneau (*joo'*-noh)
Mesabi Range (muh-*sahb'*-ih)
Monongahela R. (muh-*nahng'*-guh-*hee'*-luh)
Omaha (oh'-mah-*hah'*)
Ouachita Mts. (*wosh'*-ih-taw')
Palouse (region) (pah-*loose'*)
Phoenix (*fee'*-nix)
Provo (*proh'*-voh)
Rio Grande (*ree'*-oh *grahn'*-day)

San Joaquin R. (sann wah-*keen′*)
Spokane (spoh′-*kann′*)
Tucson (*too′*-sonn)
Valdez (val-*deez′*)
Uruguay (*yoor′*-uh-gweye′)
Montevideo (mahnt′-e-vi-*day′*-oh)
Uzbekistan (ooz-*beck′*-ih-stahn′) ("oo" as in "wood")
Amu Darya (river) (ah-moo′ *dahr′*-yuh, dahr-*yah′*)
Bukhara (buh-*kah′*-ruh)
Fergana Valley (fair-*gahn′*-uh)
Samarkand (*samm′*-ur-*kand′*; Rus., suh-mur-*kahnt′*)
Tashkent (tash-*kent′*, tahsh-)

Vanuatu (van′-oo-*ay′*-too)
Port Vila (*vee′*-la)
Venezuela (ven′-ez-*way′*-la)
Caracas (ka-*rak′*-as)
Maracaibo (mar′-a-*keye′*-boh)
Orinoco R. (oh-rih-*noh′*-koh)
Valencia (va-*len′*-see-a)
Vietnam (vee-et′-*nahm′*)
Annam (region) (uh-*namm′*)
Cochin China (*koh′*-chin)
Hanoi (ha-*noi′*)
Ho Chi Minh City (Saigon) (hoe chee *min′*; *seye′*-gahn)
Mekong R. (may-*kahng′*)

Yemen
Aden (*ah*-d'n)
S'ana (sah-*na′*)
Yugoslavia (yoo′-goh-*slahv′*-ee-a)
Belgrade (*bell′*-grade, -grahd)
Kosovo (*kaw′*-suh-voh′)
Montenegro (republic) (mahn′-tee-*nay′*-groh)

Zaire (*zeye′*-eer′)
Kananga (kay-*nahng′*-guh)
Kasai R. (kah-*seye′*)
Kinshasa (*keen′*-shah-suh)
Kisangani (kee′-sahn-*gahn′*-ih)
Lake Kivu (*kee′*-voo)
Lualaba R. (loo′-ah-*lahb′*-a)
Lubumbashi (loo-boom-*bah′*-shee)
Matadi (muh-*tah′*-dee)
Shaba (region) (*shah′*-bah)
Zambia (*zam′*-bee-a)
Kariba Dam (kah-*ree′*-buh)
Lusaka (loo-*sahk′*-a)
Zambezi R. (zamm-*bee′*-zee)
Zimbabwe (zim-*bahb′*-way)
Bulawayo (bool′-uh-*way′*-oh)
Harare (hah-*rahr′*-ee)
Que Que (*kway′ kway′*)
Wankie (*wahn′*-kee)

index

credits

Figure 1.1: (right) From *Geography for Life: National Geography Standards 1994*, National Geographic Research and Exploration, Washington, DC, pp. 34–35 **Table 1.1:** Source: *World Population Data Sheet*, Population Reference Bureau, 2007; *U.N. Human Development Report*, United Nations, 2007; *World Factbook*, CIA, 2007. **Figure 1.9:** Source: *Goode's 20th edition*, published by Rand McNally. **Figure 1.11:** (top) After Bergman, 1995; (bottom) After Rubenstein, 1995 **Table 1.3:** Source: www.aag.org. **Figure 2.1:** Adapted from NASA, "Global Tectonic Activity Map of the Earth," DTAM-1, 2002. **Figure 2.2:** Adapted from Donald Hyndman and David Hyndman. 2006. *Natural Hazards and Disasters*. Belmont: Brooks Cole, p. 16. **Figure 2.3:** Adapted with permission from UNEP/GRID, "Mean Annual Precipitation," Geneva Data Sets-Atmosphere, GNV174 (http://www.grid.unep.ch/data). **Figure 2.8:** From Tom L. McKnight and Darrel Hess, *Physical Geography: A Landscape Appreciation, Animation Edition,* 7th ed., © 2004. Reproduced by permission of Pearson Education, Inc., Upper Saddle River, NJ **Figure 2.10:** Adapted from Rand McNally's *Classroom Atlas,* 2003. **Figure 2.11:** Adapted from World Wildlife Fund ecoregions data, 1999. **Figure 2.13:** Adapted from Conservation International, "Hotspots Explorer," 2007 (http://www.biodiversity hotspots.org/xp/Hotspots). **Figure 2.16:** Cornelia Dean, "Study Sees 'Global Collapse' of Fish Species." *New York Times,* November 3, 2006, p. A16. **Figure 3.5:** Data from World Bank, 2007. **Figure 3.B:** Adapted from David W. Pearce and R. Kerry Turner, *Economics of Natural Resources and the Environment* (Baltimore: Johns Hopkins University Press, 1990), p. 348. **Figure 3.7:** Data from Population Reference Bureau, 2007 **Figure 3.8:** Data from Population Reference Bureau, 2007 **Figure 3.9:** Data from Population Reference Bureau, 2007 **Figure 3.15:** Data from Population Reference Bureau, 2007 **Figure 3.16:** Data from Population Reference Bureau, 2007 **Figure 3.21:** From *Environmental Science*, 4th edition, by Miller. © 1993. Reprinted with permission of Wadsworth, a division of Thomson Learning, Inc. **Table 4.1:** Source: *World Population Data Sheet*, Population Reference Bureau, 2007; *U.N. Human Development Report*, United Nations, 2007; *World Factbook*, CIA, 2007. **Figure 4.6:** After Hoffman, 1990 **Figure 4.7:** (left) Adapted from Rand McNally's *Classroom Atlas,* 2003; (right) Adapted from World Wildlife Fund ecoregions data, 1999. **Figure 4.8:** From Patrick Wiegand, ed., *The Oxford School Atlas* (Oxford University Press, 1997). Copyright © 1997 Oxford University Press. Reprinted by permission of Oxford University Press. **Figure 4.11:** Adapted from Bernard Comrie et al., *The Atlas of Languages,* rev. ed. (New York: Facts on File, 2003), p. 41. **Figure 4.B:** Adapted from a map in *The Economist,* Sept. 20, 1997, p. 53. © 1998 The Economist Newspaper Ltd. All rights reserved. Reprinted with permission. Further reproduction prohibited. **Table 4.2:** From Jennifer Ehrlich and Tom Vandyck, "Belgian 'Malcolm X' Seeks Office," *The Christian Science Monitor,* May 16, 2003, p. 6. Data originates from *A Guidebook on Islam and Muslims in the Wide Contemporary Europe*, edited by B. Marechal (Louvain-la-Neuve: Academia Bruylant, 2002) **Figure 4.15:** Adapted from "European Union Survey," *The Economist,* May 31, 1997, p. 5. © 1997 The Economist Newspaper Ltd. All rights reserved. Reprinted with permission. Further reproduction prohibited. **Figure 4.16:** Adapted NATO map data from EUROSTAT (Statistical Office of the European Communities), 1996; *World Factbook*, CIA, 1995. **Figure 4.1.2:** Data from Global Forest Watch, 2004 (http://www.globalforestwatch.org/english/datawarehouse/index.asp). **Figure 4.1.4:** Based on Terry G. Jordan and Bella Bychova Jordan, *The European Culture Area: A Systematic Geography* (Lanham, Md.: Rowman and Littlefield, 2002), p. 300.

Figure 4.1.5: After Hoffman, 1990, p. 245. **Figure 4.1.9:** Based on Terry G. Jordan and Bella Bychova Jordan, *The European Culture Area: A Systematic Geography* (Lanham, Md.: Rowman and Littlefield, 2002), p. 326. **Figure 4.2.22:** Adapted from maps in John Haywood, ed., *Atlas of World History* (Barnes and Noble Books, 2004) **Figure 4.2.30:** Adapted from Ian Barnes and Robert Hudson, *The History Atlas of Europe* (Macmillan Inc., 1998), p. 151. **Table 5.1:** *World Population Data Sheet*, Population Reference Bureau, 2007; *U.N. Human Development Report*, United Nations, 2007; *World Factbook*, CIA, 2007. **Figure 5.2:** (top) From Patrick Wiegand, ed., *The Oxford School Atlas* (Oxford University Press, 1997). Copyright © 1997 Oxford University Press. Reprinted by permission of Oxford University Press. **Figure 5.2:** Data from *World Factbook*, CIA, 2007. **Figure 5.4:** (left) Adapted from Rand McNally's *Classroom Atlas,* 2003; (right) Adapted from World Wildlife Fund ecoregions data, 1999. **Figure 5.6:** From Patrick Wiegand, ed., *The Oxford School Atlas* (Oxford University Press, 1997). Copyright © 1997 Oxford University Press. Reprinted by permission of Oxford University Press. **Figure 5.11:** Adapted with permission from Bernard Comrie, et al., *The Atlas of Languages,* Revised Edition (New York: Facts on File, 2003). **Figure 5.D:** Based on Robin R. Milner-Gulland, *Cultural Atlas of Russia and the Former Soviet Union* (New York: Checkmark, 1998), map X.X. **Figure 6.3:** (top) From Patrick Wiegand, ed., *The Oxford School Atlas* (Oxford University Press, 1997). Copyright © 1997 Oxford University Press. Reprinted by permission of Oxford University Press. **Figure 6.5:** (left) Adapted from Rand McNally's *Classroom Atlas,* 2003; (right) Adapted from World Wildlife Fund ecoregions data, 1999. **Figure 6.7:** From Patrick Wiegand, ed., *The Oxford School Atlas* (Oxford University Press, 1997). Copyright © 1997 Oxford University Press. Reprinted by permission of Oxford University Press. **Figure 6.17:** Bernard Comrie, et al., *The Atlas of Languages,* Revised Edition (New York: Facts on File, 2003). **Figure 6.C:** Adapted from National Geographic's *Atlas of the Middle East* (National Geographic, 2003). **Figure 6.1.1:** Adapted from John Haywood, ed., *Atlas of World History* (Barnes and Noble Books, 1997). **Figure 6.1.2:** From *The Middle East and North Africa: A Political Geography,* by A. Drysdale and G. Blake. Copyright © 1985 Oxford University Press, Inc. Used by permission of Oxford University Press, Inc. **Figure 6.1.3:** From *The Middle East and North Africa: A Political Geography,* by A. Drysdale and G. Blake. Copyright © 1985 Oxford University Press, Inc. Used by permission of Oxford University Press, Inc. **Figure 6.1.B:** Zone control data from Foundation for Middle East Peace, 2002 (www.fmep.org); security wall route from IDF maps. **Figure 6.1.7:** Held, Colbert. 1994. *Middle East Patterns*. **Figure 6.1.23:** Data from CIA maps. **Figure 6.1.32:** Ethnicity and poppy data from CIA maps. **Figure 7.3:** (top) From Patrick Wiegand, ed., The Oxford School Atlas (Oxford University Press, 1997). Copyright © 1997 Oxford University Press. Reprinted by permission of Oxford University Press. **Figure 7.5:** (left) Adapted from Rand McNally's *Classroom Atlas,* 2003.; (right): Adapted from World Wildlife Fund ecoregions data, 1999. **Figure 7.7:** From Patrick Wiegand, ed., The Oxford School Atlas (Oxford University Press, 1997). Copyright © 1997 Oxford University Press. Reprinted by permission of Oxford University Press. **Figure 7.11:** Bernard Comrie, et al., *The Atlas of Languages,* Revised Edition (New York: Facts on File, 2003). **Figure 7.F:** From David E. Sanger, "China's Big Need for Oil is High on U.S. Agenda," *The New York Times,* 4-19-06, A1. **Figure 7.17:** From "The World in 1900," in *Atlas of World History,* ed. R. R. Palmer (Chicago: Rand McNally, 1957), pp. 169–171. Used with permission. **Figure 7.1.7:** A. Smith, *Atlas of Mesozoic and Cenozoic*

Coastlines (New York: Cambridge University Press, 1994). **Figure 7.2.7:** Data from Global Forest Watch, 2004 (http://www.globalforestwatch.org/english/datawarehouse/index.asp). **Figure 7.2.9:** From Wayne Arnold, "A Gas Pipeline to the World Outside," *The New York Times,* 10-26-01, C1. Copyright © 2001 The New York Times Co. **Figure 7.3.E:** Sources: Erik Eckholm, "China Will Move Waters to Quench Thirst of Cities," *The New York Times,* 8-27-02, A1; "Water in China: In Deep," *The Economist,* 08-18-01, 31. **Figure 7.3.14:** *China Atlas of Population and Environment;* http://141.211.136.211/eng/macro/atlas2000/ybtableview.asp?ID=4). **Table 8.1:** *World Population Data Sheet*, Population Reference Bureau, 2007; *U.N. Human Development Report*, United Nations, 2007; *World Factbook*, CIA, 2007. **Figure 8.3:** (left) From Patrick Wiegand, ed., *The Oxford School Atlas* (Oxford University Press, 1997). Copyright © 1997 Oxford University Press. Reprinted by permission of Oxford University Press. **Figure 8.5(a):** (left) Adapted from Rand McNally's *Classroom Atlas*, 2003; (right) Adapted from World Wildlife Fund ecoregions data, 1999. **Figure 8.7:** *The Oxford School Atlas* (Oxford University Press, 1997). **Figure 8.9:** Robert E. Gabler, James F. Peterson, and L. Michael Trepasso, *Essentials of Physical Geography* (Philadelphia: Saunders, 1987), p. 562. **Figure 8.12:** Bernard Comrie et al., *The Atlas of Languages,* rev. ed (New York: Facts on File, 2003); settlement data from John Noble Wilford, "Seeking Polynesia's Beginnings in an Archipelago of Shards." *New York Times,* January 8, 2002, p. D1. **Figure 9.3:** (top) From Patrick Wiegand, ed., *The Oxford School Atlas* (Oxford University Press, 1997). Copyright © 1997 Oxford University Press. Reprinted by permission of Oxford University Press. **Figure 9.8:** (left) Adapted from Rand McNally's *Classroom Atlas,* 2003; (right) Adapted from World Wildlife Fund ecoregions data, 1999. **Figure 9.9:** From Patrick Wiegand, ed., *The Oxford School Atlas* (Oxford University Press, 1997). Copyright © 1997 Oxford University Press. Reprinted by permission of Oxford University Press. **Figure 9.A:** (left) Joint United Nations Program on H.I.V./AIDS. (right) Data from UNAIDS. **Figure 9.13:** Bernard Comrie, et al., *The Atlas of Languages,* Revised Edition (New York: Facts on File, 2003). **Figure 9.15:** Robert Stock, *Africa South of the Sahara,* 2nd ed. (New York: Guilford Press, 2004), p. 10. **Figure 9.16:** Adapted from John Haywood, ed., *Atlas of World History* (Barnes and Noble Books, 1997). **Figure 9.18:** Adapted from USGS country mineral maps. **Figure 9.1.1:** Robert Stock, *Africa South of the Sahara*. **Figure 9.1.B:** From "Linguistic Groups in Nigeria,", CIA map, 1979. **Table 10.1:** *World Population Data Sheet*, Population Reference Bureau, 2007; *U.N. Human Development Report*, United Nations, 2007; *World Factbook*, CIA, 2007. **Figure 10.3:** From Patrick Wiegand, ed., *The Oxford School Atlas* (Oxford University Press, 1997). Copyright © 1997 Oxford University Press. Reprinted by permission of Oxford University Press. **Figure 10.5(a):** (left) Adapted from Rand McNally's *Classroom Atlas,* 2003; (right) Adapted from World Wildlife Fund ecoregions data, 1999. **Figure 10.6:** David L. Clawson, *Latin America and the Caribbean* (Dubuque, Iowa: Brown, 1997), p. 46. **Figure 10.7:** From Patrick Wiegand, ed., *The Oxford School Atlas* (Oxford University Press, 1997). Copyright © 1997 Oxford University Press. Reprinted by permission of Oxford University Press. **Figures 10.A and B:** From *Living in the Environment,* 13th edition, by Miller. © 2004. Reprinted with permission of Brooks/Cole, a division of Thomson Learning, Inc. **Figure 10.11:** *The History Atlas of South America,* Edwin Early, ed., (Macmillan, 1998). **Figure 10.15:** Bernard Comrie, et al., *The Atlas of Languages,* Revised Edition (New York: Facts on File, 2003). **Table 10.2:** *World Factbook*, CIA, 2007. **Figure 10.23:** Adapted from *The Economist,* April 11, 1998, p. 125. **Figure 10.E:** Data from the U.N. Drug Control Program (UNDCP), 2007. **Figure 10.1.6:** G. Tyler Miller, *Living in the Environment,* 13th ed. © 2004. Reprinted with permission of Brooks/Cole, a division of Thomson Learning Inc. **Figure 10.2.14:** Adapted from *Atlas of World History* (Barnes and Noble Books, 1997). **Figure 10.2.C:** Data from Global Forest Watch, 2004 (http://www.globalforestwatch.org/english/interactive.maps/brazil.htm). **Table 11.1** *World Population Data Sheet*, Population Reference Bureau, 2007; *U.N. Human Development Report*, United Nations, 2007; *World Factbook*, CIA, 2007. **Figure 11.2:** (top) From Patrick Wiegand, ed., *The Oxford School Atlas* (Oxford University Press, 1997). Copyright © 1997 Oxford University Press. Reprinted by permission of Oxford University Press. **Figure 11.3:** Data from the U.S. Census Bureau and Statistics Canada. **Figure 11.5(a):** (left) Adapted from Rand McNally's *Classroom Atlas,* 2003; (right) Adapted from World Wildlife Fund ecoregions data, 1999. **Figure 11.6:** From Patrick Wiegand, ed., *The Oxford School Atlas* (Oxford University Press, 1997). Copyright © 1997 Oxford University Press. Reprinted by permission of Oxford University Press. **Figure 11.7:** Map adapted from Timothy Egan, "Vanishing Point: Amid Dying Towns of Rural Plains, One Makes a Stand," *The New York Times,* 12/01/03, p. A18. **Figure 11.8:** Carl Waldman, *Atlas of the North American Indian* (Facts on File, 1985). **Figure 11.12:** Adapted from John Haywood, ed., *Atlas of World History* (Barnes and Noble Books, 1997). **Figure 11.15:** Bernard Comrie, et al., *The Atlas of Languages,* Revised Edition (New York: Facts on File, 2003). **Figure 11.16:** *New Historical Atlas of Religion in America* (Oxford University Press, 2001). **Figure 11.18:** (left) Map from Alberta Geological Survey (http://www.ags.gov.ab.ca/GRAPHICS/CBM/oil_sands/alberta_oil_sands_map2.jpg). **Figure 11.1.7:** Getis, Arthur and Judith Getis. 1995. *The United States and Canada: The Land and Its People.* Dubuque, Iowa: Wm. C. Brown Publishers, p. 231. **Figure 11.1.B:** Map adapted from: Krauss, Clifford. October 10, 2005. "As Polar Ice Turns to Water, Dreams of Treasure Abound." *The New York Times*, pp. A1, 10, 11. **Table 11.2.1** *World Population Data Sheet*, Population Reference Bureau, 2007; *U.N. Human Development Report*, United Nations, 2007; *World Factbook*, CIA, 2007. **Endpapers:** front and rear, courtesy of National Geographic, © 2002.

NORTH POLAR REGION
ARCTIC OCEAN
RUSSIA
CANADA
ALASKA
GREENLAND
ATLANTIC OCEAN
GREENLAND (KALAALLIT NUNAAT)
NUNAVUT
NORTHWEST TERRITORIES
HUDSON BAY
ALBERTA
ONTARIO
QUEBEC
UNITED STATES
NORTH PACIFIC
NORTH ATLANTIC
OCEAN
MEXICO
GULF OF MEXICO
CARIBBEAN SEA
VENEZUELA
COLOMBIA
BRAZIL
BOLIVIA
ARGENTINA
CHILE
PERU
KIRIBATI
FRENCH POLYNESIA
SOUTH PACIFIC OCEAN
SOUTHWEST PACIFIC BASIN
SOUTH ATLANTIC OCEAN
ARGENTINE BASIN
BRAZIL BASIN
ANGOLA BASIN
SCOTIA SEA
WEDDELL SEA
ANTARCTICA
UNITED KINGDOM
IRELAND
FRANCE
SPAIN
PORTUGAL
ICELAND
NORWAY
NORWEGIAN SEA
ALGERIA
LIBYA
MAURITANIA
NIGER
CHAD
NIGERIA
ANGOLA
NAMIBIA
LARGEST COUNTRIES
LARGEST COUNTRIES
NATIONAL GEOGRAPHIC
NORTH AMERICA
SOUTH AMERICA
EUROPE
AFRICA
AUSTRALIA
VEGETATION AND LAND USE